Handbook of Lipid Research 8

Lipid Second Messengers

Handbook of Lipid Research

Editor: Fred Snyder
Oak Ridge Associated Universities
Oak Ridge, Tennessee

Handbook of Lipid Research 8

Lipid Second Messengers

Edited by

Robert M. Bell

Duke University
Durham, North Carolina
and Glaxo Wellcome, Inc.
Research Triangle Park, North Carolina

John H. Exton

Vanderbilt University
School of Medicine
Nashville, Tennessee

and

Stephen M. Prescott

University of Utah
Salt Lake City, Utah

Plenum Press • *New York and London*

QP
552
.L5
L57X
1996

ISSN 0163-9102

ISBN 0-306-45174-3

© 1996 Plenum Press, New York
A Division of Plenum Publishing Corporation
233 Spring Street, New York, N. Y. 10013

Printed in the United States of America

Contributors

Yoshinori Asaoka, Biosignal Research Center, Kobe University, Kobe 657, Japan

Robert M. Bell, Department of Biochemistry, Duke University Medical Center, Durham, North Carolina 27710

Stephen B. Bocckino, Sphinx Pharmaceuticals Corp., Durham, North Carolina 27707

Lewis C. Cantley, Division of Signal Transduction, Beth Israel Hospital, and Department of Cell Biology, Harvard Medical School, Boston, Massachusetts 02115

Ghassan S. Dbaibo, Department of Pediatrics, Duke University Medical Center, Durham, North Carolina 27710

Brian C. Duckworth, Department of Physiology, Tufts University, Boston, Massachusetts 02111

John H. Exton, Howard Hughes Medical Institute and Department of Molecular Physiology and Biophysics, Vanderbilt University School of Medicine, Nashville, Tennessee 37232

Yusuf A. Hannun, Department of Medicine and Cell Biology, Duke University Medical Center, Durham, North Carolina 27710

Kees Jalink, Division of Cellular Biochemistry, The Netherlands Cancer Institute, 1066 CX Amsterdam, The Netherlands

David A. Jones, The Eccles Program in Human Molecular Biology and Genetics and the Nora Eccles Harrison Cardiovascular Research and Training Institute, University of Utah, Salt Lake City, Utah 84112

Dennis C. Liotta, Department of Chemistry, Emory University, Atlanta, Georgia 30322-3050

Thomas M. McIntyre, Departments of Medicine and Biochemistry, University of Utah School of Medicine, Salt Lake City, Utah 84112

Elizabeth A. Meade, The Eccles Program in Human Molecular Biology and Genetics and the Nora Eccles Harrison Cardiovascular Research and Training Institute, University of Utah, Salt Lake City, Utah 84112

Alfred H. Merrill, Jr., Department of Biochemistry, Emory University School of Medicine, Atlanta, Georgia 30322-3050

Wouter H. Moolenaar, Division of Cellular Biochemistry, The Netherlands Cancer Institute, 1066 CX Amsterdam, The Netherlands

Yasutomi Nishizuka, Biosignal Research Center, Kobe University, and Department of Biochemistry, Kobe University School of Medicine, Kobe 650, Japan

Lina M. Obeid, Departments of Medicine and Cell Biology, Duke University Medical Center, Durham, North Carolina 27710

Stephen M. Prescott, Departments of Medicine and Biochemistry, the Nora Eccles Harrison Cardiovascular Research and Training Institute, and the Eccles Institute of Human Genetics, University of Utah School of Medicine, Salt Lake City, Utah 84112

Andrew F. G. Quest, Institute of Biochemistry, University of Lausanne, CH-1066 Lausanne, Switzerland

Daniel M. Raben, Department of Physiology, Johns Hopkins University School of Medicine, Baltimore, Maryland 21205

Ronald E. Riley, Toxicology and Mycotoxins Research Unit, U.S. Department of Agriculture, Agriculture Research Service, Athens, Georgia 30613

Yosuke Tsujishita, Biosignal Research Center, Kobe University, Kobe 657, Japan

Ralph E. Whatley, Department of Medicine, East Carolina University School of Medicine, Greenville, North Carolina 27858

Guy A. Zimmerman, Department of Medicine, University of Utah School of Medicine, Salt Lake City, Utah 84112

Preface

Lipids traditionally have been viewed as serving two functions: to form cellular membranes and serve as energy stores. These are crucial functions—for example, the formation of a membrane may have been one of the earliest events in the evolution of life since this allowed "cells" in which the internal chemical reactions could be separated from the environment. Likewise, energy storage as fat is efficient since lipids do not have to be hydrated; thus, there is a lower total weight per amount of energy stored as compared with carbohydrates. However, in the last two decades another role for lipids, or products derived from them, has taken center stage. This came with the increasing recognition that lipids can be signaling molecules.

This book deals with a variety of lipids that have been shown to be messengers. It excludes the steroid hormones, which were the earliest to be studied, but deals with all of the other classes known. Messengers generated from lipids can function either as intracellular or intercellular messengers. In each case, the signal is generated by an enzyme catalyzing the conversion of the "reservoir" lipid to the messenger. The best studied example of an intracellular messenger is diacylglycerol, which exerts its effects by activating protein kinase C, which in turn phosphorylates intracellular targets. The eicosanoids are the most thoroughly studied family of intercellular messengers and they exert their effects by binding to surface receptors on the target cell. These two examples reflect a general scheme: the intracellular messengers activate protein kinases or phosphatases, and the intercellular messengers work by binding to receptors—thus far, all utilize receptors in the G-protein-coupled family.

From this perspective, the complex lipids in cells can be viewed as a reservoir of potential signaling molecules. For example, consider the messengers that can be generated from phosphatidylcholine, which is the most prevalent phospholipid in cellular membranes. As such, it was recognized to be essential for function, like the walls of a building, but not as interesting as what was happening inside. It now is known that almost any portion of the molecule can lead to a signaling compound. For example, phospholipase D can cleave the choline to yield it and phosphatidic acid. Choline may have signaling properties and phosphatidic acid is known to activate target molecules within the cell and may also serve as an intercellular messenger. The phosphatidic acid can be acted on by phospholipase A_2 to yield lysophosphatidic acid, which is a potent mitogen. If we return to the parent molecule, a phospholipase C-catalyzed reaction yields diacylglycerol, which activates the protein kinase C pathway. Or, a phospholipase A_2 acting on phosphatidylcholine can liberate arachidonic acid, which is converted to one of many eicosanoids. The same reaction also yields lysophosphatidylcholine, which has been shown to have a variety of effects on cells. Additionally, the lysophosphatidyl choline (specifically, a form that has an ether-bond at the *sn*-1 position) is a substrate for acetylation to yield platelet-activating factor, which has diverse potent action in a variety of cells. In summary, the most commonplace of membrane phospholipids can yield an amazing variety of signaling molecules. A similar spectrum of signals can be derived from phosphatidylinositol and sphingomyelin, which suggests that all complex lipids may serve as "reservoirs" and that the number of lipid-derived signals will continue to grow.

Robert M. Bell
John H. Exton
Stephen M. Prescott

Contents

Chapter 2

Lipid Signaling for Protein Kinase C Activation

Yoshinori Asaoka, Yosuke Tsujishita, and Yasutomi Nishizuka

Chapter 3

Phosphatidic Acid

Stephen B. Bocckino and John H. Exton

Chapter 6

Bioactive Properties of Sphingosine and Structurally Related Compounds

Alfred H. Merrill, Jr., Dennis C. Liotta, and Ronald E. Riley

Chapter 7

Platelet-Activating Factor and PAF-Like Mimetics

Ralph E. Whatley, Guy A. Zimmerman, Stephen M. Prescott,
and Thomas M. McIntyre

Chapter 8

Lysophosphatidic Acid

Wouter H. Moolenaar and Kees Jalink

Chapter 9

Prostaglandins and Related Compounds: Lipid Messengers with Many Actions

Elizabeth A. Meade, David A. Jones, Guy A. Zimmerman,
Thomas M. McIntyre, and Stephen M. Prescott

Chapter 1

Diacylglycerols
Biosynthetic Intermediates and Lipid Second Messengers

Andrew F. G. Quest, Daniel M. Raben, and Robert M. Bell

1.1. Introduction

Diacylglycerols (DAGs), long-recognized intermediates of lipid biosynthesis (Bell and Coleman, 1980), are now established second messengers crucial to both short- and long-term regulation of cellular function (Bishop and Bell, 1989; Nishizuka, 1992). DAGs are formed by hydrolysis of membrane phospholipids through the action of phospholipases. These intracellular events are elicited in response to highly diverse agents impinging on the exterior surface of a cell, such as growth factors, hormones, and neurotransmitters. The location of DAG formation, the molecular species generated, and the time frame of DAG transients contribute to the versatility of this lipid-derived second messenger molecule. Extended eleva- tions of DAGs, in particular derived from phosphatidylcholine, are important in the regulation of cell growth and transformation. Several functionally distinct intracellular pools of DAG exist, in particular at the plasma membrane and in the endoplasmic reticulum (ER). While DAG in the ER is a biosynthetic intermediate, plasma membrane DAGs are important for signaling. More recent studies indicate that DAG signals are also generated in intracellular compartments, such as the nucleus, where potential downstream effector molecules (protein kinase C or cytidylyltransferase) are also present. A functional distinction for DAGs on the basis of intracellular location is, therefore, no longer immediately apparent.

Andrew F. G. Quest • Institute of Biochemistry, University of Lausanne, CH-1066 Lausanne, Switzer- land. *Daniel M. Raben* • Department of Physiology, Johns Hopkins University School of Medi- cine, Baltimore, Maryland 21205. *Robert M. Bell* • Department of Biochemistry, Duke Univer- sity Medical Center, Durham, North Carolina 27710.

Handbook of Lipid Research, Volume 8: Lipid Second Messengers, edited by Robert M. Bell *et al.* Plenum Press, New York, 1996.

Rather, metabolic channeling of DAGs is emerging as a critical determinant of their fate.

In this chapter, we summarize our current understanding of the roles for DAG in biosynthesis and signaling and discuss evidence indicating how the two functions might be linked.

1.2. Lipid Biosynthesis and Distribution

1.2.1. Properties of Lipids

Lipids are amphipathic molecules characterized by the simultaneous presence within the same molecule of a hydrophobic region, which tends to aggregate, and a hydrophilic region, which may interact stably with either the intra- or extracellular water phase. These properties allow lipids to form bilayers in the biological membranes.

Phospholipids, the main component of eukaryotic cellular membranes (Van Meer, 1989), are divided into two main classes, glycero- and sphingophospholipids. Typical values for the lipid composition of membranes from different subcellular compartments are shown in Table 1-1.

Phospholipids are not thought to be sufficiently water-soluble to permit unaided diffusion through the aqueous cytosol (Chojnacki and Dallner, 1988). Since the majority of membrane phospholipid biosynthesis occurs on the ER (Bell and Coleman, 1980; Bishop and Bell, 1988; Dawidowicz, 1987; Van Meer, 1989), at least three mechanisms have been proposed for interorganelle trafficking (J. Vance, 1991): (1) Net lipid transport within the cell may result by vesicle-mediated transport, whereby vesicles bud from the donor membrane and fuse with the

Table 1-1. Typical Lipid Composition of Rat Liver Membranes[c]

Phospholipids[a]	Mitochondrial membrane[b]	ER[b]	Plasma membrane[b]	Lysosomal membrane[b]	Nuclear membrane[c]
SPM	0.5	2.5	16.0	20.3	6.3
PC	40.3	58.4	39.3	39.7	52.2
PI	4.6	10.1	7.7	4.5	4.1
PS	0.7	2.9	9.0	1.7	5.6
PE	34.6	21.8	23.3	14.1	25.1
CL	17.8	1.1	1.0	1.0	ND[d]
LBPA	0.2			7.0	
Cholesterol/phospholipid (mole/mole)	0.03	0.08	0.40/0.76	0.49	2.3

[a]SPM, sphingomyelin; PC, phosphatidylcholine; PI, phosphatidylinositol; PS, phosphatidylserine; PE, phosphatidylethanolamine; CL, cardiolipin; LBPA, lyso bisphosphatidic acid.
[b]Van Meer (1989); data expressed as percent of total phospholipid phosphorus.
[c]Gurr et al. (1963); data expressed as percent of total phospholipid phosphorus.
[d]ND, not detected.

target membrane. This process links lipid distribution within an animal eukaryotic cell to the far better characterized targeting processes of integral membrane and secretory proteins which, with the exception of mitochondrial and peroxisomal proteins, pass through and are dispatched from the ER (Van Meer, 1989). (2) Lipids may be transported within the cell by phospholipid exchange/transfer proteins (TP). Proteins, which specifically interact with specific phospholipids, have been described for phosphatidylinositol (PI-TP) and phosphatidylcholine (PC-TP). In addition, a nonspecific lipid transfer protein, also referred to as sterol carrier protein 2, exists (see Wirtz, 1991, and references therein). (3) Finally, membrane lipid flow may occur through a collision-based mechanism between membrane compartments that are juxtaposed *in vivo*, such as ER and mitochondria (J. Vance, 1991).

In contrast to the phospholipids (Table 1-1), many intermediates in the biosynthetic pathways leading to the formation of these lipids (Fig. 1-1) or phospholipid hydrolysis products in signaling pathways (Fig. 1-2) are not limited in their distribution to bilayer structures. Monoacylglycerophospholipids such as lysophosphatidylcholine (LPC) exhibit detergentlike properties (Kawashima and Bell, 1987), while lysophosphatidic acids (LPAs) are readily water-soluble at millimolar concentrations without such deleterious effects on biomembranes (Jalink *et al.*, 1990; Moolenaar, 1994). Therefore, these lipids may equilibrate between different lipid bilayer structures. However, as is the case for phospholipids, redistribution in the absence of active transport is topographically restricted by the presence of charged head groups (Ganong and Bell, 1984; Pagano and Sleight, 1985; Sleight and Pagano, 1984, 1985). Neutral lipids, like DAG or ceramide, containing long-chain fatty acids, remain confined to membrane structures in the absence of transport proteins, but may "flip" between leaflets (Ganong and Bell, 1984; Pagano and Sleight, 1985), while smaller, uncharged molecules such as fatty acids readily equilibrate throughout the cell (Chojnacki and Dallner, 1988). Finally, some of the precursor molecules for lipid biosynthesis are not amphipathic molecules, and thus are exclusively present in the aqueous phase.

Based on their partitioning properties in the absence of proteins that facilitate transport/diffusion, intermediates and products of lipid-synthesis and -degradation (signaling) pathways are divided here into five categories. In the subsequent discussion they will be referred to using the Roman numeral groupings indicated in Fig. 1-1: (I) precursor molecules limited in distribution to the aqueous phase (e.g., choline, serine); (II) amphipathic molecules that are water miscible but not topologically restricted (e.g., free fatty acids); (III) amphipathic molecules that are water miscible but topologically restricted (e.g., LPA); (IV) amphipathic molecules present in lipid bilayers which may "flip" between the two leaflets (e.g., DAG, ceramide); and (V) amphipathic molecules that are limited in their distribution to bilayer structures and are topologically restricted (e.g., phospholipids).

The combination of these partitioning properties of cellular lipids, together with the compartmentalization of proteins involved in biosynthesis and lipid transport, will determine the lipid composition of membranes in different sub-

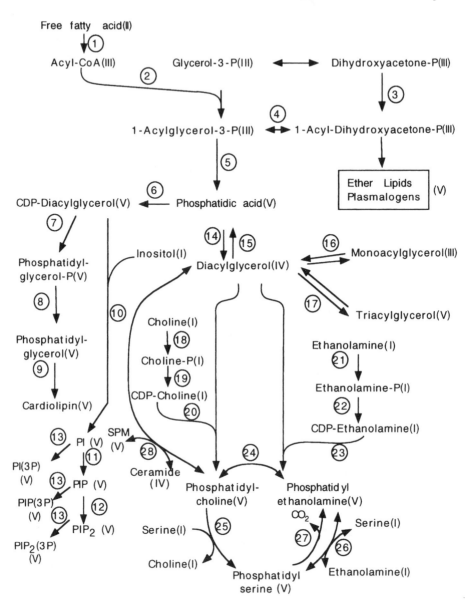

Figure 1-1. Enzymes of glycerolipid and sphingolipid biosynthesis: DAG as a versatile intermediate. Intermediates and products of lipid biosynthesis are divided into five categories (I–V) according to their partitioning properties in the absence of proteins that facilitate transport/diffusion: (I) precursor molecules limited in distribution to the aqueous phase; (II) amphipathic molecules that are water miscible but not topologically restricted; (III) amphipathic molecules that are water miscible but topologically restricted; (IV) amphipathic molecules present in lipid bilayers which may flip between the two leaflets; (V) amphipathic molecules that are limited in their distribution to bilayer structures are topologically restricted. 1, fatty acid CoA ligase (EC 6.2.1.3); 2, *sn*-glycerol 3-P acyltransferase (EC 2.3.1.15); 3, dihydroxyacetone-P acyltransferase (EC 2.3.1.42); 4, acyl(alkyl)dihydroxyacetone-P oxido-reductase (EC 1.1.1.101); 5, lysophosphatidic acid acyltransferase (EC 2.3.1.20); 6, phosphatidic acid

cellular compartments. As indicated in Table 1-1, these compositions may vary significantly, even between intracellular compartments connected by vesicular flow such as ER, Golgi, nuclear and plasma membranes. Since several possibilities for lipid exchange between membranes exist, it follows that lipid-synthesis, -degradation, and -exchange processes must be tightly regulated and coordinated within the cell.

1.2.2. Glycerolipid Biosynthesis

For an extensive discussion of phospholipid biosynthesis, the interested reader is referred to existing reviews on the topic (Bell and Coleman, 1980, 1983; Bishop and Bell, 1988; Esko and Raetz, 1983; Hannun, 1994; Van Meer, 1989). Biosynthetic pathways are illustrated (Fig. 1-1) emphasizing intracellular distribution of the enzymes (Table 1-2) as well as the nature of the substrates involved (I–V, Fig. 1-1).

During *de novo* glycerolipid synthesis, DAG is formed by the stepwise acylation of *sn*-glycerol-3-phosphate to 1-acylglycerol-3-phosphate and phosphatidic acid (PA). The enzymes in this segment of the pathway are present in microsomal (ER) and mitochondrial fractions (Table 1-2). Alternatively, acylglycerol-3-phosphate may also be generated by dihydroxyacetone phosphate acyltransferase and acyl-(alkyl) dihydroxyacetone phosphate oxidoreductase, enzymes that are present in the ER and the peroxisomes. While the overall contribution through this pathway to acylglycerol-3-phosphate synthesis is considered minor, these activities are crucial for ether lipid (plasmalogen) synthesis (Zoeller and Raetz, 1986). Acylglycerol- and acyl(alkyl)glycerol-3-phosphate intermediates synthesized in mitochondria or microsomes must return to the ER for acylation by the lysophosphatidic acid acyltransferase to yield PA, which lies at a branchpoint in mammalian glycerolipid synthesis. PA may be utilized as an intermediate for the formation of CDP-diacylglycerol by PA cytidylyltransferase present in mitochondria or on the ER.

While mitochondria have glycerol phosphate acyltransferase activity, they essentially lack LPA acyltransferase (Bell and Coleman, 1983). CDP-diacylglycerol or PA must be transported from the ER to the mitochondria to be utilized by enzymes exclusively present in the mitochondrial compartment (Table 1-2), which

---

cytidylyltransferase (CDP-diacylglycerol synthase) (EC 2.7.7.41); 7, glycerol phosphate phosphatidyltransferase (EC 2.7.8.5); 8, phosphatidylglycerol phosphate phosphatase (EC 3.1.3.27); 9, cardiolipin synthase (EC 2.7.8.-); 10, phosphatidylinositol synthase (EC 2.7.8.11); 11, phosphatidylinositol-4-kinase (EC 2.7.1.67); 12, phosphatidylinositol-4-phosphate kinase; 13, phosphatidylinositol-3-kinase; 14, phosphatidic acid phosphohydrolase (EC 3.1.3.4); 15, diacylglycerol kinase (EC 2.7.1.-); 16, monoacylglycerol acyltransferase (EC 2.3.1.22); 17, diacylglycerol acyltransferase (EC 2.3.1.20); 18, choline kinase (EC 2.7.1.32); 19, choline phosphate cytidylyltransferase (EC 2.7.7.15); 20, diacylglycerol: choline phosphotransferase (EC 2.7.8.2); 21, ethanolamine kinase (EC 2.7.1.cc); 22, ethanolamine phosphate cytidylyltransferase (EC 2.7.7.14); 23, diacylglycerol:ethanolamine phosphotransferase (EC 2.7.8.1); 24, phosphatidylethanolamine *N*-methyltransferase (EC 2.1.1.17); 25, phosphatidylcholine: serine *O*-phosphatidyltransferase; 26, phosphatidylethanolamine:serine *O*-phosphatidyltransferase; 27, phosphatidylserine decarboxylase; 28, sphingomyelin synthase.

Table 1-2. Intracellular Distribution of Enzymes Involved in Phospholipid Biosynthesis

Enzyme number	Intracellular distribution					Refs.[a]
	Cytosolic	Microsomal	Mitochondrial	Plasma membrane	Other	
1		+	+			1
2		+	+			1–3
3		+			peroxisomal	1, 2
4		+			peroxisomal	1, 2
5		+				2
6		+	+			1
7			+			1, 2
8			+			1, 2
9			+			1, 2
10		+		(+)		4, 5
11				+		6
12				+		6
13				+		7
14	+	+		+	lysosomal	1, 2
15	+	+		+	many	8
16		+				1, 9
17		+				1
18	+					2, 10
19	+	+			Golgi	10, 11
20		+				10, 11
21	+					2
22	+					2
23		+			Golgi	11
24		+			Golgi	11
25		+			Golgi	11
26	+					2
27			+			12, 13
28				+	Golgi	14, 15

[a]Table compiled using data from the following references: (1) Bell and Coleman (1980, 1983); (2) Bishop and Bell (1989); (3) Esko and Raetz (1983); (4) Imai and Gershengorn (1987); (5) Santiago *et al.* (1993); (6) Hokin (1985); (7) Majerus (1992); (8) Besterman *et al.* (1986b); (9) Bhat *et al.* (1994); (10) Pelech *et al.* (1981); (11) J. Vance and Vance (1988); (12) J. Vance (1991); (13) Hovius *et al.* (1992); (14) Van Meer (1989); (15) Van Helvoort *et al.* (1994).

generate phosphatidylglycerol phosphate, phosphatidylglycerol, and cardiolipin (Fig. 1-1), the lipids characteristic of the mitochondrial compartment (Table 1-1).

On the cytoplasmic surface of the ER, CDP-diacylglycerol is consumed to generate phosphatidylinositol (PI) by PI synthase, an enzyme that is reported also to be present at the plasma membrane (Imai and Gershengorn, 1987). Exchange of PI between ER and plasma membrane is mediated by PI-TPs (Wirtz, 1991). PI phospholipids are unique in that synthesis is not completed at the ER. Additional phosphorylation steps are required. PI-4-phosphate (PIP) and -4,5-bisphosphate (PIP_2) are formed by phosphatidylinositol and -inositol-4-phosphate kinase present at the plasma membrane (Berridge, 1987; Hokin, 1985). PI lipids phosphorylated in the 3-position of the inositol ring may also be generated at the plasma

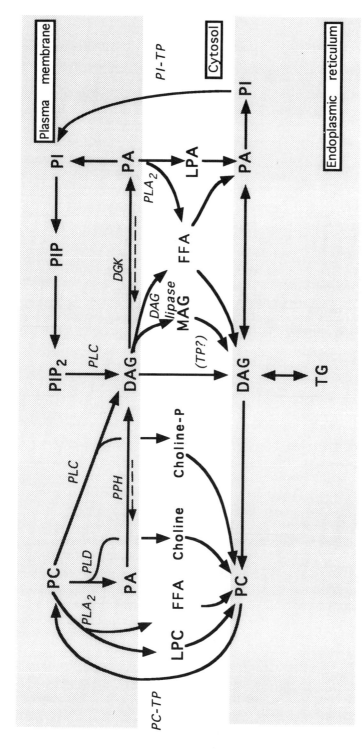

Figure 1-2. Formation and attenuation of DAG signals: lipid recycling. The events (summarized at the end of Section 1.5) leading to formation or consumption of DAG at the plasma membrane are illustrated. As a simplification, only PI and PC derivatives are considered. Events occurring at the ER are shown in greater detail in Figure 1-1.

membrane by a cytosolic PI-3-kinase in response to agonists that evoke classical phospholipid turnover at the membrane (reviewed in Majerus, 1992).

The predominant fate of PA is dephosphorylation by PA phosphohydrolase (PPH), an enzyme present at both the ER and the plasma membrane, to yield DAG which plays a central role in glycero- and sphingomyelin biosynthesis (Fig. 1-1). The reverse reaction from DAG to PA is catalyzed by diacylglycerol kinase (DGK), an enzyme that is also detected at multiple locations throughout the cell. How PPH and DGK are regulated to prevent the formation of a futile cycle between DAG and PA is not known. However, recent experiments with DGK suggest that metabolite channeling could play an important role (Van der Bend *et al.*, 1994). Both the PPH and DGK present at the plasma membrane are implicated in the control of DAG lipid second messenger levels and will be discussed again later (see Fig. 1-2).

Triacylglycerols containing three long-chain fatty acid residues, represent long-term storage molecules of lipids which are formed from DAG by a microsomal diacylglycerol acyltransferase (Bell and Coleman, 1983). Alternatively, a third short-chain C_4 fatty acid may be added by an enzyme present in microsomal preparations. Subsequent removal of the two long-chain fatty acid residues by a phospholipase(s) yields monobutyrin, a DAG-derived second messenger that stimulates both angiogenesis and vasodilation (Wilkison and Spiegelman, 1993). Also, monoacylglycerol acyltransferase, an intrinsic membrane protein of the ER, may play an important role in the regulation of glycerolipid biosynthesis because its end product is DAG. This enzyme is further implicated in the conservation of essential fatty acids released by lipase activity from tri- and diacylglycerols during physiologic periods characterized by massive lipolysis (Bhat *et al.*, 1994).

The predominant route for PC synthesis in mammalian cells is via the CDP-choline (Kennedy) pathway (D. Vance, 1990). Choline kinase, a cytosolic enzyme of which several isoforms have been described in rat liver (D. Vance, 1985), catalyzes the first step committed to PC biosynthesis. This enzyme may have a regulatory role in the synthesis of PC since mitogenic factors increase choline kinase activity in cultured fibroblasts (Warden and Friedkin, 1985). It is generally thought, however, that cytidylyltransferase, an enzyme present in both cytosolic and microsome fractions, catalyzes the rate-limiting step in PC biosynthesis (Pelech and Vance, 1984). The cytosolic form appears to represent an inactive reservoir of the enzyme and activation occurs by membrane translocation. Association of cytidylyltransferase with microsomes occurs under conditions that stimulate PC biosynthesis, such as the addition of fatty acids to hepatocytes or phorbol esters to cultured cells (Pelech and Vance, 1984) or the treatment of cells with phospholipase C (PLC) (Sleight and Kent, 1983; Wright *et al.*, 1985). Conversely, inhibition of PC biosynthesis is accompanied by partial inactivation of the microsomal cytidylyltransferase activity and redistribution of the enzyme to the cytosol (Pelech *et al.*, 1981). While phosphorylation of the cytidylyltransferase by cAMP-dependent protein kinase has been proposed to release the enzyme from the membrane, it now seems more likely that the cAMP-induced reduction in PC synthesis results from decreased levels of DAG, which limit the choline phosphotransferase reaction and thereby the rate of PC biosynthesis (Jamil *et al.*, 1992). The last step in PC biosynthesis is catalyzed by an intrinsic microsomal protein, choline phosphotransferase. A similar phosphotransferase catalyzes the synthesis of both diacyl- and alkylacyl-PC species, while a unique enzyme is involved in the *de novo* synthesis of platelet-activating factor (PAF) (Chao and Olson, 1993; Woodard *et al.*, 1987).

PC biosynthesis may also occur through successive transfer of methyl groups from *S*-adenosylmethionine to phosphatidylethanolamine (PE). PE *N*-methyltransferase is an intrinsic membrane protein present in the hepatic microsomal fraction which displays all three methyltransferase activities. The rate-limiting step is the formation of phosphatidylmonomethyl PE. In mammals, the methylation pathway is thought to be significant only in liver where the PE methyltransferase route accounts for 20–30% of PC synthesis (Ridgeway and Vance, 1987).

Once PC has been formed, the fatty acyl composition may be modified by consecutive deacylation/reacylation reactions. This process allows for the introduction of polyunsaturated fatty acids like arachidonic acid into the C-2 position of the glycerol backbone. Deacylation is catalyzed by phospholipases A_1 and A_2, which in addition to their role in lipid biosynthesis play an important role in signal transduction events involving lipid-derived second messengers, as will be discussed later. The two phospholipases are present in the cytosol fraction and may associate transiently with the membrane for catalysis (Dennis, 1994; Pete *et al.*, 1994). Phospholipases of the A_2 type are also found in secretory granules and mitochondria (Dennis, 1994). Reacylation is catalyzed by lyso PC:acyl CoA acyltransferase, an activity found in microsomal preparations (Van Heusden *et al.*, 1981). A specific acyltransferase (acetyl CoA:1-alkyl-*sn*-glycerophosphocholine acetyltransferase) is important in one of the pathways leading to PAF formation (Chao and Olson, 1993; Chilton *et al.*, 1984).

The primary route of phosphatidylserine (PS) biosynthesis in mammalian cells occurs by base exchange of free serine with the polar head group of existing phospholipids by PC:- or PE:serine *O*-phosphatidyltransferase present in microsomal and Golgi fractions (Bishop and Bell, 1989; J. Vance and Vance, 1988).

Three pathways have been described in mammalian cells for the biosynthesis of PE: (1) the CDP-ethanolamine pathway, (2) PS decarboxylation, and (3) the exchange of free ethanolamine for the head group moiety of PS or PC (Miller and Kent, 1986). PE biosynthesis by the CDP-ethanolamine route is similar to that described above for PC. The first steps in the CDP-ethanolamine pathway, phosphorylation of ethanolamine and subsequent formation of CDP-ethanolamine, are catalyzed by two cytosolic enzymes, ethanolamine kinase and CTP-ethanolamine cytidylyltransferase (D. Vance, 1985). As with the CDP-choline pathway, the second step catalyzed by the cytidylyltransferase is rate-limiting (Miller and Kent, 1986). Ethanolamine phosphotransferase, an integral microsomal membrane protein, catalyzes the formation of PE from CDP-ethanolamine (Polokoff *et al.*, 1981).

Initially, the CDP-ethanolamine pathway was considered the major route of PE synthesis in mammalian cells. However, even at physiological concentrations of ethanolamine (20 μM), the CDP-ethanolamine pathway contributes only 20–30% of the PE formed, and most mammalian cells do not require ethanolamine for growth (Miller and Kent, 1986; Voelker, 1984). Furthermore, PS decarboxylation is sufficient to support PE synthesis in the BHK-21 cell line (Voelker, 1984) and PS is the sole precursor of PE in Chinese hamster ovary (CHO) cells (Miller and Kent, 1986). Together, these observations indicate that PS decarboxylation represents the major PE synthesis pathway in mammalian cells (see also J. Vance, 1991).

1.2.3. Topography of Glycerolipid Biosynthesis

A question of principal interest is that concerning the origin of membrane asymmetry. In addition to being confined to bilayer structures in the cell, phospholipids are topographically restricted. The asymmetric distribution of phospholipids between membrane leaflets is well documented for the plasma membrane (Bretscher, 1972; Rothman and Lenard, 1977; Sleight and Pagano, 1978) and is most likely established in the ER or Golgi. Although the actual quantities of individual lipids present in inner and outer leaflet are controversial, it is generally agreed, for most biological membranes, that the two leaflets differ in their composition. In erythrocytes, the best characterized cellular system, PS, PE, and probably PI are located preferentially in the inner (cytoplasmic) leaflet, while PC and sphingomyelin (SPM) prevail in the outer (exoplasmic) leaflet of the membrane (Devaux, 1991).

While lateral diffusion of lipids within a bilayer is very rapid (10^7 sec^{-1}), the rate of protein-independent transverse diffusion, or flip-flop, is generally slow (the half-time for exchange is several hours) (Devaux, 1991). Thus, asymmetric distribution of the biosynthetic enzymes represents an intriguingly simple possibility to generate bilayer asymmetry. Indeed, studies employing proteases and impermeant mercury-dextran inhibitors revealed that the catalytic domains of most enzymes of the glycerophospholipid biosynthetic pathway, with the exception of those present in peroxisomes and mitochondria, are exposed on the cytoplasmic surface of the ER (Ballas and Bell, 1981; Ballas et al., 1984; Bell et al., 1981).

It is noteworthy that the cytoplasmic surface of the ER becomes the predominant site of glycerophospholipid biosynthesis in a cell because of the presence of essentially three biosynthetic enzymes located almost exclusively in the ER, namely, lysophosphatidic acid acyltransferase, diacylglycerol:choline and diacylglycerol:ethanolamine phosphotransferase (Table 1-2). These enzymes convert water-soluble substrates, which may equilibrate throughout the cell, to amphipathic molecules that are topologically restricted to the cytoplasmic surface of the ER (Fig. 1-1). Because of the properties of phospholipids (V, amphipathic, non-water-miscible), the ER grows by synthesis and insertion of new molecules. Exchange of phospholipids between membranes must now occur by the three possibilities initially discussed, namely, vesicle transport, protein-facilitated transport, or collision exchange (J. Vance, 1991).

The fact that PC is synthesized on the cytoplasmic leaflet of the ER, yet appears preferentially on the exoplasmic leaflet of the plasma membrane (Devaux, 1991), indicates that transverse glycerophospholipid segregation is not established exclusively by synthesis and that transbilayer mechanisms for phospholipid transport exist. Indeed, rapid transmembrane movement of the aminophospholipids PS and PE by an ATP-requiring process was demonstrated first in human red blood cells and later in other plasma membranes. It has been suggested that the same mechanism could explain the ATP requirement for the maintenance of erythrocyte membrane shape (Devaux, 1988; Seigneuret and Devaux, 1984). Rapid translocation of PC (within 45 min or less) has also been observed in microsomal preparations (Van den Besselaar *et al.*, 1978; Zilversmit and Hughes, 1977), but not across the plasma membrane (Sleight and Pagano, 1984). A PC transporter or "flippase" present in microsomal membrane preparations was implicated in this process by an ATP-independent mechanism (Bishop and Bell, 1985). Others have reconstituted a PC transporter in proteoliposomes using detergent-solubilized protein from rat liver microsomes (Backer and Dawidowicz, 1987).

Thus, while asymmetric synthesis of glycerophospholipids certainly contributes to the final asymmetry of these lipids in cellular membranes, lipid transport processes are also important.

1.2.4. Intracellular Lipid Transport

The ER is considered the major site of phospholipid biosynthesis (Bell and Coleman, 1980; Bishop and Bell, 1988; Dawidowicz, 1987; Van Meer, 1989). The Golgi apparatus has some phospholipid biosynthesis capacity for both glycero- and sphingophospholipids (Hannun, 1994; Higgins and Fieldsend, 1987; Pagano, 1988; J. Vance and Vance, 1988; Van Meer, 1989). Both the plasma and nuclear membranes appear almost devoid of such activities, which correlates with their involvement in signal transduction events. This physical separation of biosynthetic and signaling pathways prevents the formation of futile cycles, but requires that membrane flow be carefully coordinated to maintain lipid composition (Table 1-1). Membrane flow from ER/Golgi to peripheral destinations has been reviewed elsewhere (Van Meer, 1989). Here, we will consider briefly some results concerning lipid transport from the plasma membrane back to the cell interior, as these will become relevant in the discussion of sphingolipid biosynthesis and lipid-dependent signaling events.

A valuable approach to study lipid transport has utilized synthetic phospholipids containing a modified fluorescent acyl residue, N-4-nitrobenzo-2-oxa-1,3-diazole amino caproic acid (C_6NBD) at the *sn*-2 position. These lipids are spontaneously transferred to the plasma membrane of intact cells at 2°C (Struck and Pagano, 1980) and their fate in cells is followed by fluorescence microscopy. However, it should be noted that results obtained concerning lipid distribution with C_6NBD fluorescent lipid probes may not accurately reflect the distribution of endogenous lipids. These fluorescent derivatives are relatively polar in comparison to natural lipids because of the presence of the short-chain fatty acid derivative at the *sn*-2 position, which is reflected in significantly enhanced simultaneous diffusion rates (Nichols and Pagano, 1983).

Both C$_6$NBD-PE and -PC, introduced into the plasma membrane of fibroblasts at 2°C, enter the cell by endocytosis after warming to 37°C and accumulate on the luminal surface of the Golgi (Sleight and Pagano, 1984, 1985). In this case, the final destination is determined by the initial site of insertion into the plasma membrane and requires endocytosis (Pagano and Sleight, 1985; Sleight and Pagano, 1984). C$_6$NBD-PE is also transferred to the nuclear envelope and mitochondria in a manner that is independent of endocytosis but requires transmembrane movement at the plasma membrane (Pagano and Sleight, 1985; Sleight and Pagano, 1985). When cells are treated with C$_6$NBD-PA at 2°C, rapid fluorescent labeling of the mitochondria, ER, and nucleus are apparent (Pagano *et al.*, 1983; Pagano and Sleight, 1985). Uptake and intracellular translocation of PA involves the formation of DAG at the plasma membrane followed by its transbilayer movement, facilitated transport to intracellular membranes, and rephosphorylation (Pagano and Longmuir, 1985; Pagano and Sleight, 1985). When cells are incubated with C$_6$NBD-ceramide at 2°C, the initial pattern of fluorescence is similar to that seen with C$_6$NBD-PA (Lipsky and Pagano, 1983). On the basis of its properties *in vitro*, it has been proposed that the fluorescent ceramide is inserted in the outer leaflet where it undergoes transmembrane movement to the cytoplasmic face and is subsequently translocated to intracellular membranes by spontaneous or facilitated diffusion or both (Pagano and Sleight, 1985). On warming of the cells to 37°C, first the Golgi and then the plasma membrane become fluorescently labeled. During this redistribution C$_6$NBD-ceramide is metabolized to fluorescent sphingomyelin and cerebroside. Monensin blocks transfer to the plasma membrane but does not affect sphingomyelin synthesis. From these studies it was concluded that the Golgi apparatus is the intracellular site of sphingomyelin biosynthesis (Pagano, 1988; Pagano and Sleight, 1985) and not the plasma membrane as suggested by others (reviewed in Van Meer, 1989). More recently, it has been proposed that the reverse action of a sphingomyelin synthase contributes substantially to PC biosynthesis (see Fig. 1-1) at the basolateral membrane of Madin-Darby canine kidney (MDCK) cells (Van Helvoort *et al.*, 1994).

1.2.5. *Sphingolipid Biosynthesis*

Ceramide, the precursor molecule present in all sphingolipids, is synthesized in steps starting with the condensation of serine and palmitoyl-CoA, to yield 3-ketosphinganine, which is then reduced to dihydrosphingosine. Dihydroceramide formed by amide linkage of fatty acids to dihydrosphingosine is subsequently converted to ceramide containing a *trans*-4,5 double bond (Hannun, 1994). The initial enzymes in this pathway appear to reside in the ER, probably facing the cytoplasm, since the substrates palmitoyl-CoA (Polokoff and Bell, 1978) and NADPH (Takahashi and Hori, 1978) cannot penetrate microsomes. The site of introduction of the double bond to yield ceramide is not known. Incorporation of ceramide into more complex sphingolipids occurs in the Golgi apparatus (Hannun, 1994; Pagano, 1988; Van Meer, 1989). The various steps in SPM and glyco-sphingolipid synthesis occur on the luminal surface of the Golgi. Since none of the sphingolipids can flip-flop, their exclusive presence in the exoplasmic leaflet must be established during biosynthesis (Van Meer, 1989). In Fig. 1-1, only the step involving SPM synthase is indicated. An important aspect of the phosphatidyl-choline:ceramide choline phosphotransferase (sphingomyelin synthase) activity is that it could serve to regulate simultaneously ceramide and DAG levels (Hannun, 1994; Van Helvoort *et al.*, 1994). DAG produced at the Golgi may be hydrolyzed by lipases, metabolized to PA by DGK, or recycled to the ER by lipid transfer proteins (Pagano, 1988).

1.3. The Emergence of Lipid Second Messengers

1.3.1. Phosphatidylinositol Turnover

Turnover of inositol phospholipids in response to hormone stimulation was initially observed 40 years ago. When slices of pigeon pancreas were treated with acetylcholine or carbamyl choline, a roughly 10-fold overall increase in the incorporation of ^{32}P-labeled phosphate into phospholipids was seen, which correlated with enhanced secretion of amylase. Stimulation of label incorporation varied considerably between phospholipids, being 15-fold for PI, 3-fold for PA, and considerably lower for PC and PE (reviewed in Hokin, 1985). The significance of this preferential turnover of PI lipids was initially puzzling and required several years of additional research to become fully understood.

An initial insight was the observation that agonists that stimulated inositol phospholipid hydrolysis also induced intracellular mobilization of an established second messenger, calcium (Michell, 1975). Consistent with the notion that one of the hydrolysis products of PIP$_2$, inositol-1,4,5-trisphosphate (IP$_3$), was acting as a second messenger, measurements in different cell types revealed that the increase in IP$_3$ either preceded or coincided with the onset of the calcium signal (Berridge, 1987). IP$_3$ is now believed to modulate intracellular calcium levels by controlling calcium channels at both the plasma membrane and ER (Berridge and Irvine, 1989).

The second major insight related to the other hydrolysis product formed by PLC-dependent cleavage of PIP$_2$, namely, DAG. Unlike IP$_3$, which is free to diffuse throughout the cytosol, DAG remains within the plane of the lipid bilayer and must be effective there before being metabolized. In 1977, a cyclic nucleotide-independent kinase activity from brain was described, which was initially thought to be activated by proteolytic cleavage (Inoue *et al.*, 1977; Takai *et al.*, 1977), but later found to be activated reversibly by lipid-soluble membrane components in the presence of calcium (Kaibuchi *et al.*, 1981; Takai *et al.*, 1979a,c). Negatively charged phospholipids such as PS, PI, or PA effectively replaced the membrane components as kinase activators. Although PE on its own was a poor activator, it cooperatively enhanced PS-stimulated kinase activity. By contrast, the presence of PC or SPM inhibited PS-dependent kinase activity (Kaibuchi *et al.*, 1981; Schatzman *et al.*, 1983; Wise *et al.*, 1982). Thus, phospholipids prevailing on the inner leaflet of the plasma membrane were the most effective activators.

Additionally, a neutral lipid, *sn*-1,2-diacylglycerol, was found to drastically enhance the reaction velocity while simultaneously reducing the calcium requirement (Kishimoto *et al.*, 1980; Takai *et al.*, 1979a,b) of this kinase activity which became known as protein kinase C (PKC). DAG was later proposed to serve as a second messenger for PKC activation (Nishizuka, 1984). Until this point DAG had been perceived exclusively as an intermediate of glycerolipid synthesis and breakdown (Bell and Coleman, 1980).

Interest in PKC and DAG as a second messenger was additionally enhanced by the observation that a tumor-promoting phorbol ester, tetradecanoylphorbol-13-acetate (TPA), activated PKC in the presence of phospholipids. Phorbol esters induce a wide variety of cellular responses *in vitro* including altered cell proliferation, differentiation, intercellular communication, and, in the mouse skin model, tumor promotion (Blumberg, 1980, 1981). Like DAG, TPA reduced the calcium requirement. Furthermore, TPA stimulated PKC-specific phosphorylation events in platelets in the complete absence of PI breakdown (Castagna *et al.*, 1982). DAG inhibited competitively the binding of phorbol esters to PKC, indicating that the binding sites of the two activators to the PKC–lipid complex in the presence of calcium are very similar, if not identical (Sharkey and Blumberg, 1985). PKC distribution in tissues and subcellular fractions correlated with phorbol ester binding activity. The specificity of PKC activation by different phorbol esters was identical to that of the phorbol ester receptor and the latter copurified with PKC (Ashendel, 1985). The identification of PKC as the phorbol ester receptor linked intracellular DAG formation to the specific activation of PKC and implicated both in signaling events related to cell growth.

1.3.2. DAG as a Second Messenger

For the aforementioned historical reasons, much of the initial work on lipid second messengers focused on products resulting from the hydrolysis of PI polyphosphates. At least three experimental criteria were utilized to establish DAG as a second messenger: (1) signal-induced, transient accumulation of DAG; (2) characterization of the molecular "features" required for DAG to act as an activator of PKC; (3) the capacity of DAG exogenously added to cells, or produced by cell treatment with PLC, to elicit responses similar to those characterized for phorbol esters. Identifying DAG as an intracellular activator of PKC was closely linked to establishing its role as second messenger.

Signal-induced transient accumulation of DAG was observed in response to various agonists in multiple tissues, including platelets on stimulation with thrombin, collagen, or PAF, pancreas with acetylcholine, mast cells with the polycationic compound 48/80, thyroid follicles with thyrotropin, Swiss mouse 3T3 cells with platelet-derived growth factor (PDGF), aortic epithelial cells with bradykinin, pituitary cells with thyrotropin-releasing hormone, and hepatocytes with vaso-pressin (reviewed in Nishizuka, 1984).

DAG was reported to activate PKC by increasing the affinity for the essential cofactors calcium and phospholipid. Assays employing lipids as multilamellar vesicles indicated that DAGs containing at least one unsaturated fatty acid at either position 1 or 2 were most effective at activating PKC, irrespective of the chain length of the other moiety (C_2 to C_{18}). DAGs with two saturated fatty acids, such as dipalmitin or distearin, were less effective (Mori et al., 1982). The presence of a short chain in position 2 increased solubility and sn-1-oleoyl-2-acetylglycerol (OAG) stimulates rapid phosphorylation of an endogenous 40-kDa protein in intact platelets within the time frame of events elicited by thrombin (Kaibuchi et al., 1983). DAG was subsequently found to activate PKC regardless of the extent of fatty acid unsaturation or chain length, providing the DAG was sufficiently hydrophobic to partition into phospholipid vesicles or phospholipid–detergent mixed micelles (Ganong et al., 1986a,b). Thus, 1-stearoyl-2-arachidonoylglycerol, the most prominent molecular species of DAG derived from PI lipids (Hannun et al., 1986), and sn-1,2-dioctanoylglycerol (diC_8) were equally effective, but sn1,2-dibutyroylglycerol, which is very soluble, was not. DAGs with long-chain fatty acids have limited water solubility and are difficult to deliver to cells. Like OAG, diC_8 proved to be cell permeable and to activate PKC in numerous cell types (Ganong et al., 1986a).

Assays employing Triton X-100 mixed micelles revealed that PKC activation by lipids does not require a lipid bilayer, that monomeric PKC is the active species, and that stoichiometric interactions of PKC with one DAG molecule in the presence of calcium and PS are sufficient for maximum activation (Hannun et al., 1985, 1986). These results are in agreement with the findings of others who quantified PKC interactions with phorbol ester and DAG in the presence of lipid vesicles (König et al., 1985). To determine the structural features of DAG essential to activate PKC, over 20 DAG analogues were prepared and tested as PKC activators or antagonists. Both the carbonyl moieties of the oxygen esters (amides and ethers were inactive) and the 3-hydroxyl moiety are required (Ganong et al., 1986b) and activation is stereospecific (Boni and Rando, 1985; Hannun et al., 1986; Rando, 1988).

Phorbol ester-induced phosphorylation events are mimicked by the exogenous addition of cell-permeable DAGs in a large number of cells, including human platelets, lymphocytes, neutrophils, HL-60 human erythroleukemia cells, human and sheep erythrocytes, rat peritoneal mast cells, GH_3 pituitary cells, and 3T3-L1 fibroblasts. Exogenous addition of DAGs inhibits phorbol ester binding to human lymphocytes, neutrophils, A431 human epidermoid carcinoma cells, and 3T3 fibroblasts. Finally, responses induced by the exogenous addition of either DAGs or phorbol esters are mimicked by addition of bacterial PLC in human neutrophils, human T lymphocytes, Friend erythroleukemic cells, human and sheep erythrocytes, guinea pig lymphocytes, rat tracheal epithelial cells, rat anterior pituitary cells, GH_3 pituitary cells, GH_4C_1 pituitary cells, mouse B lymphocytes, and mouse 3T3 fibroblasts (reviewed in Ganong et al., 1986a).

Taken together, these results firmly established DAG as an intracellular second messenger which activates PKC by an interaction that was stoichiometric, stereospecific, and occurred in the presence of calcium and PS cofactors. However, evidence also exists suggesting that structural features of the membrane, which are influenced by the presence of DAG, might play an important role in PKC activation (Slater *et al.*, 1994; Zidovetzki and Lester, 1992). Because PKC is a target for tumor promoters and since several oncogenes resemble transmembrane signaling proteins, it was suggested that DAGs produced in response to growth factors stimulated cell proliferation and that specific oncogenes may alter DAG levels (Bell, 1986). DAG levels present in K-*ras*-transformed rat kidney cells grown at 34 and 38°C are increased by 168 and 138%, respectively, in comparison to normal rat kidney (NRK) cells. Such elevated levels of DAG are also seen in *sis*-transformed NRK cells (Preiss *et al.*, 1986) as well as in H-, K-, and N-*ras*-transformed NIH 3T3 cells (Fleischman *et al.*, 1986). The data support the hypothesis that altered levels of DAG second messengers play an important role in cellular transformation.

The IP_3-mediated increase in intracellular free calcium is thought to synergize with DAG to translocate PKC to the membrane and activate the enzyme (Berridge, 1987; Nishizuka, 1984). In addition to the well-documented ability of ligand-induced polyphosphoinositide degradation to translocate and activate PKC, DAG has been proposed to activate PKC in the absence of an increase in the free cytosolic calcium concentration (Di Virgilio *et al.*, 1984; Drust and Martin, 1985). Fatty acid analysis of the DAG produced in hepatocytes in response to hormonal agonists, such as vasopressin, epinephrine and angiotensin II, or TPA, shows that the rapid increase in DAG is only partially derived from phosphoinositides (Bocckino *et al.*, 1985; Hughes *et al.*, 1984). Although phorbol esters do not enhance phosphoinositide hydrolysis, rapid generation of DAG is seen in [^{14}C]oleate-prelabeled HL-60 cells on treatment with phorbol dibutyrate. In [^{3}H]choline-prelabeled 3T3-L1 preadipocyte cells, addition of phorbol esters, serum, or PDGF induces the rapid formation of [^{3}H]phosphocholine within 15 sec. These results imply that DAG may be generated by PLC-mediated cleavage from PC (Besterman *et al.*, 1986a). Subsequently, it has become clear that DAG is also generated from PC and PE by a pathway involving phospholipase D (PLD) and subsequent PPH activity (see Exton, 1990). Thus, glycerophospholipids serve as precursors for the formation of DAG by multiple agonist-dependent phospholipases and these events are elicited in different membranes, in particular at the plasma and nuclear membranes, as will be discussed below.

1.4.   Intracellular Formation of DAG Second Messengers

1.4.1.   Time Frame of DAG Formation

As just discussed, the role of DAGs in signal transduction cascades is now widely recognized. These lipids are increased in response to diverse signals which may lead to very different endpoints for the cell, such as mitogenesis and differentiation. Since DAGs are biosynthetic intermediates, they are also present in the uninduced cells, indicating that signaling events mediated by diglycerides require: (1) a certain threshold level of diglyceride for activation; (2) the appearance, disappearance, or change in the level of a particular molecular species on stimulation; or (3) both of the above. In addition, recent evidence supports the notion that the subcellular compartment in which DAG changes occur is likely to be a major factor in modulating the ultimate physiological response.

Three important parameters need to be considered in assessing the role of diglyceride in mediating a particular response. First, the kinetics of DAG production must be established. This is important in determining the possible relationships between DAG production and the response under investigation. Second, changes in molecular species of induced DAGs must be quantified. This will aid in establishing their potential roles as second messengers, as well as their physiological source. Finally, identification of the subcellular compartment in which DAGs are formed, and lipid topography within the compartment, are likely to be essential in establishing the mechanisms of regulation and physiological roles of DAGs.

It is now evident that the kinetics of diglyceride production depend on the nature and concentration of the agonist in question, as well as the cell type studied. For example, in IIC9 fibroblasts the kinetics of α-thrombin-induced diglyceride production is concentration dependent. High concentrations of α-thrombin (14 nM) induce a biphasic increase in diglyceride mass. The early phase peaks at 15 sec and a later phase peaks at 5 min after the addition of catalytically active α-thrombin (Wright et al., 1988). Diglyceride mass remains elevated for at least 4 h as long as catalytically active α-thrombin is present. Similar biphasic increases in diglyceride production have been observed in a number of other systems (see Exton, 1990). On the other hand, low mitogenic concentrations (2.8 pM) induce a monophasic rise in diglyceride mass (Wright et al., 1988).

In contrast to α-thrombin, the polypeptide growth factors EGF and PDGF induce monophasic increases in diglyceride mass at all mitogenic concentrations in IIC9 cells (Pessin et al., 1990; Wright et al., 1988). However, in NIH-3T3 cells, PDGF-induced DAG production is biphasic (Fukami and Takenawa, 1989). In PC12 cells, the neurotrophic factors NGF and bFGF stimulate biphasic increases in diglycerides (Altin and Bradshaw, 1990) while the muscarinic agonist carbachol stimulates a monophasic rise in diglyceride mass (Altin and Bradshaw, 1990; Horwitz, 1990). These differences are likely related to the ability of these agonists to stimulate multiple DAG-generating pathways and the particular pathway(s) activated is cell-type dependent.

Monophasic or biphasic DAG production reflect differences in the sources of this lipid. When DAG production is monophasic, it is derived largely from PC hydrolysis (Billah and Anthes, 1990; Exton, 1990). Alternatively, when DAG production occurs in two phases the transient initial phase is mainly derived from phosphoinosides while the more sustained phase is derived largely from PC (Exton, 1990; Pessin and Raben, 1989; Pessin et al., 1990; Raben et al., 1990). The physiological significance of these kinetics is not entirely clear but there is much speculation about the potential requirement for the sustained phase in modulating long-term responses including mitogenesis (Nishizuka, 1992, and see below).

1.4.2. Identification of DAG Molecular Species and Potential Sources

Identifying the molecular species of DAGs produced in response to an agonist will be important for establishing their source and potential physiological roles (see below). Essentially two parameters are used to define the molecular species of a diglyceride or phospholipid. First, while fatty acids are usually attached to the sn-2-glycerol carbon via an ester linkage, the fatty acid at sn-1 may be linked via (1) ester, (2) ether, or (3) alkenyl ether. Second, molecular species are further defined by the fatty acid pairs attached to the glycerol backbone. Bond linkage analysis may be accomplished by taking advantage of the differential stability of these bonds under alkaline and acidic conditions while fatty acid pairs may be identified by TLC, HPLC, or GC methodologies (reviewed in Leach and Raben, 1993; Raben et al., 1990).

One important question concerning agonist-induced diglycerides refers to identification of their source. There are three possible sources; (1) de novo synthesis, (2) hydrolysis of triglycerides, and (3) hydrolysis of an existing phospholipid. Differentiating among these possibilities requires analysis of released head groups, acute labeling with radioactive glycerol, and molecular species analysis.

The involvement of de novo synthesis is difficult to determine. The appearance of agonist-induced radiolabeled diglycerides in cells acutely labeled with radiolabeled glycerol is often taken as an indication of de novo synthesis. However, PA is an obligatory precursor to not only diglyceride synthesis but phospholipid synthesis as well. Therefore, it is important to establish that phospholipids were not labeled, or that the label was not "cycled through" a phospholipid in order to establish this pathway as a source of the agonist-induced diglycerides. In some selected cases, the appearance of diglyceride molecular species, known to be present only in phospholipids synthesized de novo, is useful (Ha and Thompson, 1992).

Hydrolysis of triglycerides results in the generation of both sn-1,2- and its enantiomer, sn-2,3-diglycerides. The appearance of sn-2,3-diglyceride is indicative of diglyceride formation by triglyceride hydrolysis. Since the E. coli diglyceride kinase enzyme is stereospecific for the sn-1,2 isomer, the difference in the amount of diglyceride measured by this enzyme and the amount measured by GC, which is not selective for either isomer, represent the amount of sn-2,3 present in the samples. In order to detect and quantify potentially low levels of the sn-2,3 isomer, the sn-1,2 isomer may be converted to PA followed by GC analysis of the remaining diglyceride to quantify the amount of sn-2,3 isomers.

The participation of phospholipid hydrolysis in the production of agonist-induced diglycerides may be determined by examination of the release of radiolabeled head groups from cells that had been metabolically labeled with the appropriate precursors (Wright et al., 1988), or by mass analysis of these products if possible (Kennerly, 1991; Palmer et al., 1989; Rittenhouse and Sasson, 1985). However, alternative sources and metabolism of these water-soluble components complicate the interpretation of these results. For example, while release of phosphorylcholine would suggest an activation of a PC-PLC, this metabolite may also be derived from the hydrolysis of SPM via a sphingomyelinase, a signaling pathway that is now drawing a considerable amount of attention (Bell *et al.*, 1992; Hannun, 1994). Alternatively, increases in diglyceride may activate the rate-determining enzyme in PC biosynthesis, CTP-phosphocholine cytidylyltransferase (Kent, 1990; Kolesnick and Hemer, 1990), leading to a decrease in phosphorylcholine as a result of its accelerated conversion to CDP-choline.

A more definitive approach to establishing the phospholipid precursors of induced DAGs involves an analysis of the molecular species of the DAGs and endogenous phospholipids (reviewed in Leach and Raben, 1993; Raben *et al.*, 1990). In this approach, the molecular species profile of DAGs is compared to that of the potential parent phospholipids (Leach *et al.*, 1992; Pessin and Raben, 1989; Pessin *et al.*, 1990). This analysis has proven useful in establishing the source of DAGs generated in response to mitogen in fibroblasts (Pessin and Raben, 1989; Pessin *et al.*, 1990) and in response to neurotransmitters and neutrophic factors in PC12 cells (Pessin *et al.*, 1991). Similar analyses have been used to identify the phospholipid from which induced PA is derived in other systems (Augert *et al.*, 1989; Holbrook *et al.*, 1992; Lee *et al.*, 1991).

There are two factors that may complicate interpretations about phospholipid sources based solely on molecular species analyses. First, conclusions about parent phospholipids may be compromised if there is selective hydrolysis of phospholipids containing certain molecular species. This is not likely to be a major factor because phospholipases, with the exception of the arachidonyl-selective and plasmalogen-selective intracellular phospholipase A_2s identified in myocardium (Hazen *et al.*, 1990; Wolf and Gross, 1985; reviewed in Dennis, 1994), do not appear to prefer phospholipids with any particular molecular species. It is possible, however, that the phospholipids hydrolyzed are restricted to a specific subcellular compartment in which the phospholipid molecular species profile is unique. This is particularly important when whole-cell phospholipids and DAGs are analyzed. Second, selective metabolism, such as the phosphorylation of specific species, may also complicate the analysis. This complication is illuminated by the identification of the DGKs which prefer certain DAG species such as ether-linked DAGs (Ford and Gross, 1990) or DAGs containing arachidonic acid (MacDonald *et al.*, 1988) (discussed below).

Using one or more of the above techniques, three potential phospholipid sources of agonist-induced diglyceride have been identified: PI, PC, and PE. In addition to the well-recognized role of the PI cycle, evidence for a PC cycle exists, in which PC is metabolized to PA and/or diglycerides followed by the "re-synthesis" of PC. In some systems, PC appears to be the predominant, if not exclusive, source of induced diglyceride (reviewed in Billah and Anthes, 1990; Cook and Wakelam, 1992; Exton, 1990). As with PI metabolism, it is not entirely clear that the metabolites derived from PC hydrolysis, PA and diglyceride, are used as the immediate precursors for PI or PC resynthesis. There is also compelling evidence implicating PE metabolism as a signal-transducing component in some systems. For example, phorbol esters have been shown to stimulate PE hydrolysis, possibly via a PLD-type enzyme, in certain cell types (Kiss and Anderson, 1989). Plasmalogen PEs and PEs with a vinyl ether linkage at sn-1 have also been implicated as important stores for arachidonic acid released in response to some agonists (Hazen *et al.*, 1991).

1.4.3. Compartmentalization of DAG Species

Three lines of evidence indicate that the molecular species of a DAG is important for its biological role. First, the number of DAG species actually present in biological tissues is lower than the number of possible species. At least ten fatty acids are commonly found in eukaryotic cells. Considering only that each DAG contains two fatty acids, and ignoring positional isomers involving all three glycerol carbons (sn-1,2 versus sn-2,3), 1024 molecular species are possible. However, only approximately 20–50 molecular species of DAGs are actually observed (Cook, 1991). This relatively restricted number of DAG species suggests that specific roles for selected species may indeed exist. Second, the fact that DAG molecular species are differentially metabolized, as discussed below, lends further support to the notion that the nature of the DAG is important in its ability to affect biological activities. Finally, the profile of DAGs and phospholipids differs not only between tissues but also among subcellular compartments, providing additional evidence for roles of selected molecular species.

Despite the above, there is a lack of knowledge regarding the potential roles of specific DAG species for the following reasons. First, most of the research investigating the roles for select populations of DAG species involved the activation of specific PKCs. Clearly, as this review will outline subsequently, DAGs serve other essential roles (see Section 1.6) and the effects of specific molecular species have not been examined there. Second, an important aspect that is beginning to emerge is that it may not be the precise molecular species, or population of species, that is important but rather the molecular species in the context of its microenvironment. That is, in defining a role for a particular molecular species, it may be important to understand the subcellular compartment in which the DAGs are located.

1.4.3.1. DAG Formation in the Nucleus

It was previously assumed that DAG formation by agonist-induced phospholipid hydrolysis occurred at the plasma membrane. However, quantification of induced diglyceride mass indicated that not all of the induced diglyceride could be present at the plasma membrane. In vasopressin-stimulated hepatocytes (Bocckino *et al.*, 1985; Exton, 1990) and α-thrombin-stimulated fibroblasts (Pessin and Raben, 1989; Pessin *et al.*, 1990; Wright *et al.*, 1988), the total amount of induced diglyceride represents nearly 1% of the total mass of cellular lipid. If all of this diglyceride were present in the plasma membrane, it would result in drastic and detrimental changes in the physical properties of this structure (Siegel *et al.*, 1989). In addition, in GH_3 cells stimulated with thyroid releasing hormone

(TRH), the small initial diglyceride production (possibly derived from PIPs) occurs at the plasma membrane while the larger later phase of diglyceride production (likely derived from PC) is produced in internal membranes (Martin *et al.*, 1990). These findings demonstrate that agonist-induced diglyceride production must occur at sites other than the plasma membrane.

In view of the above and the fact that any signal transduction mechanism must ultimately account for the modulation of nuclear events, the effect of agonist on nuclear DAGs became an important question. One indication that these lipids may be elevated in nuclei was derived from studies of the localization of PKC (see Section 1.6.1.3). Using immunological and biochemical analyses, this kinase has been identified in the nucleus of IIC9 fibroblasts (Leach *et al.*, 1992), liver (Masmoudi *et al.*, 1989), HL-60 cells (Hocevar and Fields, 1991), and 3T3 cells (Divecha *et al.*, 1991; Fields *et al.*, 1990; Leach *et al.*, 1989; Thomas *et al.*, 1988). While in some systems, such as liver (Masmoudi *et al.*, 1989), PKC appears to be constitutively expressed in the nucleus, in other systems, PKC is present in nuclei prepared from agonist-stimulated, but not unstimulated, cells (Divecha *et al.*, 1991; Fields *et al.*, 1990; Leach *et al.*, 1989, 1992; Thomas *et al.*, 1988). Since PKC is a physiological target for DAGs, these findings support the hypothesis that nuclear association of PKC correlates with agonist-induced elevation of nuclear DAGs.

Agonist-induced increases in nuclear DAG levels have indeed been reported. A robust, rapid, albeit transient rise in nuclear DAGs in response to α-thrombin has been observed in quiescent IIC9 cells (Jarpe *et al.*, 1994; Leach *et al.*, 1992). EGF also increases nuclear DAG levels (D. M. Raben and M. B. Jarpe, unpublished observation). The thrombin-induced rise was accompanied by an increase in the level of nuclear PKC-α (Leach *et al.*, 1992). In Swiss 3T3 cells, IGF-1-induced increases in nuclear DAGs occur concomitant with increases in nuclear PKC activity (Divecha *et al.*, 1991). Taken together, the data clearly suggest a linkage between mitogen-induced nuclear lipid metabolism, PKC activation, and cellular proliferation.

1.4.3.2. Source(s) of Induced Nuclear DAGs

Identification of the lipids that serve as the source of the induced nuclear DAGs will provide crucial information required for elucidating the enzymatic mechanisms responsible for the increase in the induced nuclear DAGs, as well as potential physiological roles. In studies with 3T3 cells, a small decrease in PIP and PIP_2 levels was observed which suggested that the hydrolysis of PIs was responsible for at least part of the increase in IGF-1-induced nuclear DAGs (Divecha *et al.*, 1991). The increase in nuclear PIs, however, did not quantitatively account for all of the induced DAGs (Divecha *et al.*, 1991), supporting the idea of another phospholipid source.

Since it is not possible to selectively radiolabel nuclear phospholipids, potential sources of the induced nuclear DAGs cannot be obtained by analysis of released water-soluble head groups from metabolically labeled cells. One approach that circumvents this problem is the analysis of DAG and phospholipid molecular species described above (Section 1.4.2). Using such an approach, it was shown in fibroblast cells that DAGs generated in response to mitogenic concentrations of α-thrombin were derived predominantly from PC (Jarpe *et al.*, 1994). Clearly, just as in whole cells, nuclear DAGs also appear to be derived largely from the hydrolysis of PI and/or PC. In this regard, it is interesting that while the molecular species of whole-cell phospholipids does not change significantly in response to α-thrombin (Pessin and Raben, 1989; Pessin *et al.*, 1990), the molecular species of nuclear PE was dramatically altered (Jarpe *et al.*, 1994). To begin with, nuclear PE profiles from quiescent cells had very little resemblance to whole-cell PE profiles (Jarpe *et al.*, 1994; Pessin and Raben, 1989; Pessin *et al.*, 1990). Quiescent nuclear PE is composed predominantly of 16:0-18:1W9 and 16:0-16:1W7 (Jarpe *et al.*, 1994). Five minutes after the addition of α-thrombin, several species with later retention times were found, with a corresponding loss of the above-mentioned species with earlier retention times (Jarpe *et al.*, 1994). These findings suggest that nuclear PE metabolism may also play a role in modulating nuclear events. It will be important to determine if the DAGs derived from these different sources serve different and specific roles.

While the above data provide strong support for the notion that nuclear diglycerides are important components of mitogenic signal transduction cascades, it is important to note that it is difficult to compare the IIC9 studies with the 3T3 studies. In the 3T3 studies, nuclei were prepared in the presence of detergents while in the IIC9 studies, nuclei were isolated without detergents. The nuclei isolated in the 3T3 studies lacked at least the outer nuclear envelope which could result in a loss of nuclear diglycerides. Examination of IIC9 nuclei isolated in the absence of detergents shows that they possess intact inner and outer membranes (Leach *et al.*, 1992). Interestingly, if IIC9 nuclei are isolated in the presence of detergent, neither diglycerides nor PKC activity can be detected at any time (M. B. Jarpe, K. L. Leach, and D. M. Raben, unpublished observations). In contrast, induced increases in nuclear diglycerides cannot be detected when 3T3 nuclei are prepared in the absence of detergent, while this is required for the detection of nuclear PKC (Divecha *et al.*, 1991). The reason for the discrepancies observed between the two systems is unclear. They may reflect basic differences between the two cell types or the quality of the nuclear preparations. This issue notwithstanding, in both studies the increase in nuclear diglycerides was rapid and mitogen dependent indicating that this response is one of the immediate agonist-induced nuclear responses.

1.4.3.3. *Molecular Mechanism(s) Involved in Agonist-Induced Nuclear DAG Formation*

The molecular mechanisms responsible for agonist-induced formation of DAGs in the nucleus have not been established. First, DAGs may be generated in one compartment, such as the plasma membrane, followed by translocation to the nucleus via one of the mechanisms discussed above (see Section 1.2.4). As mentioned, this transport must be tightly controlled to maintain lipid compositions and avoid futile cycles. Alternatively, the DAGs may be generated in the nucleus through the action of phospholipases on the endogenous nuclear phospholipids.

Such a hypothesis is supported by the observation that a number of lipid-metabolizing enzymes have been identified in this compartment (see Raben *et al.*, 1994). Indeed, enzymes involved in both the PI cycle (reviewed in Divecha *et al.*, 1993) and PC cycle (Wang *et al.*, 1993) are located in the nucleus.

1.5. Turnover of Intracellular DAG

As discussed in the previous section, DAG formation after cell stimulation with agonists has been monitored by using labeled biosynthetic precursors of phospholipids, such as arachidonic acid (Habenicht *et al.*, 1981; Kennerly *et al.*, 1979), or by quantitation of *sn*-1,2-DAGs present in crude lipid extracts (Bocckino *et al.*, 1985; Kennerly *et al.*, 1979; Preiss *et al.*, 1986). Using such approaches in platelets, thrombin and other agonists have been shown to induce a two- to threefold increase in DAG levels within 30 sec, which declines to basal levels within 5 min (Preiss *et al.*, 1986). Furthermore, in PDGF-stimulated 3T3 fibroblast cells, similar increases in DAG levels were seen after 10 min, which declined to basal levels within 50 min (Habenicht *et al.*, 1981). Thus, DAG production in cells is transient and DAG is metabolized in cells after agonist-induced formation at rates that vary considerably. In the same way that DAGs are produced at different rates, reflecting the phospholipid source and pathway employed, they may also be metabolized by a variety of mechanisms. Essentially, four types of reactions have been implicated in removal of DAG second messenger signals: (1) phosphorylation by DGK to PA; (2) lipolytic breakdown to yield monoacylglycerols and free fatty acids; (3) conversion to either PC or TG; (4) formation of bisphosphatidic acid. These pathways (Fig. 1-2) are discussed in the order indicated.

In summary (see Fig. 1-2), DAG may be formed by agonist-stimulated hydrolysis of PIP_2 or PC through the action of PLCs or PLD together with PPH. The hydrolysis products PA or DAG remain bound to the plasma membrane. Attenuation of DAG signals may occur through four pathways, of which three are shown: conversion to PA by DGK, further hydrolysis by DAG lipase and recycling in ER, transport to ER and direct recycling without further hydrolysis. DAG from PIP_2 is indicated as being preferentially recycled through PA into PI lipids either directly at the plasma membrane or after transport to ER. Transport through the cytosol to the ER from the plasma membrane may be facilitated by PLA_2_ or DAG lipase-mediated formation of lysolipids and free fatty acids which are readily miscible with aqueous phase. Transport of the phospholipids back to the plasma membrane occurs through vesicle-mediated transport or as monomers with the help of transfer proteins.

1.5.1. DAG Phosphorylation by DGKs

Mammalian DGKs are a diverse family of enzymes. Activity is present in many tissues and found in cytosolic, microsomal, and nuclear subcellular locations (MacDonald *et al.*, 1988). The best characterized DGK is an 80-kDa, cytosolic protein, which was initially purified from porcine brain (Kanoh *et al.*, 1983) and later found to be abundant in the cytosol of porcine and human lymphocytes (Yamada *et al.*, 1989). Membrane-bound DGK activities have been described in several tissues, but may be identical to better characterized cytosolic isoforms. In particular, brain membranes contain a protein that is immunologically and electrophoretically identical to the 80-kDa enzyme (Kanoh *et al.*, 1986). However, two distinct proteins of 110 and 150 kDa were recently purified to homogeneity starting from either rat brain cytosol or membranes, respectively. The 150-kDa protein is a peripheral membrane protein (Kato and Takenawa, 1990).

In addition to brain and liver DGK activities from cytosol and microsome preparations which distinguish between DAGs and ether-linked diradylglycerols (Ford and Gross, 1990), an arachidonoyl-specific DGK exists (MacDonald *et al.*, 1988). In baboon tissues, this DGK with acyl chain specificity is most abundant in testes and brain, but present at lower levels in spleen, kidney, liver, and skeletal muscle (Lemaitre *et al.*, 1990). The arachidonoyl-specific DGK was purified to homogeneity and identified as a 58-kDa, putative integral membrane protein, which is associated with the particulate fraction by both ionic and hydrophobic forces, suggesting interactions of the protein *in vivo* with other membrane or cytoskeletal proteins (Walsh *et al.*, 1994).

Since mammalian phosphoinositides are selectively enriched in arachidonate at the *sn*-1,2 position, arachidonoyl-specific DGK has been implicated in the rapid and selective clearance of the arachidonate-rich DAG pool arising from stimulus-induced phosphoinositide turnover. In support of the notion that DGKs may be linked to the phosphorylation of specific, signaling-related DAG pools, no PA formation is seen in intact Jurkat T cells and human foreskin fibroblasts when massive DAG formation is provoked by treatment of the cells with exogenous bacterial PLC. It has been suggested that DGK may associate physically with endogenous PLC and that receptor-induced DAG would be channeled from the PLC to the active site of DGK (Van der Bend *et al.*, 1994).

PA, the product derived from DAG by DGK-dependent phosphorylation, may itself be a second messenger. PA is implicated in the regulation of PKC, PI 4-kinase, PLC, adenylate cyclase, a novel PA-activated kinase, and the neutrophil oxidase. Additionally, PA can be metabolized to yield effector molecules such as free fatty acids and LPA by a PA lipase (Perry *et al.*, 1993, and references therein). Finally, PA can be converted to DAG again by PA phosphohydrolase, which is present at the plasma membrane and ER (Pagano and Longmuir, 1985; Perry *et al.*, 1993). At the plasma membrane this step is essential for agonist-dependent formation of DAG from PC through the PLD pathway (Exton, 1990, 1994). Translocation of PA phosphohydrolase to the ER is enhanced by free fatty acids and their CoA esters, which thereby trigger an increase in the synthesis of glycerolipids, especially triglycerides (Hopewell *et al.*, 1985).

1.5.2. *DAG Hydrolysis by Lipases*

A second possibility for the removal of DAG signals is by the sequential action of DAG and monoacylglycerol (MAG) lipase (Bell *et al.*, 1979; Rebbechi *et al.*,

1983). The diglyceride lipase activity associated with the particulate fraction in platelets appears sufficient to generate arachidonate for the burst of prostaglandin and thromboxane synthesis that follows thrombin stimulation of platelets (Bell *et al.*, 1979). Likewise, hydrolysis of DAG by the sequential action of DAG- and MAG-lipases yields arachidonic acid that is necessary for insulin secretion in isolated pancreatic islet membrane preparations (Konrad *et al.*, 1994).

DAGs metabolized by DAG lipase may be reutilized for lipid biosynthesis. In Swiss 3T3 cells, MAGs are converted back to phospholipids, either by an *sn*-2-acyl MAG acyltransferase activity or by an MAG kinase activity that phosphorylates preferentially *sn*-2-arachidonoyl MAG followed by a stearoyl-specific acyltransferase activity that converts *sn*-2-arachidonoyl lyso-PA into *sn*-1-stearoyl-2-arachidonoyl PA. These activities are membrane associated (Simpson *et al.*, 1991). Interestingly, a DGK activity present in pig brain cytosol fractions also exhibits MAG kinase activity (Kanoh *et al.*, 1990), suggesting that DGK may contribute to the attenuation of DAG-derived signals in several ways.

In stimulated hepatocytes [^{3}H]-PA production occurs at the plasma membrane (Seyfred and Wells, 1984b). Furthermore, the plasma membrane contains other enzymes of the PI cycle, including a PI synthase with properties distinct from the enzyme present at the ER (Imai and Gershengorn, 1987), PI and PIP kinases (Imai *et al.*, 1986; Seyfred and Wells, 1984a). DAG produced at the plasma membrane by agonist-induced PI-lipid hydrolysis may be directly recycled into PI lipids at the plasma membrane, without requiring shuttling to the ER as was initially proposed (Hokin, 1985). However, others did not detect a distinct plasma membrane PI synthase activity (Santiago *et al.*, 1993), and availability of PI lipids for PLC-dependent signaling at the plasma membrane appears to depend on a PI-TP (Thomas *et al.*, 1993).

1.5.3. DAG Recycling into Lipid Biosynthesis

DAG signals may be eliminated by conversion to either PC or TG. In experiments where NIH 3T3 fibroblasts were supplemented with exogenous DAGs containing long-chain fatty acids, the fate of individual DAG species depended strongly on the fatty acids present (Florin-Christensen *et al.*, 1992). For instance, *sn*-1,2-dioleoyl DAG is rapidly converted predominantly to PC and TG and, to a lesser extent, to MAG, fatty acids (DAG lipase pathway), or to PA and PI (DGK pathway). Thus, *sn*-1,2-dioleoyl DAG is largely utilized as an intact molecule. Both *sn*-1-stearoyl-2-myristoyl glycerol and *sn*-1- stearoyl-2-arachidonoyl glycerol produce substantial labeling of PC, but *sn*-1-stearoyl-2-myristoyl glycerol gives rise to the highest portion of TG and lowest of PA and PI, while *sn*-1-stearoyl-2-arachidonoyl glycerol yields the opposite pattern of phospholipid synthesis (Florin-Christensen *et al.*, 1992). These results are in excellent agreement with the observation that phosphorylation of monoarachidonoyl glycerol channels the glycerol backbone to PA and PI, whereas acylation to DG directs the glycerol backbone to PC and TG (Simpson *et al.*, 1991).

The two enzymes involved in PC and TG resynthesis, CDP-choline:DAG choline phosphotransferase and acyl-CoA:DAG acyltransferase, respectively, both reside in the ER (D. Vance, 1990; Table 1-2). DAGs recycled by PC/TG synthesis are likely to be translocated as such within cells from the plasma membrane to the ER. Lysosomal degradation is not seen, indicating that transport on endosomal membranes does not prevail (Florin-Christensen *et al.*, 1992). Rather, DAG translocation by TPs, which exist for different phospholipids (Wirtz, 1991), is proposed. Consistent with this suggestion, enhancement of DAG transfer by a nonspecific TP from bovine liver has been reported (Nichols and Pagano, 1983).

The observation that DAG recycling through biosynthetic pathways in the ER dominates in NIH 3T3 fibroblasts (Florin-Christensen *et al.*, 1992) could well explain the relatively slow time course of DAG formation and removal seen in Swiss 3T3 cells stimulated with PDGF (Habenicht *et al.*, 1981). In platelets, however, DAG turnover is rapid (Preiss *et al.*, 1986). Cell-permeable, short-chain DAGs (diC_8) are metabolized to the appropriate PA derivatives (DGK pathway), or, on DGK inhibition with diC_8-ethyleneglycol, to MAG by DAG lipase activity (Bishop *et al.*, 1986). Thus, in platelets DAG turnover is probably more rapid because the reduction of DAG levels occurs through processes at the plasma membrane and does not require transport to the ER.

1.5.4. DAG Incorporation into Bisphosphatidic Acid

The last possibility, which was reported in human fibroblasts, involves the PLD-mediated formation of bisphosphatidic acid from DAG and PA. Bisphosphatidic acid formation is rapid and transient and is induced by stimuli that lead to both DAG and PA formation, such as endothelin, LPA, fetal calf serum, phorbol esters, diC_8, and bacterial PLC. The formation of bisphosphatidic acid assigns for the first time a physiological role to the transphosphatidylation reaction of PLD. Furthermore, since bisphosphatidic acid derivatives are associated with endosomal/lysosomal compartments (Table 1-1), it has been speculated that bisphosphatidic acid represents a "sink" for the lipid second messengers DAG and PA, which would be degraded by the endocytic route (Van Blitterswijk and Hlikman, 1993).

1.6. Intracellular Targets for DAG

In contrast to the initial assumption that PKC is the exclusive intracellular target for both DAGs and phorbol esters, it is now apparent that several others exist. Many contain domains homologous to cysteine-rich motifs that are crucial for DAG-dependent regulation of PKC activity (Fig. 1-3).

In summary (see Fig. 1-3), although several other proteins have been identified whose activities are modulated by DAG, PKCs are still the best-characterized intracellular targets of DAG signals. Once active, PKC phosphorylates and regulates the activity of many membrane-associated proteins, including receptors, ion channels, ion pumps and also phospholipases. PKC activation may inhibit PLC, but enhance PLA_2 and PLD activity. PLD activation, together with PPH, would

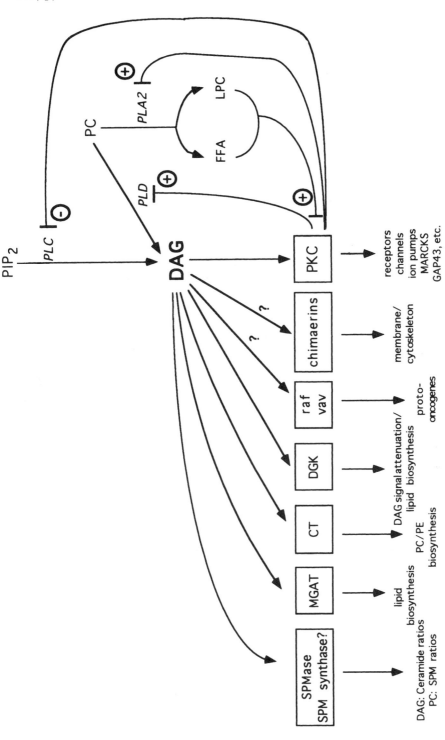

Figure 1-3. Intracellular targets for DAG signals: PKC-mediated positive feed-back prolongs DAG signals.
Aside from PKC, several other targets for DAG have been identified that are discussed in Section 1.6. In some cases the connections made are controversial (see text) as is indicated by the question marks.

increase DAG levels without the calcium transient which occurs upon PLC-dependent hydrolysis of PIP_2. Activation of PLA_2 leads to the formation of lyso-lipids and free fatty acids, which both can enhance PKC activity. Free fatty acids may reduce the calcium requirement of cPKCs. Thus, both cPKCs and nPKCs may remain active for longer periods of time, well beyond the time-frame of an initial DAG peak resulting from PLC-mediated PIP_2 hydrolysis. Cysteine-rich motifs present in PKC are crucial for DAG-dependent regulation of PKC activity. Additional proteins whose function may be modulated by DAG or other lipid second messengers exist that contain similar cysteine-rich motifs. For instance, chimaerins which are implicated in the control of membrane-cytoskeleton interactions, the proto-oncogenes *vav* and *raf*, and DGK which is important for attenuation of DAG signals and lipid biosynthesis. Thus, cysteine-rich regions are emerging as a highly versatile motif, potentially involved in mediating regulation by several lipid second messengers in addition to DAG, such as LPA, PIP_3, or ceramide. Interestingly, free fatty acids enhance the activity of several proteins in addition to PKC, only some of which possess cysteine-rich motifs, namely PPH, cytidylyltransferase, DGK and sphingomyelinase. Finally, DAG may control both PC and PE biosynthesis by modulating the activity of cholinephosphate and ethanolaminephosphate cytidylyltransferase, as well as sphingomyelinase activity.

1.6.1. Protein Kinase C

1.6.1.1. PKC Isoforms

In view of the diversity of cellular functions subject to regulation by PKC, it is not surprising that PKC is actually a family of enzymes (Nishizuka, 1988, 1992). At present the mammalian PKC family consists of 12 different PKC isoforms represented by Greek symbols: α, βI, βII, γ (calcium-dependent, cPKCs), δ, ε, η, θ (calcium-independent nPKCs), ζ, λ, ι (atypical PKCs, aPKCs), and a putative transmembrane isoform μ (Akimoto *et al.*, 1994; Dekker and Parker, 1994; Johannes *et al.*, 1994; Nishizuka, 1992; Selbie *et al.*, 1993). Two functionally distinct domains are distinguished within PKCs (Lee and Bell, 1986; Ono *et al.*, 1989b), an amino-terminal regulatory domain that interacts with lipids and a catalytic domain which is more similar to functionally comparable regions of other serine/threonine kinases (Bell and Burns, 1991; Hanks *et al.*, 1988). In the absence of the regulatory domain, PKCs are constitutively active (Inoue *et al.*, 1977; Takai *et al.*, 1977; Woodgett *et al.*, 1987). Within the regulatory domain, elements referred to as cysteine-rich regions play a crucial role in DAG/phorbol ester-dependent regulation (Burns and Bell, 1992; Ono *et al.*, 1989b; Quest *et al.*, 1994a,b).

1.6.1.2. PKC Activation

In intact cells phorbol esters elicit a rapid decrease in soluble and concomitant increase in PKC associated with the particulate fraction (Kraft and Anderson, 1983). In a reconstitution system employing red blood cell membranes, PKC translocation to the membranes and activation is stimulated by calcium and phorbol esters (May *et al.*, 1985; Wolf *et al.*, 1985). Thus, PKC activation is suggested to involve translocation of soluble, cytosolic PKC to the plasma membrane, where activation by pharmacological reagents like phorbol esters or the physiological lipid activator DAG, released in response to receptor stimulation by agonists, occurs in the presence of calcium and PS (Bell, 1986).

Within the regulatory domain, a pseudosubstrate domain is present which mimics a putative target phosphorylation site. A peptide from this region is a potent inhibitor of PKC activity (House and Kemp, 1987, 1990). Antipeptide antibodies specific for the pseudosubstrate region activate PKC in the absence of lipid cofactors (Makowsky and Rosen, 1989) and deletion of the pseudosubstrate region leads to constitutively active PKC (Pears *et al.*, 1990). Thus, PKC activation occurs by release of autoinhibitory restraints imposed on the kinase domain through the pseudosubstrate motif in the regulatory domain. Conformational changes required for PKC activation, as evidenced by exposure of the pseudosubstrate motif, are dictated by interactions with activating lipids such as PS and DAG, but do not occur by binding to nonactivating lipids (Orr *et al.*, 1992; Orr and Newton, 1994). The cysteine-rich motifs present in all PKCs play a crucial role in PKC activation by lipid interactions (reviewed in Bell and Burns, 1991; Burns and Bell, 1992; Quest and Bell, 1994a).

1.6.1.3. PKC Association/Compartmentalization

All models proposed for lipid-dependent PKC activation are an oversimplification since not all PKC is cytosolic in the unstimulated state. Even a small fraction of the cPKCs, which are mainly present in the cytosol in unstimulated cells, is associated with the particulate fraction. For nPKCs and aPKCs, this fraction is generally higher, although the extent may vary depending on the cells examined. It is not known whether all PKC associated with the particulate fraction, even in the absence of stimuli, is active (reviewed in Hug and Sarre, 1993).

PKCs are associated with other membrane compartments, in particular the nucleus. Treatment of cells with phorbol esters is thought to induce changes in cell growth and gene expression by direct activation of PKC. One explanation of how such events may be regulated by PKC is if PKC were to translocate to the nucleus. Indeed, PMA treatment of NIH 3T3 fibroblasts induces PKCα translocation to the nuclear envelope (James and Olson, 1992; Leach *et al.*, 1989). PKCα translocation to the nuclear fraction occurs also in response to agonists. Treatment of rat IIC9 fibroblasts with α-thrombin stimulates nuclear DAG levels and is accompanied by an increase in nuclear PKC activity and localization of PKCα in the nucleus (Leach *et al.*, 1992). In Swiss 3T3 cells treated with insulin-like growth factor-1, nuclear DAG levels increase with concomitant translocation of PKC to the nucleus. However, bombesin, which causes similar increases in DAG levels at the plasma membrane, has no effect on nuclear inositide levels and induces PKC translocation to postnuclear fractions (Divecha *et al.*, 1991). Furthermore, a PKCα mutant lacking the regulatory domain accumulates predominantly in the nuclear envelope fraction, and two sequence elements in the hinge and catalytic region are implicated in nuclear targeting (James and Olson, 1992).

Association with the nuclear compartment is not limited to PKCα. PKCβ is the main PKC isoform detected in preparations of rat liver nuclei (Rogue *et al.*, 1990). In renal epithelial cells, vitamin D3 enhances translocation of PKCβ to the nucleus (Simboli-Campbell *et al.*, 1994). In HL-60 cells expressing both PKCα and PKCβII, bryostatin treatment causes selective translocation of PKCβII to the nuclear membrane (Hocevar and Fields, 1991). At the nucleus, PKCβII phosphorylates directly the nuclear envelope polypeptide lamin B, which leads to solubilization of lamin B, indicative of mitotic nuclear envelope breakdown *in vitro* (Hocevar *et al.*, 1993). A nuclear-specific *in vitro* lipid or lipid metabolite is implicated in the specific activation of PKCβII (Murray *et al.*, 1994). PKCβII is translocated to the nucleus of intact cells at the G2/M phase transition, suggesting that this isoform, like p34^{cdc2}/cyclin B, is a physiological lamin kinase (Goss *et al.*, 1994). PKC-L(η) may reside permanently in the cell nucleus (Greif *et al.*, 1992), although others found this nPKC isoform located on the rough ER and outer nuclear membrane, but not the nucleus or inner nuclear membrane (Chida *et al.*, 1994).

1.6.1.4. PKC Substrates

Since PKC is also present in the particulate fraction in the absence of cellular stimuli, these interactions could contribute to the intracellular distribution/compartmentalization of PKC isoforms before activation. Alternatively, proteins called *r*eceptors for *a*ctivated *C*-*k*inase (RACKs) have recently been identified, which bind to the PKC regulatory domain, specifically after activation by lipids (Disatnik *et al.*, 1994; Mochley-Rosen *et al.*, 1990, 1991a,b; Ron *et al.*, 1994). These interactions are expected to contribute substantially to PKC substrate specificity. In experiments where the regulatory domain of PKCε was fused to the catalytic domain of PKCγ, substrate specificity was converted to that of PKCε (Pears *et al.*, 1991). In PKCη the pseudosubstrate motif mediates low histone kinase activity of the wild-type enzyme (Dekker *et al.*, 1993). Both reports underscore the role of the PKC regulatory domain in determining substrate specificity.

On activation, PKC phosphorylates a large number of proteins both *in vivo* and *in vitro* (Nishizuka, 1986; Woodgett *et al.*, 1987) and catalyzes intrapeptide phosphorylation at several sites (Flint *et al.*, 1990; Newton and Koshland, 1989). The role of autophosphorylation remains unclear, but it may be important for PKC maturation and activation (Zhang *et al.*, 1994), possibly in conjunction with PKC phosphorylation by an exogenous kinase (Cazaubon *et al.*, 1994) and/or for downregulation of PKC through calpains (Kishimoto *et al.*, 1983, 1989; discussed in Quest and Bell, 1994a).

A large number of PKC substrates have been identified *in vivo* and *in vitro* (Conn and Sweatt, 1994; Nishikawa and Hidaka, 1994; Nishizuka, 1986; Pucéat and Brown, 1994; Woodgett *et al.*, 1987), but the significance of PKC phosphorylation is often unresolved. Arbitrarily, PKC substrates are divided into five categories:

(1) receptor proteins involved in signal transduction and PKC activation (EGFR, T-cell or insulin receptor); (2) ion channels, pumps (calcium and potassium channels in heart or nicotinic acetylcholine, GABA, glutamate receptors in neurons); (3) contractile and cytoskeletal proteins (myosin light chain, vinculin, microtubule-associated proteins); (4) membrane-associated proteins (MARCKs, GAP43, pp60[src]); (5) others (glycogen phosphorylase kinase, glycogen synthase).

Many of the proteins are associated with either cell membranes, the cytoskeleton, or both. Interestingly, some proteins regulated by protein kinase C are phospholipases (see Fig. 1-2). For instance, PKC is implicated in a feedback inhibitory loop for PLC. By contrast, PKC activation stimulates both PLD and PLA_2 activity (reviewed in Lambeth, 1994). PKC is further implicated in both the regulation of PAF biosynthesis and as an effector downstream of the PAF receptor (Bussolino et al., 1994; Chao and Olson, 1993; Ninio and Joly, 1991). Free fatty acids released by PLA_2 (Tsujishita et al., 1994) activate PKCs by a mechanism distinct from that of DAGs (reviewed in Quest and Bell, 1994a) and may serve to reduce the calcium requirement of cPKCs (Nishizuka, 1992). Stimulation of PLD activity leads to the generation of DAG, in some cases without detectable phosphoinositide hydrolysis or calcium mobilization (Rosoff et al., 1988; Wright et al., 1988). In a cell-free system employing fibroblast plasma membranes, PKC was required for activation of PLD by PMA, but the effect was independent of ATP or phosphorylation (Conricode et al., 1992). In the case of PLA_2, regulation occurs through phosphorylation by mitogen-activated protein (MAP) kinase by both PKC-dependent and -independent pathways (Qiu and Leslie, 1994).

Arachidonic acid formed by PLA_2 activation has recently been identified as a mediator of SPM hydrolysis and ceramide formation in response to tumor necrosis factor alpha (TNFα) (Jayadev et al., 1994). Thus, a putative link exists between DAG formation and signaling events related to ceramide, which activates both a downstream kinase, which shares the substrate specificity of MAP kinases, and a phosphatase of the heterotrimeric protein phosphatase 2A group (Hannun, 1994; Kolesnick and Golde, 1994; Liu et al., 1994; Wolff et al., 1994). PLA_2 is further implicated in the rapid formation of LPA in the inner leaflet of platelet plasma membrane. LPA evokes multiple biological effects, including stimulation of platelet aggregation, fibroblast proliferation, and neurite retraction (reviewed in Moolenaar, 1994).

1.6.1.5. Cysteine-Rich Motifs

All PKC isoforms contain at least one cysteine-rich motif. Both cPKCs and nPKCs bind phorbol esters, are activated by DAG, and contain a tandem repeat of two cysteine-rich motifs (Nishizuka, 1992). In PKCγ both motifs are required for high-affinity phorbol dibutyrate (PDBu) binding in the presence of calcium (Quest and Bell, 1994b). PKCμ also contains two cysteine-rich motifs, but these are spaced farther apart and the isoform may not bind phorbol esters (Johannes et al., 1994). Others, however, recently cloned, expressed, and characterized a murine PKC isoform called PKD which shows 92% homogy to PKCμ and displays a serine protein kinase activity that is stimulated by PDBu and DAGs (Van Lint *et al.*, 1994). aPKCs contain only one cysteine-rich motif and also do not bind phorbol esters (Nishizuka, 1992; Selbie *et al.*, 1993).

A single cysteine-rich motif alone is sufficient for PDBu binding (Burns and Bell, 1991; Ono *et al.*, 1989b; Quest *et al.*, 1994a). Initial experiments characterizing phorbol ester and DAG binding sites in intact PKC revealed that both molecules bind to a similar site within the calcium–phospholipid–PKC complex (König *et al.*, 1985; Sharkey and Blumberg, 1985; Sharkey *et al.*, 1984). These sites are indeed highly overlapping and present within an individual cysteine-rich region of PKC, where zinc coordination is critical for this function (Quest *et al.*, 1994a,b). A 43-amino-acid sequence was defined by deletion analysis as the minimal requirement for PDBu binding (Quest *et al.*, 1994b).

PKCζ contains a single cysteine-rich region, but neither binds phorbol ester nor is activated by DAG or phorbol esters (Kazanietz *et al.*, 1994; Nakanishi and Exton, 1992; Ono *et al.*, 1989a). Likewise, the *raf* proto-oncogene product, containing only one cysteine-rich region, lacks phorbol ester binding (Ghosh *et al.*, 1994; Kazanietz *et al.*, 1994), while the results are controversial for *vav*. Expression of *vav* in *Spodoptera frugiperda* (Sf9) insect cells yielded a protein that binds neither PDBu nor bryostatin (Kazanietz *et al.*, 1994), while others identified *vav* as a protein with guanine exchange activity for *ras* that is stimulated by phorbol esters and DAG in T cells (Gulbins *et al.*, 1994). Two other proteins with just a single cysteine-rich motif, *n*-chimaerin and *unc*-13, do bind phorbol esters (Ahmed *et al.*, 1992; Maruyama and Brenner, 1991). Thus, as the experiments where individual cysteine-rich regions of PKC are expressed indicate, a single cysteine-rich motif is sufficient to confer phorbol ester binding to a protein.

However, the presence of such a motif does not automatically implicate regulation by DAG. Rather, cysteine-rich regions may represent a versatile motif capable of mediating regulation by a variety of additional lipid second messengers (Liscovitch and Cantley, 1994; Quest *et al.*, 1995; Roberts, 1994), such as ceramide for *vav* and PKCζ (Gulbins *et al.*, 1994; Lozano *et al.*, 1994), LPA for *n*-chimaerin (Ahmed *et al.*, 1993) or PIP_3 for PKCζ (Nakanishi *et al.*, 1993) as well as PKCδ,ε,η (Liscovitch and Cantley, 1994; Toker *et al.*, 1994). Furthermore, sequence elements lying within the cysteine-rich region of *raf* mediate protein interactions with GTP-*ras* (Ghosh *et al.*, 1994). Even cysteine-rich regions of cPKCs, which bind phorbol esters or DAG when expressed individually, may not be functional in the context of the regulatory domain and fulfill other functions, such as mediate protein–protein interactions (Quest and Bell, 1994b; Quest *et al.*, 1994a). Consistent with the observation that stoichiometric binding of either one molecule of phorbol ester or DAG is sufficient for maximal activation of PKC (Hannun *et al.*, 1985; Hannun and Bell, 1986; König *et al.*, 1985), only one of the two cysteine-rich regions of PKCγ is suggested to interact directly with PDBu or DAG (Quest and Bell, 1994b).

1.6.2. Chimaerins

To date three chimaerins have been identified: *n*-chimaerin in neurons, β-chimaerin in testicular germ cells, and β2-chimaerin in cerebellar granule cells

(Ahmed *et al.*, 1990; Hall *et al.*, 1990; Leung *et al.*, 1993, 1994). The amino-terminal region of *n*-chimaerin shows homology to PKCs in a single cysteine-rich region containing the consensus motif with six cysteines and two histidines (C_6H_2) (Ahmed *et al.*, 1991). In β2-chimaerin an additional unique SH2 domain is present (Leung *et al.*, 1994). The carboxy-terminal catalytic region has sequence homology to Bcr (product of breakpoint cluster region gene involved in Philadelphia chromosome translocation), *rho*-GTPase activating protein (*rho*-GAP), and *ras*-GAP-binding protein p190 (Ahmed *et al.*, 1990; Hall *et al.*, 1990; Leung *et al.*, 1993, 1994).

n-Chimaerin displays GAP activity for the p21*rho* subfamily of *ras* G proteins (Manser *et al.*, 1992) that is regulated by lipid interactions in the cysteine-rich domain (Ahmed *et al.*, 1993). Phorbol esters, but not DAG, synergize with PS and PA to activate *n*-chimaerin, while LPA, phosphoinositide lipids (PI, PIP, PIP_2), and arachidonic acid inhibit *n*-chimaerin (Ahmed *et al.*, 1993).

The *ras*-related *rho*/*rac* subfamily of small G proteins is involved in the regulation of actin cytoskeleton organization (Ridley and Hall, 1992; Ridley *et al.*, 1992). The chimaerin family of *rac*-GAPs provides a link between lipid second messengers and organization of the actin cytoskeleton in that they may allow actin microfilament structures and other *rac* effectors to sense the changes occurring at the plasma membrane (Ahmed *et al.*, 1993).

Apparently, an intricate connection exists between the cytoskeleton and signal transduction. Several proteins of the cytoskeleton associate with lipids and may thereby modulate in particular the formation of lipid second messengers. Furthermore, many proteins including PKC, implicated in different aspects of signaling, associate with the cytoskeleton (Den Hartigh *et al.*, 1992; Luna and Hitt, 1992; Pappelenbosch *et al.*, 1993; Payrastre *et al.*, 1991; Yu *et al.*, 1992; Zhang *et al.*, 1992). Although DAG has no apparent effect *in vitro* on *n*-chimaerin *rac*-GAP activity, this may not be the case for other *rho*-GAPs that have been identified by a novel overlay assay (Manser *et al.*, 1992). Indeed, the formation of new actin nucleation sites at the membrane of *Dictyostelium* amoebae is triggered by DAG in a protein-dependent, but PKC-independent manner. One of the proteins responsible for this DAG effect is a tightly bound peripheral membrane protein (Shariff and Luna, 1992). It is intriguing to speculate that proteins involved in DAG-dependent actin nucleation may be related to the aforementioned *rho*-GAPs.

1.6.3. DAG Kinase

The primary sequence of the 80-kDa cytosolic DGK obtained by molecular cloning contains two EF-hand motifs, typical of calcium-binding proteins, and two cysteine-rich, zinc finger-like motifs (Sakane *et al.*, 1990). The human counterpart of this DGK has been purified from normal white blood cells. The cDNA sequence determined also contains two cysteine-rich motifs, which are homologous to the C_6H_2 consensus of PKCs and chimaerins (Ahmed *et al.*, 1991; Schaap *et al.*, 1990). Despite the C_6H_2 consensus, both DGK cysteine-rich sequences do not bind phorbol esters after the same denaturation/renaturation procedure that yielded functional, PDBu binding PKC and *n*-chimaerin from *Escherichia coli* inclusion bodies (Ahmed *et al.*, 1991). The 80-kDa DGK is activated *in vitro* by calcium and PS and translocates reversibly to the membrane in the presence of calcium (Sakane *et al.*, 1991). DAG also enhances membrane association of affinity-purified DGK (Besterman *et al.*, 1986b). In Swiss 3T3 cells roughly 60–70% of DGK is in the cytosol while the remainder is associated with the particulate fraction. Treatment of the cells with phorbol esters enhances association with the particulate fraction, but probably does so by increasing DAG levels. Redistribution of DGK is also seen when cells are treated with cell-permeable diC_8 (Maroney and Macara, 1989). Since phosphorylation of DGK *in vitro* by both PKC and cAMP-dependent protein kinase enhance phospholipid association of DGK (Kanoh *et al.*, 1989), it remains unclear to what extent DAGs may directly modulate the intracellular distribution of DGK, for instance through interactions in the cysteine-rich regions.

1.6.4.   Vav Proto-oncogene

T-cell activation rapidly increases the *ras* p21 GTP content. Likewise, PDBu increased *ras*-GTP to comparable levels and PKC was implicated as a mediator (Downward *et al.*, 1990), although PKC-dependent phosphorylation of N-*ras* and K-*ras* has no effect on either GTP binding or the GTPase activity (Ballester *et al.*, 1987). *Vav* is a proto-oncogene expressed exclusively in hematopoietic cells that shares homology with guanine nucleotide releasing factors (Gulbins *et al.*, 1993) and contains two cysteine-rich motifs, the first conforming to the canonical pattern of C_4 zinc fingers in transcription factors (Katzav *et al.*, 1989), while the second is homologous to the C_6H_2 cysteine-rich regions of PKCs and chimaerins (Ahmed *et al.*, 1991). Tyrosine phosphorylation of *vav*, as well as guanine nucleotide release activity, are increased in Jurkat T cells after stimulation of the T-cell antigen receptor (TCR)-CD3 (Gulbins *et al.*, 1993). The guanine release activity of *vav* is also enhanced in a tyrosine phosphorylation-independent manner, by treatment of COS-1 cells expressing *vav* with either phorbol myristate-13-acetate (PMA), OAG, or type III ceramide (Gulbins *et al.*, 1994). Thus, phorbol ester-stimulated increases in *ras* p21 GTP content (Downward *et al.*, 1990) may in fact occur in a PKC-independent fashion through direct stimulation of *vav* guanine nucleotide releasing activity. However, it should be noted that others did not find PDBu binding activity in *vav* preparations from Sf9 cells (Kazanietz *et al.*, 1994). These differences may reflect limitations of the expression system employed.

1.6.5.   Raf Proto-oncogene

The *raf* proto-oncogene is an important downstream effector of *ras* in intra-cellular signaling (Avruch *et al.*, 1994). Like DGK and *vav*, *raf* contains a C_6H_2 cysteine-rich motif (Ahmed *et al.*, 1991). Membrane association of *raf* is critical for signaling and is mediated *in vivo* by interactions with GTP-*ras* (Leevers *et al.*, 1994; Stokoe *et al.*, 1994). *Raf* neither binds phorbol esters (Ghosh *et al.*, 1994; Kazanietz *et al.*, 1994) nor appears to interact with DAG (Ghosh *et al.*, 1994). In light of the discrepancies discussed for *vav*, these observations may also reflect limitations of the expression systems employed. However, the cysteine-rich region is important for association with PS/PC liposomes *in vitro*. Furthermore, the presence of a short amino-terminal segment within the cysteine-rich region of *raf* dramatically enhances association of raf to GTP-*ras in vitro* (Ghosh *et al.*, 1994). Thus, it has been suggested that the cysteine-rich motif in *raf* may be important for membrane translocation and as part of the high-affinity binding site for GTP-*ras* (Ghosh *et al.*, 1994). A role for the cysteine-rich region of *raf* in mediating interactions with *ras* is disputed by others (Avruch *et al.*, 1994).

1.6.6. CTP:Phosphocholine Cytidylyltransferase

The majority of PC in animal cells is synthesized *de novo* via the CDP-choline pathway in which CT (EC 2.7.7.15) catalyzes the rate-limiting step (see discussion of PC synthesis). CT activity is distributed between cytosol and membrane fractions (Pelech and Vance, 1989). The conversion of inactive cytosolic, to active membrane-bound CT is promoted by fatty acids and DAG, effectors of PC biosynthesis. Membrane association is also enhanced by the presence of anionic phospholipids, while aminolipids such as sphingosine are inhibitory (Feldman and Weinhold, 1987; Kalmar *et al.*, 1990; Weinhold *et al.*, 1984, 1986; Wright *et al.*, 1985). Thus, CT has several regulatory properties in common with PKC.

Phorbol esters accelerate the rate of PC synthesis from choline. TPA-stimulated PC synthesis is attributed to a two- to threefold increase in the rate of CT-catalyzed reaction and correlates with a redistribution of CT from the cytosol to ER (Pelech and Vance, 1989). Redistribution may result from TPA-stimulated activation of PLD and subsequent elevation of intracellular DAG (Utal *et al.*, 1991). In rat hepatocytes, cAMP analogues and glucagon reduce DAG levels by 30%. A highly significant correlation is seen between the cAMP-mediated decrease in DAG levels and the rate of PC biosynthesis (Jamil *et al.*, 1992). DAG levels are thought to directly regulate CT activity by stimulating membrane translocation. CT contains a motif reminiscent of a cysteine-rich region in PKC. However, a predicted amphipathic helix sequence is suggested to mediate membrane interactions (Kalmar *et al.*, 1990). Additionally, the availability of DAG limits PE synthesis in okadaic acid-treated hepatocytes (Tijburg *et al.*, 1992). These observations implicate DAG as a major regulatory component of phospholipid biosynthesis.

A strong correlation is also seen between increased CT phosphorylation and decreased PC synthesis, suggesting CT may be physiologically regulated by phosphorylation (Watkins and Kent, 1991). In intact cells, cell-free extracts, and with purified CT in reconstituted systems, CT is directly phosphorylated by cAMP-dependent protein kinase on serine which causes release of the enzyme from the particulate to the soluble fraction (Pelech and Vance, 1989). Immunofluorescence studies indicate that dephosphorylated CT may preferentially translocate to the nuclear envelope. Furthermore, it is suggested that nuclear membranes represent an important site of membrane biosynthesis. Consistent with this notion, treatment of HeLa cells with oleate stimulates CT activity and translocation to the nuclear envelope (Watkins and Kent, 1992).

The carboxy-terminal sequence of CT contains three Ser/Pro-rich consensus motifs and a total of seven Ser-Pro sites that are potential targets for phosphorylation by cyclin-dependent kinases (Jackowski, 1994; Kalmar *et al.*, 1990). Since even soluble CT is predominantly intranuclear, rather than cytosolic (Wang *et al.*, 1993), and CT phosphorylation fluctuates during the cell cycle, cyclin-dependent CT phosphorylation appears an ideally suited mechanism to coordinate phospholipid biosynthesis. A peak of PC biosynthesis is observed in the S phase of the cell cycle (Jackowski, 1994).

However, DAG levels are also known to fluctuate considerably as a function of growth conditions or growth phase (Van Veldhoven and Bell, 1988) and may therefore be important in the control of cell cycle-dependent lipid biosynthesis. Consistent with this possibility, hydrolysis of nuclear phosphoinositides is also increased in HeLa cells in the S phase of the cell cycle (York and Majerus, 1994).

1.6.7. Monoacylglycerol Acyltransferase

Monoacylglycerol acyltransferase is an integral membrane ER protein implicated in the regulation of glycerolipid biosynthesis because its end product is DAG. A microsomal monoacylglycerol acyltransferase activity from rat hepatocytes is regulated by *sn*-1,2-diacylglycerols in mixed micelle assays (Bhat *et al.*, 1994). Whether the previously discussed monoacylglycerol acyltransferase activity in Swiss 3T3 fibroblasts (see Section 1.5.3), responsible in part for attenuation of DAG signals in these cells (Simpson *et al.*, 1991), corresponds to this hepatic enzyme is not known.

1.6.8. Sphingomyelinases

DAGs, but not phorbol esters, stimulate SPM hydrolysis by both acidic and neutral sphingomyelinases. However, the acidic sphingomyelinase activity in GH_3 pituitary cell extracts is preferentially activated by addition of diC_8. This effect is suggested to be PKC-independent, since stimulation of SPM hydrolysis occurs even in cells where PKC has been downregulated by extended phorbol ester treatment (Kolesnick, 1987). A DAG-stimulated acidic sphingomyelinase is implicated in coupling the TNF-responsive PC-PLC to the formation of ceramide and induction of NF-κB activity (Schütze *et al.*, 1992). Activation of the acidic sphingomyelinase, which probably resides in acidic compartments, like lysosomes and endosomes, is suggested to occur via cointernalization of DAG with TNF/TNFR complexes (Heller and Krönke, 1994). The same receptor which generates ceramide by a neutral sphingomyelinase at the plasma membrane as part of an SPM cycle initially described in vitamin D_3-stimulated HL-60 cells (Hannun, 1994; Okazaki *et al.*, 1989) is stimulated by PLA_2-dependent release of arachidonic acid (Jayadev *et al.*, 1994).

SPM and DAG are converted to PC and ceramide by a phosphocholinetransferase reaction, operating in the opposite direction of SPM synthesis (Fig. 1-1). A plasma membrane SPM synthase that is perhaps distinct from the Golgi enzyme is implicated in both SPM-PC and DAG/ceramide interconversion. The former reaction is important for establishing epithelial cell membrane polarity, while the latter may be crucial for the regulation of these two lipid second messengers at the plasma membrane (Van Helvoort *et al.*, 1994).

1.7. Roles for Different DAG Species

Intracellular DAG signals are terminated by channeling into different recycling pathways. Thus, different DAG species could also activate some or all of the downstream effectors discussed in a very selective fashion. Since PKC represents the best-characterized target for DAGs, much of the evidence available addressing distinct roles for DAGs relates essentially to PKC.

1.7.1. Regulation of PKCs by Different DAGs

In IIC9 fibroblasts, PKC activation appeared to be determined by the DAG source. High α-thrombin concentrations (500 ng/ml) led to DAG release from both PI and PC. During the early phase of DAG production, in which phosphoinositide lipids represent the major source, PKC is activated. Low concentrations of α-thrombin (100 pg/ml) led to DAG production from PC, but did not activate PKC, as assessed by the criteria membrane translocation and phosphorylation state of the 80-kDa PKC substrate (Leach *et al.*, 1991). The source of DAGs can be distinguished by the acyl chain composition. For example, phosphoinositide-derived DAGs in liver are enriched in 1-stearoyl-2-arachidonoyl species ($\geqslant 50\%$), while this species represents only 10% of PC-derived DAGs (Patton *et al.*, 1982). However, differences in fatty acid composition of DAGs do not affect PKC activation *in vitro*, as long as the DAG species is sufficiently hydrophobic to partition into phospholipid vesicles or detergent/lipid mixed micelles (Ganong *et al.*, 1986a,b). Others observed that hydrolysis of PIP$_2$ caused translocation of calcium-dependent PKCα, and calcium-independent PKCε, while hydrolysis of PC led to translocation of PKCε only (Ha and Exton, 1993). The differences observed are explained on the basis of the absence of an increase in intracellular calcium levels when PC hydrolysis occurs rather than the result of a difference between DAG species (Exton, 1994; Ha and Exton, 1993).

Studies *in vitro*, characterizing synthetic DAGs in mixed micelle assays, suggested that the PKC was activated by DAGs, but not by ether-linked diglycerides (Ganong *et al.*, 1986a,b). Both 1-*O*-alk-1′-enyl-2-acyl-*sn*-glycerol and 1-*O*-alkyl-2-acyl-*sn*-glycerol are naturally occurring constituents of rabbit myocardium, and the former accumulates in ischemic myocardium. A study utilizing naturally occurring ether-linked diglycerides which accumulate in myocardium, neutrophils, or Martin Darby Canine kidney cells during cellular perturbations, revealed that ether-linked diglycerides, like DAGs, activate purified rat brain PKC to the same extent in the presence of PS vesicles. However, in mixed-micelle assays, the ether-linked diglycerides were less effective. Furthermore, in assays employing PS vesicles, significantly higher calcium concentrations were required for ether-linked diglycerides to activate PKC and this calcium dependence varied between cPKCs (Ford *et al.*, 1989).

These observations highlight a number of problems arising in the character-ization of lipid cofactors for PKC activation: (1) The results obtained *in vitro* may vary dramatically depending on the assay employed. Clearly DAGs display differ-ences with respect to the nature of the lipid phase in which they may function as activators (Slater *et al.*, 1994; Zidovetzki and Lester, 1992). (2) In PKC phospho-rylation assays, the substrate protein is a major determinant of lipid and calcium cofactor requirements for activation. These may vary from no PS or calcium requirement as is the case for protamine, to a requirement for PS only with myelin basic protein or for PS and calcium with histones (Epand, 1994). (3) The efficacy of a lipid cofactor may depend on the presence of other cofactors (divalent cations, other phospholipids, free fatty acids), ionic strength (Hannun and Bell, 1990), or the critical micellar concentration (free fatty acids; Khan *et al.*, 1992). (4) Finally, whether a lipid cofactor is required or not may depend strongly on the manner in which PKC itself is present (intracellular, homogenate, pure prepara-tion containing several isoforms, individual isoforms). These points listed here for DAG-dependent activation of PKC apply to the analysis of lipid cofactor require-ments for any protein.

1.7.2.   Regulation of Cell Growth by DAGs

Since PKC is the major intracellular phorbol ester receptor and phorbol esters bind to the same site as DAGs, both PKCs and DAGs are strongly implicated in playing a major role in the regulation of cell growth and differentiation. The question has arisen whether DAGs from phosphoinositides or PC are essential for proliferation.

1.7.2.1.   DAGs in Mitogenesis

Microinjection of NIH 3T3 cells with antibodies against PIP_2 abolished mito-genic response to both PDGF and bombesin (Matuoka *et al.*, 1988). EGF, fibroblast growth factor, insulin, and serum also stimulate nuclear [^{3}H]thymidine incor-poration, but this effect is not blocked by the same antibodies (Matuoka *et al.*, 1988). In hamster (NIL) fibroblasts, α-thrombin stimulates ^{32}P incorporation into PI and turnover of phosphoinositides. The concentration dependence of throm-bin stimulation correlated with initiation of cell proliferation. Neomycin, which binds to PIP and PIP_2 and thus inhibits PLC, blocks thrombin-dependent initia-tion of DNA synthesis and cell proliferation (Carney *et al.*, 1985). Microinjection of antibodies to PLCγ inhibited serum-induced DNA synthesis (Smith *et al.*, 1990).

Inidividual mitogenic agents are often not fully effective alone but require the synergistic action of additional agents to elicit a full mitogenic response (Rozengurt, 1986). However, serum contains several mitogenic agents. The inhibi-tion of serum-induced mitogenesis by the anti-PLCγ antibody indicates that all agents present in serum lead to mitogenesis through pathways in which PLCγ is crucial. In light of results discussed below, this exclusive role for derivatives of phosphoinositides in mitogenic responses is questionable.

Stimulation of PC hydrolysis by growth factors, such as PDGF and EGF, has been observed in various cells (Exton, 1990). PC is hydrolyzed to yield DAGs by either the action of PLCs or the consecutive action of PLD and PA phosphohydrolase. In many cases, agonists that induce PC breakdown also stimulate PC resynthesis by activation of CTP:phosphocholine cytidylyltransferase (Billah and Anthes, 1990; Exton, 1990; Pelech and Vance, 1989). Independent of whether DAG is liberated from PC by a PLC- or PLD-dependent pathway, it is produced without the simultaneous production of IP_3 and accompanying increase in free intracellular calcium concentration.

Intracellular DAG formation induced in response to many agonists is biphasic. The initial rise in DAG correlates with IP_3 formation and the fatty acid composition of DAG released is similar to that of phosphoinositides, suggesting that early DAG peaks are formed by PLC-dependent cleavage of PIP_2. Two lines of evidence indicate that sustained elevations of DAG result from cleavage of other phospholipids. First, DAG accumulates to levels exceeding the mass of inositol phosphates in a manner that is kinetically distinct from IP_3 formation (Uhing et al., 1989). Second, phorbol esters (Agwu et al., 1989; Billah et al., 1989), growth factors like EGF and PDGF (Exton, 1990; Pessin and Raben, 1989; Wright et al., 1988), the ras oncogene product (Lacal et al., 1987c; Morris et al., 1989) and certain interleukins (Rosoff et al., 1988) induce the formation of DAG without simultaneous phosphoinositide hydrolysis.

A mutant PDGF was expressed in CHO cells, which induced PLC-dependent PIP_2 hydrolysis, but did not sustain a mitogenic response (Escobedo and Williams, 1988). Other PDGF receptor mutants, which are unable to bind PI-specific PLC, still induce mitogenesis on ligand binding (Seedorf et al., 1992). The presence of genestein, a tyrosine kinase inhibitor, blocked PDGF-induced DNA synthesis, receptor autophosphorylation, PLC activation, and the intracellular accumulation of IP_3 and calcium in C3H10T1/2 mouse fibroblasts. After removal of genestein, addition of PDGF induced DNA synthesis in the absence of PLC activation (Hill et al., 1990).

In IIC9 fibroblasts, DAG release on stimulation with high concentrations of α-thrombin (500 ng/ml) is biphasic, with an initial peak at 15 sec and a second peak at 5 min. In cells pretreated with chymotrypsin, both α-thrombin and EGF lead to the formation of only the second DAG peak, in the absence of detectable inositol phosphate production (Wright et al., 1988).

In summary, these results clearly indicate that PIP_2 hydrolysis and DAG production are not required for mitogenesis, but rather sustained elevations of DAG resulting from PC hydrolysis are more important.

1.7.2.2. DAGs in Oncogenesis

Genes involved in the transduction of signals required for the control of normal cell proliferation are commonly altered in the neoplastic process (Bishop, 1991). Many examples of such proto-oncogenes, which give rise to oncogenes when altered, can be growth factors, growth factor receptors, or proteins like *ras* or *src*, which are linked to postreceptor signaling events (Barbacid, 1987; Pawson and Hunter, 1994). Cell transformation through the presence of the oncogenes *ras*, *src*, or *sis* correlates with elevated levels of DAG (Diaz-Lavada *et al.*, 1990; Fleischman *et al.*, 1986; Preiss *et al.*, 1986). The following discussion will focus mainly on *ras* since it is one of the most commonly mutated oncogenes in human cancer (Barbacid, 1987).

Ras belongs to the family of small G proteins which modulate a wide variety of cellular functions. The p21^{ras} protein controls regulatory pathways critical for normal proliferation and neoplastic transformation of fibroblasts, for the maturation of *Xenopus laevis* oocytes as well as for the differentiation of rat pheochromocytoma PC12 cells (Bourne *et al.*, 1990; McCormick, 1994). Recently, considerable progress has been made in understanding how *ras* is involved in such a multitude of processes. *Ras* is a crucial component in the signaling cascade linking growth factor receptors at the surface to activation of transcription in the nucleus. For example, EGF receptor phosphorylation on tyrosines leads, through the intermediate proteins GRB2 and SOS, to increased levels of GTP-bound *ras* at the plasma membrane, which then activates a downstream cascade including *raf*, MAP kinase kinase, and MAP kinase (reviewed in Blennis, 1993; Crews and Erickson, 1993; Marshall, 1994; Maruta and Burgess, 1994).

In *ras*-transformed cells, elevated DAG levels are seen consistently and stem most likely from the hydrolysis of PC and PE (Fleischman *et al.*, 1986; Lacal *et al.*, 1987a,b; Lacal, 1990; Preiss *et al.*, 1986; Wolfman and Macara, 1987). Microinjection of the inhibitory anti-*ras* monoclonal antibody 259 into NIH 3T3 cells inhibits transformation and proliferation by oncogenes with tyrosine kinase activity (Smith *et al.*, 1986). This antibody did not prevent activation of PLC or PLA$_2$, suggesting that they are not downstream effectors activated by *ras* (Yu *et al.*, 1988). Alternatively, microinjection of *ras* into *X. laevis* oocytes induced biphasic phosphocholine release and concomitant increase in choline kinase activity. The second phase correlated with DAG production indicating that *ras* may control both PC biosynthesis and degradation (Lacal, 1990). Activation of a PC-specific PLC may be all that is required for *ras*-induced germinal vesicle breakdown (GVBD) in *X. laevis* oocytes. Injection of PC-specific PLC from *Bacillus cereus* causes GVBD. Injection of antibodies against PC-specific PLC prevent *ras*-induced GVBD (Garcia de Herreros *et al.*, 1991).

One PKC isoform, PKCζ, appears crucial for GVBD in *X. laevis* oocytes induced by *ras*, insulin, or PLC treatment (Dominguez *et al.*, 1992). This is difficult to reconcile with observations suggesting that PKCζ neither binds nor is down-regulated by phorbol esters, nor is activated by either phorbol esters or DAGs (Nakanishi and Exton, 1992; Ono *et al.*, 1989a; Ways *et al.*, 1992). However, PKCζ is activated by PIP$_3$ (Nakanishi *et al.*, 1993), a PI lipid arising through activation of PI(3)-kinase (Fig. 1-1), which binds directly to and is activated by GTP-*ras* (Rodriguez-Viciana *et al.*, 1994).

Mitogenic responses to *ras* appear to rely on PC hydrolysis in fibroblasts. Expression of a dominant negative H-*ras* mutant (H-*ras* Asn-17), which preferentially binds GDP versus GTP, inhibits NIH 3T3 fibroblast proliferation in response to various peptide growth factors and phorbol ester treatment. Treatment of cells expressing H-*ras* Asn-17 with exogenous PC-specific PLC overcomes inhibition of mitogenesis (Cai *et al.*, 1992). PC-specific PLC from *Bacillus cereus* effectively mimics *ras* and *src* transformation of NIH 3T3 fibroblasts, an effect that is not seen with PI-specific PLC from *Bacillus thuringiensis* (Diaz-Laviada *et al.*, 1990). These results indicate that PC hydrolysis is a target of *ras* during transduction of growth factor-initiated mitogenic signals.

PKC is crucial for *ras*-dependent signaling in fibroblasts. Stimulation of mitogenesis in quiescent Swiss 3T3 cells by microinjection of H-*ras* p21 is markedly reduced when PKC is downregulated by chronic phorbol ester treatment. The mitogenic response is reconstituted by coinjection of H-*ras* p21 and PKC, indicating that functional PKC is required for the mitogenic activity of H-*ras* p21 (Lacal *et al.*, 1987b). Downregulation of PKC by prolonged phorbol ester treatment prevented the ability of *ras* introduced into Swiss 3T3 cells by scrape-loading to elevate DAG (from PC) levels (Price *et al.*, 1989).

Alternatively, *ras* appears necessary for certain aspects of PKC function. For instance, phorbol ester-stimulated mitogenesis requires *ras*, although the immediate target is probably PKC (Cai *et al.*, 1992; Yu *et al.*, 1988). Thus, the precise relationship between *ras* and PKC remains unclear.

These contradictory results concerning the positioning of *ras* relative to PKCs may be explained because PKCs are not the sole targets of phorbol esters. Thus, clarification may arise through the identification of proteins that bind phorbol esters and modulate *ras*-nucleotide binding. One possibility is the protein *vav*, which has been discussed (Gulbins *et al.*, 1994). Alternatively, *ras* may appear downstream of PKC in the signaling cascade, if direct activation of PKC modulates the activity of an element upstream of *ras*. The recent discovery that PKC-dependent phosphorylation inhibits a protein tyrosine phosphatase PTP-PEST (Garton and Tonks, 1994) provides just one of many possibilities. This may also explain why phorbol ester treatment of cells increases tyrosine phosphorylation of certain proteins (Gilmore and Martin, 1983).

1.8. Summary

Glycerolipids are synthesized on the cytoplasmic surface of the ER. A "PC-flippase" activity in the ER is implicated in establishing the preferential presence of PC on the exoplasmic surface of the plasma membrane (Bishop and Bell, 1989). Ceramide is synthesized on the cytoplasmic side of the ER, but subsequent steps leading to the formation of SPM and more complex sphingolipids occur on the luminal surface of the Golgi, explaining the exclusive presence of these lipids on the exoplasmic leaflet of the plasma membrane (Hannun, 1994; Van Meer, 1989). Thus, the asymmetric distribution of sphingolipids, preferentially in the outer membrane leaflet, reflects more closely the asymmetry of their biosynthesis in the Golgi, than is the case for the biosynthesis of glycerolipids. The composition of individual leaflets, in turn, determines the types of lipid hydrolyzing activities required to generate DAG second messengers in response to receptor activation by agonists.

DAGs are formed from the glycerophospholipids PI, PC, and PE, by agonist-stimulated pathways involving PLCs and PLDs at several subcellular locations. These factors determine the rate at which DAG signals are generated in a cell. Several possibilities for removal of DAG signals exist, which are utilized to widely varying extents in different cells. The kinetics of DAG formation and removal together account for the time frame and transience of DAG signals. The fatty acid composition of the DAG species, which depends on the phospholipid DAG was derived from, channels a DAG molecule into a particular recycling pathway and thereby determines how rapidly DAG signals disappear within the cell. DAG signals resulting from the hydrolysis of PI lipids are accompanied by an increase in intracellular calcium, which is not seen on PC or PE hydrolysis. This presence of an additional second messenger released simultaneously with DAG influences the cellular response observed.

PKCs are the best-characterized targets for DAGs. PKC activation occurs through lipid interactions within the regulatory domain. PKC responds to agonist-induced changes in DAG levels at the plasma membrane and elicits a wide variety of responses related to cell growth and differentiation. PKC isoforms vary in their intracellular distribution and contribute thereby to the propagation of DAG-related signals throughout the cell. In particular, PKCβII may be activated in response to lipid (DAG?) signals in the nucleus. The possibly exclusive localization of PKCη in the ER indicates that this isoform could play an important role in the regulation of lipid biosynthesis in epithelial cells.

PKC activation through DAG results in the formation of secondary lipid hydrolysis products, including again DAG, which prolong PKC activation. Such long-term responses are essential for cell proliferation and differentiation (Nishi-zuka, 1992). Lipid-second messengers derived from both glycerophospholipids and sphingophospholipids may be formed as a result of an initial DAG signal.

The cysteine-rich regions in PKCs, identified as the elements responsible for DAG-dependent PKC activation, are present in several proteins, some of which are phorbol ester binding proteins and also regulated by DAG. Thus, caution is required in the interpretation of effects of phorbol esters or cell-permeable DAGs *in vivo* as being solely the result of interactions with PKC. Furthermore, cysteine-rich regions mediate regulation by lipid second messengers other than DAG as well as protein–protein interactions. These regions represent a highly versatile protein motif implicated in responses to multiple lipid-derived second messengers.

DAG levels are elevated in transformed cells, fluctuate depending on the growth conditions, and may play an important role in cell cycle-dependent control of lipid biosynthesis. Pools of DAG exist both at the plasma membrane and in the nuclear/ER compartments which are important for signaling. Regulation of cytidylyltransferase activity by DAG and phosphorylation, are both potentially important determinants in the control of PC biosynthesis. The initially proposed distinction between DAGs involved in signaling events at the plasma membrane and DAGs involved in biosynthesis at the ER is no longer obvious.

Methods of analysis represent the major limitation to improving our understanding of intracellular DAG functions. Currently we may at best study synchronized cell populations. The results obtained reflect overall changes in DAG levels. The existence of multiple pools of DAG, which may change simultaneously within any given cell, indicates that current analytical methods need to be refined. Thus, a major step toward improving our understanding of intracellular DAG functions will be taken when the means are developed to study DAG changes and location of DAG pools in individual cells.

ACKNOWLEDGMENTS.   We are grateful to Elaine Bardes for assistance with figures and tables. This work was supported by Swiss National Science Foundation grant 3100-040477.94 (A.F.G.Q.) and National Institutes of Health grants HL39086 (D.M.R.) and GM38737 (R.M.B.).

References

Agwu, D. E., McPhail, L. C., Chabot, M. C., Daniel, L. W., Wykle, R. L., and McCall, C. E., 1989, Choline-linked phosphoglycerides. A source of phosphatidic acid and diglycerides in stimulated neutrophils, *J. Biol. Chem.* **264**:1405–1413.

Ahmed, S., Kozma, R., Monfries, C., Hall, C., Lim, H. H., and Smith, P., 1990, Human brain n-chimaerin cDNA encodes a novel phorbol ester receptor, *Biochem. J.* **272**:767–773.

Ahmed, S., Kozma, R., Lee, J., Monfries, C., Harden, N., and Lim, L., 1991, The cysteine-rich domain of human proteins, neuronal chimaerin, protein kinase C and diacylglycerol kinase binds zinc, *Biochem. J.* **280**:233–241.

Ahmed, S., Murutama, I. N., Kozma, R., Lee, J., Brenner, S., and Lim, L., 1992, The *Caenorhabditis elegans unc*-13 gene product is a phospholipid-dependent high affinity phorbol ester receptor, *Biochem. J.* **287**:995–999.

Ahmed, S., Lee, J., Kozma, R., Best, A., Monfries, C., and Lim, L., 1993, A novel functional target for tumor-promoting phorbol esters and lysophosphatidic acid. The p-21*rac*-GTPase activating protein *n*-chimaerin, *J. Biol. Chem.* **268**:10709–10712.

Akimoto, K., Mizuno, K., Osada, S.-i. Harai, S.-i., Tanuma, S.-i., Suzuki, K., and Ohno, S., 1994, A new member of the third class in the protein kinase C family, PKCλ, expressed predominantly in an undifferentiated mouse embryonal carcinoma cell line and also in many tissues and cells, *J. Biol. Chem.* **269**:12677–12683.

Altin, J., and Bradshaw, R. A., 1990, The production of 1,2-diacylglycerol in PC12 cells by nerve growth factor and basic fibroblast growth factor, *J. Neurochem.* **54**:1666–1676.

Ashendel, C. L., 1985, The phorbol ester receptor: A phospholipid-regulated kinase, *Biochim. Biophys. Acta* **822**:219–242.

Augert, G., Bocckino, S. B., Blackmore, P. F., and Exton, J. H., 1989, Hormonal stimulation of diacylglycerol formation in hepatocytes. *J. Biol. Chem.* **264**:21689–21698.

Avruch, J., Zhang, X.-F., and Kyriakis, J. M., 1994, Raf meets ras: completing the framework of a signal transduction pathway, *Trends Biochem. Sci.* **19**:279–283.

Backer, J. M., and Dawidowicz, E. A., 1987, Reconstitution of a phospholipid flippase from rat liver microsomes, *Nature* **327**:341–343.

Ballas, L. M., and Bell, R. M., 1981, Topography of glycerolipid synthetic enzymes. Synthesis of phosphatidylserine, phosphatidylinositol and glycerolipid intermediates occurs on the cytoplasmic surface of rat liver microsome vesicles. *Biochim. Biophys. Acta* **665**:586–595.

Ballas, L. M., Lazarow, P. B., and Bell, R. M., 1984, Glycerolipid synthetic capacity of rat liver peroxisomes, *Biochim. Biophys. Acta* **795**:297–300.

Ballester, R., Furth, M. E., and Rosen, O. M., 1987, Phorbol ester- and protein kinase C-mediated phosphorylation of the cellular Kirsten *ras* gene product, *J. Biol. Chem.* **262**:2688–2695.

Barbacid, M., 1987, *Ras* genes, *Ann. Rev. Biochem.* **56:**779–827.

Bell, R. L., Kennerly, D. A., Stanford, N., and Majerus, P. W., 1979, Diglyceride lipase: a pathway for arachidonate release from human platelets, *Proc. Natl. Acad. Sci. USA* **76:**3238–3241.

Bell, R. M., 1986, Protein kinase C activation by diacylglycerol second messengers, *Cell* **45:**631–632.

Bell, R. M., and Burns, D. J., 1991, Lipid activation of protein kinase C, *J. Biol. Chem.* **266:**4661–4664.

Bell, R. M., and Coleman, R. A., 1980, Enzymes of glycerolipid synthesis in eukaryotes, *Annu. Rev. Biochem.* **49:**459–487.

Bell, R. M., and Coleman, R. A., 1983, Enzymes of triacylglycerol formation in mammals, in: *The Enzymes*, Vol. 16 (P. D. Boyer, ed.), pp. 87–111, Academic Press, New York.

Bell, R. M., Ballas, L. M., and Coleman, R. A., 1981, Lipid topogenesis, *J. Lipid Res.* **22:**391–403.

Bell, R., Burns, D., Okazaki, T., and Hannun, Y., 1992, Network of signal transduction pathways involving lipids: Proteins kinase C-dependent and independent pathways, *Adv. Exp. Med. Biol.* **318:**275–284.

Berridge, M. J., 1987, Inositol trisphosphate and diacylglycerol: Two interacting second messengers, *Annu. Rev. Biochem.* **56:**159–193.

Berridge, M. J., and Irvine, R. F., 1989, Inositol phosphates and cell signalling, *Nature* **341:**197–205.

Besterman, J. M., Duronio, V., and Cuatrecasas, P., 1986a, Rapid formation of diacylglycerol from phosphatidylcholine: A pathway for generation of a second messenger, *Proc. Natl. Acad. Sci. USA* **83:**6785–6789.

Besterman, J. M., Pollenz, R. S., Booker, E. L., Jr., and Cuatrecasas, P., 1986b, Diacylglycerol-induced translocation of diacylglycerol kinase: Use of affinity-purified enzyme in a reconstitution system, *Proc. Natl. Acad. Sci. USA* **83:**9378–9382.

Bhat, B. G., Wong, P., and Coleman, R. A., 1994, Hepatic monoacylglycerol acyltransferase is regulated by *sn*-1,2-diacylglycerol and by specific lipids in Triton X-100/phospholipid-mixed micelles, *J. Biol. Chem.* **269:**13172–13178.

Billah, M. M., and Anthes, J. C., 1990, The regulation and cellular functions of phosphatidylcholine hydrolysis, *Biochem. J.* **269:**281–291.

Billah, M. M., Pai, J.-K., Mullmann, T. J., Egan, R. W., and Siegel, M. I., 1989, Regulation of phospholipase D in HL-60 granulocytes. Activation by phorbol esters, diglyceride and calcium ionophore via protein kinase C-independent mechanisms, *J. Biol. Chem.* **264:**9069–9076.

Bishop, J. M., 1991, Molecular themes in oncogenesis, *Cell* **64:**235–248.

Bishop, W. R., and Bell, R. M., 1985, Assembly of the endoplasmic reticulum phospholipid bilayer: The phosphatidylcholinetransporter, *Cell* **42:**51–60.

Bishop, W. R., and Bell, R. M., 1988, Functions of diacylglycerol in glycerolipid metabolism, signal transduction and cellular transformation, *Oncogene Res.* **2:**205–208.

Bishop, W. R., and Bell, R. M., 1989, Assembly of phospholipids into cellular membranes: Biosynthesis, transmembrane movement and intracellular translocation, *Annu. Rev. Cell Biol.* **4:**576–610.

Bishop, W. R., Ganong, B. R., and Bell, R. M., 1986, Attenuation of *sn*-1,2-diacylglycerol second messengers by diacylglycerol kinase. Inhibition by diacylglycerol analogs *in vitro* and in human platelets, *J. Biol. Chem.* **261:**6993–7000.

Blennis, J., 1993, Signal transduction via MAP kinases: Proceed at your own RSK, *Proc. Natl. Acad. Sci. USA* **90:**5889–5892.

Blumberg, P. M., 1980, *In vitro* studies on the mode of action of phorbol esters, potent tumor promoters: Part 1, *CRC Crit. Rev. Toxicol.* **8:**153–197.

Blumberg, P. M., 1981, *In vitro* studies on the mode of action of phorbol esters, potent tumor promoters: Part 2, *CRC Crit. Rev. Toxicol.* **8:**199–234.

Bocckino, S. B., Blackmore, P. F., and Exton, J. H., 1985, Stimulation of 1,2-diacylglycerol accumulation in hepatocytes by vasopressin, epinephrine and angiotensin II, *J. Biol. Chem.* **260:**14201–14207.

Boni, L. T., and Rando, R. R., 1985, The nature of protein kinase C activation by physically defined phospholipid vesicles and diacylglycerols, *J. Biol. Chem.* **260:**10819–10825.

Bourne, H. R., Sanders, D. A., and McCormick, F., 1990, The GTPase superfamily: A conserved switch for diverse cell functions, *Nature* **348:**125–132.

Bretscher, M. S., 1972, Asymmetrical lipid bilayer structure for biological membranes, *Nature New Biol.* **236:**11–12.

Burns, D. J., and Bell, R. M., 1991, Protein kinase C contains two phorbol ester binding domains, J. Biol. Chem. **266:**18330–18338.

Burns, D. J., and Bell, R. M., 1992, Lipid regulation of protein kinase C, in: *Protein Kinase C: Current Concepts and Future Perspectives* (D. Lester and R. M. Epand, eds.), pp. 25–40, Horwood, Chichester, England.

Bussolino, F., Silvagno, F., Garbarino, G., Costamagna, C., Sanavio, F., Arese, M., Soldi, R., Aglietta, M., Pescarmona, G., Camussi, G., and Bosia, A., 1994, Human endothelial cells are targets for platelet-activating factor (PAF). Activation of α and β protein kinase C isoenzymes in endothelial cells stimulated by PAF, *J. Biol. Chem.* **269:**2877–2886.

Cai, H., Erhardt, P., Szeberenyl, J., Diaz-Meco, M. T., Johansen, T., Moscat, J., and Cooper, G. M., 1992, Hydrolysis of phosphatidylcholine is stimulated by *ras* proteins during mitogenic signal transduction, *Mol. Cell. Biol.* **12:**5329–5335.

Carney, D. H., Scott, D. L., Gordon, E. A., and LaBelle, E. F., 1985, Phosphoinositides in mitogenesis: Neomycin inhibits thrombin-stimulated phosphoinositide turnover and inhibition of cell proliferation, *Cell* **42:**479–488.

Castagna, M., Takai, Y., Kaibuchi, K., Sano, K., Kikkawa, U., and Nishizuka, Y., 1982, Direct activation of calcium-activated, phospholipid-dependent protein kinase by tumor-promoting phorbol esters, *J. Biol. Chem.* **257:**7847–7851.

Cazaubon, S., Bornancin, F., and Parker, P. J., 1994, Threonine-497 is a critical site for permissive activation of protein kinase Cα, *Biochem. J.* **301:**443–448.

Chao, W., and Olson, M. S., 1993, Platelet-activating factor: Receptors and signal transduction, *Biochem. J.* **292:**617–629.

Chida, K., Sagara, H., Suzuki, Y., Murakami, A., Osada, S.-i., Ohno, S., Hirosawa, K., and Kuroki, T., 1994, The η isoform of protein kinase C is localized on rough endoplasmic reticulum, *Mol. Cell. Biol.* **14:**3782–3790.

Chilton, F. H., Ellis, J. M., Olson, S. C., and Wykle, R. L., 1984, 1-O-alkyl-2-arachidonoyl-*sn*-glycero-3-phosphocholine. A common source of platelet-activating factor and arachidonate in human polymorphonuclear leukocytes, *J. Biol. Chem.* **259:**12014–12019.

Chojnacki, T., and Dallner, G., 1988, The biological role of dolichol, *Biochem. J.* **251:**1–9.

Conn, P. J., and Sweatt, J. D., 1994, Protein kinase C in the nervous system, in: *Protein Kinase C* (J. F. Kuo, ed.), pp. 199–235, Oxford University Press, London.

Conricode, K. M., Brewer, K. A., and Exton, J. H., 1992, Activation of phospholipase D by protein kinase C. Evidence for a phosphorylation-independent mechanism, *J. Biol. Chem.* **267:**7199–7202.

Cook, H. W., 1991, Fatty acid desaturation and chain elongation in eukaryotes, in: *Biochemistry of Lipids, Lipoproteins and Membranes* (D. E. Vance and J. Vance, eds.), pp. 141–169, Benjamin/Cummings, Menlo Park, CA.

Cook, S. J., and Wakelam, M. J. O., 1992, Stimulated phosphatidylcholine hydrolysis as a signal transduction pathway in mitogenesis, *Cell. Signal.* **3:**273–282.

Crews, C. M., and Erickson, R. L., 1993, Extracellular signals and reversible phosphorylation: What to MEK of it all, *Cell* **74:**215–217.

Dawidowicz, E. A., 1987, Dynamics of membrane lipid metabolism and turnover, *Annu. Rev. Biochem.* **56:**43–61.

Dekker, L. V., and Parker, P. J., 1994, Protein kinase C—A question of specificity, *Trends Biochem. Sci.* **19:**73–77.

Dekker, L. V., McIntyre, P., and Parker, P. J., 1993, Mutagenesis of the regulatory domain of rat protein kinase C-η. A molecular basis for restricted histone kinase activity, *J. Biol. Chem.* **268:**19498–19504.

Den Hartigh, J. C., van Bergen en Hengegowen, P. M. P., Verkleij, A. J., and Boonstra, J., 1992, The EGF receptor is an actin-binding protein, *J. Cell Biol.* **119:**349–355.

Dennis, E. A., 1994, Diversity of group types, regulation and function of phospholipase A_2, *J. Biol. Chem.* **269:**13057–13060.

Devaux, P. F., 1988, Phospholipid flippases, *FEBS Lett.* **234:**8–12.

Devaux, P. F., 1991, Static and dynamic lipid asymmetry in cell membranes, *Biochemistry* **30:**1163–1173.

Diaz-Laviada, I., Larrodera, P., Diaz-Meco, M., Cornet, M. E., Guddal, P. H., Johansen, T., and Moscat,

J., 1990, Evidence for a role of phosphatidylcholine-hydrolysing phospholipase C in the regulation of protein kinase C by *ras* and *scr* oncogenes, *EMBO J.* **9:**3907–3912.

Disatnik, M.-H., Hernandez-Sotomayer, S. M. T., Jones, G., Carpenter, G., and Mochley-Rosen, D., 1994, Phospholipase C-γl binding to intracellular receptors for activated protein kinase C, *Proc. Natl. Acad. Sci. USA* **91:**559–563.

Divecha, N., Banfic, H., and Irvine, R. F., 1991, The polyphosphoinositide cycle exists in the nuclei of Swiss 3T3 cells under the control of a receptor (for IGF-1) in the plasma membrane, and stimulation of the cycle increases nuclear diacylglycerol and apparently induces translocation of protein kinase C to the nucleus, *EMBO J.* **10:**3207–3214.

Divecha, N., Banfie, H., and Irvine, R., 1993, Inositides and the nucleus and inositides in the nucleus, *Cell* **74:**405–407.

Di Virgilio, F., Lew, D. P., and Pozzan, T., 1984, Protein kinase C activation of physiological processes in human neutrophils at vanishingly small cytosolic Ca^{2+} levels, *Nature* **310:**691–693.

Dominguez, I., Diaz-Meco, M. T., Municio, M. M., Berra, E., Garcia de Herreros, A., Cornet, M. E., Sanz, L., and Moscat, J., 1992, Evidence for a role of protein kinase C subspecies in maturation of *Xenopus laevis* oocytes, *Mol. Cell. Biol.* **12:**3776–3783.

Downward, J., Graves, J. D., Warne, P. H., Rayter, S., and Cantrell, D. A., 1990, Stimulation of p21^{ras} upon T-cell activation, *Nature* **346:**719–723.

Drust, D. S., and Martin, T. F. J., 1985, Protein kinase C translocates from cytosol to membrane upon hormone activation: Effects of thyrotropin-releasing hormone in GH$_3$ cells, *Biochem. Biophys. Res. Commun.* **128:**531–537.

Epand, R. M., 1994, *In vitro* assays of protein kinase C activity, *Anal. Biochem.* **218:**241–247.

Escobedo, J. A., and Williams, L. T., 1988, A PDGF receptor domain essential for mitogenesis but not for many other responses to PDGF, *Nature* **335:**85–87.

Esko, J. D., and Raetz, C. R. H., 1983, Synthesis of phospholipids in animal cells, in: *The Enzymes*, Vol. 16 (P. D. Boyer, ed.), pp. 208–253, Academic Press, New York.

Exton, J. H., 1990, Signaling through phosphatidylcholine breakdown, *J. Biol. Chem.* **265:**1–4.

Exton, J. H., 1994, Messenger molecules derived from membrane lipids, *Curr. Opin. Cell Biol.* **6:**226–229.

Feldman, D. A., and Weinhold, P. A., 1987, CTP:phosphorylcholine cytidylyltransferase from rat liver. Isolation and characterization of the catalytic subunit, *J. Biol. Chem.* **262:**9075–9081.

Fields, A. P., Tyler, G., Kraft, A. S., and May, W. S., 1990, Role of nuclear protein kinase C in the mitogenic response to platelet-derived growth factor, *J. Cell Sci.* **96:**107–114.

Fleischman, L. F., Chahwala, S. B., and Cantley, L., 1986, Ras-transformed cells: Altered levels of phosphatidylinositol-4,5-bisphosphate and catabolites, *Science* **231:**407–410.

Flint, A. J., Palachini, R. D., and Koshland, D. E., Jr., 1990, Autophosphorylation of protein kinase C at three separate regions of its primary sequence, *Science* **249:**408–411.

Florin-Christensen, J., Florin-Christensen, M., Delfino, J. M., Stegmann, T., and Rasmussen, H., 1992, Metabolic fate of plasma membrane diacylglycerols in NIH 3T3 fibroblasts, *J. Biol. Chem.* **267:**14783–14789.

Ford, D. A., and Gross, R. W., 1990, Differential metabolism of diradylglycerol molecular subclasses and molecular species by rat brain diglyceride kinase, *J. Biol. Chem.* **265:**12280–12286.

Ford, D. A., Miyake, R., Glaser, P. E., and Gross, R. W., 1989, Activation of protein kinase C by naturally occurring ether-linked diglycerides, *J. Biol. Chem.* **264:**13818–13824.

Fukami, K., and Takenawa, T., 1989, Quantitative changes in polyphosphoinositides, 1,2-diacylglycerol and inositol 1,4,5-triphosphate by platelet-derived growth factor and prostaglandin F$_2$ α, *J. Biol. Chem.* **264:**14985–14989.

Ganong, B. R., and Bell, R. M., 1984, Transmembrane movement of phosphatidylglycerol and diacylglycerol sulfhydryl analogues, *Biochemistry* **23:**4977–4983.

Ganong, B. R., Loomis, C. R., Hannun, Y. A., and Bell, R. M., 1986a, Regulation of protein kinase C by lipid cofactors, in: *Cell Membranes: Methods and Reviews* (E. Elson, W. Frazier, and L. Glaser, eds.), Plenum Press, New York.

Ganong, B. R., Loomis, C. R., and Bell, R. M., 1986b, Specificity and mechanism of protein kinase C activation by *sn*-1,2-diacylglycerols, *Proc. Natl. Acad. Sci. USA* **83:**1184–1188.

Garcia de Herreros, A., Dominguez, I., Diaz-Meco, M. T., Giaziani, G., Cornet, M. E., Guddal, P. H., Johansen, T., and Moscat, J., 1991, Requirement of phospholipase C catalyzed hydrolysis of phosphatidylcholine for maturation of *Xenopus laevis* oocytes in response to insulin and *ras* p21, *J. Biol. Chem.* **266:**6825–6829.

Garton, A. J., and Tonks, N. K., 1994, PTP-PEST: A protein tyrosine phosphatase regulated by serine phosphorylation, *EMBO J.* **13:**3763–3771.

Ghosh, S., Xie, W. Q., Quest, A. F. G., Mabrouk, G. M., Strum, J. C., and Bell, R. M., 1994, The cysteine-rich region of *raf*-1 kinase contains zinc, translocates to liposomes, and is adjacent to a segment that binds GTP-*ras*, *J. Biol. Chem.* **269:**10000–10007.

Gilmore, T., and Martin, S., 1983, Phorbol ester and diacylglycerol induce protein phosphorylation at tyrosine, *Nature* **306:**487–490.

Goss, V. L., Hocevar, B. A., Thompson, L. J., Stratton, C. A., Burns, D. J., and Fields, A. P., 1994, Identification of nuclear βII-protein kinase C as a mitotic lamin kinase, *J. Biol. Chem.* **269:**19074–19080.

Greif, H., Ben-chaim, J., Shimon, T., Bechor, E., Eldar, H., and Livneh, E., 1992, The protein kinase C-L(η) gene product is localized in the cell nucleus, *Mol. Cell Biol.* **12:**1304–1311.

Gulbins, E., Coggeshall, K. M., Baier, G., Katzov, S., Burn, P., and Altman, A., 1993, Tyrosine kinase-stimulated guanine-nucleotide exchange of *vav* in T-cell activation, *Science* **260:**822–825.

Gulbins, E., Coggeshall, K. M., Baier, G., Telford, D., Langlet, C., Baier-Bitterlich, G., Bonnefoy-Berard, N., Burn, P., Wittinghofer, A., and Altman, A., 1994, Direct stimulation of *vav* guanine nucleotide exchange activity for *ras* by phorbol esters and diglycerides, *Mol. Cell. Biol.* **14:**4749–4758.

Gurr, M. I., Finean, J. B., and Hawthorne, J. N., 1963, The phospholipids of liver-cell fractions: The phospholipid composition of the liver-cell nucleus, *Biochim. Biophys. Acta* **70:**406–416.

Ha, K. S., and Exton, J. H., 1993, Differential translocation of protein kinase C isozymes by thrombin and platelet-derived growth factor. A possible function for phosphatidylcholine-derived diacylglycerol, *J. Biol. Chem.* **268:**10534–10539.

Ha, K. S., and Thompson, G. A., 1992, Biphasic changes in the level and composition of Dunaliella salina plasma membrane diacylglycerols following hypoosmotic shock, *Biochemistry* **31:**596–603.

Habenicht, A. J. R., Glomset, J. A., King, W. C., Nist, C., Mitchell, C. D., and Ross, R., 1981, Early changes in phosphatidylinositol and arachidonic acid metabolism in quiescent Swiss 3T3 cells stimulated to divide by platelet-derived growth factor, *J. Biol. Chem.* **256:**12329–12335.

Hall, C., Monfries, C., Smith, P., Lim, H. H., Kozma, R., Ahmed, S., Vanniasingham, V., Leung, T., and Lim, L., 1990, Novel human brain cDNA encoding a 34000 Mr protein *n*-chimaerin, related to both the regulatory domain of protein kinase C and BCR, the product of the breakpoint cluster region gene, *J. Mol. Biol.* **211:**11–16.

Hanks, S. K., Quinn, A. M., and Hunter, T., 1988, The protein kinase C family: Conserved features and deduced phylogeny of the catalytic domains, *Science* **241:**45–52.

Hannun, Y. A., 1994, The sphingomyelin cycle and second messenger function of ceramide, *J. Biol. Chem.* **269:**3125–3128.

Hannun, Y. A., and Bell, R. M., 1986, Phorbol ester binding and activation of protein kinase C on Triton X-100 mixed micelles containing phosphatidylserine, *J. Biol. Chem.* **261:**9341–9347.

Hannun, Y. A., and Bell, R. M., 1990, Rat brain protein kinase C: Kinetic analysis of substrate dependence, allosteric regulation, and autophosphorylation, *J. Biol. Chem.* **265:**2962–2972.

Hannun, Y. A., Loomis, C. R., and Bell, R. M., 1985, Activation of protein kinase C by Triton X-100 mixed micelles containing diacylglycerol and phosphatidylserine, *J. Biol. Chem.* **260:**10039–10043.

Hannun, Y. A., Loomis, C. R., and Bell, R. M., 1986, Protein kinase C activation in mixed micelles, *J. Biol. Chem.* **261:**7184–7190.

Hazen, S. L., Stuppy, R. J., and Gross, R. W., 1990, Purification and characterization of canine myocardial cytosolic phospholipase A$_2$: A calcium-independent phospholipase with absolute *sn*-2 regiospecificity for diradylglycerophospholipids, *J. Biol. Chem.* **265:**10622–10630.

Hazen, S. L., Ford, D. A., and Gross, R. W., 1991, Activation of a membrane-associated phospholipase A$_2$ during myocardial ischemia which is highly selective for plasmalogen substrate, *J. Biol. Chem.* **266:**5629–5633.

Heller, R. A., and Krönke, M., 1994, Tumor necrosis factor receptor-mediated signaling pathways, *J. Cell Biol.* **126:**5–9.

Higgins, J. A., and Fieldsend, J. K., 1987, Phosphatidylcholine synthesis for incorporation into membranes or for secretion as plasma lipoproteins by Golgi membranes of rat liver, *J. Lipid Res.* **28:** 268–278.

Hill, T. D., Dean, N. D., Mordan, L. J., Lau, A. F., Kanemitsu, M. Y., and Boynton, A. L., 1990, PDGF-induced activation of phospholipase C is not required for induction of DNA synthesis, *Science* **248:**1660–1663.

Hocevar, B. A., and Fields, A. P., 1991, Selective translocation of βII-protein kinase C to the nucleus of human promyelocytic (HL60) leukemia cells, *J. Biol. Chem.* **266:**28–33.

Hocevar, B. A., Burns, D. J., and Fields, A. P., 1993, Identification of protein kinase C (PKC) phosphorylation sites on human lamin B. Potential role of PKC in nuclear lamina structural dynamics, *J. Biol. Chem.* **268:**7545–7552.

Hokin, L. E., 1985, Receptors and phosphoinositide second messengers, *Annu. Rev. Biochem.* **54:** 205–235.

Holbrook, P. G., Pannell, L. K., Murata, Y., and Daly, J. W., 1992, Molecular species analysis of a product of phospholipase D activation: Phosphatidylethanol is formed from phosphatidylcholine in phorbol ester and bradykinin-stimulated PC12 cells, *J. Biol. Chem.* **267:**16834–16840.

Hopewell, R., Martin-Sanz, P., Martin, A., Saxton, J., and Brindley, D. N., 1985, Regulation of the translocation of phosphatidate phosphohydrolase between cytosol and the endoplasmic reticulum of rat liver. Effects of unsaturated fatty acids, spermine, nucleotides, albumin and chlorpromazine, *Biochem. J.* **232:**485–491.

Horwitz, J., 1990, Carbachol and bradykinin increase the production of diacylglycerol from sources other than inositol-containing phospholipids in PC12 cells, *J. Neurochem.* **54:**983–991.

House, C., and Kemp, B. E., 1987, Protein kinase C contains a pseudosubstrate prototype in the regulatory domain, *Science* **238:**1726–1728.

House, C., and Kemp, B. E., 1990, Protein kinase C pseudosubstrate prototype: Structure–function relationships, *Cell. Signal.* **2:**187–190.

Hovius, R., Faber, B., Brigot, B., Nicolay, K., and de Kruijff, B., 1992, On the mechanism of mitochondrial decarboxylation of phosphatidylserine, *J. Biol. Chem.* **267:**16790–16795.

Hug, H., and Sarre, T. F., 1993, Protein kinase C isozymes: Divergence in signal transduction, *Biochem. J.* **291:**329–343.

Hughes, B. P., Rye, K.-A., Pickford, L. B., Barritt, G. J., and Chalmers, A. H., 1984, A transient increase in diacylglycerols is associated with the action of vasopressin on hepatocytes, *Biochem. J.* **222:**535–540.

Imai, A., and Gershengorn, M. C., 1987, Independent phosphatidylinositol synthesis in pituitary plasma membrane and endoplasmic reticulum, *Nature* **325:**726–728.

Imai, A., Rebbechi, M. J., and Gershengorn, M. C., 1986, Differential regulation by phosphatidyl-inositol-4,5-bisphosphate of pituitary plasma membrane and cytosolic phosphoinositide kinases, *Biochem. J.* **240:**341–348.

Inoue, M. A., Kishimoto, A., Takai, Y., and Nishizuka, Y., 1977, Studies on a cyclic nucleotide-independent protein kinase and its proenzyme in mammalian tissue II, *J. Biol. Chem.* **252:**7610–7616.

Jackowski, S., 1994, Coordination of membrane phospholipid synthesis in the cell cycle, *J. Biol. Chem.* **269:**3858–3867.

Jalink, K., van Corven, E. J., and Moolenaar, W. H., 1990, Lysophosphatidic acid, but not phosphatidic acid, is a potent Ca^{2+}-mobilizing stimulus for fibroblasts. Evidence for extracellular site of action, *J. Biol. Chem.* **265:**12232–12239.

James, G., and Olson, E., 1992, Deletion of the regulatory domain of protein kinase Cα exposes regions in the hinge and catalytic domains that mediate nuclear targeting, *J. Cell Biol.* **116:**863–874.

Jamil, H., Utal, A. K., and Vance, D. E., 1992, Evidence that cyclic AMP-induced inhibition of phosphatidylcholine biosynthesis is caused by a decrease in cellular diacylglycerol levels in cultured rat hepatocytes, *J. Biol. Chem.* **267:**1752–1760.

Jarpe, M. B., Leach, K. L., and Raben, D. M., 1994, α-Thrombin-induced nuclear *sn*-1,2-diacylglycerols are derived from phosphatidylcholine hydrolysis in cultured fibroblasts, *Biochemistry* **33:**526–534.

Jayadev, S., Linardic, C. M., and Hannun, Y. A., 1994, Identification of arachidonic acid as a mediator of sphingomyelin hydrolysis in response to tumor necrosis factor α, *J. Biol. Chem.* **269:**5757–5763.

Johannes, F. J., Prestle, J., Eis, S., Oberhagemann, P., and Pfizenmaier, K., 1994, PKCμ is a novel, atypical member of the protein kinase C family, *J. Biol. Chem.* **269**:6140–6148.

Kaibuchi, K., Takai, Y., and Nishizuka, Y., 1981, Cooperative roles of various membrane phospholipids on the activation of calcium-activated, phospholipid-dependent protein kinase, *J. Biol. Chem.* **256**:7146–7149.

Kaibuchi, K., Takai, Y., Sawamura, M., Hoshijima, M., Fujikura, T., and Nishizuka, Y., 1983, Synergistic functions of protein phosphorylation and calcium mobilization in platelet activation, *J. Biol. Chem.* **258**:6701–6704.

Kalmar, G. B., Kay, R. J., Lachance, A., Aebersold, R., and Cornell, R. B., 1990, Cloning and expression of rat liver CTP:phosphocholine cytidylyltransferase: An amphipathic protein that controls phosphatidylcholine synthesis, *Proc. Natl. Acad. Sci. USA* **87**:6029–6033.

Kanoh, H., Kondoh, H., and Ono, T., 1983, Diacylglycerol kinase from pig brain. Purification and phospholipid dependencies, *J. Biol. Chem.* **258**:1767–1774.

Kanoh, H., Iwata, T., Ono, T., and Suzuki, T., 1986, Immunological characterization of *sn*-1,2-diacylglycerol and *sn*-2-monoacylglycerol kinase from pig brain, *J. Biol. Chem.* **261**:5597–5602.

Kanoh, H., Yamada, K., Sakane, F., and Imaizumi, T., 1989, Phosphorylation of diacylglycerol kinase *in vitro* by protein kinase C, *Biochem. J.* **258**:455–462.

Kanoh, H., Yamada, K., and Sakane, F., 1990, Diacylglycerol kinase: A key modulator of signal transduction? *Trends Biochem. Sci.* **15**:47–50.

Kato, M., and Takenawa, T., 1990, Purification and characterization of membrane-bound and cytosolic forms of diacylglycerol kinase from rat brain, *J. Biol. Chem.* **265**:794–800.

Katzav, S., Martin-Zanca, D., and Barbacid, M., 1989, VAV, *EMBO J.* **8**:2283–2290.

Kawashima, Y., and Bell, R. M., 1987, Assembly of the endoplasmic reticulum and phospholipid bilayer. Transporters for phosphatidylcholine and metabolites, *J. Biol. Chem.* **262**:16495–16502.

Kazanietz, M. G., Bustelo, X. R., Barbacid, M., Kolch, W., Mischak, H., Wong, G., Pettit, G. R., Bruns, J. D., and Blumberg, P. M., 1994, Zinc finger domains and phorbol ester pharmacophore. Analysis of binding to mutated form of protein kinase Cζ and the *vav* and c-*raf* proto-oncogene products, *J. Biol. Chem.* **269**:11590–11594.

Kennerly, D. A., 1991, Quantitative analysis of water-soluble products of cell-associated phospholipase-C and phospholipase-D catalyzed hydrolysis of phosphatidylcholine, *Methods Enzymol.* **197**:191–197.

Kennerly, D. A., Parker, C. W., and Sullivan, T. J., 1979, Use of diacylglycerol kinase to quantitate picomole levels of 1,2- diacylglycerol, *Anal. Biochem.* **98**:123–131.

Kent, C., 1990, Regulation of phosphatidylcholine biosynthesis, *Prog. Lipid Res.* **29**:87–105.

Khan, W. A., Blobe, G. C., and Hannun, Y. A., 1992, Activation of protein kinase C by oleic acid, *J. Biol. Chem.* **267**:20878–20886.

Kishimoto, A., Takai, Y., Mori, T., Kikkawa, U., and Nishizuka, Y., 1980, Activation of calcium and phospholipid-dependent protein kinase by diacylglycerol, its possible relation to phosphatidylinositol turnover, *J. Biol. Chem.* **255**:2272–2276.

Kishimoto, A., Kajikawa, N., Shiota, M., and Nishizuka, Y., 1983, Proteolytic activation of calcium-activated, phospholipid-dependent protein kinase C by calcium-dependent neutral protease, *J. Biol. Chem.* **258**:1156–1164.

Kishimoto, A., Mikawa, K., Hashimoto, K., Yasuda, I., Tanaka, S., Tominga, M., Kuroda, T., and Nishizuka, Y., 1989, Limited proteolysis of protein kinase C subspecies by calcium-dependent neutral protease (calpain), *J. Biol. Chem.* **264**:4088–4092.

Kiss, Z., and Anderson, W. B., 1989, Phorbol ester stimulates the hydrolysis of phosphatidylethanolamine in leukemic HL-60, NIH3T3 and baby hamster kidney cells, *J. Biol. Chem.* **264**:1483–1487.

Kolesnick, R. N., 1987, 1,2-Diacylglycerols but not phorbol esters stimulate sphingomyelin hydrolysis in GH₃ pituitary cells, *J. Biol. Chem.* **262**:16759–16762.

Kolesnick, R. N., and Golde, D. W., 1994, The sphingomyelin pathway in tumor necrosis factor and interleukin-1 signaling, *Cell* **77**:325–328.

Kolesnick, R. N., and Hemer, M. R., 1990, Physiologic 1,2-diacylglycerol levels induce protein kinase C-independent translocation of a regulatory enzyme, *J. Biol. Chem.* **265**:10900–10904.

König, B., Di Nitto, P. A., and Blumberg, P. M., 1985, Stoichiometric binding of diacylglycerol to the phorbol ester receptor, *J. Cell. Biochem.* **29**:37–44.

Konrad, R. J., Major, C. D., and Wolf, B. A., 1994, Diacylglycerol hydrolysis to arachidonic acid is necessary for insulin secretion from isolated pancreatic islets: Sequential actions of diacylglycerol and monoacylglycerol lipases, *Biochemistry* **33:**13284–13294.

Kraft, A. S., and Anderson, W. B., 1983, Phorbol esters increase the amount of Ca^{2+}, phospholipid-dependent protein kinase associated with the plasma membrane, *Nature* **301:**621–623.

Lacal, J. C., 1990, Diacylglycerol production in Xenopus laevis oocytes after microinjection of p21^{ras} proteins in a consequence of activation of phosphatidylcholine metabolism, *Mol. Cell. Biol.* **10:**333–340.

Lacal, J. C., De la Peña, P., Moscat, J., Garcia-Barreno, P., Anderson, P. S., and Aaronson, S. A., 1987a, Rapid stimulation of diacylglycerol production in *Xenopus* oocytes by microinjection of H-*ras*-p21, *Science* **238:**533–536.

Lacal, J. C., Fleming, T. P., Warren, B. S., Blumberg, P. M., and Aaronson, S. A., 1987b, Involvement of functional protein kinase C in the mitogenic response to H-*ras* oncogene product, *Mol. Cell. Biol.* **7:**4146–4149.

Lacal, J. C., Moscat, J., and Aaronson, S. A., 1987c, Novel source of 1,2-diacylglycerol elevated in cells transformed by Ha-*ras* oncogene, *Nature* **330:**269–272.

Lambeth, J. D., 1994, Receptor-regulated phospholipases and their generation of lipid-mediators, which activate PKC, in: *Protein Kinase C* (J. K. Kuo, ed.), pp. 121–170, Oxford University Press, London.

Leach, K. L., and Raben, D. M., 1993, α-Thrombin-stimulated 1,2-diacylglycerol formation: The relationship between phospholipid hydrolysis and protein kinase C activation, *Neuroscience* **3:**120–132.

Leach, K. L., Powers, E. A., Ruff, V. A., Jaken, S., and Kaufman, S., 1989, Type 3 protein kinase C localization to the nuclear envelope of phorbol ester-treated NIH 3T3 cells, *J. Cell Biol.* **109:**685–695.

Leach, K. L., Ruff, V. A., Wright, T. M., Pessin, M. S., and Raben, D. M., 1991, Dissociation of protein kinase C activation and *sn*-1,2-diacylglycerol formation. Comparison of phosphatidylinositol- and phosphatidylcholine-derived diglycerides in α-thrombin-stimulated fibroblasts, *J. Biol. Chem.* **266:**3215–3221.

Leach, K. L., Ruff, V. A., Jarpe, M. B., Adams, L. D., Fabbro, D., and Raben, D. M., 1992, α-Thrombin stimulates nuclear diglyceride levels and differential localization of protein kinase C isozymes in IIC9 cells, *J. Biol. Chem.* **267:**21816–21822.

Lee, M.-H., and Bell, R. M., 1986, The lipid-binding, regulatory domain of protein kinase C, *J. Biol. Chem.* **261:**14867–14870.

Lee, C., Fisher, S. K., Agranoff, B. W., and Hajra, A. K., 1991, Quantitative analysis of molecular species of diacylglycerol and phosphatide formed upon muscarinic activation of human SK-N-Sh neuroblastoma cells, *J. Biol. Chem.* **266:**22837–22846.

Leevers, S. J., Paterson, H. F., and Marshall, C. J., 1994, Requirement for *ras* in raf activation is overcome by targeting raf to the plasma membrane, *Nature* **369:**411–414.

Lemaitre, R. N., King, W. C., MacDonald, M. L., and Glomset, J. A., 1990, Distribution of distinct arachidonoyl-specific and non-specific isoenzymes of diacylglycerol kinase in baboon (*Papio cynocephalus*) tissues, *Biochem. J.* **266:**291–299.

Leung, T., How, B.-E., Manser, E., and Lim, L., 1993, Germ cell β-chimaerin, a new GTPase-activating protein for p21*rac*, is specifically expressed during the acrosomal assembly stage in rat testes, *J. Biol. Chem.* **268:**3813–3816.

Leung, T., How, B.-E., Manser, E., and Lim, L., 1994, Cerebellar β2-chimaerin, a GTPase-activating protein for p21*ras*-related *rac* is specifically expressed in granule cells and has a unique N-terminal SH$_2$ domain, *J. Biol. Chem.* **269:**12888–12892.

Lipsky, N. G., and Pagano, R. E., 1983, Sphingolipid metabolism in cultured fibroblasts. Microscopic and biochemical studies employing a fluorescent ceramide analogue, *Proc. Natl. Acad. Sci. USA* **80:**2608–2612.

Liscovitch, M., and Cantley, L., 1994, Lipid second messengers, *Cell* **77:**329–334.

Liu, J., Mathias, S., Yang, Z., and Kolesnick, R. N., 1994, Renaturation and tumor necrosis factor-α stimulation of a 97-kDa ceramide-activated protein kinase, *J. Biol. Chem.* **269:**3047–3052.

Lozano, J., Berra, E., Municio, M. M., Diaz-Meco, M. T., Dominguez, I., Sanz, L., and Moscat, J., 1994, Protein kinase Cζ isoform is critical for κ B-dependent promoter activation by sphingomyelinase, *J. Biol. Chem.* **269:**19200–19202.

Luna, E. J., and Hitt, A. L., 1992, Cytoskeleton–plasma membrane interactions, *Science* **258:**955–964.

McCormick, F., 1994, Activators and effectors of *ras* p21 proteins, *Curr. Opin. Genet. Dev.* **4:**71–76.

MacDonald, M. L., Mack, K. F., Richardson, C. N., and Glomset, J. A., 1988, Regulation of diacylglycerol kinase reaction in Swiss 3T3 cells. Increased phosphorylation of endogenous diacylglycerol and decreased phosphorylation of didecanoylglycerol in response to platelet-derived growth factor, *J. Biol. Chem.* **263:**1575–1583.

Majerus, P. W., 1992, Inositol phosphate biochemistry, *Annu. Rev. Biochem.* **61:**225–250.

Makowsky, M., and Rosen, O. M., 1989, Complete activation of protein kinase C by antipeptide antibody directed against the pseudosubstrate prototype, *J. Biol. Chem.* **264:**16155–16159.

Manser, E., Leung, T., Monfries, C., Teo, M., Hall, C., and Lim, L., 1992, Diversity and versatility of GTPase activating proteins for the p21rho subfamily of ras G proteins detected by a novel overlay assay, *J. Biol. Chem.* **267:**16025–16028.

Maroney, A. C., and Macara, I. G., 1989, Phorbol ester-induced translocation of diacylglycerol kinase from cytosol to the membrane in Swiss 3T3 fibroblasts, *J. Biol. Chem.* **264:**2537–2544.

Marshall, C. J., 1994, MAP kinase kinase kinase, MAP kinase kinase, and MAP kinase, *Curr. Opin. Genet. Dev.* **4:**82–89.

Martin, T. F. J., Hsieh, K.-P., and Porter, B. W., 1990, The sustained second phase of hormone-stimulated diacylglycerol accumulation does not activate protein kinase C in GH3 cells, *J. Biol. Chem.* **265:**7623–7631.

Maruta, H., and Burgess, A. W., 1994, Regulation of the ras signalling network, *BioEssays* **16:**489–496.

Maruyama, I. N., and Brenner, S., 1991, A phorbol ester/diacylglycerol-binding protein encoded by the unc-13 gene of Caenorhabditis elegans, *Proc. Natl. Acad. Sci. USA* **88:**5729–5733.

Masmoudi, A., Labourdette, G., Mersel, M., Huang, F. L., Huang, K.-P., Vincendon, G., and Malviya, A. N., 1989, Protein kinase C located in rat liver nuclei: Partial purification and biochemical and immunochemical characterization, *J. Biol. Chem.* **264:**1172–1179.

Matuoka, K., Fukami, K., Nakanishi, O., Kawai, S., and Takenawa, T., 1988, Mitogenesis in response to PDGF and bombesin abolished by microinjection of antibody to PIP$_2$, *Science* **239:**640–643.

May, W. S., Jr., Sahyoun, N., Wolf, M., and Cuatrecasas, P., 1985, Role of intracellular calcium mobilization in the regulation of protein kinase C-mediated membrane processes, *Nature* **317:**549–551.

Michell, R. H., 1975, Inositol phospholipids and cell surface receptor function, *Biochim. Biophys. Acta* **415:**81–147.

Miller, M. A., and Kent, C., 1986, Characterization of the pathways for phosphatidylethanolamine biosynthesis in Chinese hamster ovary mutant and parental cell lines, *J. Biol. Chem.* **261:**9753–9761.

Mochley-Rosen, D., Heinrich, C. J., Cheever, L., Khaner, H., and Simpson, P. C., 1990, A protein kinase C isoform is translocated to cytoskeletal elements in activation, *Cell Regul.* **1:**689–706.

Mochley-Rosen, D., Khaner, H., and Lopez, J., 1991a, Identification of intracellular receptor proteins for activated protein kinase C, *Proc. Natl. Acad. Sci. USA* **88:**3997–4000.

Mochley-Rosen, D., Khaner, H., Lopez, J., and Smith, B. L., 1991b, Intracellular receptors for activated protein kinase C, *J. Biol. Chem.* **266:**14866–14868.

Moolenaar, W. H., 1994, LPA: A novel lipid second messenger with diverse biological actions, *Trends Cell Biol.* **4:**213–219.

Mori, T., Takai, Y., Yu, B., Takahashi, J., Nishizuka, Y., and Fujikura, T., 1982, Specificity of fatty acyl moieties of diacylglycerol for the activation of calcium-activated, phospholipid-dependent protein kinase, *J. Biochem.* **91:**427–431.

Morris, J. D. H., Price, B., Lloud, A. C., Self, A. J., Marshall, C. J., and Hall, A., 1989, Scrape-loading of Swiss 3T3 cells with ras protein rapidly activates protein kinase C in the absence of phosphoinositide hydrolysis, *Oncogene* **4:**27–31.

Murray, N. R., Burns, D. J., and Fields, A. P., 1994, Presence of a βII protein kinase C-selective nuclear membrane activation factor in human leukemia cells, *J. Biol. Chem.* **269:**21385–21390.

Nakanishi, H., and Exton, J. H., 1992, Purification and characterization of the ζ isoform of protein kinase C from bovine kidney, *J. Biol. Chem.* **267:**16347–16354.

Nakanishi, H., Brewer, K. A., and Exton, J. H., 1993, Activation of the ζ isozyme of protein kinase C by phosphatidylinositol-3,4,5-trisphosphate, *J. Biol. Chem.* **268:**13–16.

Newton, A. C., and Koshland, D. E., Jr., 1987, Protein kinase C autophosphorylates by an intrapeptide reaction. J. Biol. Chem. **262:**10185–10188.

Nichols, J. W., and Pagano, R. E., 1983, Resonance energy transfer assay of protein-mediated lipid transfer between vesicles, *J. Biol. Chem.* **258:**5368–5371.

Ninio, E., and Joly, F., 1991, Transmembrane signaling and Paf-acether biosynthesis, *Lipids* **26:**1034–1037.

Nishikawa, M., and Hidaka, H., 1994, Protein kinase C in smooth muscle, in: *Protein Kinase C* (J. F. Kuo, ed.), pp. 236–248, Oxford University Press, London.

Nishizuka, Y., 1984, The role of protein kinase C in cell surface signal transduction and tumor promotion, *Nature* **308:**693–695.

Nishizuka, Y., 1986, Studies and perspectives of protein kinase C, *Science* **233:**305–312.

Nishizuka, Y., 1988, The molecular heterogeneity of protein kinase C and its implications for cellular regulation, *Nature* **334:**661–665.

Nishizuka, Y., 1992, Intracellular signaling by hydrolysis of phospholipids and activation of protein kinase C, *Science* **258:**607–614.

Okazaki, T., Bell, R. M., and Hannun, Y. A., 1989, Sphingomyelin turnover induced by vitamin D_3 in HL-60 cells. Role in cell differentiation, *J. Biol. Chem.* **264:**19076–19080.

Ono, Y., Fujii, T., Ogita, K., Kikkawa, U., Igarashi, K., and Nishizuka, Y., 1989a, Protein kinase Cζ from rat brain: Its structure, expression and properties, *Proc. Natl. Acad. Sci. USA* **86:**3099–3103.

Ono, Y., Fujii, T., Igarashi, K., Kuno, T., Tanaka, C., Kikkawa, U., and Nishizuka, Y., 1989b, Phorbol ester binding to protein kinase C requires a cysteine-rich zinc finger-like sequence, *Proc. Natl. Acad. Sci. USA* **86:**4868–4871.

Orr, J. W., and Newton, A. C., 1994, Intrapeptide regulation of protein kinase C, *J. Biol. Chem.* **269:**8383–8387.

Orr, J. W., Keranen, L. M., and Newton, A. C., 1992, Reversible exposure of the pseudosubstrate domain of protein kinase C by phosphatidylserine and diacylglycerol, *J. Biol. Chem.* **267:**15263–15266.

Pagano, R. E., 1988, What is the fate of diacylglycerol produced at the Golgi apparatus? *Trends Biochem. Sci.* **13:**202–205.

Pagano, R. E., and Longmuir, K. J., 1985, Phosphorylation, transbilayer movement and facilitated transport of diacylglycerol are involved in the uptake of a fluorescent analog of phosphatidic acid by cultured fibroblasts, *J. Biol. Chem.* **260:**1909–1916.

Pagano, R. E., and Sleight, R. G., 1985, Defining lipid transport pathways in animal cells, *Science* **229:**1051–1057.

Pagano, R. E., Longmuir, K. J., and Martin, O. C., 1983, Intracellular translocation and metabolism of a fluorescent phosphatidic acid analogue in cultured fibroblasts, *J. Biol. Chem.* **258:**2034–2040.

Palmer, S., Hughes, K. T., Lee, D. Y., and Wakelam, M. J. O., 1989, Development of a novel Ins(1,4,5)-P3-specific binding assay. Its use to determine the intracellar concentration of Ins(1,4,5)P3 in unstimulated and vasopressin-stimulated rat hepatocytes, *Cell. Signal.* **1:**147–156.

Pappelenbosch, M. P., Tertoolen, L. G. J., Hage, W. J., and de Laat, S. W., 1993, Epidermal growth factor-induced actin remodeling is regulated by 5-lipoxygenase and cyclooxygenase products, *Cell* **74:**565–575.

Patton, G. M., Fasulo, J. M., and Robins, S. J., 1982, Separation of phospholipids and individual molecular species of phospholipids by high-performance liquid chromatography, *J. Lipid Res.* **23:**190–196.

Pawson, T., and Hunter, T., 1994, Signal transduction and growth control in normal and cancer cells, *Curr. Opin. Genet. Dev.* **4:**1–4.

Payrastre, B., van Bergen en Henegouwen, P. M. P., Breton, M., den Hartigh, J. C., Plantavid, M., Verkleij, A. J., and Boonstra, J., 1991, Phosphoinositide kinase, diacylglycerol kinase and phospholipase C activities associated with the cytoskeleton: Effect of epidermal growth factor, *J. Cell Biol.* **115**:121–128.

Pears, C. J., Kour, G., House, C., Kemp, B. E., and Parker, P. J., 1990, Mutagenesis of the pseudosubstrate site of protein kinase C leads to activation, *Eur. J. Biochem.* **194**:89–94.

Pears, C. J., Schaap, D., and Parker, P. J., 1991, The regulatory domain of protein kinase C-ε restricts the catalytic domain specificity, *Biochem. J.* **276**:257–260.

Pelech, S. L., and Vance, D. E., 1984, Regulation of PC biosynthesis, *Biochim. Biophys. Acta* **779**:217–251.

Pelech, S. L., and Vance, D. E., 1989, Signal transduction via phosphatidylcholine cycles, *Trends Biochem. Sci.* **14**:28–30.

Pelech, S. L., Pritchard, P. H., and Vance, D. E., 1981, cAMP analogues inhibit phosphatidylcholine biosynthesis in cultured rat hepatocytes, *J. Biol. Chem.* **256**:8283–8286.

Perry, D. K., Stevens, V. L., Widlanski, T. S., and Lambeth, J. D., 1993, A novel ectophosphatidic acid phosphohydrolase activity mediates activation of neutrophil superoxide generation by exogenous phosphatidic acid, *J. Biol. Chem.* **268**:25302–25310.

Pessin, M. S., and Raben, D. M., 1989, Molecular species analysis of 1,2-diglycerides stimulated by α-thrombin in cultured fibroblasts, *J. Biol. Chem.* **264**:8729–8738.

Pessin, M. S., Baldassare, J. J., and Raben, D. M., 1990, Molecular species analysis of mitogen-stimulated 1,2-diglyceride in fibroblasts: Comparison of α-thrombin, epidermal growth factor, *J. Biol. Chem.* **265**:7959–7966.

Pessin, M. S., Altin, J. G., Jarpe, M., Tansley, F., Bradshaw, R. A., and Raben, D. M., 1991, Carbachol stimulates a different phospholipid metabolism than nerve growth factor and basic fibroblast growth factor in PC12 cells, *Cell Regul.* **2**:383–390.

Pete, M. J., Ross, A. H., and Exton, J. H., 1994, Purification and properties of phospholipase A$_1$ from bovine brain, *J. Biol. Chem.* **269**:19494–19500.

Polokoff, M. A., and Bell, R. M., 1978, Limited palmitoyl-CoA penetration into microsomal vesicles as evidenced by a highly latent ethanol acyltransferase activity, *J. Biol. Chem.* **253**:7173–7178.

Polokoff, M. A., Wing, D. C., and Raetz, C. R. H., 1981, Isolation of somatic cell mutant defective in the biosynthesis of phosphatidylethanolamine, *J. Biol. Chem.* **256**:7687–7690.

Preiss, J., Loomis, C. R., Bishop, W. R., Stein, R., Niedel, J. E., and Bell, R. M., 1986, Quantitative measurement of *sn*-1,2-diacylglycerols present in platelets, hepatocytes and *ras*- and *sis*-transformed normal rat kidney cells, *J. Biol. Chem.* **261**:8597–8600.

Price, B. D., Morris, J. D. H., Marshall, C. J., and Hall, A., 1989, Stimulation of phosphatidylcholine hydrolysis, diacylglycerol release, and arachidonic acid production by oncogenic ras is a consequence of protein kinase C activation, *J. Biol. Chem.* **264**:16638–16643.

Pucéat, M., and Brown, J. H., 1994, Protein kinase C in heart, in: *Protein Kinase C* (J. F. Kuo, ed.), pp. 249–268, Oxford University Press, London.

Quest, A. F. G., and Bell, R. M., 1994a, The molecular mechanism of protein kinase C regulation by lipids, in: *Protein Kinase C* (J. F. Kuo, ed.), Oxford University Press, London.

Quest, A. F. G., and Bell, R. M., 1994b, The regulatory region of protein kinase C: Studies of phorbol ester binding to individual and combined functional segments expressed as glutathione-S-transferase fusion proteins indicate a complex mechanism of regulation by phospholipids, phorbol esters and divalent cations, *J. Biol. Chem.* **269**:20000–20012.

Quest, A. F. G., Bardes, E. S. G., and Bell, R. M., 1994a, A phorbol ester binding domain of protein kinase Cγ. High affinity binding to a glutathione-S-transferase/Cys2 fusion protein, *J. Biol. Chem.* **269**:2953–2960.

Quest, A. F. G., Bardes, E. S. G., and Bell, R. M., 1994b, A phorbol ester binding domain of protein kinase Cγ. Deletion analysis of the Cys2 domain defines a minimal 43-amino acid peptide, *J. Biol. Chem.* **269**:2961–2970.

Quest, A. F. G., Ghosh, S., Xie, W. Q., and Bell, R. M., 1995, Diacylglycerol second messengers: Molecular switches in growth control, in: *Proceedings of the 3rd International Eicosanoids and Other Bioactive Lipids in Cancer, Inflammation and Radiation Injury Conference*, in press.

Qiu, Z.-H., and Leslie, C. C., 1994, Protein kinase C-dependent and -independent pathways of mitogen-activated protein kinase activation in macrophages by stimuli that activate phospholipase A$_2$, *J. Biol. Chem.* **269**:19480–19487.

Raben, D. M., Pessin, M. S., Rangan, L. A., and Wright, T. M., 1990, Kinetic and molecular species analyses of mitogen-induced increases in diglyceride: Evidence for stimulated hydrolysis of phosphoinositides and phosphatidylcholine, *J. Cell. Biochem.* **44:**117–125.

Raben, D. M., Jarpe, M. B., and Leach, K. L., 1994, Nuclear lipid metabolism in NEST: nuclear envelope signal transduction, *J. Membr. Biol.* **142:**1–7.

Rando, R. R., 1988, Regulation of protein kinase C activity by lipids, *FASEB J.* **2:**2348–2355.

Rebbechi, M. J., Kolesnick, R. N., and Gershengorn, M. C., 1983, Thyrotropin-releasing hormone stimulates rapid loss of phosphatidylinositol and its conversion to 1,2-diacylglycerol and phosphatidic acid in rat mammotropic pituitary cells. Association with calcium mobilization and prolactin secretion, *J. Biol. Chem.* **258:**227–234.

Ridgeway, N. D., and Vance, D. E., 1987, Purification of phosphatidylethanolamine-N-methyltransferase from rat liver, *J. Biol. Chem.* **262:**17231–17239.

Ridley, A. J., and Hall, A., 1992, The small GTP-binding protein *rho* regulates the assembly of focal adhesions and actin stress fibers in response to growth factors, *Cell* **70:**389–399.

Ridley, A. J., Paterson, H. F., Johnston, C. L., Diekmann, D., and Hall, A., 1992, The small GTP-binding protein *rac* regulates growth factor-induced membrane ruffling, *Cell* **70:**401–410.

Rittenhouse, S. E., and Sasson, J. P., 1985, Mass changes in myoinositol triphosphate in human platelets stimulated by thrombin. Inhibitory effects of phorbol esters, *J. Biol. Chem.* **260:**8657–8660.

Roberts, M. F., 1994, First thoughts on lipid second messengers, *Trends Cell Biol.* **4:**219–223.

Rodriguez-Viciana, P., Warne, P. H., Dhand, R., Vanhaesebroeck, B., Gout, I., Fry, M. J., Waterfield, M. D., and Downward, J., 1994, Phosphatidylinositol-3-OH kinase as a direct target of *ras*, *Nature* **370:**527–532.

Rogue, P., Labourette, G., Masmoudi, A., Yoshida, Y., Huang, F. L., Huang, K.-P., Zwiller, J., Vincendon, G., and Malviya, A. N., 1990, Rat liver nuclei protein kinase C is the isozyme type II, *J. Biol. Chem.* **265:**4161–4165.

Ron, D., Chen, C.-H., Caldwell, J., Jamieson, L., Orr, E., and Mochley-Rosen, D., 1994, Cloning of an intracellular receptor for protein kinase C: A homology to the β subunit of G proteins, *Proc. Natl. Acad. Sci. USA* **91:**839–843.

Rosoff, P. M., Savage, N., and Dinarello, C. A., 1988, Interleukin-1 stimulates diacylglycerol production in T lymphocytes by a novel mechanism, *Cell* **54:**73–81.

Rothman, J. E., and Lenard, J., 1977, Membrane asymmetry, *Science* **195:**743–753.

Rozengurt, E., 1986, Early signals in the mitogenic response, *Science* **254:**161–166.

Sakane, F., Yamada, K., Kanoh, H., Yokoyama, C., and Tanabe, T., 1990, Porcine diacylglycerol kinase has zinc finger and E-F hand motifs, *Nature* **344:**345–348.

Sakane, F., Yamada, K., Imai, S.-i., and Kanoh, H., 1991, Porcine 80-kDa diacylglycerol kinase is a calcium-binding and calcium-phospholipid-dependent enzyme and undergoes calcium-dependent translocation, *J. Biol. Chem.* **266:**7096–7100.

Santiago, O. M., Rosenberg, L. I., and Monaco, M. E., 1993, Organization of the phosphoinositide cycle. Assessment of inositol transferase activity in purified plasma membranes, *Biochem. J.* **290:**179–183.

Schaap, D., de Widt, J., van der Wal, J., Vanderkerkhove, J., van Damme, J., Gussow, D., Ploegh, H. L., van Blitterswijk, W. J., and van der Bend, R. L., 1990, Purification, cDNA cloning and expression of human diacylglycerol kinase. *FEBS Lett.* **275:**151–158.

Schatzman, R. C., Raynor, R. L., Fritz, R. B., and Kuo, J. F., 1983, Purification to homogeneity, characterization and monoclonal antibodies of phospholipid-sensitive Ca^{2+}-dependent protein kinase from spleen. *Biochem. J.* **209:**435–443.

Schütze, S., Potthoff, K., Macleidt, T., Berkovic, D., Wiegmann, K., and Krönke, M., 1992, TNF activates NF-κB by phosphatidylcholine-specific phospholipase C-induced acidic sphingomyelin breakdown. *Cell* **71:**765–776.

Seedorf, K., Millauer, B., Kostka, G., Schlessinger, J., and Ullrich, A., 1992). Differential effects of carboxy-terminal sequence deletions on platelet-derived growth factor receptor signaling activities and interactions with cellular substrates. *Mol. Cell. Biol.* **12:**4347–4356.

Seigneuret, M., and Devaux, P. F., 1984, ATP-dependent asymmetric distribution of spin-labeled phospholipids in the erythrocyte membrane: relation to shape changes. *Proc. Natl. Acad. Sci. USA* **81**:3751–3755.

Selbie, L. A., Schmitz-Peiffer, C., Sheng, Y., and Biden, T. J., 1993, Molecular cloning and characterization of PKCι, an atypical isoform of protein kinase C derived from insulin-secreting cells. *J. Biol. Chem.* **268**:24296–24302.

Seyfred, M. A., and Wells, W. W., 1984b, Subcellular site and mechanism of vasopressin-stimulated hydrolysis of phosphoinositides in rat hepatocytes, *J. Biol. Chem.* **259**:7666–7672.

Shariff, A., and Luna, E. J., 1992, Diacylglycerol-stimulated formation of actin nucleation sites at the plasma membranes, *Science* **256**:245–247.

Sharkey, N. A., and Blumberg, P. M., 1985, Kinetic evidence that 1,2-diolein inhibits phorbol ester binding via a competitive mechanism, *Biochem. Biophys. Res. Commun.* **133**:1051–1056.

Sharkey, N. A., Leach, K. L., and Blumberg, P. A., 1984, Competitive inhibition by diacylglycerol of specific phorbol ester binding, *Proc. Natl. Acad. Sci. USA* **81**:607–610.

Siegel, D. P., Banschbach, J., Alford, D., Ellens, H., Lis, L. J., Quinn, P. J., Yeagel, P. L., and Benz, J., 1989, Physiological levels of diacylglycerols in phospholipid membrane induce fusion and stabilize inverted phases, *Biochemistry* **28**:3703–3709.

Simboli-Campbell, M., Gagnon, A. M., Franks, D. J., and Welsh, J. E., 1994, 1,25-Dihydroxyvitamin D$_3$ translocates protein kinase Cβ to the nucleus and enhances plasma membrane association of protein kinase Cα in renal epithelial cells, *J. Biol. Chem.* **269**:3257–3264.

Simpson, C. M. F., Itabe, H., Reynolds, C. N., King, W. C., and Glomset, J. A., 1991, Swiss 3T3 cells preferentially incorporate *sn*-2-arachidonoyl monoacylglycerol into *sn*-1-stearoyl-2-arachidonoyl phosphatidylinositol, *J. Biol. Chem.* **266**:15902–15909.

Slater, S. J., Kelly, M. B., Taddeo, F. J., Ho, C., Rubin, E., and Stubbs, C. D., 1994, The modulation of protein kinase C activity by membrane lipid bilayer structure, *J. Biol. Chem.* **269**:4866–4871.

Sleight, R. G., and Kent, C., 1983, Regulation of phosphatidylcholine biosynthesis in mammalian cells, *J. Biol. Chem.* **258**:831–835.

Sleight, R. G., and Pagano, R. E., 1978, Phospholipid asymmetry in LM cell plasma membrane derivatives: Polar head group and acyl chain distributions, *Biochemistry* **17**:332–338.

Sleight, R. G., and Pagano, R. E., 1984, Transport of a fluorescent phosphatidylcholine analog from the plasma membrane to the Golgi apparatus, *J. Cell Biol.* **99**:742–751.

Sleight, R. G., and Pagano, R. E., 1985, Transbilayer movement of a fluorescent phosphatidylethanolamine analog across the plasma membrane of cultured mammalian cells, *J. Biol. Chem.* **260**:1146–1154.

Smith, M. R., DeGudicibus, S. J., and Stacy, D. W., 1986, Requirement for c-ras proteins during viral oncogene transformation, *Nature* **320**:540–543.

Smith, M. R., Liu, Y.-L., Kim, H., Rhee, S. G., and Kung, H.-F., 1990, Inhibition of serum- and ras-stimulated DNA synthesis by antibodies to phospholipase C, *Science* **247**:1074–1077.

Stokoe, D., Macdonald, S. G., Cadwallder, K., Symons, M., and Hancock, J. F., 1994, Activation of *raf* as a result of recruitment to the plasma membrane, *Science* **264**:1463–1467.

Struck, D. K., and Pagano, R. E., 1980, Insertion of fluorescent phospholipids into the plasma membrane of a mammalian cell, *J. Biol. Chem.* **255**:5404–5410.

Takahashi, T., and Hori, S. H., 1978, Intramembraneous localization of rat liver microsomal hexose-6-phosphate dehydrogenase membrane permeability to its substrates, *Biochim. Biophys. Acta* **524**:262–276.

Takai, Y., Kishimoto, A., Inoue, M., and Nishizuka, Y., 1977, Studies on a cyclic nucleotide-independent protein kinase and its proenzyme in mammalian tissue. I. Purification and characterization of an active enzyme from bovine cerebellum, *J. Biol. Chem.* **252**:7603–7609.

Takai, Y., Kishimoto, A., Iwasa, Y., Kawahara, Y., Mori, T., and Nishizuka, Y., 1979a, Calcium-dependent activation of a multifunctional protein kinase by membrane phospholipids, *J. Biol. Chem.* **254**:3692–3695.

Takai, Y., Kishimoto, A., Kikkawa, U., Mori, T., and Nishizuka, Y., 1979b, Unsaturated diacylglycerol as a possible messenger for the activation of a calcium-activated phospholipid-independent protein kinase system, *Biochem. Biophys. Res. Commun.* **91**:1218–1224.

Takai, Y., Kishimoto, A., Iwasa, Y., Kawahari, Y., Mori, T., and Nishizuka, Y., 1979c, A role of membranes in the activation of a new multifunctional protein kinase system, *J. Biochem.* **86**:575–578.

Thomas, G. M. H., Cunningham, E., Fensome, A., Ball, A., Totty, N. F., Truong, O., Hsuan, J. J., and Cockcroft, S., 1993, An essential role for phosphatidylinositol transfer protein in phospholipase C-mediated inositol lipid signaling, *Cell* **74:**919–928.

Thomas, T. P., Talwar, H. V., and Anderson, W. B., 1988, Phorbol ester-mediated association of protein kinase C to the nuclear fraction in NIH-3T3 cells, *Cancer Res.* **48:**1910–1919.

Tijburg, L. B., Vermeulen, P. S., Schmitz, M. G., and van Golde, L. M., 1992, Okadaic acid inhibits phosphatidylethanolamine biosynthesis in rat hepatocytes, *Biochem. Biophys. Res. Commun.* **182:**1226–1231.

Toker, A., Meyer, M., Reddy, K. K., Falck, J. R., Aneja, R., Aneja, S., Parra, A., Burns, D. J., Ballas, L. M., and Cantley, L. C., 1994, Activation of protein kinase C family members by the novel polyphospho-inositides PtdIns-3,4-P_2 and PtdIns-3,4,5-P_3, *J. Biol. Chem.* **269:**32358–32367.

Tsujishita, Y., Asaoka, Y., and Nishizuka, Y., 1994, Regulation of phospholipase A_2 in human leukemia cell lines: Its implication for intracellular signaling, *Proc. Natl. Acad. Sci. USA* **91:**6274–6278.

Uhing, R. J., Prpic, V., Hollenbach, P. W., and Adams, D. O., 1989, Involvement of protein kinase C in platelet-activating factor-stimulated diacylglycerol accumulation in murine peritoneal macro-phages, *J. Biol. Chem.* **264:**9224–9230.

Utal, A. K., Jamil, H., and Vance, D. E., 1991, Diacylglycerol signals the translocation of CTP:choline-phosphate cytidylyltransferase in HeLa cells treated with 12-O-tetradecanoyl phorbol-13-acetate, *J. Biol. Chem.* **266:**24084–24091.

Valverde, A. M., Sinnett-Smith, J., Van Lint, J., and Rozengurt, E., 1994, Molecular cloning and characterization of protein kinase D: A target for diacylglycerol and phorbol esters with a distinctive catalytic domain, *Proc. Natl. Acad. Sci. USA* **91:**8572–8576.

Van Blitterswijk, W. J., and Hlikman, H., 1993, Rapid attenuation of receptor-induced diacylglycerol and phosphatidic acid by phospholipase D-mediated transphosphatidylation: Formation of bis-phosphatidic acid, *EMBO J.* **12:**2655–2662.

Vance, D. E., 1985, Phospholipid metabolism in eukaryotes, in: *Biochemistry of Lipids and Membranes* (D. E. Vance and J. E. Vance, eds.), pp. 242–270, Benjamin/Cummings, Menlo Park, CA.

Vance, D. E., 1990, Phosphatidylcholine metabolism: Masochistic enzymology, metabolic regulation and lipoprotein assembly, *Biochem. Cell Biol.* **68:**1151–1165.

Vance, J. E., 1991, Newly made phosphatidylserine and phosphatidylethanolamine are preferentially translocated between rat liver mitochondria and endosome reticulum, *J. Biol. Chem.* **266:**89–97.

Vance, J. E., and Vance, D. E., 1988, Does rat liver Golgi have the capacity to synthesize phospholipids for lipoprotein secretion? *J. Biol. Chem.* **263:**5898–5909.

Van den Besselaar, A. M., De Kruijff, B., van den Bosch, H., and van Deenen, L. L. M., 1978, Phosphatidylcholine mobility in liver microsomal membranes, *Biochim. Biophys. Acta* **863:**193–204.

Van der Bend, R. L., de Widt, J., Hilkman, H., and van Blitterswijk, W. J., 1994, Diacylglycerol kinase in receptor-stimulated cells converts its substrate in a topologically restricted manner, *J. Biol. Chem.* **269:**4098–4102.

Van Helvoort, A., van't Hoff, W., Ritsema, T., Sandra, A., and van Meer, G., 1994, Conversion of diacylglycerol to phosphatidylcholine on the basolateral surface of epithelial (Mardin-Darby canine kidney) cells. Evidence for the reverse action of a sphingomyelin synthase, *J. Biol. Chem.* **269:**1763–1769.

Van Heusden, G. P. H., Noteborn, H. P. J. M., and van den Bosch, H., 1981, Selective utilization of palmitoyl lysophosphatidylcholine in the synthesis of disaturated phosphatidycholine in rat lung. A combined *in vitro* and *in vivo* approach, *Biochim. Biophys. Acta* **664:**49–60.

Van Lint, J., Sinnett-Smith, J., and Rozengurt, E., 1995, Expression and characterization of PKD, a phorbol ester and diacylglycerol-stimulated serine protein kinase, *J. Biol. Chem.* **270:**1455–1461.

Van Meer, G., 1989, Lipid traffic in animal cells, *Annu. Rev. Cell Biol.* **5:**247–255.

Van Veldhoven, P. P., and Bell, R. M., 1988, Effect of harvesting methods, growth conditions and growth phase on diacylglycerol levels in cultured human adherent cells, *Biochim. Biophys. Acta* **959:**185–196.

Voelker, D. R., 1984, Phosphatidylserine functions as the major precursor of phosphatidylethanol-amine in cultured BHK-21 cells, *Proc. Natl. Acad. Sci. USA* **81:**2669–2673.

Walsh, J. P., Suen, R., Lemaitre, R. N., and Glomset, J. A., 1994, Arachidonoyl-diacylglycerol kinase from bovine testes. Purification and properties, *J. Biol. Chem.* **269**:21155–21164.

Wang, Y., Sweitzer, T. D., Weinhold, P. A., and Kent, C., 1993, Nuclear localization of soluble CTP:phosphocholine cytidylyltransferase, *J. Biol. Chem.* **268**:5899–5904.

Warden, C. H., and Friedkin, M., 1985, Regulation of choline kinase activity and phosphatidylcholine biosynthesis by mitogenic growth factors in 3T3 fibroblasts, *J. Biol. Chem.* **260**:6006–6011.

Watkins, J. D., and Kent, C., 1991, Regulation of CTP:phosphocholine cytidylyltransferase activity and subcellular location by phosphorylation in Chinese hamster ovary cells. The effect of phospholipase C treatment, *J. Biol. Chem.* **266**:21113–21117.

Watkins, J. D., and Kent, C., 1992, Immunolocalization of membrane-associated CTP:phosphocholine cytidylyltransferase in phosphocholine-deficient Chinese hamster ovary cells, *J. Biol. Chem.* **267**:5686–5692.

Ways, D. K., Cook, P. P., Webster, C., and Parker, P., 1992, Effect of phorbol ester on protein kinase C-ζ, *J. Biol. Chem.* **267**:4799–4805.

Weinhold, P. A., Rounsifer, M. E., Williams, S. E., Brubaker, P. G., and Feldman, D. A., 1984, CTP:phosphorylcholine cytidylyltransferase in rat lung. The effect of free fatty acids on the translocation of activity between microsomes and cytosol, *J. Biol. Chem.* **259**:10315–10321.

Weinhold, P. A., Rounsifer, M. E., and Feldman, D. A., 1986, The purification and characterization of CTP:phosphorylcholine cytidylyltransferase from rat liver, *J. Biol. Chem.* **261**:5104–5110.

Wilkison, W. O., and Spiegelman, B. M., 1993, Biosynthesis of the vasoactive lipid monobutyrin. Central role of diacylglycerol, *J. Biol. Chem.* **268**:2844–2849.

Wirtz, K. W. A., 1991, Phospholipid transfer proteins, *Annu. Rev. Biochem.* **60**:73–99.

Wise, B. C., Raynor, R. L., and Kuo, J. F., 1982, Phospholipid-sensitive calcium-dependent protein kinase from heart: Purification and general properties, *J. Biol. Chem.* **257**:8481–8488.

Wolf, M., Le Vinett, H., May, W. S., Jr., Cuatrecasas, P., and Sayhoun, N., 1985, A model for intracellular translocation of protein kinase C involving synergism between Ca^{2+} and phorbol esters, *Nature* **317**:546–549.

Wolf, R. A., and Gross, R. W., 1985, Identification of neutral active phospholipase C which hydrolyzes choline glycerophospholipids and plasmalogen selective phospholipase A_2 in canine myocardium, *J. Biol. Chem.* **260**:7295–7303.

Wolff, R. A., Dobrowsky, R. T., Bielawska, A., Obeid, L. M., and Hannun, Y. A., 1994, Role of ceramide-activated protein phosphatase in ceramide-mediated signal transduction, *J. Biol. Chem.* **269**:19605–19609.

Wolfman, A., and Macara, I. G., 1987, Elevated levels of diacylglycerol and decreased phorbol ester sensitivity in ras-transformed fibroblasts, *Nature* **235**:359–361.

Woodard, D. S., Lee, T., and Snyder, F., 1987, The final step in the *de novo* biosynthesis of platelet-activating factor, *J. Biol. Chem.* **262**:2520–2527.

Woodgett, J. R., Hunter, T., and Gould, K. L., 1987, Protein kinase C and its role in cell growth, in: *Cell Membranes: Methods and Reviews* (E. Elson, W. Frazier, and L. Glaser, eds.), pp. 215–340, Plenum Press, New York.

Wright, P. S., Morand, J. N., and Kent, C., 1985, Regulation of phosphatidylcholine biosynthesis in Chinese hamster ovary cells by reversible membrane association of CTP:phosphocholine cytidylyltransferase, *J. Biol. Chem.* **260**:7919–7926.

Wright, T. M., Rangan, L. A., Shin, H. S., and Raben, D. M., 1988, Kinetic analysis of 1,2-diacylglycerol mass levels in cultured fibroblasts. Comparison of stimulation by α-thrombin and epidermal growth factor, *J. Biol. Chem.* **263**:9374–9380.

Yamada, M. A., Sakane, F., and Kanoh, H., 1989, Immunoquantitation of 80 kDa diacylglycerolkinase in pig and human lymphocytes and several other cells, *FEBS Lett.* **244**:402–406.

York, J. D., and Majerus, P. W., 1994, Nuclear phosphatidylinositols decrease during S-phase of the cell cycle in HeLa cells, *J. Biol. Chem.* **269**:7847–7850.

Yu, C.-L., Tsai, M.-H., and Stacey, D. W., 1988, Cellular *ras* activity and phospholipid metabolism, *Cell* **52**:63–71.

Yu, F.-X., Sun, H.-Q., Janmey, P. A., and Yin, H. L., 1992, Identification of a phosphoinositide-binding sequence in an actin monomer-binding domain of gelsolin, *J. Biol. Chem.* **267**:14616–14621.

Zhang, J., Fry, M. J., Waterfield, M. D., Jaken, S., Liao, L., Fox, J. E. B., and Rittenhouse, S. E., 1992, Activated phosphoinositide 3-kinase associates with membrane cytoskeleton in thrombin-exposed platelets, *J. Biol. Chem.* **267:**4686–4692.

Zhang, J., Wang, L., Schwartz, J., Bond, R. W., and Bishop, W. R., 1994, Phosphorylation of Thr[642] is an early event in the processing of newly synthesized protein kinase CβI and is essential for its activation, *J. Biol. Chem.* **269:**19578–19584.

Zidovetzki, R., and Lester, D. S., 1992, The mechanism of activation of protein kinase C: A biophysical perspective, *Biochim. Biophys. Acta* **1134:**261–272.

Zilversmit, D. B., and Hughes, M. E., 1977, Extensive exchange of rat liver microsomal phospholipids, *Biochim. Biophys. Acta* **469:**99–110.

Zoeller, R. A., and Raetz, C. R. H., 1986, Isolation of animal cell mutants deficient in plasmalogen biosynthesis and peroxisome assembly, *Proc. Natl. Acad. Sci. USA* **83:**5170–5174.

Chapter 2

Lipid Signaling for Protein Kinase C Activation

Yoshinori Asaoka, Yosuke Tsujishita, and Yasutomi Nishizuka

2.1. Introduction

Stimulation of receptors or opening of ion channels initiates a cascade of intracellular events through a highly organized network of cell signaling pathways. The hydrolysis of phosphatidylinositol 4,5-bisphosphate (PIP_2) by phospholipase C (PLC) produces inositol 1,4,5-trisphosphate (IP_3) and diacylglycerol (DAG). IP_3 increases intracellular Ca^{2+} and DAG activates protein kinase C (PKC), both of which synergistically act to elicit a wide variety of cellular responses. The appearance of DAG derived from inositol phospholipids (PI), however, is transient and insufficient for the sustained activation of PKC, which is essential for long-term cellular responses such as growth and differentiation (Berry *et al.*, 1990; William *et al.*, 1990; Aihara *et al.*, 1991; Asaoka *et al.*, 1991). For sustained activation of this enzyme, several mechanisms have been discussed (Nishizuka, 1995). For instance, phosphatidylcholine (PC) is hydrolyzed by phospholipase D (PLD) and the resulting phosphatidic acid (PA) is dephosphorylated to produce DAG (Billah and Anthes, 1990; Exton, 1990; Liscovitch, 1992). In addition, phospholipase A_2 (PLA_2) is activated by most of the agonists that induce PI hydrolysis (Axelrod *et al.*, 1988). Arachidonic acid induces many physiological and pathological processes after being converted to various eicosanoids. Other products of the PC hydrolysis catalyzed by PLA_2, various *cis*-unsaturated fatty acids and lysophospholipids, appear to potentiate PKC activation, thereby contributing to cell signaling, at least partly through the PKC pathway (Fig. 2-1; Asaoka *et al.*, 1992a; Nishizuka, 1992). Although the biochemical mechanism of receptor-mediated activation of PLD

Yoshinori Asaoka and Yosuke Tsujishita • Biosignal Research Center, Kobe University, Kobe 657, Japan. *Yasutomi Nishizuka* • Biosignal Research Center, Kobe University, and Department of Biochemistry, Kobe University School of Medicine, Kobe 650, Japan.

Handbook of Lipid Research, Volume 8: Lipid Second Messengers, edited by Robert M. Bell *et al.* Plenum Press, New York, 1996.

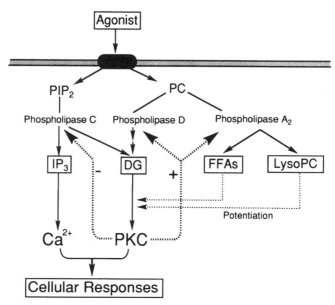

Figure 2-1. Schematic representation of signal-induced phospholipid degradation. (Adapted from Nishizuka, 1992.)

and PLA$_2$ remains largely unclear, this chapter will describe the signal-induced degradation of various membrane phospholipids that play a role in transmitting information from extracellular signals across the cell membrane.

2.2. Transient and Sustained Elevation of Diacylglycerol

When cell surface receptors are stimulated, PLC is immediately activated and DAG is produced from PI, most rapidly from PIP$_2$. This DAG molecule disappears quickly. Early experiments using resting T lymphocytes as a model system showed that repeated additions of a membrane-permeant DAG can mimic a single dose of a phorbol ester for the induction of T-lymphocyte activation (Fig. 2-2). Under the experimental conditions employed, a major portion of this exogenous DAG is metabolized rapidly, whereas the phorbol ester is hardly metabolizable and remains in the membrane for a prolonged period of time (Asaoka *et al.*, 1991). This observation suggests that the sustained level of DAG is a prerequisite for long-term cellular responses such as T-lymphocyte activation.

Following the initial transient increase, the DAG level often increases again, when cells are stimulated by long-acting signals, particularly some growth factors. This second wave of DAG begins with a relatively slow onset and persists longer. The fatty acid composition of the DAG molecule suggests that it results from the hydrolysis of PC (Billah and Anthes, 1990; Exton, 1990). Both DAG molecules

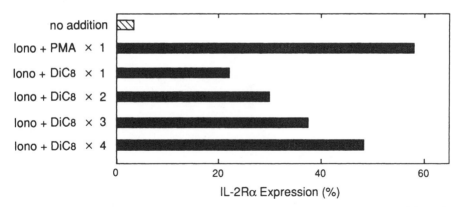

Figure 2-2. Activation of T lymphocytes by repeated additions of a membrane-permeant DAG. Resting T lymphocytes (5×10^5 cells/ml) were stimulated with either DiC_8 or phorbol ester in the presence of ionomycin. DiC_8 (25 μM) was added repeatedly (1–4 times with an interval of 3 hr) in the presence of ionomycin (0.25 μM) as indicated. A single dose of PMA (10 nM) and ionomycin (1 μM) was added where specified ("Iono + PMA"). Cells were collected 16 hr after the first addition and the expression of the α subunit of interleukin-2 receptor was measured. Detailed experimental conditions are given in Asaoka *et al.* (1991). DiC_8, 1,2-dioctanoylglycerol; PMA, phorbol 12-myristate 13-acetate; Iono, ionomycin; IL-2Rα, α-subunit of interleukin-2 receptor.

derived from PC and from PI are capable of activating PKC *in vitro* (Zeisel, 1993). It has been proposed that the DAG derived from PI hydrolysis is rapidly phosphorylated to PA by the action of DAG kinase, and converted again to PI, whereas the DAG produced from PC hydrolysis is not preferentially phosphorylated by DAG kinase but is degraded slowly by DAG lipase (Ford and Gross, 1990; Lee *et al.*, 1991; Florin-Christensen *et al.*, 1992). This slow degradation of the DAG that is produced from PC possibly contributes to the sustained elevation of DAG. Obviously, the level of DAG depends on the balance of its formation and degradation. It is possible that DAG kinase and DAG lipase are regulated for the maintenance of the DAG level. Fatty acids are indeed inhibitory for mammalian DAG lipases (Farooqui *et al.*, 1989).

PLC enzymes thus far purified from mammalian tissues catalyze the hydrolysis of PI but not PC (Dennis *et al.*, 1991), although the occurrence of PLC that is reactive with PC has often been proposed (Wolf and Gross, 1985; Cai *et al.*, 1992; Xu *et al.*, 1993; Rao and Mufson, 1994). This proposal is based primarily on the observation that phosphocholine is produced soon after receptor stimulation. However, no evidence is available unequivocally indicating the occurrence of PC-reactive PLC in mammalian tissues. Phosphocholine may likely be produced by consecutive actions of PLA_2 and lysophospholipase, followed by phosphodiesterase. Choline kinase may also produce phosphocholine from choline. This enzyme is active in mammalian tissues (Macara, 1989; Teegarden *et al.*, 1990; Cuadrado *et al.*, 1993). In addition, sphingomyelinase may contribute to produce phosphocholine. Several lines of evidence strongly suggest that PLD plays a role in the production of DAG from PC to sustain PKC activation (Billah and Anthes, 1990; Exton, 1990; Liscovitch, 1992).

2.3.   *Phospholipase D for Sustained Protein Kinase C Activation*

PLD hydrolyzes PC to produce PA and choline. In the presence of primary alcohols, such as methanol and ethanol, this enzyme catalyzes transphosphatidylation to produce phosphatidylalcohol rather than PA (Chalifour *et al.*, 1980). Although PA itself has been proposed to act as a second messenger for mitogenic signaling in certain cell types (Boarder, 1994), it is generally thought that PA produced from PC by PLD action is converted to DAG by the action of a phosphomonoesterase, thereby activating PKC (Billah and Anthes, 1990; Exton, 1990; Cockcroft, 1992; Liscovitch, 1992). Despite ubiquitous distribution, this enzyme has not been solubilized or purified from any mammalian tissue or cell type.

Early observations in intact cell systems have repeatedly shown that PLD is activated by a wide variety of extracellular signals, notably by PKC activators such as phorbol esters and membrane-permeant DAG molecules, occasionally synergistically with a Ca^{2+} ionophore. Certain PKC isoforms, especially the α and βI isoforms, are proposed to play a role in the PLD activation (Pai *et al.*, 1991; Pachter *et al.*, 1992; Eldar *et al.*, 1993; Balboa *et al.*, 1994). It is suggested that tyrosine phosphorylation plays a role in the activation of PLD, since pervanadate, a tyrosine phosphatase inhibitor, enhances PLD activity, whereas several tyrosine kinase inhibitors such as genistein prevent the enzyme activation (Bourgoin and Grinstein, 1992; Uings *et al.*, 1992; Kusner *et al.*, 1993).

In permeabilized cells and cell-free preparations, both hydrolysis and transphosphatidylation reactions by PLD are activated by GTP-γ-S (Bocckino *et al.*, 1987; Geny and Cockcroft, 1992; Kanoh *et al.*, 1993). A cytosolic soluble protein factor, identified as the ADP-ribosylation factor (ARF), is needed for this PLD activation (Brown *et al.*, 1993; Cockcroft *et al.*, 1994). ARF, a member of the family of small GTP-binding proteins, plays a role in the membrane trafficking, such as membrane targeting and fusion of secretory vesicles for release and exocytosis (De Matteis *et al.*, 1993). The biochemical mechanism of PLD activation, however, remains unclear.

2.4.   *Phospholipase A₂ for Protein Kinase C Activation*

PLA_2 is ubiquitously expressed in mammalian tissues, and several secretory and cytosolic enzymes have been identified (Dennis *et al.*, 1991; Kudo *et al.*, 1993; Dennis, 1994). Based on the substrate specificity, the enzymes may be divided into two groups. One group preferentially cleaves arachidonic acid, whereas the other group cleaves various fatty acids nonselectively from the *sn*-2 position of phospholipids. Thus far, the arachidonic acid-selective enzymes have been emphasized, because this fatty acid is the rate-limiting precursor of prostaglandins and leukotrienes. *cis*-Unsaturated fatty acids, however, are shown to exhibit several biological actions, such as modulation of ion channel properties (Petrou *et al.*, 1993), Ca^{2+} pump (Cardoso and De Meis, 1993), membrane fusion during exocytosis (Creutz, 1992), and enhancement of long-term potentiation of synaptic processes (Linden *et al.*, 1987; Bliss and Collingridge, 1993).

The finding that PKC is activated by *cis*-unsaturated fatty acids (McPhail *et al.*, 1984) was soon followed by the observation that various *cis*-unsaturated fatty acids such as oleic, linoleic, linolenic, arachidonic, and docosahexaenoic acids greatly enhance the DAG- or phorbol ester-dependent activation of PKC, at the basal level of Ca^{2+} concentration (Seifert *et al.*, 1988; Shinomura *et al.*, 1991; Chen and Murakami, 1992). Neither saturated fatty acid nor *trans*-unsaturated fatty acid was effective for enhancing this PKC activation. In human platelets, the release of serotonin, which is induced by either a membrane-permeant DAG or a phorbol ester, is significantly potentiated by *cis*-unsaturated fatty acids (Fig. 2-3; Yoshida *et al.*, 1992). *cis*-Unsaturated fatty acids also enhance the DAG- or phorbol ester-dependent phosphorylation of a specific endogenous PKC substrate, pleckstrin. Fatty acid alone is inactive unless DAG or phorbol ester is present. Kinetic analysis with the fluorescent Ca^{2+} indicator fura-2 suggests that *cis*- unsaturated fatty acids

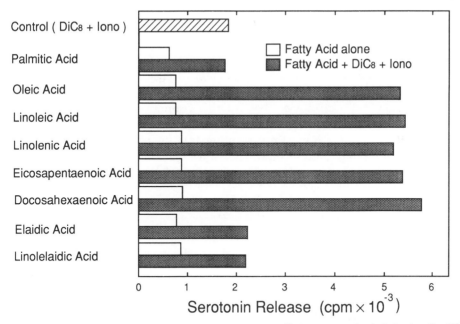

Figure 2-3. Effect of various fatty acids on serotonin release. [^{14}C]-Serotonin-loaded platelets (1×10^8 cells/ml) were stimulated for 1 min in the presence of various fatty acids (50 μM) with (shaded bars) or without (open bars) DiC_8 (25 μM) plus ionomycin (0.8 μM). Control (hatched bar) represents the serotonin release induced by DiC_8 (25 μM) plus ionomycin (0.8 μM) in the absence of fatty acids. Detailed experimental conditions are given in Yoshida *et al.* (1992). DiC_8, 1,2-dioctanoylglycerol; Iono, ionomycin.

markedly increase an apparent affinity of PKC activation for Ca^{2+}, thereby causing nearly full cellular responses at the basal Ca^{2+} concentration. Consistent with the observation from *in vitro* studies, saturated and *trans*-unsaturated fatty acids are inactive for this platelet response. Synergistic action of *cis*-unsaturated fatty acid and DAG or phorbol ester was observed also for the differentiation of HL-60 cells to macrophages, as measuring the expression of cell surface marker CD11b (Fig. 2-4) as well as the appearance of phagocytic activity (Asaoka *et al.*, 1993). Analogously, the glutamate release from isolated nerve terminals, which is induced by a membrane-permeant DAG or phorbol ester, is also potentiated by *cis*-unsaturated fatty acids (Herrero *et al.*, 1992). In this study, *cis*-unsaturated fatty acid alone was inactive, unless the nerve terminals were stimulated by a membrane-permeant DAG, a phorbol ester, or a physiological ligand, *trans*-1-amino-cyclo-penthyl-1,3-dicarboxylate.

2-Lysophosphatidylcholine (lysoPC), the other product of PC hydrolysis by PLA_2, shows membrane-lytic activity and is toxic to the cell (Weltzien, 1979). Normally, lysoPC, once produced, is rapidly degraded further or acylated to produce PC again. LysoPC itself exhibits some biological activities such as enhancement of sustained Na^+ current (Undrovinas *et al.*, 1992), chemotaxis (Quinn *et al.*, 1988), smooth muscle relaxation (Saito *et al.*, 1988), and induction of epidermal growth factor-like growth factor in monocytes (Nakano *et al.*, 1994). When added to intact cells together with a membrane-permeant DAG or a phorbol ester, lysoPC potentiates significantly subsequent cellular responses. For example, lysoPC greatly enhances the activation of human resting T lymphocytes that is induced by a membrane-permeant DAG plus a Ca^{2+} ionophore (Fig. 2-5; Asaoka *et al.*, 1992b). Similarly, differentiation of HL-60 cells to macrophages, which is induced by a membrane-permeant DAG or a phorbol ester, is potentiated by lysoPC (Fig. 2-4; Asaoka *et al.*, 1993). Other lysophospholipids are practically inactive except for lysophosphatidylethanolamine, which was slightly active as lysoPC. LysoPC alone was inert unless either DAG or phorbol ester was added, suggesting that lysoPC interacts with the PKC signaling pathway. In cell-free

Figure 2-4. Stimulatory actions of lysoPC, linoleic acid, and venom PLA_2 on CD11b expression induced by a single addition of PMA. HL-60 cells (2.0×10^5 cells/ml) were treated with PMA (1 nM) in the presence of lysoPC (50 μM), linoleic acid (50 μM), or *Clotalus adamanteus* PLA_2 (50 units/ml). In control experiments, the cells were incubated with PMA alone as indicated. After a 24-hr incubation, the expression of CD11b was measured with a flow cytometer. The fluorescence intensity (logarithmic scale) is represented on the horizontal axis and the cell number is on the vertical axis. Detailed experimental conditions are given in Asaoka *et al.* (1993). PMA, phorbol 12-myristate 13-acetate; LA, linoleic acid.

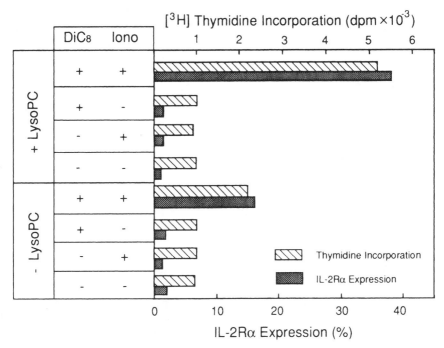

Figure 2-5. Effect of lysoPC on T-lymphocyte activation induced by DiC$_8$ plus ionomycin. Resting T lymphocytes (5×10^5 cells/ml) were stimulated with DiC$_8$ (50 μM) plus ionomycin (0.5 μM) and lysoPC (50 μM), as indicated. After a 16-hr incubation, the α subunit of interleukin-2 receptor expression (shaded bars) and [^{3}H]thymidine incorporation (hatched bars) were determined. Detailed experimental conditions are given in Asaoka *et al.* (1992b). DiC$_8$, 1,2-dioctanoylglycerol; Iono, ionomycin; IL-2Rα, α subunit of interleukin-2 receptor.

enzymatic reactions, lysoPC enhances the DAG-dependent PKC activation at micromolar concentrations (Oishi *et al.*, 1988; Sasaki *et al.*, 1993), but the mechanism underlying the biological action of lysoPC observed in intact cells remains unknown.

2.5. Possible Link of Receptor to Phospholipase A$_2$ Activation

Mitogenic signals that induce activation of PLC and PLD frequently cause the release of arachidonic acid. The mechanism of this receptor-mediated activation of PLA$_2$ has been discussed repeatedly (Dennis *et al.*, 1991; Cockcroft, 1992; Nishizuka, 1992; Kudo *et al.*, 1993). In cultured rat thyroid cells, PLA$_2$ is activated by stimulation of α$_1$-adrenergic receptor that is coupled to a pertussis toxin-sensitive G-protein (Burch *et al.*, 1986). In bovine rod outer segments, the release of arachidonic acid is shown to be mediated by the βγ subunits of transducin

(Axelrod *et al.*, 1988). In rabbit platelets, histamine stimulates PLA_2 probably through H_1 receptors (Murayama *et al.*, 1990).

A cytosolic PLA_2 with a mass of 85 kDa, which is expressed in many tissues, has been isolated and its primary structure elucidated (Clark *et al.*, 1991; Rehfeldt *et al.*, 1991; Sharp *et al.*, 1991). This enzyme is activated by the phosphorylation of its specific seryl residue by mitogen-activated protein kinases. Subsequent binding of this enzyme to membranes in a Ca^{2+}-dependent manner produces its catalytic activity (Nemenoff *et al.*, 1993; Kramer *et al.*, 1993; Lin *et al.*, 1993). Arachidonic acid is preferentially cleaved by this enzyme. Since mitogen-activated protein kinases are located downstream from tyrosine kinases as well as from PKC, this cytosolic PLA_2 is regulated indirectly by receptor stimulation.

However, the mechanism of the receptor-mediated activation of PLA_2 enzymes, which exhibit nonselective cleavage of various *cis*-unsaturated fatty acids, remains unclear. Several Ca^{2+}-independent cytosolic PLA_2 enzymes have been described (Kudo *et al.*, 1993; Dennis, 1994). Two Ca^{2+}-independent enzymes of 39 and 110 kDa have been partially purified from bovine brain cytosol, but their specificity toward fatty acids to be cleaved is unknown (Hirashima *et al.*, 1992). Ca^{2+}-independent cytosolic PLA_2 of 40 kDa has also been purified from canine myocardium, but the enzyme shows a preference for arachidonoyl-containing phospholipids (Hazen *et al.*, 1990). This enzyme seems to be activated by ATP. A similar enzyme has been purified from the cytosol of a murine macrophagelike cell line (Ackermann *et al.*, 1994). The regulatory mechanism of these Ca^{2+}-independent cytosolic PLA_2 enzymes is unknown.

Permeabilized human leukemia HL-60 and U-937 cells, suspended in an acidic or alkaline medium, release various *cis*-unsaturated fatty acids, most abundantly oleic and arachidonic acids (Tsujishita *et al.*, 1994). Concomitant production of lysophospholipids suggests that PLA_2 may play a major role in this fatty acid release reaction. The release of fatty acids particularly in an alkaline medium is considerably accelerated by GTP-γ-S as well as by aluminum fluoride, suggesting a potential role of a trimeric G-protein. The reaction is activated also by vanadate, and inhibited by genistein, suggesting the involvement of tyrosine phosphorylation. This fatty acid release reaction is further affected by both PKC activators and inhibitors, but is insensitive to Ca^{2+} concentrations. In contrast, the fatty acid release in an acidic medium is absolutely dependent on 10^{-7} M Ca^{2+}, while neither vanadate, GTP-γ-S, nor aluminum fluoride is capable of potentiating this release reaction (Table 2-1). All of the available evidence is indirect, but suggests that several Ca^{2+}-insensitive cytosolic, arachidonic acid-nonselective types of PLA_2 enzymes are linked to receptors for their activation (Tsujishita *et al.*, 1994).

2.6. Protein Kinase C Activation by Lipid Mediators

There is more than one PKC, and multiple discrete isoforms are expressed in mammals. These isoforms have subtly different enzymological properties, differential tissue expression, and specific intracellular localization (Asaoka *et al.*,

Table 2-1. Potential Regulatory Mechanism of Putative PLA_2 Enzymes in Human Leukemia Cells[a]

Regulation	PLA$_2$	
	Acidic pH	Alkaline pH
Ca^{2+}	0.2–2 × 10^{-7} M	−
G-protein mechanism		
GTP-γ-S	−	+
Aluminum fluoride	−	+ +
Phosphorylation mechanism		
Tyrosine kinase?		
Vanadate + ATP	−	+ +
Vanadate + ATP + GTP-γ-S	−	+ + +
Vanadate + ATP + genistein	−	−

[a]Detailed explanations are given in Tsujishita et al. (1994).

1992a). To date, 11 isoforms have been identified in mammalian tissues, although some of these isoforms do not show typical characteristics of the classical PKC enzymes in their mode of activation (Table 2-2).

The enzymes may be divided into three groups (Nishizuka, 1992). Group A consists of four classical, conventional PKC (cPKC) isoforms (α, βI, βII, and γ) that emerged from the initial screening (Nishizuka, 1988). Group B consists of five new, novel PKC (nPKC) isoforms [δ, ε, η(L), θ, and μ] subsequently found. Group C consists of two atypical PKC (aPKC) isoforms [ζ and λ(ι)] more recently characterized. The cPKC enzymes have four conserved (C_1 to C_4) and five variable (V_1 to V_5) regions. The C_1 region is a membrane-binding domain having two

Table 2-2. PKC Isoforms in Mammalian Tissues

	Subspecies	Amino acid residues	Ca^{2+} and lipid activators	Tissue expression
cPKC	α	672	Ca^{2+}, DAG, PS, FFAs, lysoPC	Universal
	βI	671	"	Some tissues
	βII	673	"	Many tissues
	γ	697	"	Brain only
nPKC	δ	673	DAG, PS	Universal
	ε	737	DAG, PS, FFA, PIP$_3$	Brain and others
	η (L)	683	DAG, PS, PIP$_3$, cholesterol sulfate	Skin, lung, heart
	θ	707	?	Muscle, T cells, etc.
	μ	912	?	Kidney cells
aPKC	ζ	592	PS, FFA, PIP$_3$?	Universal
	λ (ι)	587	?	Many tissues

[a]The activators for each isoform are determined with calf thymus Hl histone and bovine myelin basic protein as model phosphate acceptors. PS, phosphatidylserine; FFAs, cis-unsaturated fatty acids; for other abbreviations, see text.

tandem repeats of a cysteine-rich zinc finger-like motif. The C_2 region appears to be related to the Ca^{2+} sensitivity of the enzyme. This region has a Ca^{2+}-binding domain called CaLB that is also present in cytosolic 85-kDa PLA_2 (Clark et al., 1991), GTPase activating protein (Vogel et al., 1988), the γ isoform of PLC (Stahl et al., 1988), and synaptotagmin (Perin et al., 1990). This domain interacts with phospholipids in a Ca^{2+}-dependent fashion, and presumably plays a role in its translocation to membranes on increase in the Ca^{2+} concentration. The C_3 region contains the catalytic site. The C_4 region seems to be necessary for recognition of the substrate to be phosphorylated. The cPKC isoforms are activated by Ca^{2+}, phosphatidylserine, and either DAG or phorbol ester, and this activation is enhanced further by cis-unsaturated fatty acids and lysoPC as described above.

The nPKC isoforms, which lack the C_2 region, do not require Ca^{2+} for their activation. The enzymes are activated by micelles composed of phosphatidylserine and either DAG or phorbol ester. The ε isoform is activated by cis-unsaturated fatty acids (Koide et al., 1992). On the other hand, the δ isoform is activated slightly by cis-unsaturated fatty acids in the absence of DAG, whereas its DAG-dependent activation is inhibited by the fatty acids (Ogita et al., 1992). The δ, ε, and η(L) isoforms are activated by phosphatidylinositol 3,4,5-triphosphate (PIP_3). The η(L) isoform is activated by cholesterol sulfate (Ikuta et al., 1994). The δ and ε isoforms are phosphorylated in native tissues, showing doublet or triplet on gel electrophoresis (Koide et al., 1992; Ogita et al., 1992). The δ isoform is phosphorylated at tyrosyl residues that is remarkably enhanced by treatment of cells with phorbol ester (Denning et al., 1993; Li et al., 1994a,b). The effect of tyrosine phosphorylation on enzymatic activity is controversial. The signal transduction leading to tyrosine phosphorylation is unclear. Recently, the physical association of Bruton tyrosine kinase with PKC was proposed (Yao et al., 1994). The pleckstrin homology domain (Haslam et al., 1993) of this tyrosine kinase appears to interact with PKC.

The aPKC enzymes, which also lack the C_2 region, have only one cysteine-rich zinc finger-like motif. The enzymes are dependent on phosphatidylserine, but not sensitive to DAG, phorbol ester, or Ca^{2+} (Ono et al., 1989). The ζ isoform is activated by cis-unsaturated fatty acids as well as by PIP_3 and partially by PIP_2 (Nakanishi et al., 1993). The ι isoform (Selbie et al., 1993) and the λ isoform (Akimoto et al., 1994) subsequently isolated from insulin-secreting cells and embryonic carcinoma cells, respectively, appear to be variants of a single entity. These isoforms show the highest amino acid sequence identity with the ζ isoform. The signal that activates the aPKC enzymes remains unknown.

2.7. Coda

The enzymological analysis outlined above suggests that the members of the PKC family respond differently to various combinations of lipids including phosphatidylserine, DAG, cis-unsaturated fatty acids and lysoPC, and hence the pattern of activation of the PKC isoforms may vary in extent, duration, and intracellular localization. In fact, the PKC isoforms so far examined show distinct tissue distribution and specific intracellular localization (Nishizuka, 1988; Tanaka and Nishizuka, 1994). Despite extensive studies, little is known about specific functions of the individual PKC isoforms.

Recently, several PKC-binding proteins have been identified from mammalian tissues and some of them cloned and structurally identified. These proteins, termed RACKs (receptors for activated C-kinase), may act as receptors for the activated forms of PKC (Hyatt *et al.*, 1994; Ron *et al.*, 1994). It has been reported that the binding of these proteins to PKC is phospholipid-dependent, making the active enzyme possible to associate with a specific site on the membrane (Mochly-Rosen *et al.*, 1991; Chapline *et al.*, 1993; Liao *et al.*, 1994). However, specificity of these proteins to each of the PKC isoforms has not been clarified.

When PKC was found to be activated by DAG, PI hydrolysis was thought to be the sole mechanism for the production of DAG which is needed for PKC activation (Nishizuka, 1984). It is now becoming clearer that various products of signal-induced hydrolysis of membrane phospholipids may also play roles in transmitting information from various extracellular signals across the membrane. Dynamic intracellular phospholipid metabolism in response to receptor stimulation is possibly crucial to understanding the precise mechanism of the signal transduction for well-organized cellular responses.

ACKNOWLEDGMENTS. The authors are grateful to Ms. Sachiko Nishiyama for her skillful secretarial assistance. The investigation in this laboratory has been supported by a special research grant from the Ministry of Education, Science and Culture, Japan; the Sankyo Foundation of Life Science; the Sankyo Research Laboratories; the Yamanouchi Foundation for Research on Metabolism Disorders; Merck Sharp & Dohme Research Laboratories; the Terumo Life Science Foundation; and the Osaka Cancer Research Fund.

References

Ackermann, E. J., Kempner, E. S., and Dennis, E. A., 1994, Ca^{2+}-independent cytosolic phospholipase A_2 from macrophage-like $P388D_1$ cells, *J. Biol. Chem.* **269**:9227–9233.

Aihara, H., Asaoka, Yoshida, K., and Nishizuka, Y., 1991, Sustained activation of protein kinase C is essential to HL-60 cell differentiation to macrophage, *Proc. Natl. Acad. Sci. USA* **88**:11062–11066.

Akimoto, K., Mizuno, K., Osada, S., Hirai, S., Tanuma, S., Suzuki, K., and Ohno, S., 1994, A new member of the third class in the protein kinase C family, PKCλ, expressed dominantly in an undifferentiated mouse embryonal carcinoma cell line and also in many tissues and cells, *J. Biol. Chem.* **269**:12677–12683.

Asaoka, Y., Oka, M., Yoshida, K., and Nishizuka, Y., 1991, Metabolic rate of membrane-permeant diacylglycerol and its relation to human resting T-lymphocyte activation, *Proc. Natl. Acad. Sci. USA* **88**:8681–8685.

Asaoka, Y., Nakamura, S., Yoshida, K., and Nishizuka, Y., 1992a, Protein kinase C, calcium and phospholipid degradation, *Trends Biochem. Sci.* **17**:414–417.

Asaoka, Y., Oka, M., Yoshida, K., Sasaki, Y., and Nishizuka, Y., 1992b, Role of lysophosphatidylcholine in T-lymphocyte activation: Involvement of phospholipase A_2 in signal transduction through protein kinase C, *Proc. Natl. Acad. Sci. USA* **89**:6447–6451.

Asaoka, Y., Yoshida, K., Sasaki, Y., and Nishizuka, Y., 1993, Potential role of phospholipase A$_2$ in HL-60 cell differentiation to macrophages induced by protein kinase C activation, *Proc. Natl. Acad. Sci. USA* **90**:4917–4921.

Axelrod, J., Burch, R. M., and Jelsema, C. L., 1988, Receptor-mediated activation of phospholipase A$_2$ via GTP-binding proteins: Arachidonic acid and its metabolites as second messengers, *Trends Neurosci.* **11**:117–123.

Balboa, M. A., Firestein, B. L., Godson, C., Bell, K. S., and Insel, P. A., 1994, Protein kinase C α mediates phospholipase D activation by nucleotides and phorbol ester in Madin-Darby canine kidney cells, *J. Biol. Chem.* **269**:10511–10516.

Berry, N., Ase, K., Kishimoto, A., and Nishizuka, Y., 1990, Activation of resting human T cells requires prolonged stimulation of protein kinase C, *Proc. Natl. Acad. Sci. USA* **87**:2294–2298.

Billah, M. M., and Anthes, J. C., 1990, The regulation and cellular functions of phosphatidylcholine hydrolysis, *Biochem. J.* **269**:281–291.

Bliss, T. V. P., and Collingridge, G. L., 1993, A synaptic model of memory: Long-term potentiation in the hippocampus, *Nature* **361**:31–39.

Boarder, M. R., 1994, A role for phospholipase D in control of mitogenesis, *Trends Pharmacol. Sci.* **15**:57–62.

Bocckino, S. B., Blackmore, P. F., Wilson, P. B., and Exton, J. H., 1987, Phosphatidate accumulation in hormone-treated hepatocyte via a phospholipase D mechanism, *J. Biol. Chem.* **262**:15309–15315.

Bourgoin, S., and Grinstein, S., 1992, Peroxides of vanadate induce activation of phospholipase D in HL-60 cells. Role of tyrosine phosphorylation, *J. Biol. Chem.* **267**:11908–11916.

Brown, H. A., Gutowski, S., Moomaw, C. R., Slaughter, C., and Sternweis, P. C., 1993, ADP-ribosylation factor, a small GTP-dependent regulatory protein, stimulates phospholipase D activity, *Cell* **75**:1137–1144.

Burch, R. M., Luini, A., and Axelrod, J., 1986, Phospholipase A$_2$ and phospholipase C are activated by distinct GTP-binding proteins in response to α$_1$-adrenergic stimulation in FRTL5 thyroid cells, *Proc. Natl. Acad. Sci. USA* **83**:7201–7205.

Cai, H., Erhardt, P., Szeberényi, J., Diaz-Meco, M. T., Johansen, T., Moscat, J., and Cooper, G. M., 1992, Hydrolysis of phosphatidylcholine is stimulated by Ras proteins during mitogenic signal transduction, *Mol. Cell Biol.* **12**:5329–5335.

Cardoso, C. M., and De Meis, L., 1993, Modulation by fatty acids of Ca^{2+} fluxes in sarcoplasmic reticulum vesicles, *Biochem. J.* **296**:49–52.

Chalifour, R. J., Taki, T., and Kanfer, J. N., 1980, Phosphatidylglycerol formation via trans-phosphatidylation by rat brain extracts, *Can. J. Biochem.* **58**:1189–1196.

Chapline, C., Ramsay, K., Klauck, T., and Jaken, S., 1993, Interaction cloning of protein kinase C substrates, *J. Biol. Chem.* **268**:6858–6861.

Chen, S. G., and Murakami, K., 1992, Synergistic activation of type III protein kinase C by *cis*-fatty acid and diacylglycerol, *Biochem. J.* **282**:33–39.

Clark, J. D., Lin, L.-L., Kriz, R. W., Ramesha, C. S., Sultzman, L. A., Lin, A. Y., Milona, N., and Knopf, J. L., 1991, A novel arachidonic acid-selective cytosolic PLA$_2$ contains a Ca^{2+}-dependent translocation domain with homology to PKC and GAP, *Cell* **65**:1043–1051.

Cockcroft, S., 1992, G-protein-regulated phospholipases D, and A$_2$-mediated signalling in neutrophils, *Biochim. Biophys. Acta* **1113**:135–160.

Cockcroft, S., Thomas, G. M. H., Fensome, A., Geny, B., Cunningham, E., Gout, I., Hiles, L., Totty, N. F., Truong, O., and Hsuan, J. J., 1994, Phospholipase D: A downstream effector of ARF in granulocytes, *Science* **263**:523–526.

Creutz, C. E., 1992, The annexins and exocytosis, *Science* **258**:924–931.

Cuadrado, A., Carnero, A., Dolfi, F., Jiménez, B., and Lacal, J. C., 1993, Phosphorylcholine: A novel second messenger essential for mitogenic activity of growth factors, *Oncogene* **8**:2959–2968.

De Matteis, M. A., Santini, G., Kahn, R. A., Di Tullio, G., and Luini, A., 1993, Receptor and protein kinase C-mediated regulation of ARF binding to the Golgi complex, *Nature* **364**:818–821.

Denning, M. F., Dlugosz, A. A., Howett, M. K., and Yasupa, S. H., 1993, Expression of an oncogenic *ras*^{Ha} gene in murine keratinocytes induces tyrosine phosphorylation and reduced activity of protein kinase C δ, *J. Biol. Chem.* **268**:26079–26081.

Dennis, E. A., 1994, Diversity of group types, regulation, and function of phospholipase A$_2$, *J. Biol. Chem.* **269:**13057–13060.

Dennis, E. A., Rhee, S. G., Billah, M. M., and Hannun, Y. A., 1991, Role of phospholipases in generating lipid second messengers in signal transduction, *FASEB J.* **5:**2068–2077.

Eldar, H., Ben-Av, P., Schmidt, U.-S., Livneh, E., and Liscovitch, M., 1993, Up-regulation of phospholipase D activity induced by overexpression of protein kinase C-α, *J. Biol. Chem.* **268:**12560–12564.

Exton, J. H., 1990, Signaling through phosphatidylcholine breakdown, *J. Biol. Chem.* **265:**1–4.

Farooqui, A. A., Rammohan, K. W., and Horrocks, L. A., 1989, Isolation, characterization, and regulation of diacylglycerol lipases from the bovine brain, *Ann. N.Y. Acad. Sci.* **559:**25–36.

Florin-Christensen, J., Florin-Christensen, M., Delfino, J. M., Stegmann, T., and Rasmussen, H., 1992, Metabolic fate of plasma membrane diacylglycerols in NIH 3T3 fibroblasts, *J. Biol. Chem.* **267:**14783–14789.

Ford, D. A., and Gross, R. W., 1990, Differential metabolism of diradylglycerol molecular subclasses and molecular species by rabbit brain diglyceride kinase, *J. Biol. Chem.* **265:**12280–12286.

Geny, B., and Cockcroft, S., 1992, Synergistic activation of phospholipase D by protein kinase C and G-protein-mediated pathways in streptolysin O-permeabilized HL-60 cells, *Biochem. J.* **284:**531–538.

Haslam, R. J., Koide, H. B., and Hemmings, B. A., 1993, Pleckstrin domain homology, *Nature* **363:**309–310.

Hazen, S. L., Stuppy, R. J., and Gross, R. W., 1990, Purification and characterization of canine myocardial cytosolic phospholipase A$_2$, *J. Biol. Chem.* **265:**10622–10630.

Herrero, I., Miras-Portugal, M. T., and Sánchez-Prieto, J., 1992, Activation of protein kinase C by phorbol esters and arachidonic acid required for the optimal potentiation of glutamate exocytosis, *J. Neurochem.* **59:**1574–1577.

Hirashima, Y., Farooqui, A. A., Mills, J. S., and Horrocks, L. A., 1992, Identification and purification of calcium-independent phospholipase A$_2$ from bovine brain cytosol, *J. Neurochem.* **59:**708–714.

Hyatt, S. L., Liao, L., Chapline, C., and Jaken, S., 1994, Identification and characterization of α-protein kinase C binding proteins in normal and transformed REF52 cells, *Biochemistry* **33:**1223–1228.

Ikuta, T., Chida, K., Tajima, O., Matsuura, Y., Iwamori, M., Ueda, Y., Mizuno, K., Ohno, S., and Kuroki, T., 1994, Cholesterol sulfate, a novel activator for the η isoform of protein kinase C, *Cell Growth Differ.* **5:**943–947.

Kanoh, H., Kanaho, Y., and Nozawa, Y., 1993, Requirement of adenosine 5′-triphosphate and Ca^{2+} for guanosine 5′-triphosphate-binding protein-mediated phospholipase D activation in rat pheochromocytoma PC 12 cells, *Neurosci. Lett.* **151:**146–149.

Koide, H., Ogita, K., Kikkawa, U., and Nishizuka, Y., 1992, Isolation and characterization of the ε subspecies of protein kinase C from rat brain, *Proc. Natl. Acad. Sci. USA* **89:**1149–1153.

Kramer, R. M., Roberts, E. F., Manetta, J. V., Hyslop, P. A., and Jakubowski, J. A., 1993, Thrombin-induced phosphorylation and activation of Ca^{2+}-sensitive cytosolic phospholipase A$_2$ in human platelets, *J. Biol. Chem.* **268:**26796–26804.

Kudo, I., Murakami, M., Hara, S., and Inoue, K., 1993, Mammalian non-pancreatic phospholipases A$_2$, *Biochim. Biophys. Acta* **117:**217–231.

Kusner, D. J., Schomisch, S. J., and Dubyak, G. R., 1993, ATP-induced potentiation of G-protein-dependent phospholipase D activity in cell-free system from U-937 promonocytic leukocytes, *J. Biol. Chem.* **267:**19973–19982.

Lee, C., Fisher, S. K., Agranoff, B. W., and Hajra, A. K., 1991, Quantitative analysis of molecular species of diacylglycerol and phosphatidate formed upon muscarinic receptor activation of human SK-N-SH neuroblastoma cells, *J. Biol. Chem.* **266:**22837–22846.

Li, W., Mischak, H., Yu, J.-C., Wang, L.-H., Mushinski, J. F., Heidaran, M. A., and Pierce, J. H., 1994a, Tyrosine phosphorylation of protein kinase C-δ in response to its activation, *J. Biol. Chem.* **269:**2349–2352.

Li, W., Yu, J.-C., Michieli, P., Beeler, J. F., Ellmore, N., Heidaran, M. A., and Pierce, J. H., 1994b, Stimulation of the platelet-derived growth factor β receptor signaling pathway activates protein kinase C-δ, *Mol. Cell Biol.* **14:**6727–6735.

Liao, L., Hyatt, S. L., Chapline, C., and Jaken, S., 1994, Protein kinase C domains involved in interactions with other proteins, *Biochemistry* **33**:1229–1233.

Lin, L.-L., Wartmann, M., Lin, A. Y., Knopf, J. L., Seth, A., and Davis, R. J., 1993, cPLA$_2$ is phosphorylated and activated by MAP kinase, *Cell* **72**:269–278.

Linden, D. J., Sheu, F.-S., Murakami, K., and Routtenberg, A., 1987, Enhancement of long-term potentiation by *cis*-unsaturated fatty acid: Relation to protein kinase C and phospholipase A$_2$, *J. Neurosci.* **7**:3783–3792.

Liscovitch, M., 1992, Crosstalk among multiple signal-activated phospholipases, *Trends Biochem. Sci.* **17**:393–399.

Macara, I. C., 1989, Elevated phosphocholine concentration in *ras*-transformed NIH 3T3 cells arises from increased choline kinase activity, not from phosphatidylcholine breakdown, *Mol. Cell Biol.* **9**:325–328.

McPhail, L. C., Clayton, C. C., and Snyderman, R., 1984, A potential second messenger role for unsaturated fatty acids: Activation of Ca^{2+}-dependent protein kinase, *Science* **224**:622–625.

Mochly-Rosen, D., Khaner, H., Lopez, J., and Smith, B. L., 1991, Intracellular receptors for activated protein kinase C. Identification of a binding site for the enzyme, *J. Biol. Chem.* **266**:14866–14868.

Murayama, T., Kajiyama, Y., and Nomura, Y., 1990, Histamine-stimulated and GTP-binding proteins-mediated phospholipase A$_2$ activation in rabbit platelets, *J. Biol. Chem.* **265**:4290–4295.

Nakanishi, H., Brewer, K. A., and Exton, J. H., 1993, Activation of the ζ isozyme of protein kinase C by phosphatidylinositol 3,4,5-trisphosphate, *J. Biol. Chem.* **268**:13–16.

Nakano, T., Raines, E. W., Abraham, J. A., Klagsbrun, M., and Ross, R., 1994, Lysophosphatidylcholine upregulates the level of heparin-binding epidermal growth factor-like growth factor mRNA in human monocytes, *Proc. Natl. Acad. Sci. USA* **91**:1069–1073.

Nemenoff, R. A., Winitz, S., Qian, N.-X., Putten, V. V., Johnson, G. L., and Heasley, L. E., 1993, Phosphorylation and activation of a high molecular weight form of phospholipase A$_2$ by p42 microtubule-associated protein 2 kinase and protein kinase C, *J. Biol. Chem.* **268**:1960–1964.

Nishizuka, Y., 1984, The role of protein kinase C in cell surface signal transduction and tumour promotion, *Nature* **308**:693–698.

Nishizuka, Y., 1988, The molecular heterogeneity of protein kinase C and its implication for cellular regulation, *Nature* **334**:661–665.

Nishizuka, Y., 1992, Intracellular signaling by hydrolysis of phospholipids and activation of protein kinase C, *Science* **258**:607–614.

Nishizuka, Y., 1995, Protein kinase C and lipid signaling for sustained cellular responses, *FASEB J.* **9**:484–496.

Ogita, K., Miyamoto, S., Yamaguchi, K., Koide, H., Fujisawa, N., Kikkawa, U., Sahara, S., Fukami, Y., and Nishizuka, y., 1992, isolation and characterization of δ-subspecies of protein kinase C from rat brain, *Proc. Natl. Acad. Sci. USA* **89**:1592–1596.

Oishi, K., Raynor, R. L., Charp, P. A., and Kuo, J. F., 1988, Regulation of protein kinase C by lysophospholipids, *J. Biol. Chem.* **263**:6865–6871.

Ono, Y., Fujii, T., Ogita, K., Kikkawa, U., Igarashi, K., and Nishizuka, Y., 1989, Protein kinase C ζ subspecies from rat brain: Its structure, expression, and properties, *Proc. Natl. Acad. Sci. USA* **86**:3099–3103.

Pachter, J. A., Pai, J.-K., Mayer-Ezell, R., Petrin, J. M., Dobek, E., and Bishop, W. R., 1992, Differential regulation of phosphoinositide and phosphatidylcholine hydrolysis by protein kinase C-β1 over-expression, *J. Biol. Chem.* **267**:9826–9830.

Pai, J.-K., Pachter, J. A., Weinstein, B., and Bishop, W. R., 1991, Overexpression of protein kinase C β1 enhances phospholipase D activity and diacylglycerol formation in phorbol ester-stimulated rat fibroblasts, *Proc. Natl. Acad. Sci. USA* **88**:598–602.

Perin, M. S., Fried, V. A., Mignery, G. A., Jahn, R., and Südhof, T. C., 1990, Phospholipid binding by a synaptic vesicle protein homologous to the regulatory region of protein kinase C, *Nature* **345**:260–263.

Petrou, S., Ordway, R. W., Singer, J. J., and Walsh, J. V., Jr., 1993, A putative fatty acid-binding domain of the NMDA receptor, *Trends Biochem. Sci.* **18**:41–42.

Quinn, M. T., Parthasarathy, S., and Steinberg, D., 1988, Lysophosphatidylcholine: A chemotactic factor for human monocytes and its potential role in atherogenesis, *Proc. Natl. Acad. Sci. USA* **85:**2805–2809.

Rao, P., and Mufson, R. A., 1994, Human interleukin-3 stimulates a phosphatidylcholine specific phospholipase C and protein kinase C translocation, *Cancer Res.* **54:**777–783.

Rehfeldt, W., Hass, R., and Goppelt-Struebe, M., 1991, Characterization of phospholipase A₂ in monocytic cell lines, *Biochem. J.* **276:**631–636.

Ron, D., Chen, C.-H., Caldwell, J., Jamieson, L., Orr, E., and Mochly-Rosen, D., 1994, Cloning of an intracellular receptor for protein kinase C: A homolog of the β subunit of G proteins, *Proc. Natl. Acad. Sci. USA* **91:**839–843.

Saito, T., Wolf, A., Menon, N. K., Saeed, M., and Bing, R. J., 1988, Lysolecithins as endothelium-dependent vascular smooth muscle relaxants that differ from endothelium-derived relaxing factor (nitric oxide), *Proc. Natl. Acad. Sci. USA* **85:**8246–8250.

Sasaki, Y., Asaoka, Y., and Nishizuka, Y., 1993, Potentiation of diacylglycerol-induced activation of protein kinase C by lysophospholipids, subspecies difference, *FEBS Lett.* **320:**47–51.

Seifert, R., Schächtele, C., Rosenthal, W., and Schultz, G., 1988, Activation of protein kinase C by *cis*- and *trans*-fatty acids and its potentiation by diacylglycerol, *Biochem. Biophys. Res. Commun.* **154:** 20–26.

Selbie, L. A., Schmitz-Peiffer, C., Sheng, Y., and Biden, T. J., 1993, Molecular cloning and characterization of PKCι, an atypical isoform of protein kinase C derived from insulin-secreting cells, *J. Biol. Chem.* **268:**24296–24302.

Sharp, J. D., White, D. L., Chiou, X. G., Goodson, T., Gamboa, G. C., McClure, D., Burgett, S., Hoskins, J., Skatrud, P. L., Sortsman, J. R., Becker, G. W., Kang, L. H., Roberts, E. F., and Kramer, R. M., 1991, Molecular cloning and expression of human Ca²⁺-sensitive cytosolic phospholipase A₂, *J. Biol. Chem.* **266:**14850–14853.

Shinomura, T., Asaoka, Y., Oka, M., Yoshida, K., and Nishizuka, Y., 1991, Synergistic action of diacylglycerol and unsaturated fatty acid for protein kinase C activation: Its possible implications, *Proc. Natl. Acad. Sci. USA* **88:**5149–5153.

Stahl, M. L., Ferenz, C. R., Kelleher, K. L., Kriz, R. W., and Knopf, J. L., 1988, Sequence similarity of phospholipase C with non-catalytic region of *src*, *Nature* **332:**269–272.

Tanaka, C., and Nishizuka, Y., 1994, The protein kinase C family for neuronal signaling, *Annu. Rev. Neurosci.* **17:**551–567.

Teegarden, D., Taparowsky, E. J., and Kent, C., 1990, Altered phosphatidylcholine metabolism in C3H 10T1/2 cells transfected with the Harvey-ras oncogene, *J. Biol. Chem.* **265:**6042–6047.

Tsujishita, Y., Asaoka, Y., and Nishizuka, Y., 1994, Regulation of phospholipase A₂ in human leukemia cell lines: Its implication for intracellular signaling, *Proc. Natl. Acad. Sci. USA* **91:**6274–6278.

Uings, I. J., Thompson, N. T., Randall, R. W., Spacey, G. D., Bonser, R. W., Hudson, A. T., and Garland, L. G., 1992, Tyrosine phosphorylation is involved in receptor coupling to phospholipase D but not phospholipase C in the human neutrophil, *Biochem. J.* **281:**597–600.

Undrovinas, A. I., Fleidervish, I. A., and Makielski, J. C., 1992, Inward sodium current at resting potentials in single cardiac myocytes induced by the ischemic metabolite lysophosphatidylcholine, *Circ. Res.* **71:**1231–1241.

Vogel, U. S., Dixon, R. A. F., Schaber, M. D., Diehl, R. E., Marshall, M. S., Scolnick, E. M., Sigal, I. S., and Gibbs, J. B., 1988, Cloning of bovine GAP and its interaction with oncogenic *ras* p21, *Nature* **335:**90–93.

Weltzien, H. U., 1979, Cytolytic and membrane-perturbing properties of lysophosphatidylcholine, *Biochim. Biophys. Acta* **559:**259–287.

William, F., Wagner, F., Karin, M., and Kraft, A. S., 1990, Multiple doses of diacylglycerol and calcium ionophore are necessary to activate AP-1 enhancer activity and induce markers of macrophage differentiation, *J. Biol. Chem.* **265:**18166–18171.

Wolf, R. A., and Gross, R. W., 1985, Identification of neutral active phospholipase C which hydrolyzes choline glycerophospholipids and plasmalogen selective phospholipase A₂ in canine myocardium, *J. Biol. Chem.* **260:**7295–7303.

Xu, X.-X., Tessner, T. G., Rock, C. O., and Jackowski, S., 1993, Phosphatidylcholine hydrolysis and c-*myc* expression are in collaborating mitogenic pathways activated by colony-stimulating factor 1, *Mol. Cell Biol.* **13:**1522–1533.

Yao, L., Kawakami, Y., and Kawakami, T., 1994, The pleckstrin homology domain of Bruton tyrosine kinase interacts with protein kinase C, *Proc. Natl. Acad. Sci. USA* **91:**9175–9179.

Yoshida, K., Asaoka, Y., and Nishizuka, Y., 1992, Platelet activation by simultaneous actions of diacylglycerol and unsaturated fatty acids, *Proc. Natl. Acad. Sci. USA* **89:**6443–6446.

Zeisel, S. H., 1993, Choline phospholipids: Signal transduction and carcinogenesis, *FASEB J.* **7:** 551–557.

Chapter 3

Phosphatidic Acid

Stephen B. Bocckino and John H. Exton

3.1. Introduction

Many agonists increase the concentration of phosphatidic acid (PA) in their target cells. This was initially thought to reflect the action of diacylglycerol kinase on the 1,2-diacylglycerol (DAG) generated by the hydrolysis of phosphoinositides (PI) by phospholipase C (PLC) (Berridge, 1984). However, later experiments indicated that the agonists promoted the formation of DAG and PA from another phospholipid and this was subsequently identified as phosphatidylcholine (PC) in most cell types (Exton, 1990, 1994; Billah and Anthes, 1990).

In initial studies, it was assumed that the formation of DAG and PA from PC was also related to the activation of a PLC with subsequent conversion of DAG to PA by DAG kinase, and there is evidence that this occurs in some cell types. However, more detailed experiments with many cell types and agonists later indicated that the formation of PA from PC often preceded that of DAG, implying the activation of a phospholipase D (PLD) with subsequent formation of DAG from PA resulting from the action of PA phosphohydrolase (Exton, 1990, 1994; Billah and Anthes, 1990). Conclusive proof that the agonists activated a PLD came when it was demonstrated that they activated the transphosphatidylation reaction, which is considered to be specific for PLD.

It is now recognized that a large number of agonists including growth factors, cytokines, hormones, and neurotransmitters promote the cellular accumulation of DAG and PA through two mechanisms. The first is by stimulating the β or γ isozymes of PI-specific phospholipase C (PI-PLC) and the second is by stimulating PLD isozymes that act on PG in most instances (Table 3-1). Although the temporal sequence may vary, PI-PLC activation usually occurs very rapidly, but is transient, whereas PLD activation is slower, but more sustained.

Stephen B. Bocckino • Sphinx Pharmaceuticals Corp., Durham, North Carolina 27707.   **John H. Exton** • Howard Hughes Medical Institute and Department of Molecular Physiology and Biophysics, Vanderbilt University School of Medicine, Nashville, Tennessee 37232.

Handbook of Lipid Research, Volume 8: Lipid Second Messengers, edited by Robert M. Bell *et al.* Plenum Press, New York, 1996.

Table 3-1. Agonist-Induced Hydrolysis of Phosphatidylcholine by Phospholipase D

Cell type or tissue	Agonists and other agents
Hematopoietic	
Neutrophils	fMLP, C5a, GM-CSF, PMA, OAG, A23187
HL-60 promyelocytic leukemia	fMLP, ATP, PMA, OAG, A23187
U937	fMLP, ATP, PAF, C5a, PMA
Macrophages	PAF, PMA, A23187
Mast cells	Antigen, IL3, PMA, A23187
RBL-2H3 basophilic leukemia	Antigen, PMA
Monocytes	PMA
Eosinophils	C5a, PMA, A23187
Platelets	Thrombin, collagen, PMA, A23187
Jurkat/T lymphocytes	Antigen, PMA
HEL erythroleukemia	Thrombin, PAF, $PGE_{1,2}$, ADP, A23187
Fibroblasts	
Swiss 3T3	Bombesin, vasopressin, PDGF, $PGF_{2\alpha}$, EGF, FGF
NIH 3T3	EGF, PDGF, $PGF_{2\alpha}$
BALB/c-3T3	PDGF, PGE_2
REF 52	Vasopressin, PMA
IIC9	Thrombin, EGF, PDGF
CCL39	Thrombin, FGF, PMA
C3H10T1/2	PMA
HeLa	PMA, A23187
HEF	PMA
Rat 1	Epinephrine, endothelin
Human	Bradykinin, PMA
BHK	FGF, PMA
Nerve	
NG108-15 neuroblastomaXglioma	Carbachol, PMA, DiC_8
1321N1 astrocytoma	Carbachol, PMA
Cerebral cortex	Carbachol, epinephrine
Hippocampus	Metabotropic excitatory amino acids
Astrocytes	PMA
PC12 pheochromocytoma	Carbachol, bradykinin, PMA
SK-N-MC neuroblastoma	Endothelin
Sympathetic neurons	Depolarization, carbachol, 5-hydroxytryptamine
Endocrine/exocrine	
Pancreas	Bombesin, CCK, carbachol, PMA, A23187
Ovarian granulosa	GnRH, PMA
Adrenal glomerulosa	Angiotensin II, carbachol
GH3 pituitary tumor	TRH
αT3-1 gonadotrop	GnRH
Epithelial	
Hepatocytes	Vasopressin, angiotensin II, epinephrine, ATP, PMA, A23187
MDCK kidney	Epinephrine, bradykinin, PMA
A431 breast carcinoma	EGF, PMA
Cardiovascular	
Heart	Carbachol, epinephrine
Vascular smooth muscle	Angiotensin II, vasopressin, PMA, A23187

(*continued*)

Table 3-1. Agonist-Induced Hydrolysis of Phosphatidylcholine by Phospholipase D
(continued)

Cell type or tissue	Agonists and other agents
A10 aortic smooth muscle	Vasopressin
A7r5 vascular smooth muscle	Vasopressin
Endothelial	Thrombin, bradykinin, FGF, ATP, PMA, A23187
Mesangial	Angiotensin II, endothelin, vasopressin
Other	
Human amnion	Bradykinin
Retina	Light
MC3T3-E1 osteoblast	$PGF_{2\alpha}$
Uterine decidua	PMA
Leydig cells	Vasopressin

The physiological role of PI hydrolysis is undoubtedly to induce a rapid rise in cytosolic Ca^{2+} resulting from the production of inositol trisphosphate, which induces Ca^{2+} release from intracellular stores, which leads in turn to Ca^{2+} influx. However, the physiological significance of the relatively small amount of DAG that is released by PI hydrolysis is obscure, although it appears to cause the translocation and activation of typical protein kinase C (PKC) isozymes (Dekker and Parker, 1994). It probably plays a role in the negative feedback that is observed when many PI-linked receptors are activated (for references, see Bird *et al.*, 1993), but it is unlikely that this is its only function. The physiological function(s) of PC hydrolysis is even less understood. DAG derived from PC is capable of activating Ca^{2+}-independent PKC isozymes (Ha and Exton, 1993a) and is probably responsible for the longer-term effects of agonists mediated PKC (Exton, 1990, 1994; Billah and Anthes, 1990). Many roles for PA have been proposed (Exton, 1994), but support for these roles in intact cell experiments is limited.

This chapter will describe the enzymes that produce and degrade PA and also the mechanisms involved in their regulation, with emphasis on PLD, DAG kinase, and PA phosphohydrolase. The signal transduction mechanisms for agonists that alter cellular PA levels will also be discussed, as will be the physiological roles proposed for PA.

3.2. Enzymes Producing and Degrading Phosphatidic Acid

PA levels are determined by the activities of the synthetic enzymes, PLD, diacylglycerol kinase (DGK), and lysophosphatidate acyltransferase (LPAAT), and of the degradative enzymes, PA phosphohydrolase (PPH) and phospholipase A_2 (PLA_2) (Fig. 3-1). Many of these enzymes have not been purified or extensively characterized and their role in signal transduction is yet to be determined. In this section, the characteristics of these enzymes are discussed.

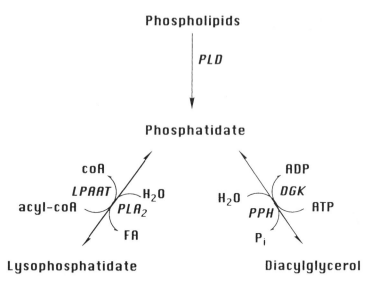

Figure 3-1. Pathways of phosphatidic acid formation and metabolism.

3.2.1. Phospholipase D

PLD catalyzes the hydrolysis of phospholipid to produce PA and a free head group. PLD has been assayed by measurement of the PA (Saito and Kanfer, 1975; Bocckino et al., 1987a; Pai et al., 1988) or the released head group (Hattori and Kanfer, 1984; Davitz et al., 1989). Since these products can be generated by enzymes other than PLD, one must be cautious in the interpretation of these assays. PLD also catalyzes a reaction, termed transphosphatidylation, in which aliphatic alcohols react with the phospholipid to form a phosphatidylalcohol and the free head group (Dawson, 1967) (Fig. 3-2). Transphosphatidylation is thought to be uniquely catalyzed by PLD, and measurement of phosphatidylalcohol formation has been used as an alternative PLD assay, especially in intact cells. Incubation with radiolabeled alcohol (Chalifour and Kanfer, 1982; Randall et al., 1990) or lipid precursors (Kobayashi and Kanfer, 1987; Pai et al., 1988) and subsequent purification of the phosphatidylalcohol by thin-layer chromatography is a commonly employed method; a mass measurement has also been described (Bocckino et al., 1987b).

Mammalian PLD activities hydrolyzing various substrates have been identified in many tissues and subcellular fractions, but purification to homogeneity has proved an elusive goal. It seems likely that multiple PLDs, activated by different signaling pathways, generate phosphatidate as a second messenger but the elucidation of their individual roles will require their purification.

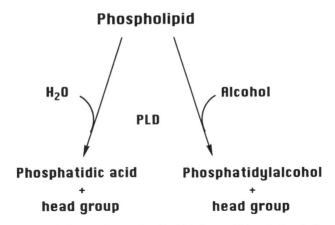

Phospholipid

H₂O

Alcohol

PLD

Phosphatidic acid
+
head group

Phosphatidylalcohol
+
head group

Figure 3-2. Hydrolytic and transphosphatidylation activities of phospholipase D.

3.2.1.1. *Glycosylphosphatidylinositol Phospholipase D (GPI-PLD)*

Over 100 proteins, including alkaline phosphatase, decay-accelerating factor, and acetylcholinesterase, have been reported to be anchored to the outer face of the plasma membrane via a glycosylphosphatidylinositol (GPI) group (reviewed in Cross, 1990). The GPI-PLDs hydrolyze the GPI anchor to release phosphatidate and a soluble protein linked to a complex glycan (Fig. 3-3). GPI-PLD has been purified from bovine serum (K.-S. Huang *et al.*, 1990), and from human plasma (Davitz *et al.*, 1989), blood being an abundant source. A GPI-PLD cDNA isolated from a bovine liver library has been cloned and expressed in COS cells (Scallon *et al.*, 1991). The calculated M_r of the recombinant GPI-PLD is 90,200, similar to the M_r values observed by SDS-PAGE of the enzyme purified from human plasma (M_r=110,000; Davitz *et al.*, 1989) and bovine serum (M_r= 100,000; K.-S. Huang *et al.*, 1990). The GPI-PLD from bovine serum migrates with an apparent M_r of 200,000 on gel filtration under nondenaturing conditions and may be a homodimer. The plasma GPI-PLD is inhibited by 1,10-phenanthroline and by EGTA, indicating a possible requirement for divalent cation for activity or in regulation (K.-S. Huang *et al.*, 1991).

Purified GPI-PLD is inactive against PC or PI substrates, and does not hydrolyze GPI substrates in an environment of intact membrane, being active only when the substrate is solubilized with detergent. This lack of activity against membrane-bound substrate may be explained by the potent inhibition of purified bovine serum GPI-PLD observed with phosphatidate and lysophosphatidate (Low and Huang, 1993), which are present in the plasma membrane. It is also possible that the physiological substrates of the serum enzyme are not GPI-anchored proteins. Metz *et al.* (1994) have described a GPI-PLD activity present in HeLa cells that

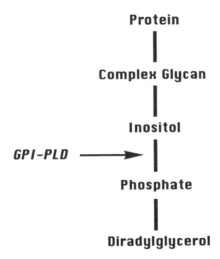

Figure 3-3. Site of action of glycosylphosphatidylinositol phospholipase D.

freed GPI-anchored proteins from cells under physiological conditions. The cellular origin of the serum enzyme and its relationship to cellular GPI-PLD have not been determined. A GPI-PLD activity has been identified in liver plasma membranes that is separable from the blood plasma enzyme by anion exchange chromatography (Heller, *et al.* 1992). GPI-PLDs are expressed in multiple tissues, including pancreatic islets of Langerhans (Metz *et al.*, 1991), mast cells (Metz *et al.*, 1992), stratified squamous epithelium (Xie *et al.*, 1993), and myeloid cells (Xie and Low, 1994).

3.2.1.2. Phosphoinositide Phospholipase D (PI-PLD)

Balsinde *et al.* (1988) have described a PLD activity in the postnuclear fraction of human neutrophils that hydrolyzed PI but not PC. Balsinde and Mollinedo (1990) later identified a PI-PLD activity in human plasma. This activity was assayed in the presence of 1 mM Ca^{2+}, and was completely inhibited in the presence of 2 mM EGTA. A PLD activity hydrolyzing [³H]stearoyl-PI has been identified in the cytosol of Madin-Darby canine kidney cells (C. Huang *et al.*, 1992). This PLD activity was activated 2.6-fold by 1 mM Ca^{2+}, but the relevance of such a high concentration of calcium to the regulation of an enzyme in the cytosol, where the calcium concentration is submicromolar, is questionable.

3.2.1.3. Phosphatidylcholine Phospholipase D (PC-PLD)

PLD activity acting on PC has been observed in many cell types and in broken cell preparations. There appear to be multiple PC-PLD activities present in different subcellular locations (Table 3-2), but little is known about their func-

Table 3-2. PC-PLD Activities in Subcellular Fractions

Tissue or cell	Localization	Reference
Brain	Particulate	Saito and Kanfer (1975)
	Microsomes	Chalifour and Kanfer (1980)
	Synaptic membrane	Hattori and Kanfer (1984)
	Cytosol	Wang *et al.* (1991)
	Plasma membrane	Kanfer and McCartney (1994)
Lung	Microsomes	Chalifour and Kanfer (1980)
	Cytosol	Wang *et al.* (1991)
HL-60 granulocytes	Particulate+cytosol	Anthes *et al.* (1991)
	Particulate	Brown *et al.* (1993)
U937 cells	Particulate+cytosol	Kusner *et al.* (1993)
Neutrophils	Plasma membrane	Gelas *et al.* (1989)
	Plasma membrane + cytosol	Olson *et al.* (1991)
Eosinophils	Cytosol	Kater *et al.* (1976)
Liver	Plasma membranes	Bocckino *et al.* (1987a)
	Cytosol	Wang *et al.* (1991)
	Mitochondria	Rakhimov *et al.* (1989)
Fibroblasts	Particulate	Conricode *et al.* (1992)
MDCK cells	Particulate	C. Huang *et al.* (1992)
Heart	Cytosol	Wang *et al.* (1991)
	Sarcolemma, mitochondria, sarcoplasmic reticulum	Panagia *et al.* (1991)
Kidney, spleen, thymus	Cytosol	Wang *et al.* (1991)
Amnion	Microsomes	Inamori *et al.* (1993)

tions. Membrane-bound PC-PLD activities have been described frequently, but have been resistant to purification as a result of loss of activity on solubilization with detergent (Bocckino and Exton, 1992). The PC-PLDs from brain and lung are exceptions, and significant progress has been made in their purification and characterization.

The PC-PLD from brain microsomes is activated by the detergents Miranol H2M and taurodeoxycholate and by unsaturated fatty acids (Chalifour and Kanfer, 1982). Taki and Kanfer (1979) partially purified PLD from a lyophilized rat brain powder to a specific activity of 120 nmole/hr per mg protein using the detergent Miranol H2M. The partially purified enzyme hydrolyzed both PC (K_m = 0.75 mM) and phosphatidylethanolamine (PE) (K_m = 0.91 mM) with a pH optimum of 6.0. This PLD did not require calcium for activity in contrast to plant PLDs (but was stimulated 65–80% by 5 mM calcium) and was completely inhibited by FeCl$_3$ or PCMB at 5 mM. Using Triton X-100, Kobayashi and Kanfer (1991) were able to purify PC-PLD from rat brain microsomes 9-fold to a specific activity of 901 nmole/hr per mg protein. Using the same procedure, they purified PC-PLD from lung microsomes 3.8-fold to a specific activity of 1151 nmole/hr per mg protein. Further attempts at purification using a variety of chromatographic media were unsuccessful.

Okamura and Yamashita (1994) have purified a PC-PLD from pig lung microsomes to apparent homogeneity using the detergent heptylthioglucoside to solubilize the enzyme. After seven chromatographic steps, the enzyme was purified to a specific activity of 3670 nmole/min per mg protein and migrated as a single protein band of M_r of 190,000. The purified PLD hydrolyzed PC, but LPC, PE and PI were not substrates. Calcium (1 mM) and magnesium (2 mM) stimulated enzyme activity 2.5 and 1.7-fold respectively, but neither cation was required for activity. The purified enzyme was activated by unsaturated fatty acids (arachidonate>oleate>linoleate), but the saturated fatty acids stearate and palmitate were inhibitory.

In contrast to the above studies, Brown *et al.* (1993) were able to solubilize PLD from HL60 cell membranes by treatment with 400 mM NaCl. Subsequent assay was conducted using mixed vesicles of PE/phosphatidylinositol 4,5-bisphosphate/PC in a molar ratio of 16:1.4:1. Phosphatidylinositol 4,5-bisphosphate was a potent activator, with phosphatidylinositol 4-phosphate and phosphatidylserine being much less effective (less than 10% and 1%, respectively). Phosphatidylinositol 4,5-bisphosphate may function as a cofactor for PLD and its omission from the assay mix may explain the failure of previous workers to extract PLD from membranes with salt (Kobayashi and Kanfer, 1991; Bocckino and Exton, 1992). Brown *et al.* (1995) have partially purified the PLD from porcine brain and shown that it was stimulated by ARF (ADP-ribosylation factor) and by cytosolic components. The purified PLD had a specific activity of 160 nmol/min per mg protein when assayed in the presence of guanine nucleotides, ARF, the cytosolic components, and the lipid mix including phosphatidylinositol 4,5-bisphosphate. This PLD activity was calculated to have an apparent molecular mass of 95,000 Da.

Massenburg and coworkers (1994) have separated two forms of PC-PLD activity found in rat brain membranes by chromatography over a heparin column. One form required sodium oleate for activity and the other required phosphatidylinositol 4,5-bisphosphate and was activated by ARF and guanine nucleotide. ARF1, ARF2 and ARF3 were active, and myristoylated ARFs were more effective than nonmyristoylated ARFs. The oleate-dependent form was not activated by phosphatidylinositol 4,5-bisphosphate or by ARF and guanine nucleotide. The two PC-PLD forms appear to be different enzymes under very different controls with physiological functions yet unknown.

A soluble PLD was purified from human eosinophils by Kater *et al.* (1976) to a specific activity of 194,000 nmole/hr per mg protein. The purified enzyme had a molecular weight of 50,000 by gel filtration and a pH optimum of 4.5–6.0, suggesting a lysosomal origin. Isoelectric focusing demonstrated that the PLD had a pI of 5.8–6.2 and that at least four contaminating bands of protein with no PLD activity were present at other isoelectric points. The eosinophil PLD both hydrolyzed PC and inactivated platelet activating factor, subsequently shown to be 1-*O*-alkyl-2-acetyl-*sn*-glycero-3-phosphocholine (reviewed in Hanahan, 1986).

Cytosolic PC-PLD activities have been observed in many other tissues by Wang *et al.* (1991). Lung was the richest source of this enzyme with a specific activity of 77 pmole/hr per mg protein. The lung cytosolic PLD was purified to 976 pmole/hr per mg protein by anion exchange chromatography. The enzyme had a pH optimum of 6.0–6.5, was activated by high concentrations of Ca^{2+} (65–90% by 10 mM Ca^{2+}), and did not require Mg^{2+} for activity. It hydrolyzed PC with an apparent K_m of 1.45 mM. The relationship of this PLD to the membrane forms and its relative importance are not known, but it should be pointed out that unpurified membrane PC-PLD has a specific activity of 60–100 nmole/hr per mg protein (Bocckino *et al.*, 1987a; Kobayashi and Kanfer, 1991), approximately 1000-fold higher than the crude cytosolic enzyme.

The substrate specificity of PC-PLD has, of necessity, been assessed in intact cells and in mixtures that may contain multiple lipases. PC-PLD is known to hydrolyze a variety of substrates, including 1-*O*-alkyl-, 1-*O*-alky-1'-enyl-, and l-acyl-PC (Pai *et al.*, 1988; Daniel *et al.*, 1993), Möhn *et al.* (1992) devised an ingenious approach using endogenous acyl-coenzyme A synthetase and acyl-coenzyme A:lysolipid acyltransferase to transfer radioactive fatty acids to phospholipid in native synaptic membranes. Oleate-stimulated neutral PLD present in the synaptic membranes hydrolyzed the labeled PC in the order : 2-oleoyl-PC > 2-myristoyl-PC > 2-palmitoyl-PC > 2-arachidinoyl-PC. Horwitz and Davis (1993) added potential substrate PCs along with sodium oleate to brain microsomes and found that phosphatidylbutanol formation, a measure of PLD activity, followed the order: egg yolk PC > 1-palmitoyl-2-oleoyl-PC > 1-stearoyl-2-oleoyl-PC > 1-stearoyl-2-arachidonoyl-PC. It appears from the results of these two studies that highly unsaturated PC may not be a good substrate for the brain PLD.

3.2.1.4. *Phosphatidylethanolamine Phospholipase D (PE-PLD)*

Kiss and Anderson (1989a,b, 1990) have described hydrolysis of PE by PLD in cultured cells. In isolated membranes from NIH 3T3 cells, both adenine and guanine nucleotides stimulated PE-PLD activity (Kiss and Anderson, 1990). The enzyme(s) responsible for PE hydrolysis has not been isolated.

3.2.1.5. *Phospholipase D Inhibitors*

Various inhibitors have been described for the PLDs, but a potent and specific inhibitor has not been discovered. As described above, PA, lyso-PA, and lipid A potently inhibited GPI-PLD (Low and Huang, 1993), but have other well-known biological activities. Fluoride ion inhibited rat liver plasma membrane PC-PLD over the concentration range of 1–10 mM (Bocckino *et al.*, 1987a; Bocckino and Exton, 1992), but stimulated PA production in isolated liver cells (Bocckino and Exton, 1992). Ethanol is a competitive inhibitor of PLD via the transphosphatidylation reaction and has been used at high concentration (200–400 mM) to examine the role of PA generated by PLD (e.g., Huang and Cabot, 1990b). Wortmannin (Reinhold *et al.*, 1990) and carbobenzyloxy-leucine-tyrosine-chloromethylketone (Kessels *et al.*, 1991a) appear to inhibit receptor coupling to PLD in neutrophils, but are not PLD inhibitors *per se*.

Aminoglycoside antibiotics inhibited neural PLD activity (Liscovitch *et al.*, 1991), perhaps by binding to PA (Brown *et al.*, 1993). Gratas and Powis (1993) have identified suramin (IC$_{50}$= 15 μM), the azo-suramin analogue NSC 79741 (58 μM), and the aminosteroid U-73,122 (78 μM) as inhibitors of solubilized rat brain PC-PLD. These compounds have been shown previously to inhibit PI-PLC (Seewald *et al.*, 1989; Powis *et al.*, 1991, 1992).

3.2.2. Lysophosphatidate Acyltransferase

Lyso-PA acyltransferase catalyzes the transfer of a fatty acyl group from acyl-coA to the 2 position of lyso-PA to form phosphatidate. The enzyme is found in microsomes and has been purified 7.5-fold from this source (Yamashita *et al.*, 1981). Rapid increases in lyso-PA acyltransferase activity have been observed on treatment of guinea pig parotid glands with isoproterenol or carbachol (Söling *et al.*, 1989). Treatment of isolated guinea pig parotid microsomes with protein kinases increased the specific activity of the acyltransferase (calcium–calmodulin-dependent protein kinase II > cyclic AMP-dependent protein kinase ≫ PKC). Protein phosphatases were shown to reverse the activation by calcium–calmodulin protein kinase. Protein phosphatase 2a also reversed the stable activation of the acyltransferase observed in microsomes isolated from carbachol-treated parotid. The authors concluded that protein phosphorylation–dephosphorylation could be responsible for rapid changes in the activity of the acyltranferase. Interleukin-1 has also been shown to produce a stable activation of microsomal lyso-PA acyltransferase in human mesangial cells (Bursten *et al.*, 1991).

3.2.3. Diacylglycerol Kinase

DAG kinase catalyzes the transfer of a phosphate group from ATP to DAG to form PA. Multiple forms of DAG kinase have been identified in various tissues and three forms have been cloned. An 83-kDa kinase has been cloned from porcine thymus (Sakane *et al.*, 1990), human lymphocytes (Schaap *et al.*, 1990), and rat brain (Goto *et al.*, 1992). This kinase had two EF hand domains (Ca^{2+}-binding domains), and the purified enzyme exhibited high-affinity Ca^{2+} binding and a requirement for calcium and phosphatidylserine (PS) for maximal activity. The enzyme also had two zinc fingers (cysteine-rich zinc-binding domains). Although this domain has been implicated in the binding of phorbol esters to protein kinase C, the purified 83-kDa DAG kinase did not bind phorbol esters (reviewed in Kanoh *et al.*, 1993). A 90-kDa DAG kinase has been cloned from rat brain (Goto and Kondo, 1993), and an 86-kDa kinase has been cloned from a hepatoma library (Kai *et al.*, 1994). These enzymes were highly homologous to the 83-kDa enzyme, and all three enzymes had two sets of EF hands and zinc fingers. Clear differences among the enzymes exist, however, in regulation and in tissue distribution. The 83-kDa enzyme was activated by calcium and PS and was expressed highly in T lymphocytes and in oligodendrocytes (Goto *et al.*, 1992). The 90-kDa kinase also required calcium, but was expressed predominantly in neurons (Goto and Kondo, 1993). The 86-kDa kinase required PS, but not calcium, for activity and was expressed predominantly in retina (Kai *et al.*, 1994). These differences in regulation and distribution, along with the existence of other probable members of the DAG kinase family (reviewed in Kanoh *et al.*, 1993), and the signaling functions of both substrate and product suggest an important regulatory role for this enzyme family.

Inhibitors of DAG kinase with varied chemical structures have been described. R 59 022 ((6-[2-(4-fluorophenyl)phenyl-methylene]-1-piperidinyl)ethyl)-7-methyl-5-*H*-thiazolo-[3,2-*a*]-pyrimidin-5-one; de Chaffoy de Courcelles *et al.*, 1985) and its more potent analogue, R 59 949 ((3-[2-94-[bis(4-fluorophenyl)-methylene]-1-piperidinyl)ethyl)-2,3-dihydro-2-thioxo-4[1*H*]-quinazolinone; de Chaffoy de Courcelles *et al.*, 1989), have been used frequently to dissect the functions of DAG kinase in cellular processes (reviewed in de Chaffoy de Courcelles, 1990). The DAG analogues dioctanolethylene glycol and 1-monooleoylglycerol (Bishop *et al.*, 1986) inhibited pig brain DAG kinase competitively with respect to DAG. The PC analogue 1-*O*-hexadecyl-2-*O*-methyl-*sn*-glycerol-3-phosphocholine and its metabolite 1-*O*-hexadecyl-2-*O*-methyl-*sn*-glycerol (Salari *et al.*, 1993) inhibited DAG kinase in WEHI-3B cells. The high-molecular-weight DAG kinase from brain was inhibited by short-chain ceramides, acting competitively with DAG (Younes *et al.*, 1992). Lysine polymers have also been shown to inhibit the kinase (Jeng *et al.*, 1988).

3.2.4. Phosphatidate Phosphohydrolase

PA phosphodydrolase catalyzes the hydrolysis of the phosphomonoester linkage in PA yielding DAG and inorganic phosphate. This connection between two lipid second messengers has received much attention and the PC-PLD-PA phosphohydrolase pathway may generate a significant portion of the agonist-induced DAG in some cell types (Billah *et al.*, 1989a; Huang and Cabot, 1990b). The enzyme can be assayed by the release of phosphate from PA, but the presence of phospholipase A and phosphatases in cell extracts would also yield phosphate (Martin *et al.*, 1991). In the presence of these contaminating enzymes, PA phosphohydrolase is assayed by following the generation of DAG from PA, sometimes in the presence of a lipase inhibitor such as tetrahydrolipstatin (Jamal *et al.*, 1991). Two forms of PA phosphohydrolase in rat liver have been described: a Mg^{2+}-dependent activity (PAP-1) that translocates between cytosol and endoplasmic reticulum and is inhibited by *N*-ethyl maleimide (NEM), and a Mg^{2+}-independent activity in the plasma membrane (PAP-2) that is not sensitive to NEM (Jamal *et al.*, 1991).

The Mg^{2+}-dependent PA phosphohydrolase (PAP-1) (reviewed in Brindley, 1988; Martin *et al.*, 1994) is thought to function in the synthesis of triacylglycerols, PE, and PC. It has been shown to translocate from the cytosol to the endoplasmic reticulum, where it is active, in response to PA, acyl-coA esters, and fatty acids. The enzyme has been partially purified from the soluble fraction of rat liver to a specific activity of 10.3 μmole/min per mg protein (Butterwith *et al.*, 1984). The purified enzyme displayed complex kinetics with inhibition by its substrate, PA, at concentrations greater than 0.4 mM. The purified enzyme required Mg^{2+} for activity, but was inhibited by concentrations of the cation greater than 0.5 mM. The phosphohydrolase was inhibited by the thiol reagent NEM and by butane-2,3-dione and cyclohexane-1,2-dione, which react with arginine.

The plasma membrane Mg^{2+}-independent PA phosphohydrolase (PAP-2) may function in cell signaling (Kanoh *et al.*, 1993). An NEM-insensitive PA phosphohydrolase is enriched in rat brain synaptosomes, consistent with this signaling function (Fleming and Yeaman, 1995). This enzyme preferentially hydrolyzed PA with short, saturated acyl chains. A PA phosphohydrolase has been purified from porcine thymus membranes to apparent homogeneity with a specific activity of 15 nmol/min per mg protein (Kanoh *et al.*, 1992a). The purified enzyme had a M_r of 83,000 by SDS-PAGE, and an apparent M_r of 218,000 by gel chromatography. The authors suggested that binding of Triton X-100 to the phosphohydrolase might explain this discrepancy in apparent molecular size. The purified phosphohydrolase did not require Mg^{2+} and was not inhibited by NEM and may be similar to the rat liver plasma membrane enzyme described by Jamal *et al.* (1991). The enzyme was inhibited by its product, DAG, but not by other lipids tested.

Inhibition of PA phosphohydrolase by a number of amphiphilic cations has been described. Mepyramine, fenfluramine, norfenfluramine, hydroxyethylnorfenfluramine, *N*-(2-benzyloxyethyl)norfenfluramine, cinchocaine, chlorpromazine, and demethylimipramine were shown to inhibit the rat liver enzyme with IC_{50} values of 0.2–0.9 mM (Brindley and Bowley, 1975). These compounds inhibit many other enzymes at these concentrations. Koul and Hauser (1987) later found that chlorpromazine, desmethylimipramine, and propranolol inhibited rat brain PA phosphohydrolase. Propranolol has since been widely used as an inhibitor of PA phosphohydrolase, but at the concentrations typically used (50–200 μM), the compound not only acts as a β-adrenergic antagonist, but also inhibits PKC (Sozzani *et al.*, 1992). The rat liver cytosolic PA phosphohydrolase was inhibited by 1,2-diacyl-*sn*-glycero-3-phosphorothioate (Bonnel *et al.*, 1989). Sphingosine and other sphingoid bases inhibited PA phosphohydrolase (Mullmann *et al.*, 1991), but these compounds also inhibit PKC (Hannun *et al.*, 1986) and activate PLD (Lavie and Liscovitch, 1990).

3.2.5. Phospholipase A₂

PLA$_2$ catalyzes the hydrolysis of the ester linkage at the *sn*-2 position to form a lysolipid and a free fatty acid. Billah *et al.* (1981) have described a platelet PLA$_2$ acting specifically on PA to produce lyso-PA. This pathway is of interest not only because of the importance of eicosanoids in platelet function, but also because it produces lyso-PA, a lipid with diverse signaling functions (reviewed in Moolenaar *et al.*, 1994). The platelet PA-PLA$_2$ was particulate, had a pH optimum of 7.0, and required Ca^{2+} for activity, with maximal activity observed at 7 μM Ca^{2+}. Quinacrine inhibited the phospholipase. A PLA$_2$ that is specific for phosphatidate has been purified from rat brain (Thomson and Clark, 1995). This enzyme had an apparent M_r of 58,000, an optimal pH of 6.0 and was calcium-independent.

A PLA$_1$ activity hydrolyzing PA has been identified in sheep pancreas and in pancreatic secretions (Dawson *et al.*, 1982). The enzyme was inhibited by Ca^{2+} and by Mg^{2+} and had a pH optimum of 6.2. Since this phospholipase apparently is secreted, it probably does not function in the breakdown of intracellular phosphatidate. PLA_1 has also been purified to apparent homogeneity from bovine brain (Pete *et al.*, 1994). Like the pancreatic enzyme, it hydrolyzes PA at pH 7 and the reaction is inhibited by Mg^{2+} (Pete *et al.*, 1995).

3.3. Agonist Regulation of Phospholipase D

The concept that many growth factors, cytokines, hormones, neurotransmitters, and related agonists stimulate PC hydrolysis by activating PLD in their target cells is now widely accepted. Table 3-1 lists the mammalian cell types that are presently known to exhibit this response, and also the natural agonists and other agents that elicit the response. It is evident from the list that the phenomenon is very widespread and is therefore likely to subserve some important physiological function(s).

Inspection of the types of agonists listed in Table 3-1 indicates that some of them act through receptors linked to heterotrimeric G proteins, e.g., vasopressin, carbachol, angiotensin II, epinephrine, bombesin, ATP, bradykinin, formyl-Met-Leu-Phe (fMLP), thrombin, gonadotropin-releasing hormone (GnRH), endothelin, and prostaglandins, whereas others act through receptors that exhibit intrinsic tyrosine kinase activity, e.g., platelet-derived growth factor (PDGF), epidermal growth factor (EGF), fibroblast growth factor (FGF), or promote the activation of *src*-related tyrosine kinases, e.g., T-cell antigen receptor and IgE receptor ($Fc_\epsilon R$). These observations point to roles for G proteins and tyrosine phosphorylation in the regulation of PLD. Table 3-1 also illustrates that in most cell lines, the tumor-promoting phorbol ester PMA also activates the enzyme and this is often true of the Ca^{2+} ionophore A23187. These findings indicate potential roles of PKC and Ca^{2+} in the regulation of the enzyme.

Because no regulated isoform of PLD has been purified to homogeneity or cloned, it is not known if G proteins, tyrosine kinases, PKC, or Ca^{2+} can directly affect the enzyme. Information concerning its regulation is derived from studies with intact or permeabilized cells or isolated plasma membranes. Because of this, the mechanisms of its control are known in far less detail than for PI-PLC.

3.3.1. Role of Heterotrimeric G Proteins in the Regulation of Phospholipase D

As indicated above, many agonists that are known to act through heterotrimeric G proteins activate PC-PLD in many cell types. Further evidence for the involvement of these G proteins comes from the fact that the response is inhibited by pertussis toxin in some instances (Agwu *et al.*, 1989; Kanaho *et al.*, 1991; Xie *et al.*, 1991; MacNulty *et al.*, 1992) and by the observation that certain agonists enhance the stimulatory effect of GTP analogues on PLD in isolated membranes or permeabilized cells (Bocckino *et al.*, 1987a; Martin and Michaelis, 1989; Qian and

Drewes, 1989; Liscovitch and Eli, 1991). As noted below, there is now evidence for the involvement of small (low M_r) GTP-binding proteins (SMGs) in the regulation of PLD. Thus, the stimulation of PLD by GTP analogues can no longer be taken as evidence of the involvement of heterotrimeric G proteins. Since NaF inhibits the activity of certain PLDs (Bocckino *et al.*, 1987a), this reagent also cannot be used to indicate the involvement of heterotrimeric G proteins.

The mechanism(s) by which G protein-linked agonists activate PLD may be direct or indirect. Because purified or recombinant forms of the enzyme are not available, the possibility of direct regulation by one or more of the known G proteins cannot be tested by reconstitution or other experiments. On the other hand, there is much suggestive evidence that the regulation may be indirect. The most plausible indirect mechanism is one in which the agonists activate G proteins that are coupled to PI-PLC (Fig. 3-4). The DAG produced by PI hydrolysis transiently activates PKC isozymes which then activate PC-PLD. The evidence that PKC regulates PC-PLD will be presented below, but other support for the scheme will be described here. First, in the vast majority of case, G protein-linked agonists that activate PC-PLD also activate PI-PLC. Second, the activation of PI-PLC precedes that of PC-PLD. Third, downregulation of PKC induced by prolonged treatment of cells with phorbol ester usually impairs agonist-induced PC-PLD activation (for references, see Section 3.3.3). Although PKC inhibitors often produce a similar impairment, their lack of specificity for PKC and their ability to

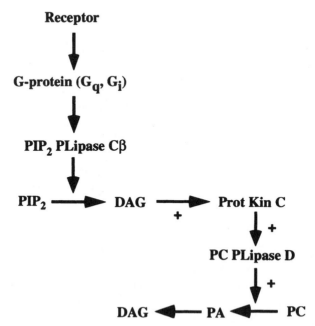

Figure 3-4. Proposed role of protein kinase C in activation of phospholipase D by agonists acting through G proteins.

produce other effects, e.g., sphingosine and staurosporine, render suspect any conclusions based on their use alone.

Although it is likely that the major mechanism by which G protein-linked agonists stimulate PC-PLD involves activation of PI-PLC and PKC (Fig. 3-4), there are some situations where other mechanisms may operate. Examples are when PC-PLD activation is observed in the apparent absence of PI hydrolysis, as will be discussed in Section 3.3.2. Other examples are the effects of agonists on PC-PLD observed in isolated membranes in the presence of GTPγS (Bocckino *et al.*, 1987a; Qian and Drewes, 1989). However, it is likely that PKC was present in these preparations since stimulatory effects of PMA were also observed. With respect to the requirement for ATP, this nucleotide was the only agonist observed to be stimulatory in the studies of Bocckino *et al.* (1987), suggesting that it might have acted both as a purinergic agonist and as a phosphorylating agent. As noted in Section 3.3.3, PKC may also activate PC-PLD by a nonphosphorylating mechanism.

Interestingly, the pertussis toxin sensitivity of agonist activation of PC-PLD in different cell lines parallels that for agonist activation of PI-PLC, suggesting the involvement of the same types of G proteins. It is now generally accepted that the pertussis toxin-insensitive mechanism of PI-PLC activation involves G proteins of the G_q family acting via their α subunits, whereas the pertussis toxin-sensitive mechanism involves certain members of the G_i family acting via their βγ subunits (Exton, 1993). The G_q α subunits activate mainly the β1 and β3 isozymes of PI-PLC, whereas the G_i βγ subunits activate mainly the β2 and β3 isozymes. Experimental modulation of the cellular levels and activities of specific G proteins and PLC and PKC isozymes should help define the role of these proteins as mediators of activation of PC-PLD by certain agonists.

3.3.2. Role of Tyrosine Phosphorylation in the Regulation of Phospholipase D

Many growth factors activate PC-PLD in fibroblasts and other cell types as illustrated in Table 3-1. The effects of these agonists are inhibited by genistein and other tyrosine kinase inhibitors (Cook and Wakelam, 1992; Ahmed *et al.*, 1994; Lee *et al.*, 1994; Pettit *et al.*, 1994; E.-J. Yeo and J. H. Exton, unpublished observations), implying, as expected, the involvement of the tyrosine kinase activity of their receptors. Activation of the T-cell antigen receptor and of the mast cell Fc_ε receptor for IgE also results in PLD activation (Stewart *et al.*, 1991; Dinh and Kennerly, 1991; Lin and Gilfillan, 1992). Tyrosine kinase inhibitors abolish the response in the RBL-2H3 mast cell line (Kumada *et al.*, 1993; Lin *et al.*, 1994) and, in both cell types, it is likely that it is secondary to the activation of nonreceptor tyrosine kinases of the Src family (Weiss, 1993). In support of this, activation of PC-PLD has been observed in fibroblasts transformed by active Src (Song *et al.*, 1991; Wyke *et al.*, 1992).

A role for a tyrosine kinase(s) in the activation of PLD by mechanisms involving G proteins is indicated by studies in permeabilized U937 promonocytic leukocytes (Dubyak *et al.*, 1993; Kusner *et al.*, 1993). In these cells, ATP markedly augments the stimulation of PLD by GTPγS, and the effect of ATP is mimicked by vanadate/H_2O_2 and blocked by tyrosine kinase inhibitors. Based on these and other findings, the authors concluded that a tyrosine kinase(s) plays a secondary or supplementary role in G protein regulation of PLD in these cells. Tyrosine kinase inhibitors also inhibit the activation of PLD by thrombin in intact platelets and by GTPγS in permeabilized platelets (Martinson *et al.*, 1994), consistent with the preceding conclusion. In permeabilized HL60 cells, Bourgoin and Grinstein (1992) also found a strong correlation between the effects of vanadyl hydroperoxide on tyrosine phosphorylation and PLD activation. The nature of the G proteins and tyrosine kinases involved in these effects needs to be defined.

In fibroblasts infected with Rous sarcoma virus, i.e., expressing v-Src activity, GTPγS stimulated PLD after permeabilization, whereas GDPβS inhibited the enzyme (Jiang *et al.*, 1994). ATP enhanced the effect of GTPγS and the enhancement was blocked by the tyrosine kinase inhibitor herbimycin A. The failure of NaF or pertussis or cholera toxin to mimic or block the action of GTPγS suggests the involvement of an SMG.

As indicated above for heterotrimeric G proteins, the absence of purified or expressed PC-PLD isozymes precludes testing the possibility that growth factor receptor tyrosine kinases or Src-like tyrosine kinases directly phosphorylate and activate PLD. Although this possibility cannot be discounted, most of the reports of growth factor and cytokine activation of the enzyme can be explained as being secondary to the activation of γ isozymes of PI-PLC and subsequently of PKC isozymes (Fig. 3-5). Thus, in the majority of cases, stimulation of PI hydrolysis precedes PC breakdown such that the accumulation of DAG is biphasic (Exton, 1990; Billah and Anthes, 1990; Cook and Wakelam, 1991b). Studies utilizing PKC inhibitors or PKC downregulation are also generally supportive of a major role for this enzyme (for references, see Section 3.3.3).

However, it must be noted that there have been several examples of agonists activating PC-PLD without the apparent involvement of PI hydrolysis. In some of these cases, it is not clear that PC hydrolysis was related to activation of PLD rather than PLC (Rosoff *et al.*, 1988; Wright *et al.*, 1988). In other cases, the dissociation seems more clear, since choline and PA or phosphatidylbutanol accumulation was observed, but inositol phosphate formation or Ca^{2+} mobilization either was not detected or was greatly reduced (Cook and Wakelam, 1992; Ha and Exton, 1993a; Ahmed *et al.*, 1994; Balboa *et al.*, 1994). In these cases, it could be postulated that there was a short-lived, localized hydrolysis of PI that was sufficient to trigger the activation of PKC and PLD, but often not large enough to result in measurable changes in whole cell Ca^{2+} and PI metabolism. This possibility is supported by experiments with the PKC inhibitor Ro-31-8220, in which a barely detectable increase in inositol triphosphate induced by EGF in Swiss 3T3 cells became a marked response in the presence of the inhibitor (Yeo and Exton, 1995). An enhancing effect of PKC inhibitors on EGF-induced inositol phosphate formation has also been observed in A431 cells and appears to involve a mechanism distal to EGF binding (Wahl and Carpenter, 1988).

An alternative approach to exploring the role of PI-PLC and PKC in growth

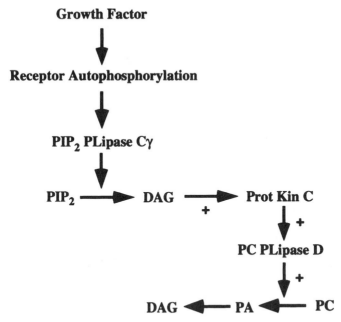

Figure 3-5. Proposed role of protein kinase C in activation of phospholipase D by growth factors.

factor stimulation of PC-PLD is to employ cells transfected with mutant growth factor receptors. Mutations (Tyr→Phe) have been made in the autophosphorylation sites of the PDGF receptor that alter its ability to interact with the different cytosolic proteins that are responsible for transducing the signal (Kashishian *et al.*, 1992; Kazlauskas *et al.*, 1992; Fantl *et al.*, 1992; Valius *et al.*, 1993; Nishimura *et al.*, 1993; Valius and Kazlauskas, 1993). The proteins interact at different autophosphorylation sites because their SH2 domains recognize specific phosphopeptide sequences (Fantl *et al.*, 1992; Cooper and Kashishian, 1993; Songyang *et al.*, 1993).

Using TRMP cells expressing PDGF receptors with Tyr→Phe mutations in residues 740, 751, 771, 1009, and 1021, which result in selective loss of coupling to the 85-kDa regulatory subunit of PI 3-kinase, Ras GAP, PI-PLCγ, or a 64-kDa protein that is an SH2-containing phosphotyrosine phosphatase (Fig. 3-6) (Valius and Kazlauskas, 1993), it has been found that only those receptors that couple to and activate PI-PLCγ are capable of activating PC-PLD (Yeo *et al.*, 1994). For example, cells expressing receptors that display no kinase activity (Leu→Arg mutation of residue 635) or have a single Tyr→Phe mutation at residue 1021 (selective loss of coupling to PI-PLCγ) do not show activation of PC-PLD in response to PDGF. On the other hand, cells with receptors (PLC+) in which there are Tyr→Phe mutations at residues 740, 751, 771, 1009 but not 1021 (i.e., which couple only to PI-PLC) do show the PLD response (Fig. 3-7). These results argue

**Autophosphorylation Sites for Interactions
of the PDGF β Receptor**

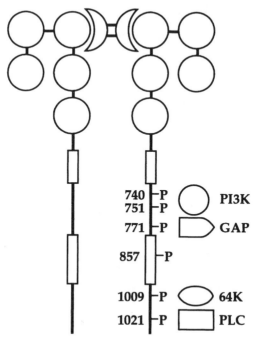

Figure 3-6. Autophosphorylation sites in the PDGF β receptor and the proteins that interact with them.

strongly that, for this agonist and cell line, activation of PC-PLD is entirely dependent on activation of PI-PLC.

Another approach to examining the dependence of PLD activation on PI-PLC activity has been utilized by Lee *et al.* (1994). These workers measured PLD activation in response to PDGF in NIH 3T3 fibroblasts overexpressing PLC-γ1 and in control cells. Activation of PLD was 10-fold greater in the overexpressing cells. Pretreatment with staurosporine or genistein greatly decreased the response, and downregulation of PKC by PMA pretreatment abolished it. Thus, in accord with the preceding findings, these data strongly implicate PI-PLC and PKC in the activation of PLD by PDGF.

3.3.3. Role of Protein Kinase C in the Regulation of Phospholipase D

Table 3-1 illustrates that PMA is capable of activating PC-PLD in very many cell types. This indication that PKC plays a major role in the regulation of the enzyme is reinforced by many studies in which inhibition and/or downregulation of PKC results in partial or complete abolition of agonist-induced PC-PLD activation.

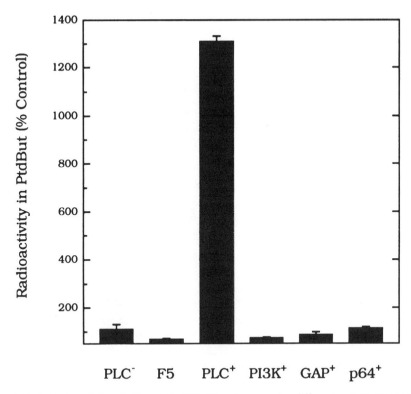

Figure 3-7. Activation of phospholipase D by PDGF in cells expressing different mutant receptors that selectively couple to phosphoinositide phospholipase or other proteins. Data from Yeo *et al.* (1994) by permission of the authors and publisher.

Examples of the effects of PKC downregulation and PKC inhibitors have been given in previous reviews (Exton, 1990, 1994; Billah and Anthes, 1990), but these will be discussed more specifically here. Except for a few reports (Lassègue *et al.*, 1991; Freeman *et al.*, 1995), downregulation of PKC by prolonged treatment of cells with PMA or phorbol dibutyrate has been shown to completely block the ability of G protein-linked agonists such as angiotensin II, bradykinin, bombesin, carbachol, PAF, GnRH, vasopressin, endothelin, and $PGF_{2\alpha}$ to activate PC-PLD in a variety of cell types, as measured by choline release or formation of PA or phosphatidylalcohol (Muir and Murray, 1987; Liscovitch and Amsterdam, 1989; Martin *et al.*, 1989; Martinson *et al.*, 1989; Price *et al.*, 1989; Uhing *et al.*, 1989; MacNulty *et al.*, 1990; Cook and Wakelam, 1991b; van Blitterswijk *et al.*, 1991; Vingaard and Hansen, 1991; Ben-Av *et al.*, 1993). Likewise, this treatment abolished or attenuated the PC-PLD response to growth factors, e.g., EGF and PDGF, in most instances (Fig. 3-8) (Besterman *et al.*, 1986a; Price *et al.*, 1989; Plevin *et al.*, 1991a; Ben-Av *et al.*, 1993; Yeo and Exton, 1995; see Kaszkin *et al.*, 1992).

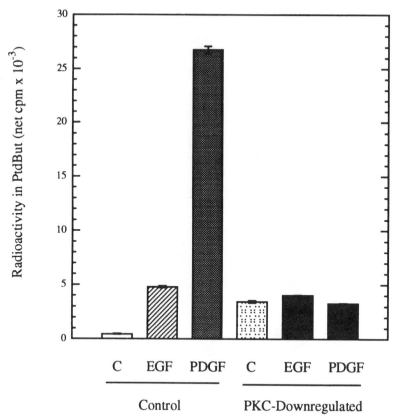

Figure 3-8. Effect of downregulation of protein kinase C on the activation of phospholipase D by EGF and PDGF.

A number of agents have been employed to inhibit PKC. Some of these lack specificity insofar as they inhibit other protein kinases, e.g., H7, or affect other enzymes (staurosporine and sphingosine). Nevertheless, studies with a variety of these inhibitors (H7, staurosporine, sphingosine, 1-*O*-hexadecyl-2-*O*-methyl glycerol, sangivamycin, calphostin C, bisindolylmaleimide, chelerythrine, Ro-31-8220) generally support the involvement of PKC in the actions of both G protein-linked agonists and growth factors on PC-PLD (Uhing *et al.*, 1989; Huang and Cabot, 1990a; Cook *et al.*, 1991; van Blitterswijk *et al.*, 1991; Purkiss and Boarder, 1992; Kester *et al.*, 1992; Pfeilschifter and Huwiler, 1993; Ben-Av *et al.*, 1993; Gustavsson *et al.*, 1994; Balboa *et al.*, 1994; Yeo and Exton, 1995). In some cases where negative results were reported, the inhibitor (H7) was only partly effective against PMA-stimulated PC-PLD (Liscovitch and Amsterdam, 1989; Llahi and Fain, 1992). In others where staurosporine was used (K.-S. Huang *et al.*, 1991; Kanoh *et al.*, 1992b), the stimulatory effect of this compound on PLD was not corrected for. However, in some instances, the failure of the more specific inhibitor Ro-31-8220 to affect agonist activation of PC-PLD (MacNaulty *et al.*, 1992; Uings *et al.*, 1992) is difficult to explain in the absence of more detailed studies.

In summary, the findings reported above support the partial or complete dependence of agonist regulation of PC-PLD on PKC in many cell types. In some cases where contrary findings have been presented, the data are limited or inadequately controlled. However, it must be recognized that there are also studies in which PLD and PKC activities are dissociated. For example, PKC inhibitors often have greater effects on the stimulation of PLD by phorbol esters than by agonists. Although this could represent the existence of PKC-independent mechanisms, it may also be related to differences in the potency or efficacy of the inhibitors against specific PKC isozymes, which may be differentially activated by phorbol esters and DAG derived from endogenous phospholipids. In support of the latter explanation, it has been found that a commonly used PKC inhibitor (Ro-31-8220) is more potent on the Ca^{2+}-dependent α and β isozymes than on the Ca^{2+}-independent ϵ isozyme, and is virtually inactive against PKCζ (Yeo and Exton, 1995).

Despite the abundance of evidence pointing to a role for PKC in the regulation of PC-PLD by many agonists, there have been only a few studies where the enzyme has been added directly to isolated membranes containing PC-PLD activity. These studies utilized membranes from CCL-39 fibroblasts, HL60 cells or neutrophils and showed that addition of purified PKC stimulated choline release and phosphatidylpropranol or phosphatidylethanol formation in a PMA-or phorbol 12,13-dibutyrate-dependent manner (Conricode *et al.*, 1992; Ohguchi *et al.*, 1995; Lopez *et al.*, 1995). The most surprising result to emerge from these studies utilizing fibroblast was that the activation of PC-PLD by PKC appeared to be independent of ATP. This was demonstrated by the fact that depletion of ATP with ATPase did not impair PLD activation in the CCL-39 membranes, but abolished the ability of PKC to undergo autophosphorylation or phosphorylate an exogenous substrate (Conricode *et al.*, 1992). In later work it was shown that only the α and β Ca^{2+}-dependent isozymes of PKC were capable of activating PC-PLD in this system (Conricode *et al.*, 1994). The possible involvement of a phosphorylation-independent mechanism of PMA activation of PLD in HL60 membranes was also suggested by the observation that concentrations of Ro-31-8425 that completely blocked PKC activity did not affect PLD activation (Ohguchi *et al.*, 1995). These results indicate that Ca^{2+}-dependent isozymes of PKC are capable of activating PC-PLD, perhaps directly, but that the activation does not necessarily require phosphorylation.

In contrast to the findings with fibroblasts described above, studies with neutrophils support a phosphorylation-dependent mechanism of control of PLD by PKC (Lopez *et al.*, 1995). The PKC substrate that was involved in the activation of the phospholipase was localized to the plasma membrane, and the potency order of PKC isozymes was $\beta_1 > \alpha > \gamma$, with the novel and atypical isozymes being ineffective.

Overexpression of the α and β_1 isozymes of PKC in fibroblast cell lines has been shown to increase the activation of PC-PLD by endothelin, thrombin, and

PDGF (Pai *et al.*, 1991; Pachter *et al.*, 1992; Eldar *et al.*, 1993), whereas depletion of PKCα by antisense methods reduced the activation of the enzyme by purinergic agonists in MDCK kidney cells (Balboa *et al.*, 1994). Using treatment with phorbol ester for various times to selectively downregulate different PKC isozymes, Pfeilschifter and Huwiler (1993) concluded that PKCϵ was responsible for angiotensin II stimulation of PC-PLD in mesangial cells. In summary, these studies implicate Ca^{2+}-dependent PKC isozymes in the regulation of PC-PLD, but delineation of the role of Ca^{2+}-independent isoforms requires further work.

In most studies of the regulation of PLD by PKC in cells, the focus has been on the plasma membrane. However, phosphatidylethanol has been shown to accumulate also in fractions enriched in mitochondria and endoplasmic reticulum in fibroblasts incubated with phorbol ester and ethanol (Edwards & Murray, 1995). Whether the lipid is produced in these fractions as a result of PLD activity or is translocated there from the plasma membrane is unknown. Isolated nuclei from MDCK cells have also been shown to have a PLD activity that is activated by PKC (Balboa *et al.*, 1995). These and other studies suggest that PLD is present in other cell structures besides the plasma membrane.

In addition to the many reports showing that short-term activation of PKC in cells is associated with increased PLD activity, there is evidence that increased activity of PKC can induce long-term negative regulation of the enzyme. For example, in many cells, chronic downregulation of PKC by phorbol esters leads to enhanced activity of PLD (e.g., Plevin *et al.*, 1991a; Balbao *et al.*, 1994; Yeo *et al.*, 1994; Yeo and Exton, 1995). On the other hand, transfection of COS-1 cells with wild-type and constitutively activated PKCα for 48 hr results in inhibition of PMA-stimulated PLD activity (McKinnon and Parker, 1994). The basis for these effects has not been explored, but they could be related to PKC acting as negative regulator of the synthesis of a PLD isozyme(s).

3.3.4. Role of Calcium Ions in the Regulation of Phospholipase D

There is also much evidence that Ca^{2+} ions play a role in agonist regulation of PC-PLD. Several studies have shown that addition of the Ca^{2+} ionophores A23187 or ionomycin stimulates PC-PLD activity, as measured by formation of PA or phosphatidylethanol (PEth) or choline release (Bocckino *et al.*, 1987a; Polverino and Barritt, 1988; Martinson *et al.*, 1989; Billah *et al.*, 1989b; Halenda and Rehm, 1990; R. Huang *et al.*, 1991; Peterson and Walter, 1992; Lin and Gilfillan, 1992; Guillemain and Rossignol, 1992; Gustavsson *et al.*, 1994). Formation of DAG from PC, as demonstrated by isotopic labeling, is also stimulated by the Ca^{2+} ionophores in some cell lines (Augert *et al.*, 1989b; Peterson and Walter, 1992). On the other hand, depletion of cellular Ca^{2+} by chelators EGTA and BAPTA resulted in inhibition of the activation of PC-PLD by G protein-linked and other agonists (Halenda and Rehm, 1990; Kessels *et al.*, 1991b; C. Huang *et al.*, 1992; Lin and Gilfillan, 1992; Wu *et al.*, 1992; Balboa *et al.*, 1994). Chelation of extracellular Ca^{2+} also decreased DAG production from PC stimulated by vasopressin in hepatocytes (Augert *et al.*, 1989b) and by carbachol in 1321N1 astrocytoma cells (Martinson *et al.*, 1989). Interestingly, there have been no reports that have explored the role of Ca^{2+} in growth factor stimulation of PC-PLD.

A major question relating to the role of Ca^{2+} is whether it is merely required for the activation of PKC. As alluded to in Section 3.3, the PKC isozymes that are initially activated in response to PI hydrolysis are of the Ca^{2+}-dependent type. In addition, these PKC isozymes have been shown to mediate PC-PLD activation. Thus, the Ca^{2+} dependence of PC-PLD activation could be attributed to the involvement of Ca^{2+}-dependent PKC isozymes, in part at least. With respect to the possibility that Ca^{2+} ions have a direct stimulatory effect on PC-PLD, most studies of the enzyme *in vitro* indicate that it is stimulated by Ca^{2+}, but only at concentrations much higher than those reported in the cytosol (see Section 3.2.1.1). Other indirect mechanisms of Ca^{2+} control are possible.

3.3.5. Role of Low-Molecular-Weight G Proteins in the Regulation of Phospholipase D

One of the most exciting recent developments in PLD research is the recognition that the enzyme is stimulated by members of the ARF and Rho families of SMGs. It was recognized by early investigators that the stimulatory effect of GTPγS on PLD in membranes from neutrophils or HL-60 cells required the addition of cytosol (Anthes *et al.*, 1991; Olson *et al.*, 1991). In addition, it was found that the ability of GTPγS to stimulate PLD in streptolysin O-permeabilized HL-60 cells declined with time as cytosolic proteins leaked from the cells (Geny and Cockcroft, 1992; Geny *et al.*, 1993). The loss of activation could be restored by adding brain cytosol to the permeabilized cells, and the reconstituting factor was partially purified and shown to be a protein of 16 kDa (Geny *et al.*, 1993). In a remarkable example of parallel research, the groups of Cockcroft and Sternweis identified the factor as ARF, an SMG that was first recognized as an ADP-ribosylation factor required for the action of cholera toxin on the α subunit of the G protein G_S and later shown to be involved in Golgi vesicular transport (Moss and Vaughan, 1993).

Using HL-60 membranes and an improved assay for PC-PLD, Brown *et al.* (1993) purified the stimulatory factor from bovine brain cytosol and showed it to be a polypeptide with an apparent M_r of 21,000. Partial amino acid sequence analysis revealed the protein to be a mixture of ARF1 and ARF3. As expected, the factor also stimulated the ADP-ribosylation of $G_{s\alpha}$. Stimulation of PLD activity was also exhibited by preparations of recombinant ARF, and myristoylated ARF was much more potent than nonmyristoylated ARF (Brown *et al.*, 1993, 1995). In an independent, but concordant, study, Cockcroft *et al.* (1994) also purified the stimulatory factor from bovine brain cytosol using cytsol-depleted HL-60 cells and measuring PLD by either labeled choline release or PEth formation. They also found it to be a low-M_r protein with sequences corresponding to ARF1 and ARF3. Recombinant ARF1 was also effective.

It is not clear from these studies if the ARFs interact directly with the PLD. Although, Brown *et al.* (1993) initially showed that a partially purified preparation of the enzyme could be markedly stimulated by ARF purified from bovine brain

Effect of GTPγS-Activated
Recombinant ARF 1 and Bovine Brain ARF
on HL 60 PC PLD

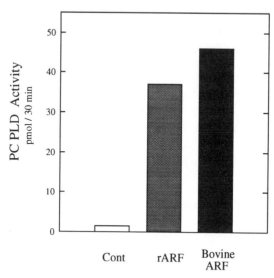

Figure 3-9. Activation of phospholipase D partially purified from HL-60 cell membranes by GTPγS in the presence of recombinant ARF1 (rARF) or ARF purified from bovine brain. Data replotted from Brown *et al.* (1993).

and by recombinant ARF1 (Fig. 3-9), recent evidence indicates that a cytosolic factor is required in addition to ARF to stimulate the enzyme (Lambeth *et al.*, 1995; Bourgoin *et al.*, 1995; Singer *et al.*, 1995). Two groups have identified this as a 50 kDa protein (Lambeth *et al.*, 1995; Bourgoin *et al.*, 1995), but its nature and function remain unknown. Although it is generally accepted that plasma membranes contain an ARF-responsive PLD, Golgi-enriched membranes have also been found to have a PLD that is activated by ARF. The activation is sensitive to brefeldin A (Ktiskakis *et al.*, 1995). Activation of PLD by ARF could involve the formation of a ternary complex, but proof of this requires the purification and identification of the PLD and cytosolic protein.

Surprisingly, ARF is not the only SMG to be implicated in the regulation of PC-PLD. In a study of the effects of GTPγS on the enzyme in neutrophil lysates, Bowman *et al.* (1993) found a requirement for protein factors in both cytosol and plasma membrane. In reconstitution experiments, the GTPγS-binding protein was found to be membrane-associated. A low Mg^{2+} requirement for GTPγS binding and a failure of AlF_4^- to activate suggested the involvement of an SMG. This idea was supported by the finding that a nonspecific stimulator of GDP dissociation from SMGs (GDS) enhanced the ability of GTP to stimulate the PLD (Bowman *et al.*, 1993). When a more specific reagent was employed, namely, an inhibitor of GDP dissociation from Rho family members (Rho GDI), it was found to inhibit the stimulatory effect of GTPγS. These findings thus implicated a membrane-associated SMG of the Rho family in the regulation of the enzyme.

The findings of Bowman *et al.* (1993) have been confirmed by Ohguchi *et al.* (1995) and in our laboratory using membranes from HL-60 cells or rat liver plasma membranes. In addition, GTPγS-dependent stimulation of PC-PLD has been observed on addition of recombinant RhoA, Rac1, or CDC42 to membranes pretreated with RhoGDI (Malcolm *et al.*, 1994; Siddiqi *et al.*, 1995; Ohguchi *et al.*, 1995). Of the three SMGs tested, RhoA was the most effective (Fig. 3-10). ARF was ineffective in the liver preparation, but produced a large stimulation in both membranes and cytosol from HL-60 cells. In HL60 cell membranes and using a partially purified preparation of PLD from brain membranes, a synergistic effect of RhoA and ARF was observed on PLD activity (Siddiqi *et al.*, 1995; Singer *et al.*, 1995).

It is presently unclear why PC-PLD is under dual control by SMGs, but one possibility is that different isozymes are involved and that a soluble enzyme is regulated by ARF, whereas a membrane-associated enzyme is controlled by Rho. Another important issue is how the regulation of PC-PLD by the two types of SMG

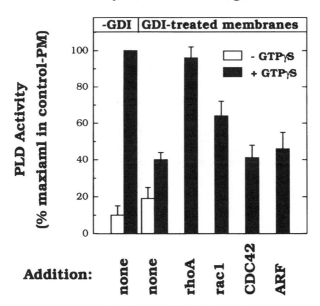

Figure 3-10. Effects of rhoA, rac1, CDC42, and ARF added to rat liver plasma membranes previously treated with RhoGDI. In the absence of GTPγS, none of the SMGs was effective in restoring PLD activity. Data from Malcolm *et al.* (1994) by permission of the authors and publisher.

fits into the scheme(s) by which the enzyme is regulated by G protein- and tyrosine kinase-linked agonists. Thus, it needs to be established whether or not the SMGs mediate the effects of certain agonists and, if so, how the agonists activate the SMGs. Activation involves the conversion of the SMG from a GDP-bound to a GTP-bound form, and agonists could induce this by altering the activity of proteins that stimulate the GTPase activity of the SMG (GAPs) or that promote or inhibit the dissociation of GDP (GDSs and GDIs). At present there is no information on whether agonists regulate the activities of ARF or Rho species in cells, although studies on actin polymerization in fibroblasts support the idea that certain agonists promote the activation of Rho or Rac (Ridley and Hall, 1992; Ridley *et al.*, 1992).

Several studies have indicated an interaction between G proteins and PKC in the regulation of PLD. Thus, GTPγS and phorbol ester or ATP act synergistically to stimulate PLD in permeabilized platelets and HL-60 cells (Xie and Dubyak, 1991; Coorssen and Haslam, 1993; Ohguchi *et al.*, 1995). The effect of ATP was not mimicked by a nonhydrolyzable analogue, implying that a phosphotransferase reaction was involved, whereas the action of GTPγS was inhibited by GDP and GDPβS, suggesting the involvement of a G protein (Xie and Dubyak, 1991). The nature of the G protein is unknown, but it is probably an SMG of the Rho family (Ohguchi *et al.*, 1995). In neutrophils, phorbol ester pretreatment enhances the activation of PLD by GTPγS plus ARF in the permeabilized cells (Whatmore *et al.*, 1994). The stimulation by GTPγS plus ARF is also increased by the addition of Mg ATP, but it is not clear that this is related to PKC activity.

The regulation of PLD by the Ras gene product *ras* p21 remains controversial. Early reports showed that PC hydrolysis was enhanced in *ras*-transformed cells, but the effect was attributed to enhanced PC-PLC activity (for references, see Exton, 1994). However, Carnero *et al.* (1994a,b) have presented evidence for the activation of PC-PLD in NIH3T3 fibroblasts transformed by *ras* and in *Xenopus* oocytes injected with *ras* p21 protein. The activation of PLD in the fibroblasts was shown by increases in PA and choline, and the formation of phosphatidylbutanol in the presence of butanol, but there were increases in DAG and phosphorylcholine, suggesting that PC-PLC was also activated. In the *Xenopus* experiments, DAG was increased in addition to PA, again suggesting activation of both enzymes. In both types of experiment, propranol reduced the increase in DAG, suggesting that it was derived in part from PA.

3.4. Regulation of Diacylglycerol Kinase and Phosphatidate Phosphohydrolase

3.4.1. Regulation of Diacylglycerol Kinases

As described in Section 3.2.3, DAG kinase exists in several isozymic forms which differ in their tissue distribution, molecular weight, cellular location, and biochemical properties (Kanoh *et al.*, 1993). Phorbol ester and DAG have been shown to induce the translocation of DAG kinase from the cytosol to the membrane *in vitro* and in intact cells (Besterman *et al.*, 1986b; Maroney and Macara, 1989). However, there was no evidence that these agents altered the activity of DAG kinase in these experiments. Phosphorylation of the 80-kDa porcine form of the enzyme by PKC and cAMP-dependent protein kinase *in vitro* has been reported, but little change in activity was detected (Kanoh *et al.*, 1989). Phosphorylation of the 86-kDa human enzyme by PKCα and PKCε and the activated EGF receptor, when these are expressed in COS-7 cells, has been reported (Schaap *et al.*, 1993). The expressed DAG kinase is also phosphorylated in response to forskolin plus isobutylmethylxanthine, but not ionomycin (Schaap *et al.*, 1993), suggesting its phosphorylation by cAMP-dependent protein kinase, but not Ca^{2+}-calmodulin-dependent protein kinase. No change in activity was detected with either EGF or cAMP stimulation. However, phosphorylation of the enzyme by PKC has been shown to increase its association with phosphatidylserine (PS) vesicles, resulting in its stabilization (Kanoh *et al.*, 1989).

The 80-kDa porcine DAG kinase sequence contains two Cys-rich Zn finger and two EF hand motifs (Sakane *et al.*, 1990). As expected from the presence of the EF hand motifs, the enzyme binds Ca^{2+} with high affinity (Sakane *et al.*, 1991). Addition of Ca^{2+} produced a small increase in activity which was enhanced by PS. This indicated that Ca^{2+}-dependent interaction with membrane phospholipids could be important in the regulation of DAG kinase. In support of this, Ca^{2+} was observed to cause membrane translocation of the enzyme in thymocyte homogenates (Sakane *et al.*, 1991) although no increase in membrane-associated enzyme activity could be detected. Surprisingly, the enzyme did not bind phorbol ester despite the presence of the Zn finger motifs (Ahmed *et al.*, 1991; Kanoh *et al.*, 1993) and deletion of the C-terminal half of the enzyme, including the Zn fingers, did not abolish PS-dependent activity (Kanoh *et al.*, 1993).

The above results indicate a role for DAG and Ca^{2+} in the regulation of DAG kinase through membrane translocation. Although translocation does not appear to activate the enzyme, its movement from the cytosol to the site of its substrate could result in more phosphorylation of DAG. The question of whether phosphorylation alters the activity of DAG kinase is still open, but PKC can promote the incorporation of up to 1 mole of phosphate per mole of enzyme and induce its membrane association, which could result in greater DAG phosphorylation, as discussed above. To date, there have been no demonstrations of the phosphorylation of DAG kinase in response to natural agonists in intact cells. There has been a single report of hormonal regulation of the enzyme. Rider and Baquet (1988) reported that treatment of rats with vasopressin or perfusion of rat livers with the hormone increased the activity of plasma membrane DAG kinase twofold.

Inhibitors of DAG kinase promote the accumulation of DAG in thrombin-stimulated platelets (de Chaffoy de Courcelles *et al.*, 1985; Bishop *et al.*, 1986), implying a significant metabolism of DAG by this enzyme. However, it is not clear that the enzyme is an important determinant of DAG levels in other cells. In fact, there is evidence that DAG lipase is the major route of DAG metabolism in most cells (Welsh and Cabot, 1987; Severson and Hee-Cheong, 1989; Asaoka *et al.*, 1991; Chuang *et al.*, 1993; Florin-Christensen *et al.*, 1993). In Swiss 3T3 cells, PDGF stimulated the phosphorylation of cellular DAG, but the increase in PA rose in parallel with the accumulation of DAG, suggesting that it was related to an increase in substrate rather than activation of DAG kinase (MacDonald *et al.*, 1988). In most cells, the accumulation of PA induced by agonists appears to be mainly related to activation of PLD, except during the initial phase when some PA is formed from DAG derived from PI hydrolysis.

3.4.2. Regulation of Phosphatidate Phosphohydrolase

As described in Section 2.4, PA phosphohydrolase exists in two forms: one form (PAP-1) translocates from the cytosol to the endoplasmic reticulum and is involved in glycerolipid synthesis, whereas the other (PAP-2) is located in the plasma membrane and is presumably involved in signal transduction by the PC-PLD pathway (Jamal *et al.*, 1991; Day and Yeaman, 1992). This discussion will be focused mainly on PAP-2, which is distinguished from PAP-1 by its insensitivity to NEM and certain other specific amino acid reagents and to thermal denaturation (Jamal *et al.*, 1991). The activity of PAP-2 was not altered by short- or long-term treatment of rat hepatocytes with vasopressin, glucagon, insulin, glucocorticoids, or triiodothyronine (Jamal *et al.*, 1991). In contrast, PAP1 activity was increased by glucagon and dexamethasone, and the increases were reversed by insulin. Sphingosine has been observed to inhibit PA phosphohydrolase activity in several cell types (Lavie *et al.*, 1990; Mullmann *et al.*, 1991; Jamal *et al.*, 1991; Gomez-Muñoz *et al.*, 1992b). However, this effect is probably unrelated to its action on PKC since it was not mimicked by staurosporine or K252a (Mullmann *et al.*, 1991; Gomez-Muñoz *et al.*, 1992b).

Translocation of PAP-1 from the cytosol to the endoplasmic reticulum is induced in liver by low concentrations of unsaturated fatty acids and influenced by a number of other agents, and the translocation may be related to the control of glycerolipid synthesis (Gomez-Muñoz *et al.*, 1992a,b; Day and Yeaman, 1992). However, PAP-2 is located in the plasma membrane and does not undergo translocation (Gomez-Muñoz *et al.*, 1992a,b). Siddiqui and Exton (1992) observed that oleate and other unsaturated fatty acids stimulated the PA phosphohydrolase activity of liver plasma membranes incubated *in vitro*. These results suggest that unsaturated fatty acids may play a role in the regulation of the hydrolase. However, *in vivo* evidence for this is presently lacking. Bursten *et al.* (1991) have reported that interleukin 1 stimulates PA phosphohydrolase activity in mesangial cells by a mechanism involving a pertussis toxin-sensitive G protein, but confirmation of these intriguing findings is lacking. The PAP-2 form of the enzyme is decreased in fibroblasts transformed by the *ras* and *fps* oncogenes, resulting in increased PA/ DAG ratio in stimulated cells (Martin *et al.*, 1993). The molecular basis for the effect is unknown.

3.5. Agonist Regulation of Cellular Phosphatidate Levels

As expected from Table 3-1, a large number of agonists increase the level of PA in their target cells. In most cases this has been demonstrated by prelabeling the phospholipids with [³H]myristic or other fatty acids that are preferentially incorporated into PC. In some cases, [³H]arachidonic acid has been used alone or in combination with myristic acid labeled with a different isotope. Arachidonic acid is incorporated into all of the phospholipids, but usually labels PI to a greater extent than the others. In other experiments, PC has been specifically labeled using either [³H]alkyl-lyso-glycero-phosphocholine or alkyl-lyso-glycero[³²P]-phosphocholine.

In general, the effects of agonists on the time course of PA accumulation depend on the relative extent to which it is formed from the phosphorylation of DAG derived from PI hydrolysis versus PLD action on PC. In cells where there is significant PI hydrolysis, there is an initial peak of DAG accumulation that is detectable as early as 10 sec and usually reaches a maximum between 30 and 60 sec. The level then declines or plateaus before rising to a larger second peak between 2 and 10 min which arises from PC. This is the pattern typically seen in fibroblasts when the cells are labeled with [³H]arachidonic acid and stimulated with PDGF, EGF, or G protein-linked agonists such as thrombin, vasopressin, bradykinin, or endothelin (Plevin *et al.*, 1991a; Cook and Wakelam, 1992; Huang and Cabot, 1990a; van Blitterswijk *et al.*, 1991; van der Bend *et al.*, 1992). In these experiments, the accumulation of labeled PA is slower than that of DAG (Huang and Cabot, 1990; van Blitterswijk *et al.*, 1991; van der Bend *et al.*, 1992), reflecting the fact that PA arising from PI is formed by the sequential action of PI-PLC and DAG kinase and not by PLD.

In experiments with various cell types in which the phospholipids have been labeled with [³H]myristic or other long-chain saturated fatty acids, a different temporal pattern of PA and DAG formation is usually seen, namely, the increase in labeled PA is more rapid than that of DAG (Martin and Michaelis, 1989; Martinson *et al.*, 1989; Huang and Cabot, 1990a; Dinh and Kennerly, 1991; Moehren *et al.*, 1994). This reflects the fact that DAG arising from PC is formed mainly by the combined action of PLD and PA phosphohydrolase rather than by PLC. The peak of DAG formed from PC hydrolysis in response to agonists in myristate-labeled cells corresponds in time to the second peak of DAG formation in cells labeled with [³H]arachidonic acid, in which a significant fraction of the labeled fatty acid is incorporated into PC as well as PI.

In studies where isotopic alkyl-lyso-glycero-phosphocholine has been used to selectively label PC, the most comprehensive analyses have been made in neutrophils and the results have been uniform. Labeled alkyl-PA or alkyl-PEth formation in response to agonists is seen within 10 or 30 sec, whereas the appearance of alkyl-acylglycerol is delayed for 30 sec or 1 min (Pai *et al.*., 1988; Billah *et al.*, 1989a; Agwu *et al.*, 1989a; Tyagi *et al.*, 1991; Chabot *et al.*, 1992). Delayed production of [³H]alkyl-acylglycerol from [³H]alkyl PC has also been observed in other cell types (Augert *et al.*, 1989b; Kondo *et al.*, 1992). These findings are consistent with the rapid activation of PC-PLD and the subsequent formation of DAG by PA phosphohydrolase.

In a large number of studies, the production of phosphatidylalcohols has been used to define the time course of activation of PLD. Although this does not define the phospholipid that is hydrolyzed, the concurrent release of choline that occurs in most cells points to PC. Formation of labeled or unlabeled PEth or phosphatidylbutanol in response to G protein-linked agonists such as vasopressin, ATP, carbachol, and thrombin in a variety of cells prelabeled with [^{3}H]myristate or [^{3}H]palmitate is usually detectable within 15 or 30 sec (Bocckino et al., 1987b; Martin and Michaelis, 1989; Martinson et al., 1990; Halenda and Rehm, 1990; Xie et al., 1991; Cook et al., 1991; Plevin and Wakelam, 1992; Kanoh et al., 1992a; Moehren et al., 1994; Sugiyama et al., 1994). The formation of the phosphatidylalcohol occurs as rapidly as [^{3}H]alkyl-PA formation from [^{3}H]alkyl-PC in neutrophils stimulated with f-Met-Leu-Phe, consistent with the activation of PLD within this time frame. Phosphatidylalcohol formation in fibroblasts stimulated with EGF or PDGF occurs on a slightly slower time scale than observed for G protein-linked agonists (Plevin et al., 1991a; Cook and Wakelam, 1992; Wright et al., 1990; Ben-Av et al., 1993). This implies a slightly slower mechanism of activation of PLD by growth factors.

The changes in PA and DAG mass induced by agonists in various cell types reflect the time courses and magnitudes of PC and PI hydrolysis. In those cells where there is a significant initial peak of DAG accumulation, PI hydrolysis is substantial, at least during the first few minutes. This is exemplified by the action of thrombin on fibroblasts or platelets (Wright et al., 1988; Nakashima et al., 1991; Ha and Exton, 1993a), the effects of f-Met-Leu-Phe and platelet-activating factor on neutrophils (Truett et al., 1988; Tou et al., 1991; Tyagi et al., 1991), and the stimulation by cross-linking of the IgE receptor in mast cells (Lin et al., 1992). On the other hand, in those situations where there is little or no discernible accumulation of DAG during this phase, PI hydrolysis is small or short-lived and PC is the major source of this lipid. Examples of this pattern of DAG accumulation are hepatocytes stimulated with vasopressin (Bocckino et al., 1987a), macrophages treated with platelet-activating factor (Uhing et al., 1989), certain fibroblasts exposed to EGF or PDGF (Wright et al., 1988; Pessin et al., 1990; Fukami and Takenawa, 1992; Ha and Exton, 1993a), and PC12 cells incubated with carbachol (Horwitz, 1990).

There have been very few chemical measurements of PA in cells. In hepatocytes, neutrophils, and HEL cells treated with vasopressin, f-Met-Leu-Phe, or platelet-activating factor and thrombin, respectively, PA mass rises more rapidly than DAG, consistent with the rapid activation of PLD in these cells (Cockcroft and Allan, 1984; Bocckino et al., 1987a; Agwu et al., 1989b; Halenda and Rehm, 1990; Tou et al., 1991; Allan and Exton, 1993; Fukami and Takenawa, 1992). There is hydrolysis of PI in these cells (Verghese et al., 1987; Augert et al., 1989a; Halenda and Rehm, 1990; Allan and Exton, 1993) but it appears to be of small magnitude during the initial period of stimulation. During the later stages of stimulation of these cells, the concentrations of DAG and PA become comparable. Since information on PA mass is lacking in other cell types, its changes relative to DAG can only be inferred from labeling studies. Most of these studies suggest that the levels of these two lipids are similar or differ by less than threefold at the later stages of stimulation. In the great majority of studies, utilizing either mass or isotopic measurements, the maximum increase in total cellular PA that has been observed has been two- to threefold over basal. This is similar to the increase in DAG, in most cases.

There have been several studies of the molecular species of DAG produced in response to agonists in several cell lines (Kennerly, 1987, 1990; Auguert et al., 1989b; Pessin et al., 1990; Lee et al., 1991; Divecha et al., 1991; Sebaldt et al., 1992). These support the conclusion that the first peak originates from a mixture of PI and PC, whereas the second peak comes mainly from PC, in agreement with the labeling studies described above and some earlier reports that studied the fatty acid composition of DAG (Banschbach et al., 1981; Bocckino et al., 1985). However, there have been only a few studies of the molecular species of PA in agonist-stimulated cells (Kennerly, 1987; Lee et al., 1991; Divecha et al., 1991) and only three reports of the changes in the fatty acid composition of PA (Broekman et al., 1981; Cockcroft and Allan, 1984; Bocckino et al., 1987a).

The study of Lee et al. (1991) examined the molecular species of PA in neuroblastoma cells stimulated with carbachol for 5 min and found that, whereas the PA of unstimulated cells resembled PC, that in the stimulated cells resembled PI. Studies of cells incubated with $^{32}P_i$ indicated a rapid increase in PA labeling, consistent with phosphorylation of DAG. Divecha et al. (1991) reported similar findings in PA molecular species in Swiss 3T3 cells before and after stimulation with bombesin. Kennerly (1987) also observed that the molecular species of PA generated by DAG kinase in mast cells stimulated for 4 min were enriched in arachidonic acid, suggesting their origin from PI. In a study of the fatty acid composition of PA in platelets stimulated with thrombin, Broekman et al. (1981) also obtained evidence that PA arose initially from PI, whereas it later came from other sources.

In contrast to the preceding findings, studies of the fatty acid composition of PA in neutrophils stimulated with f-Met-Leu-Phe (Cockcroft and Allan, 1984) and hepatocytes stimulated with vasopressin (Bocckino et al., 1985, 1987b) indicated that the newly formed PA was mainly derived from sources other than PI, with the most likely candidate being PC. These differences probably reflect the different cell types and agonists studied. In addition, it should be recognized that comparisons in this type of study are always made with the molecular species and fatty acid composition of total cellular PI or PC, and that PA and DAG could arise from discrete pools of these phospholipids.

There has been only one analysis of the molecular species of PEth (Holbrook et al., 1992). This utilized analysis by fast atom bombardment–mass spectrometry to show that, in PC12 cells stimulated with bradykinin or PMA, the molecular species of PEth found were identical to those in PC, confirming that PC was the major substrate of the activated PLD.

In summary, the various studies of the changes in PA and DAG in cells utilizing mass measurements and a variety of isotopic means to label different phospholipids generally support the following scheme of growth factor and G protein-linked agonist action: There is rapid initial activation of PI hydrolysis that may be small and difficult to detect by whole cell analyses in some instances. This generates DAG in the plasma membrane that is detected as the first peak of DAG accumulation in many cells. In these cells, part of the DAG may be converted to PA by DAG kinase. The increase in DAG in the membrane results in activation of Ca^{2+}-dependent PKC isozymes, and this activation may or may not be detectable by translocation studies. Activation of PKC leads by an unknown mechanism to activation of PC-PLD which generates PA in the membrane. This is converted to DAG by PA phosphohydrolase, accounting for the second peak of DAG accumulation.

As a consequence of PKC activation, there is often downregulation of agonist-stimulated PI hydrolysis. This accounts for the drop or plateau in the DAG level between the first and second peaks that is frequently observed. However, despite the decrease in DAG formation from PI as a result of downregulation, activation of PC-PLD may persist. This may be because the DAG formed from PC via the sequential action of PLD and PA phosphohydrolase is able to activate Ca^{2+}-independent PKC isozymes which can feed back to activate PC-PLD. This positive feedback mechanism could explain the observation that PC-PLD activation and DAG production from PC can be maintained for a long time in certain situations (e.g., Wright *et al.*, 1990; Pessin *et al.*, 1990).

It should be emphasized that the scheme proposed above to explain the changes in DAG and PA in agonist-stimulated cells does not exclude other mechanisms. In particular, there is strong evidence that PC can also be hydrolyzed by a PLC in certain cells (Slivka *et al.*, 1988; Martinson *et al.*, 1989; Pfeffer *et al.*, 1990; Huang and Cabot, 1990b; Choudhury *et al.*, 1991; van Blitterswijk *et al.*, 1991; Pettit *et al.*, 1994). In addition, PE hydrolysis can occur in some cells (Kiss and Anderson, 1989a,b, 1990). Finally, agonists may act by mechanisms involving other signals or intermediary proteins besides or in addition to PKC, e.g., G proteins, Ca^{2+}, tyrosine kinases.

3.6. Actions of Phosphatidic Acid

One of the greatest areas of deficiency in understanding the physiological role of PA is the definition of its true intracellular targets. PA is widely accepted as a likely second messenger, but there is no convincing information about its physiological targets. One of the first suggestions was that it promoted Ca^{2+} influx into cells. This was based on experiments in which it was added exogenously to cells (Salmon and Honeyman, 1980; Putney *et al.*, 1980; Ohsako and Deguchi, 1981) or to liposomes containing a Ca^{2+} indicator (Serhan *et al.*, 1981). However, the latter results were challenged (Holmes and Yoss, 1983) and it now seems likely that the findings with the cells were the result of the activation of cell surface receptors for lysophosphatidic acid (lyso-PA), which induce intracellular Ca^{2+} release and subsequent Ca^{2+} influx (Moolenaar *et al.*, 1992). Mobilization of internal Ca^{2+} in response to PA or lyso-PA preparations has been shown in several studies and this was associated with PI breakdown (Murayama and Ui, 1987; Moolenaar *et al.*, 1986; Van Corven *et al.*, 1989; Jalink *et al.*, 1990; McGhee and Shoback, 1990). PI hydrolysis has also been observed in other cell types (Knauss *et al.*, 1990; Plevin *et al.*, 1991b).

PA has been reported to inhibit adenylate cyclase by a pertussis toxin-sensitive (G_i) mechanism (Proll *et al.*, 1985; Murayama and Ui, 1987). As in the case of PI hydrolysis and subsequent Ca^{2+} mobilization, the effect may have resulted from contaminating lyso-PA since this lipid also inhibits adenylate cyclase (van Corven *et al.*, 1989) and its actions are blocked by pertussis toxin (van Corven *et al.*, 1989, 1993).

PA has been observed to stimulate arachidonic acid release from fibroblasts, implying activation of PLA_2 (Murayama and Ui, 1987). This is also seen with lyso-PA as agonist (van Corven *et al.*, 1989) and since the effect is potentiated by a GTP analogue in permeabilized cells, it might also involve a G protein. Stimulation of PC-PLD by PA or lyso-PA has also been reported (Ha and Exton, 1993b; Ha *et al.*, 1994; van der Bend *et al.*, 1992).

In addition to the reported effects of PA added to intact cells, which may be caused by traces of lyso-PA, there have been many observations of direct effects of PA on various enzymes, e.g., PI-PLC (Jackowski and Rock, 1989; Jones and Carpenter, 1993), PI-4P kinase (Moritz *et al.*, 1992), a protein tyrosine phosphatase (Zhao *et al.*, 1993), phosphorylase *b* kinase (Negami *et al.*, 1986), and other protein kinases (Bocckino *et al.*, 1991; Jones *et al.*, 1994; Khan *et al.*, 1994; McPhail *et al.*, 1995). There have also been reports of PA and lyso-PA effects on the proteins that regulate the function of certain LMGs. Thus, PA was reported to inhibit the GTPase-activating protein (GAP) for Ras (Tsai *et al.*, 1988) and to stimulate *n*-chimaerin, which is a GAP for Rac (Ahmed *et al.*, 1993). At the present time, it is difficult to fit these *in vitro* observations into a coherent picture of cellular PA action. For example, it seems more likely that the effects of exogenous PA and lyso-PA on PI-PLC are mediated by receptors and G proteins and not by PA *per se*, and that the activation of phosphorylase *b* kinase is related to the rise in cytosolic Ca^{2+}. With respect to the effects of PA on Ras-GAP, the concentrations of lipid required are very high, and the molecular species of PA that are active are very different from those that induce mitogenesis (Yu *et al.*, 1988).

Many reports have pointed to a role for PA in the respiratory burst in neutrophils, which is activated by inflammatory or phagocytic stimuli and is responsible for producing O_2 which is dismutated to H_2O_2. The reactive O_2 species are bactericidal and act as mediators of inflammation. The respiratory burst is caused by activation of NADPH oxidase, which is a multicomponent enzyme complex. There have been several studies of the effects of PA on the enzyme in cell-free and intact cell systems. Qualliotine-Mann *et al.* (1993) have shown that addition of PA to membrane plus cytosolic fractions from neutrophils produces a small stimulation of the oxidase. This is greatly increased by the addition of DAG, which by itself has little effect. This study expands on the earlier report of the *in vitro* stimulation of the enzyme by PA alone (Agwu *et al.*, 1991). In this report, the PA and DAG levels of intact neutrophils were also modulated by f-Met-Leu-Phe, propranolol, and exogenous PLA_1 and the results indicated that the NADPH oxidase activity correlated better with the level of PA than that of

DAG. Bauldry *et al.* (1992) examined the changes in PA, DAG, and O_2^- production in permeabilized neutrophils and also noted a close correlation between oxidase activation and PA production. Other reports have implicated both PA and DAG in regulation of the enzyme (referenced in Bauldry *et al.*. 1992, and Qualliotine-Mann *et al.*, 1993).

Stimulatory effects of PA on NADPH oxidase in intact and broken-cell neutrophil preparations have been reported by other workers (Bellavite *et al.*, 1988; Ohtsuka *et al.*, 1989; Rossi *et al.*, 1990; Mitsuyama *et al.*, 1993; Perry *et al.*, 1993). However, it should be recognized that there is also evidence for a role of DAG and that the effects of exogenous PA could be due to its conversion to DAG by a PA phosphohydrolase ecto-enzyme (Perry *et al.*, 1993). As illustrated by the findings of Qualliotine-Mann *et al.* (1993), both PA and DAG could play a role in regulation of NADPH oxidase. In an interesting development in this area, McPhail *et al.* (1995) have provided evidence that the activation of the oxidase by PA plus DAG involves phosphorylation of the p47-*phox* component.

An intriguing observation that was madè by early investigators is that PA added to fibroblasts is mitogenic (Moolenaar *et al.*, 1986; Yu *et al.*, 1988; Imagawa *et al.*, 1989; Knauss *et al.*, 1990; van Corven *et al.*, 1992; Fukami and Takenawa, 1992). Although lyso-PA is now known to be mitogenic also (van Corven *et al.*, 1989, 1992, 1993; Fukami and Takenawa, 1992; Tigyi *et al.*, 1994), there is evidence that PA exerts its action independently of lyso-PA in some cases (van Corven *et al.*, 1992; Fukami and Takenawa, 1992). The mechanism(s) by which PA and lyso-PA exert their stimulatory effect on DNA synthesis is undefined. As described above, both lipids elicit activation of PI-PLC, PLA_2, and PC-PLD, and thus produce many of the signals thought to be involved in the mitogenic action of other agonists. In common with these agonists, lyso-PA stimulates mitogen-activated protein kinase (MAP kinase) through a mechanism involving activation of Ras (Howe and Marshall, 1993; Hordijk *et al.*, 1994). MAP kinase(s) can phosphorylate certain transcription factors and is thus thought to be important in the control of transcription.

Lyso-PA also stimulates protein tyrosine phosphorylation in fibroblasts (Hordijk *et al.*, 1994; Seufferlein and Rozengurt, 1994). The proteins that are phosphorylated, besides MAP kinases, are the focal adhesion kinase p125[FAK], the focal adhesion protein paxillin, and several other unidentified proteins (Kumagai *et al.*, 1993b; Seufferlein and Rozengurt, 1994; Hordijk *et al.*, 1994). The tyrosine phosphorylation of p125[FAK], paxillin, and the other proteins is blocked by cytochalasin D and PDGF, both of which disrupt the network of actin filaments (Seufferlein and Rozengurt, 1994), implying involvement of the cytoskeleton in the phosphorylation. Further to this point, Kumagai *et al.* (1993a) have reported that inactivation of the SMG Rho by ADP-ribosylation induced by the C3 exotoxin of *C. botulinum* attenuates the protein tyrosine phosphorylation induced by lyso-PA. The involvement of Rho in the action of lyso-PA to induce focal adhesion and stress fiber formation (actin polymerization) in fibroblasts was earlier reported by Ridley and Hall (1992). A role for PC-PLD and PA formation in the effects of thrombin and lyso-PA on actin polymerization in fibroblasts has been proposed by

Ha and Exton (1993b) and Ha *et al.* (1994). However, the relationship between PLD activation and the actions of Rho and p125[FAK] in the assembly of focal adhesions and actin stress fibers is presently unclear.

The physiological role of lyso-PA is discussed at length in another chapter in this monograph. However, as alluded to above, many of the reported effects of PA in cells are probably largely attributable to the presence of lyso-PA in the PA preparations used or possibly the formation of lyso-PA by secreted PLA_2. This makes it difficult to define the physiological actions of PA *per se* from such experiments. Another potential complication is the presence of a PA phospho-hydrolase which converts exogenous PA to DAG (Perry *et al.*, 1993) and the possible presence of a membrane receptor for PA which could lead to the generation of intracellular signals, as in the case of lyso-PA. A further problem could arise from the rapid intracellular metabolism of any exogenous PA that crossed the membrane bilayer, i.e., it may be difficult to achieve a physiologically significant increase in PA in the inner leaflet or to dissociate the effects of PA *per se* from that of its metabolites.

For the above reasons, the most productive approach to determining the physiological role of PA may be to define the intracellular proteins that bind PA in a selective manner and/or whose activity is selectively changed by PA. The functions of such proteins, if known, should relate to the physiological actions of agonists that activate PLD.

References

Agwu, D., McPhail, L. C., Chabot, M. C., Daniel, L. W., Wykle, R. L., and McCall, C. E., 1989a, Choline-linked phosphoglycerides. A source of phosphatidic acid and diglycerides in stimulated neutrophils, *J. Biol. Chem.* **264**:1405.

Agwu, D. E., McPhail, L. C., Wykle, R. L., and McCall, C. E., 1989b, Mass determination of receptor-mediated accumulation of phosphatidate and diglycerides in human neutrophils measured by Coomassie blue staining and densitometry, *Biochem. Biophys. Res. Commun.* **159**:79.

Agwu, D. E., McPhail, L. C., Sozzani, S., Bass, D. A., and McCall, C. E., 1991, Phosphatidic acid as a second messenger in human polymorphonuclear leukocytes. Effects on activation of NADPH oxidase, *J. Clin. Invest.* **88**:531.

Ahmed, S., Kozma, R., Lee, J., Monfries, C., Harden, N., and Lim, L., 1991, The cysteine-rich domain of human proteins, neuronal chimaerin, proteinkinase C and diacylglycerol kinase binds zinc. Evidence for the involvement of a zinc-dependent structure in phorbol ester binding, *Biochem. J.* **280**:233.

Ahmed, S., Lee, J., Kozma, R., Best, A., Monfries, C., and Lim, L., 1993, A novel functional target for tumor-promoting phorbol esters and lysophosphatidic acid. The p21*rac*-GTPase activating protein *n*-chimaerin, *J. Biol. Chem.* **268**:10709.

Ahmed, A., Plevin, R., Shoaibi, M. A., Fountain, S. A., Ferriani, R. A., and Smith, S. K., 1994, Basic FGF activates phospholipase D in endothelial cells in the absence of inositol-lipid hydrolysis, *Am. J. Physiol.* **266**:C206.

Allan, C., and Exton, J. H., 1993, Quantification of inositol phospholipid breakdown in isolated rat hepatocytes, *Biochem. J.* **290**:865.

Anthes, J. C., Wang, P., Siegel, M. I., Egan, R. W., and Billah, M. M., 1991, Granulocyte phospholipase D is activated by a guanine nucleotide dependent protein factor, *Biochem. Biophys. Res. Commun.* **175**:236.

Asaoka, Y., Oka, M., Yoshida, K., and Nishizuka, Y., 1991, Metabolic rate of membrane-permeant diacylglycerol and its relation to human resting T-lymphocyte activation, *Proc. Natl. Acad. Sci. USA* **88:**8681.

Augert, G., Blackmore, P. F., and Exton, J. H., 1989a, Changes in the concentration and fatty acid composition of phosphoinositides induced by hormones in hepatocytes, *J. Biol. Chem.* **264:**2574.

Augert, G., Bocckino, S. B., Blackmore, P. F., and Exton, J. H., 1989b, Hormonal stimulation of diacylglycerol formation in hepatocytes. Evidence for phosphatidylcholine breakdown, *J. Biol. Chem.* **264:**21689.

Balboa, M. A., Balsinde, J., Dennis, E. A., and Insel, P. A., 1995, A phospholipase D-mediated pathway for generating diacylglycerol in nuclei from Madin-Darby canine kidney cells, *J. Biol. Chem.* **270:**11738.

Balboa, M. A., Firestein, B. L., Godson, C., Bell, K. S., and Insel, P. A., 1994, Protein kinase C_α mediates phospholipase D activation by nucleotides and phorbol ester in Madin-Darby canine kidney cells. Stimulation of phospholipase D is independent of activation of polyphosphoinositide-specific phospholipase C and phospholipase A_2, *J. Biol. Chem.* **269:**10511.

Balsinde, J., and Mollinedo, F., 1990, Phosphatidylinositol hydrolysis by human plasma phospholipase D, *FEBS Lett.* **259:**237.

Balsinde, J., Diez, E., and Mollinedo, F., 1988, Phosphatidylinositol-specific phospholipase D: A pathway for generation of a second messenger, *Biochem. Biophys. Res. Commun.* **154:**502.

Banschbach, M. W., Geison, R. L., and Hokin-Neaverson, M., 1981, Effects of cholinergic stimulation of levels and fatty acid composition of diacylglycerols in mouse pancreas, *Biochim. Biophys. Acta* **663:**34.

Bauldry, S. A., Elsey, K. L., and Bass, D. A., 1992, Activation of NADPH oxidase and phospholipase D in permeabilized human neutrophils. Correlation between oxidase activation and phosphatidic acid production, *J. Biol. Chem.* **267:**25141.

Bellavite, P., Corso, F., Dusi, S., Grzeskowiak, M., Della-Bianca, V., and Rossi, F., 1989, Activation of NADPH-dependent superoxide production in plasma membrane extracts of pig neutrophils by phosphatidic acid, *J. Biol. Chem.* **263:**8210.

Ben-Av, P., Eli, Y., Schmidt, U.-S., Tobias, K. E., and Liscovitch, M., 1993, Distinct mechanisms of phospholipase D activation and attenuation utilized by different mitogens in NIH-3T3 fibroblasts, *Eur. J. Biochem.* **215:**455.

Berridge, M. J., 1984, Inositol trisphosphate and diacylglycerol as second messengers, *Biochem. J.* **220:**345.

Besterman, J. M., Duronio, V., and Cuatrecasas, P., 1986a, Rapid formation of diacylglycerol from phosphatidylcholine: A pathway for generation of a second messenger, *Proc. Natl. Acad. Sci. USA* **83:**6785.

Besterman, J. M., Pollenz, R. S., Booker, E. L., Jr., and Cuatrecasas, P., 1986b, Diacylglycerol-induced translocation of diacylglycerol kinase: Use of affinity-purified enzyme in a reconstitution system, *Proc. Natl. Acad. Sci. USA* **83:**9378.

Billah, M. M., and Anthes, J. C., 1990, The regulation and cellular functions of phosphatidylcholine hydrolysis, *Biochem. J.* **269:**281.

Billah, M. M., Lapetina, E., and Cuatrecasas, P., 1981, Phospholipase A_2 activity specific for phosphatidic acid. A possible mechanism for the production of arachidonic acid in platelets, *J. Biol. Chem.* **256:**5399.

Billah, M. M., Eckel, S., Mullmann, T. J., Egan, R. W., and Siegel, M. I., 1989a, Phosphatidylcholine hydrolysis by phospholipase D determines phosphatidate and diglyceride levels in chemotactic peptide stimulated human neutrophils. Involvement of phosphatidate phosphohydrolase in signal transduction, *J. Biol. Chem.* **264:**17069.

Billah, M. M., Pai, J.-K., Mullmann, T. J., Egan, R. W., and Siegel, M. I., 1989b, Regulation of phospholipase D in HL-60 granulocytes. Activation by phorbol esters, diglyceride, and calcium ionophore via protein kinase C-independent mechanisms, *J. Biol. Chem.* **264:**9069.

Bird, G. St. J., Rossier, M. F., Obie, J. F., and Putney, J. W., Jr., 1993, Sinusoidal oscillations in intracellular calcium requiring negative feedback by protein kinase C, *J. Biol. Chem.* **268:**8425.

Bishop, W. R., Ganong, B. R., and Bell, R. M., 1986, Attenuation of *sn*-1,2-diacyglycerol second messengers by diacylglycerol kinase. Inhibition by diacylglycerol analogs *in vitro* and in human platelets, *J. Biol. Chem.* **261**:6993.

Bocckino, S. B., and Exton, J. H., 1992, Phosphatidylcholine metabolism in signal transduction, in: *Cellular and Molecular Mechanisms of Inflammation*, Vol. 3, Part A (C. C. Cochrane and M. A. Gimbrone, Jr., eds.), pp. 89–114, Academic Press, San Diego.

Bocckino, S. B., Blackmore, P. F., and Exton, J. H., 1985, Stimulation of 1,2-diacylglycerol accumulation in hepatocytes by vasopressin, epinephrine and angiotensin II, *J. Biol. Chem.* **260**:14201.

Bocckino, S. B., Blackmore, P. F., Wilson, P. B., and Exton, J. H., 1987a, Phosphatidate accumulation in hormone-treated hepatocytes via a phospholipase D mechanism, *J. Biol. Chem.* **262**:15309.

Bocckino, S. B., Wilson, P. B., and Exton, J. H., 1987b, Ca^{2+}-mobilizing hormones elicit phosphatidylethanol formation via phospholipase D activation, *FEBS Lett.* **255**:201.

Bocckino, S. B., Wilson, P. B., and Exton, J. H., 1991, Phosphatidate-dependent protein phosphorylation, *Proc. Natl. Acad. Sci. USA* **88**:6210.

Bonnel, S., Lin, Y.-P., Kelley, M. J., Carman, G. M., and Eichberg, J., 1989, Interactions of thiophosphatidic acid with enzymes which metabolize phosphatidic acid. Inhibition of phosphatidic acid phosphatase and utilization by CDP-diacylglycerol synthase, *Biochim. Biophys. Acta* **1005**:289.

Bourgoin, S., and Grinstein, S., 1992, Peroxides of vanadate induce activation of phospholipase D in HL-60 cells. Role of tyrosine phosphorylation, *J. Biol. Chem.* **267**:11908.

Bourgoin, S., Harbour, D., Desmarais, Y., Takai, Y., and Beaulieu, A., 1995, Low molecular weight GTP-binding proteins in HL-60 granulocytes. Assessment of the role of ARF and of a 50-kDa cytosolic protein in phospholipase D activation, *J. Biol. Chem.* **270**:3172.

Bowman, E. P., Uhlinger, D. J., and Lambeth, J. D., 1993, Neutrophil phospholipase D is activated by a membrane-associated Rho family small molecular weight GTP-binding protein, *J. Biol. Chem.* **268**:21509.

Brindley, D. N., 1988, General introduction and phosphatidate phosphohydrolase in the liver, in: *Phosphatidate Phosphohydrolase*, CRC Series in Enzyme Biology (D. N. Brindley, ed.), Vol. 1, pp. 1–77, CRC Press, Boca Raton, FL.

Brindley, D. N., and Bowley, M., 1975, Drugs affecting the synthesis of glycerides and phospholipids in rat liver. The effects of clofibrate, halofenate, fenfluramine, amphetamine, cinchocaine, chlorpromazine, demethylimipramine, mepyramine and some of their derivatives, *Biochem. J.* **148**:461.

Broekman, M. J., Ward, J. W., and Marcus, A. J., 1981, Fatty acid composition of phosphatidylinositol and phosphatidic acid in stimulated platelets. Persistence of arachidonyl-stearyl structure, *J. Biol. Chem.* **256**:8271.

Brown, H. A., Gutowski, S., Kahn, R. A., and Sternweis, P. C., 1995, Partial purification and characterization of Arf-sensitive phospholipase D from porcine brain, *J. Biol. Chem.* **270**:14935.

Brown, H. A., Gutowski, S., Moomaw, C. R., Slaughter, C., and Sternweis, P. C., 1993, ADP-ribosylation factor, a small GTP-dependent regulatory protein, stimulates phospholipase D activity, *Cell* **75**:1137.

Bursten, S. L., Harris, W. E., Bomsztyk, K., and Lovett, D., 1991, Interleukin-1 rapidly stimulates lysophosphatidate acyltransferase and phosphatidate phosphohydrolase activities in human mesangial cells, *J. Biol. Chem.* **266**:20732.

Butterwith, S. C., Hopewell, R., and Brindley, D. N., 1984, Partial purification and characterization of a soluble phosphatidate phosphohydrolase of rat liver, *Biochem. J.* **220**:825.

Carnero, A., Cuadrado, A., del Peso, L., and Lacal, J. C., 1994a, Activation of type D phospholipase by serum stimulation and *ras*-induced transformation in NIH3T3 cells, *Oncogene* **9**:1387.

Carnero, A., Dolfi, F., and Lacal, J. C., 1994b, *ras*-p21 activates phospholipase D and A2, but not phospholipase C or PKC, in *Xenopus laevis* oocytes, *J. Cell. Biochem.* **54**:478.

Chabot, M. C., McPhail, L. C., Wykle, R. L., Kennerly, D. A., and McCall, C. E., 1992, Comparison of diglyceride production from choline-containing phosphoglycerides in human neutrophils stimulated with *N*-formylmethionyl-leucylphenylalanine, ionophore A23187 or phorbol 12-myristate 13-acetate, *Biochem. J.* **286**:693.

Chalifour, R. J., and Kanfer, J. N., 1980, Microsomal phospholipase D of rat brain and lung tissues, *Biochem. Biophys. Res. Commun.* **96:**742.

Chalifour, R., and Kanfer, J. N., 1982, Fatty acid activation and temperature perturbation of rat brain microsomal phospholipase D, *J. Neurochem.* **39:**299.

Choudhury, G. G., Sylvia, V. L., and Sakaguchi, A. Y., 1991, Activation of a phosphatidylcholine-specific phospholipase C by colony stimulating factor 1 receptor requires tyrosine phosphorylation and a guanine nucleotide-binding protein, *J. Biol. Chem.* **266:**23147.

Chuang, M., Lee, M. W., Zhao, D., and Severson, D. L., 1993, Metabolism of a long-chain diacylglycerol by permeabilized A10 smooth muscle cells, *Am. J. Physiol.* **265:**C927.

Cockcroft, S., and Allan, D., 1984, The fatty acid composition of phosphatidylinositol, phosphatidate and 1,2-diacylglycerol in stimulated human neutrophils, *Biochem. J.* **222:**557.

Cockcroft, S., Thomas, G. M. H., Fensome, A., Geny, B., Cunningham, E., Gout, I., Hiles, I., Totty, N. F., Truong, O., and Hsuan, J. J., 1994, Phospholipase D: A downstream effector of ARF in granulocytes, *Science* **263:**523.

Conricode, K. M., Brewer, K. A., and Exton, J. H., 1992, Activation of phospholipase D by protein kinase C. Evidence for a phosphorylation-independent mechanism, *J. Biol. Chem.* **267:**7199.

Conricode, K. M., Smith, J. L., Burns, D. J., and Exton, J. H., 1994, Phospholipase D activation in fibroblast membranes by the α and β isoforms of protein kinase C, *FEBS Lett.* **342:**149.

Cook, S. J., and Wakelam, M. J. O., 1991a, Stimulated phosphatidylcholine hydrolysis as a signal transduction pathway in mitogenesis, *Cell. Signal.* **3:**273.

Cook, S. J., and Wakelam, M. J. O., 1991b, Hydrolysis of phosphatidylcholine by phospholipase D is a common response to mitogens which stimulate inositol lipid hydrolysis in Swiss 3T3 fibroblasts, *Biochim. Biophys. Acta* **1092:**265.

Cook, S. J., and Wakelam, M. J. O., 1992, Epidermal growth factor increases sn-1,2,-diacylglycerol levels and activates phospholipase D-catalysed phosphatidylcholine breakdown in Swiss 3T3 cells in the absence of inositol-lipid hydrolysis, *Biochem. J.* **285:**247.

Cook, S. J., Briscoe, C. P., and Wakelam, M. J. O., 1991, The regulation of phospholipase D activity and its role in sn-1,2-diradylglycerol formation in bombesin- and phorbol 12-myristate 13-acetate stimulated Swiss 3T3 cells, *Biochem. J.* **280:**431.

Cooper, J. A., and Kashishian, A., 1993, In vivo binding properties of SH2 domains from GTPase-activating protein and phosphatidylinositol 3-kinase, *Mol. Cell. Biol.* **13:**1737.

Coorssen, J. R., and Haslam, R. J., 1993, GTPγS and phorbol ester act synergistically to stimulate both Ca^{2+}-independent secretion and phospholipase D activity in permeabilized human platelets. Inhibition by BAPTA and analogues, *FEBS Lett.* **316:**170.

Cross, G. M. C., 1990, Glycolipid anchoring of plasma membrane proteins, *Annu. Rev. Cell Biol.* **6:**1.

Daniel, L. W., Huang, C., Strum, J. C., Smitherman, P. K., Greene, D., and Wykle, R. L., 1993, Phospholipase D hydrolysis of choline phosphoglycerides is selective for the alkyl-linked subclass of Madin-Darby canine kidney cells, *J. Biol. Chem.* **268:**21519.

Davitz, M. A., Hom, J., and Schenkman, S., 1989, Purification of a glycosyl-phosphatidylinositol-specific phospholipase D from human plasma, *J. Biol. Chem.* **264:**13760.

Dawson, R. M. C., 1967, The formation of phosphatidylglycerol and other phospholipids by the transferase activity of phospholipase D, *Biochem. J.* **102:**205.

Dawson, R. M. C., Irvine, R., Hirasawa, K., and Hemington, N. L., 1982, Hydrolysis of phosphatidylinositol by pancreas and pancreatic secretions, *Biochim. Biophys. Acta* **710:**212.

Day, C. P., and Yeaman, S. J., 1992, Physical evidence for the presence of two forms of phosphatidate phosphohydrolase in rat liver, *Biochim. Biophys. Acta* **1127:**87.

de Chaffoy de Courcelles, D., 1990, The use of diacylglycerol kinase inhibitors for elucidating the roles of protein kinase C, *Adv. Second Messenger Phosphoprotein Res.* **24:**491.

de Chaffoy de Courcelles, D., Roevens, P., and Van Belle, H., 1985, R 59 022, a diacylglycerol kinase inhibitor. Its effect on diacylglycerol and thrombin-induced C kinase activation in the intact platelet, *J. Biol. Chem.* **260:**15762.

de Chaffoy de Courcelles, D., Roevens, P., Van Belle, H., Kennis, L., Somers, Y., and De Clerck, F., 1989, The role of endogenously formed diacylglycerol in the propagation and termination of platelet activation. A biochemical and functional analysis using the novel diacylglycerol kinase inhibitor, R 59 949, *J. Biol. Chem.* **264:**3274.

Dekker, L. V., and Parker, P. J., 1994, Protein kinase C—a question of specificity, *Trends Biochem. Sci.* **19:**73.

Dinh, T. T., and Kennerly, D. A., 1991, Assessment of receptor-dependent activation of phosphatidylcholine hydrolysis by both phospholipase D and phospholipase C, *Cell Regul.* **2:**299.

Divecha, N., Lander, D. J., Scott, T. W., and Irvine, R. F., 1991, Molecular species analysis of 1,2-diacylglycerols and phosphatidic acid formed during bombesin stimulation of swiss 3T3 cells, *Biochim. Biophys. Acta* **1093:**184.

Dubyak, G. R., Schomisch, S. J., Kusner, D. J., and Xie, M., 1993, Phospholipase D activity in phagocytic leucocytes is synergistically regulated by G-protein- and tyrosine kinase-based mechanisms, *Biochem. J.* **292:**121.

Edwards, Y. S., and Murray, A. W., 1995, Accumulation of phosphatidylalcohol in cultured cells: use of subcellular fractionation to investigate phospholipase D activity during signal transduction, *Biochem. J.* **308:**473.

Eldar, H., Ben-Av, P., Schmidt, U.-S., Livneh, E., and Liscovitch, M., 1993, Up-regulation of phospholipase D activity induced by overexpression of protein kinase C-α. Studies in intact Swiss/3T3 cells and in detergent-solubilized membranes *in vitro*, *J. Biol. Chem.* **268:**12560.

Exton, J. H., 1990, Signaling through phosphatidylcholine breakdown, *J. Biol. Chem.* **265:**1.

Exton, J. H., 1993, Phosphoinositide phospholipases and G-proteins in hormone action, *Ann. Rev. Physiol.* **56:**349.

Exton, J. H., 1994, Phosphatidylcholine breakdown and signal transduction, *Biochim. Biophys. Acta* **1212:**26.

Fantl, W. J., Escobedo, J. A., Martin, G. A., Turck, C. W., del Rosario, M., McCormick, F., and Williams, L. T., 1992, Distinct phosphotyrosines on a growth factor receptor bind to specific molecules that mediate different signaling pathways, *Cell* **69:**413.

Fleming, I. N., and Yeaman, S. J., 1995, Subcellular distribution of N-ethylmaleimide-sensitive and -insensitive phosphatidic acid phosphohydrolase in rat brain, *Biochim. Biophys. Acta* **1254:**161.

Florin-Christensen, J., Florin-Christensen, M., Delfino, J. M., and Rasmussen, H., 1993, New patterns of diacylglycerol metabolism in intact cells, *Biochem. J.* **289:**783.

Freeman, E. J., Chisolm, G. M., and Tallant, E. A., 1995, Role of calcium and protein kinase C in the activation of phospholipase D by angiotensin II in vascular smooth muscle cells, *Arch. Biochem. Biophys.* **319:**84.

Fukami, K., and Takenawa, T., 1992, Phosphatidic acid that accumulates in platelet-derived growth factor-stimulated Balb/c 3T3 cells is a potential mitogenic signal, *J. Biol. Chem.* **267:**10988.

Gelas, P., Ribbes, G., Record, M., Terce, F., and Chap, H., 1989, Differential activation by fMet-Leu-Phe and phorbol ester of a plasma membrane phosphatidylcholine-specific phospholipase D in human neutrophil, *FEBS Lett.* **251:**213.

Geny, B., and Cockcroft, S., 1992, Synergistic activation of phospholipase D by protein kinase C- and G-protein-mediated pathways in streptolysin O-permeabilized HL60 cells, *Biochem. J.* **284:**531.

Geny, B., Fensome, A., and Cockcroft, S., 1993, Rat brain cytosol contains a factor which reconstitutes guanine-nucleotide-binding-protein-regulated phospholipase-D activation in HL60 cells previously permeabilized with streptolysin O, *Eur. J. Biochem.* **215:**389.

Gomez-Muñoz, A., Hamza, E. H., and Brindley, D. N., 1992a, Effects of sphingosine, albumin and unsaturated fatty acids on the activation and translocation of phosphatidate phosphohydrolases in rat hepatocytes, *Biochim. Biophys. Acta* **1127:**49.

Gomez-Muñoz, A., Hatch, G. M., Martin, A., Jamal, Z., Vance, D. E., and Brindley, D. N., 1992b, Effects of okadaic acid on the activities of two distinct phosphatidate phosphohydrolases in rat hepatocytes, *FEBS Lett.* **301:**103.

Goto, K., and Kondo, H., 1993, Molecular cloning and expression of a 90 kDa diacylglycerol that predominantly localizes in neurons, *Proc. Natl. Acad. Sci. USA* **90:**7598.

Goto, K., Watanabe, M., Kondo, H., Yuasa, H., Sakane, F., and Kanoh, H., 1992, Gene cloning, sequence, expression and in situ localization of 80 kDa diacylglycerol kinase specific to oligodendrocyte of rat brain, *Brain Res. Mol. Brain Res.* **16:**75.

Gratas, C., and Powis, G., 1993, Inhibition of phospholipase D by agents that inhibit cell growth, *Anticancer Res.* **13:**1239.

Guillemain, I., and Rossignol, B., 1992, Evidence for receptor-linked activation of phospholipase D in rat parotid glands, *FEBS Lett.* **313:**489.

Gustavsson, L., Moehren, G., Torres-Marquez, M. E., Benistant, C., Rubin, R., and Hoek, J. B., 1994, The role of cytosolic Ca^{2+}, protein kinase C, and protein kinase A in hormonal stimulation of phospholipase D in rat hepatocytes, *J. Biol. Chem.* **269:**849.

Ha, K.-S., and Exton, J. H., 1993a, Differential translocation of protein kinase C isozymes by thrombin and platelet-derived growth factor. A possible function for phosphatidylcholine-derived diacylglycerol, *J. Biol. Chem.* **268:**10534.

Ha, K.-S., and Exton, J. H., 1993b, Activation of actin polymerization by phosphatidic acid derived from phosphatidylcholine in IIC9 fibroblasts, *J. Cell Biol.* **123:**1789.

Ha, K.-S., Yeo, E.-J., and Exton, J. H., 1994, Lysophosphatidic acid activation of phosphatidylcholine-hydrolysing phospholipase D and actin polymerization by a pertussis toxin-sensitive mechanism, *Biochem. J.* **303:**55.

Halenda, S. P., and Rehm, A. G., 1990, Evidence for the calcium-dependent activation of phospholipase D in thrombin-stimulated human erythroleukaemia cells, *Biochem. J.* **267:**479.

Hanahan, D. J., 1986, Platelet-activating factor: A biologically active phosphoglyceride, *Annu. Rev. Biochem.* **55:**483.

Hannun, Y. A., Loomis, C. R., Merrill, A. H., and Bell, R. M., 1986, Sphingosine inhibition of protein kinase C activity and of phorbol dibutyrate binding in vitro and in human platelets, *J. Biol. Chem.* **261:**12604.

Hattori, H., and Kanfer, J., 1984, Synaptosomal phospholipase D: Potential role in providing choline for acetylcholine synthesis, *Biochem. Biophys. Res. Commun.* **124:**945.

Heller, M., Bieri, S., and Brodbeck, U., 1992, A novel form of glycosylphosphatidylinositol-anchor converting activity with a specificity of a phospholipase D in mammalian liver membranes, *Biochim. Biophys. Acta* **1109:**109.

Holbrook, P. G., Pannell, L. K., Murata, Y., and Daly, J. W., 1992, Molecular species analysis of a product of phospholipase D activation. Phosphatidylethanol is formed from phosphatidylcholine in phorbol ester- and bradykinin-stimulated PC12 cells, *J. Biol. Chem.* **267:**16834.

Holmes, R. P., and Yoss, N. L., 1983, Failure of phosphatidic acid to translocate Ca^{2+} across phosphatidylcholine membranes, *Nature* **305:**637.

Hordijk, P. L., Verlaan, I., van Corven, E. J., and Moolenaar, W. H., 1994, Protein tyrosine phosphorylation induced by lysophosphatidic acid in Rat-1 fibroblasts. Evidence that phosphorylation of MAP kinase is mediated by the G_i-$p21^{ras}$ pathway, *J. Biol. Chem.* **269:**645.

Horwitz, J., 1990, Carbachol and bradykinin increases the production of diacylglycerol from sources other than inositol-containing phospholipids in PC12 cells, *J. Neurochem.* **54:**983.

Horwitz, J., and Davis, L. L., 1993, The substrate specificity of brain microsomal phospholipase D, *Biochem. J.* **295:**793.

Howe, L. R., and Marshall, C. J., 1993, Lysophosphatidic acid stimulates mitogen-activated protein kinase activation via a G-protein-coupled pathway requiring $p21^{ras}$ and $p74^{raf-1}$, *J. Biol. Chem.* **268:**20717.

Huang, C., and Cabot, M. C., 1990a, Vasopressin-induced polyphosphoinositide and phosphatidylcholine degradation in fibroblasts. Temporal relationship for formation of phospholipase C and phospholipase D hydrolysis products, *J. Biol. Chem.* **265:**17468.

Huang, C., and Cabot, M. C., 1990b, Phorbol esters stimulate the accumulation of phosphatidate, phosphatidylethanol and diacylglycerol in three cell types, *J. Biol. Chem.* **265:**14858.

Huang, C., Wykle, R. L., Daniel, L. W., and Cabot, M. C., 1992, Identification of phosphatidylcholine-selective and phosphatidylinositol-selective phospholipases D in Madin-Darby canine kidney cells, *J. Biol. Chem.* **267:**16859.

Huang, K.-S., Li, S., Fung, W.-J., Holmes, J. D., Reik, L., Pan, Y.-C. E., and Low, M. G., 1990, Purification and characterization of glycosyl-phosphatidylinositol-specific phospholipase D, *J. Biol. Chem.* **265:**17738.

Huang, K.-S., Li, S., and Low, M. G., 1991, Glycosylphosphatidylinositol-specific phospholipase D, *Methods Enzymol.* **197:**567.

Huang, R., Kucera, G. L., and Rittenhouse, S. E., 1991, Elevated cytosolic Ca^{2+} activates phospholipase D in human platelets, *J. Biol. Chem.* **266:**1652.

Imagawa, W., Bandyopadhyay, G. K., Wallace, D., and Nandi, S., 1989, Phospholipids containing polyunsaturated fatty acyl groups are mitogenic for normal mouse mammary epithelial cells in serum-free primary cell culture, *Proc. Natl. Acad. Sci. USA* **86:**4122.

Inamori, K., Sagawa, N., Kasegawa, M., Itoh, H., Ueda, H., Kobayashi, F., Ihara, Y., and Mori, T., 1993, Identification and partial characterization of phospholipase D in the human amniotic membrane, *Biochem. Biophys. Res. Commun.* **191:**1270.

Jackowski, S., and Rock, C. O., 1989, Stimulation of phosphatidylinositol 4,5-bisphosphate phospholipase C activity by phosphatidic acid, *Arch. Biochem. Biophys.* **268:**516.

Jalink, K., van Corven, E. J., and Moolenaar, W. H., 1990, Lysophosphatidic acid, but not phosphatidic acid, is a potent Ca^{2+}-mobilizing stimulus for fibroblasts. Evidence for an extracellular site of action, *J. Biol. Chem.* **265:**12232.

Jamal, Z., Martin, A., Gomez-Muñoz, A., and Brindley, D. N., 1991, Plasma membrane fractions from rat liver contain a phosphatidate phosphohydrolase distinct from that in the endoplasmic reticulum and cytosol, *J. Biol. Chem.* **266:**2988.

Jeng, I. M., Klemm, N., and Wu, C. Z., 1988, Inhibition of rat diacylglycerol kinase activity by synthetic L-lysine containing polypeptides, *Biochem. Int.* **16:**853.

Jiang, H., Alexandropoulos, K., Song, J., and Foster, D. A., 1994, Evidence that v-Src-induced phospholipase D activity is mediated by a G protein, *Mol. Cell. Biol.* **14:**3676.

Jones, G. A., and Carpenter, G., 1993, The regulation of phospholipase C-γ1 by phosphatidic acid. Assessment of kinetic parameters, *J. Biol. Chem.* **268:**20845.

Jones, L. G., Ella, K. M., Bradshaw, C. D., Gause, K. C., Dey, M., Wisehart-Johnson, A. E., Spivey, E. C., and Meier, K. E., 1994, Activation of mitogen activated protein kinases and phospholipase D in A7r5 vascular smooth muscle cells, *J. Biol. Chem.* **269:**23790.

Kai, M., Sakane, F., Imai, S.-I., Wada, I., ad Kanoh, H., 1994, Molecular cloning of a diacylglycerol kinase isozyme predominantly expressed in human retina with a truncated and inactive enzyme expression in most other human cells, *J. Biol. Chem.* **269:**18492.

Kanfer, J. N., and McCartney, D., 1994, Phospholipase D activity of isolated rat brain plasma membranes, *FEBS Lett.* **337:**251.

Kanoh, H., Yamada, K., Sakane, F., and Imaizumi, T., 1989, Phosphorylation of diacylglycerol kinase *in vitro* by protein kinase C, *Biochem. J.* **258:**455.

Kanoh, H., Imai, S., Yamada, K., and Sakane, F., 1992a, Purification and properties of phosphatidic acid phosphatase from porcine thymus membranes, *J. Biol. Chem.* **267:**25309.

Kanoh, H., Kanaho, Y., and Nozawa, Y., 1992b, Pertussis toxin-insensitive G protein mediates carbachol activation of phospholipase D in rat pheochromocytoma PC12 cells, *J. Neurochem.* **59:**1786.

Kanoh, H., Sakane, F., Imai, S.-I., and Wada, I., 1993, Diacylglycerol kinase and phosphatidic acid phosphatase—enzymes metabolizing lipid second messengers, *Cell. Signal.* **5:**494.

Kashishian, A., Kazlauskas, A., and Cooper, J. A., 1992, Phosphorylation sites in the PDGF receptor with different specificities for binding GAP and P13 kinase *in vivo*, *EMBO J.* **11:**1373.

Kaszkin, M., Seidler, L., Kast, R., and Kinzel, V., 1992, Epidermal-growth-factor-induced production of phosphatidylalcohols by HeLa cells and A431 cells through activation of phospholipase D, *Biochem. J.* **287:**51.

Kater, L. A., Goetzl, E. J., and Austen, K. F., 1976, Isolation of human eosinophil phospholipase D, *J. Clin. Invest.* **57:**1173.

Kazlauskas, A., Kashishian, A., Cooper, J. A., and Valius, M., 1992, GTPase-activating protein and phosphatidylinositol 3-kinase bind to distinct regions of the platelet-derived growth factor receptor β subunit, *Mol. Cell. Biol.* **12:**2534.

Kennerly, D. A., 1987, Diacylglycerol metabolism in mast cells. Analysis of lipid metabolic pathways using molecular species analysis of intermediates, *J. Biol. Chem.* **262:**16305.

Kennerly, D. A., 1990, Phosphatidylcholine is a quantitatively more important source of increased 1,2-diacylglycerol than is phosphatidylinositol in mast cells, *J. Immunol.* **144:**3912.

Kessels, G. C. R., Gervaix, A., Lew, P. D., and Verhoeven, A. J., 1991a, The chymotrypsin inhibitor carbobenzyloxy-leucine-tyrosine-chloromethylketone interferes with phospholipase D activation induced by formyl-methionyl-leucyl-phenylalanine in human neutrophils, *J. Biol. Chem.* **266**:15870.

Kessels, G. C. R., Roos, D., and Verhoeven, A. J., 1991b, fMet-leu-phe-induced activation of phospholipase D in human neutrophils. Dependence on changes in cytosolic free Ca^{2+} concentration and relation with respiratory burst activation, *J. Biol. Chem.* **265**:23152.

Kester, M., Simonson, M. S., McDermott, R. G., Baldi, E., and Dunn, M. J., 1992, Endothelin stimulates phosphatidic acid formation in cultured rat mesangial cells: Role of a protein kinase C-regulated phospholipase D, *J. Cell. Physiol.* **150**:578.

Khan, W. A., Blobe, G. C., Richards, A. L., and Hannun, Y. A., 1994, Identification, partial purification, and characterization of a novel phospholipid-dependent and fatty acid-activated protein kinase from human platelets, *J. Biol. Chem.* **269**:9729.

Kiss, Z., and Anderson, W. B., 1989a, Phorbol ester stimulates the hydrolysis of phosphatidylethanolamine in leukemic HL-60, NIH 3T3, and baby hamsters kidney cells, *J. Biol. Chem.* **264**:1483.

Kiss, Z., and Anderson, W. B., 1989b, Alcohols selectively stimulate phospholipase D-mediated hydrolysis of phosphatidylethanolamine in NIH 3T3 cells, *FEBS Lett.* **257**:45.

Kiss, Z., and Anderson, W. B., 1990, ATP stimulates the hydrolysis of phosphatidylethanolamine in NIH 3T3 cells, *J. Biol. Chem.* **265**:7345.

Knauss, T. C., Jaffer, F. E., and Abboud, H. E., 1990, Phosphatidic acid modulates DNA synthesis, phospholipase C, and platelet-derived growth factor mRNAs in cultured mesangial cells. Role of protein kinase C, *J. Biol. Chem.* **265**:14457.

Kobayashi, M., and Kanfer, J. N., 1987, Phosphatidylethanol formation via transphosphatidylation by rat brain synaptosomal phospholipase D, *J. Neurochem.* **48**:1597.

Kobayashi, M., and Kanfer, J. N., 1991, Solubilization and purification of rat tissue phospholipase D, *Methods Enzymol.* **197**:575.

Kondo, T., Konishi, F., Inui, H., and Inagami, T., 1992, Diacylglycerol formation from phosphatidylcholine in angiotensin II-stimulated vascular smooth muscle cells, *Biochem. Biophys. Res. Commun.* **187**:1460.

Koul, O., and Hauser, G., 1987, Modulation of rat brain cytosolic phosphatidate phosphohydrolase: Effect of cationic amphiphilic drugs and divalent cations, *Arch. Biochem. Biophys.* **253**:453.

Ktistakis, N. T., Brown, H. A., Sternweis, P. C., and Roth, M. G., 1995, Phospholipase D is present on Golgi-enriched membranes and its activation by ADP-ribosylation factor is sensitive to brefeldinA, *Proc. Natl. Acad. Sci. USA* **92**:4952.

Kumada, T., Miyata, H., and Nozawa, Y., 1993, Involvement of tyrosine phosphorylation in IgE receptor-mediated phospholipase D activation in rat basophilic leukemia (RBL-2H3) cells, *Biochem. Biophys. Res. Commun.* **191**:1363.

Kumagai, N., Morii, N., Fujisawa, K., Nemoto, Y., and Narumiya, S., 1993a, ADP-ribosylation of rho p^{21} inhibits lysophosphatidic acid-induced protein tyrosine phosphorylation and phosphatidylinositol 3-kinase activation in cultured Swiss 3T3 cells, *J. Biol. Chem.* **268**:24535.

Kumagai, N., Morii, N., Fujisawa, K., Yoshimasa, T., Nakao, K., and Narumiya, S., 1993b, Lysophosphatidic acid induces tyrosine phosphorylation and activation of MAP-kinase and focal adhesion kinase in cultured Swiss 3T3 cells, *FEBS Lett.* **329**:273.

Kusner, D. J., Schomisch, S. J., and Dubyak, G. R., 1993, ATP-induced potentiation of G-protein-dependent phospholipase D activity in a cell-free system from U937 promonocytic leukocytes, *J. Biol. Chem.* **268**:19973.

Lassègue, B., Alexander, R. W., Clark, M., and Griendling, K. K., 1991, Angiotensin II-induced phosphatidylcholine hydrolysis in cultured vascular smooth-muscle cells, *Biochem. J.* **276**:19.

Lavie, Y., and Liscovitch, M., 1990, Activation of phospholipase D by sphingoid bases in NG108-15 neural-derived cells, *J. Biol. Chem.* **265**:3868.

Lavie, Y., Piterman, O., and Liscovitch, M., 1990, Inhibition of phosphatidic acid phosphohydrolase activity by sphingosine. Dual action of sphingosine in diacylglycerol signal termination, *FEBS Lett.* **277**:7.

Lee, C., Fisher, S. K., Agranoff, B. W., and Hajra, A. K., 1991, Quantitative analysis of molecular species of diacylglycerol and phosphatidate formed upon muscarinic receptor activation of human SK-N-SH neuroblastoma cells, *J. Biol. Chem.* **266:**22837.

Lee, Y. H., Kim, H. S., Pai, J.-K., Ryu, S. H., and Suh, P.-G., 1994, Activation of phospholipase D induced by platelet-derived growth factor is dependent upon the level of phospholipase C-γ1, *J. Biol. Chem.* **269:**26842.

Lin, P., and Gilfillan, A. M., 1992, The role of calcium and protein kinase C in the IgE-dependent activation of phosphatidylcholine-specific phospholipase D in a rat mast (RBL 2H3) cell line, *Eur. J. Biochem.* **207:**163.

Lin, P., Fung, W.-J. C., and Gilfillan, A. M., 1992, Phosphatidylcholine-specific phospholipase D-derived 1,2-diacylglycerol does not initiate protein kinase C activation in the RBL 2H3 mast-cell line, *Biochem. J.* **287:**325.

Lin, P., Fung, W.-J. C., Li, S., Chen, T., Repetto, B., Huang, K.-S., and Gilfillan, A. M., 1994, Temporal regulation of the IgE-dependent 1,2-diacylglycerol production by tyrosine kinase activation in a rat (RBL 2H3) mast-cell line, *Biochem. J.* **299:**109.

Liscovitch, M., and Amsterdam, A., 1989, Gonadotropin-releasing hormone activates phospholipase D in ovarian granulosa cells. Possible role in signal transduction, *J. Biol. Chem.* **264:**11763.

Liscovitch, M., and Eli, Y., 1991, Ca^{2+} inhibits guanine nucleotide-activated phospholipase D in neural-derived NG108-15 cells, *Cell Regul.* **2:**1011.

Liscovitch, M., Chalifa, V., Danin, M., and Eli, Y., 1991, Inhibition of neural phospholipase D by aminoglycoside antibiotics, *Biochem. J.* **279:**319.

Liscovitch, M., Chalifa, V., Pertile, P., Chen, C.-S., and Cantley, L. C., 1994, Novel function of phosphatidylinositol 4,5-bisphosphate as a cofactor for brain membrane phospholipase D, *J. Biol. Chem.* **269:**21403.

Llahi, S., and Fain, J. N., 1992, α_1-Adrenergic receptor-mediated activation of phospholipase D in rat cerebral cortex, *J. Biol. Chem.* **267:**3679.

Lopex, I., Burns, D. J., and Lambeth, J. D., 1995, Regulation of phospholipase D by protein kinase C in human neutrophils. Conventional isoforms of protein kinase C phosphorylate A phospholipase D-related component in the plasma membrane, *J. Biol. Chem.* **270:**19465.

Low, M. G., and Huang, K.-S., 1993, Phosphatidic acid, lysophosphatidic acid and lipid A are inhibitors of glycosylphosphatidylinositol-specific phospholipase D. Specific inhibition of a phospholipase by product analogues? *J. Biol. Chem.* **268:**8480.

MacDonald, M. L., Mack, K. F., Richardson, C. N., and Glomset, J. A., 1988, Regulation of diacylglycerol kinase reaction in Swiss 3T3 cells. Increased phosphorylation of endogenous diacylglycerol and decreased phosphorylation of didecanoylglycerol in response to platelet derived growth factor, *J. Biol. Chem.* **263:**1575.

McGhee, J. G., and Shoback, D. M., 1990, Effects of phosphatidic acid on parathyroid hormone release, intracellular free Ca^{2+}, and inositol phosphates in dispersed bovine parathyroid cells, *Endocrinology* **126:**899.

McKinnon, M., and Parker, P. J., 1994, Phospholipase-D activation can be negatively regulated through the action of protein kinase C, *Biochim. Biophys. Acta* **1222:**109.

McPhail, L. C., Qualliotine-Mann, D., and Waite, K. A., 1995, Cell-free activation of neutrophil NADPH oxidase by a phosphatidic acid-regulated protein kinase, *Proc. Natl. Acad. Sci. USA* **92:**7931.

MacNulty, E., Plevin, R., and Wakelam, M. J. O., 1990, Stimulation of the hydrolysis of phosphatidylinositol 4,5-bisphosphate and phosphatidylcholine by endothelin, a complete mitogen for Rat-1 fibroblasts, *Biochem. J.* **272:**761.

MacNulty, E. E., McClue, S. J., Carr, I. C., Jess, T., Wakelam, M. J. O., and Milligan, G., 1992, α_2-C10 adrenergic receptors expressed in rat 1 fibroblasts can regulated both adenylylcyclase and phospholipase D-mediated hydrolysis of phosphatidylcholine by interacting with pertussis toxin-sensitive guanine nucleotide-binding proteins, *J. Biol. Chem.* **267:**2149.

Malcolm, K. C., Ross, A. H., Qiu, R.-G., Symons, M., and Exton, J. H., 1994, Activation of rat liver phospholipase D by the small GTP binding protein rhoA, *J. Biol. Chem.* **269:**25951.

Maroney, A. C., and Macara, I. G., 1989, Phorbol ester-induced translocation of diacylglycerol kinase from the cytosol to the membrane in Swiss 3T3 fibroblasts, *J. Biol. Chem.* **264:**2537.

Martin, A., Gomez-Muñoz, A., Jamal, Z., and Brindley, D. N., 1991, Characterization and assay of phosphatidate phosphohydrolase, *Methods Enzymol.* **197:**553.

Martin, A., Gomez-Muñoz, A., Waggoner, D. W., Stone, J. C., and Brindley, D. N., 1993, Decreased activities of phosphatidate phosphohydrolase and phospholipase D in *ras* and tyrosine kinase (fps) transformed fibroblasts, *J. Biol. Chem.* **266:**23924.

Martin, A., Gomez-Muñoz, A., Duffy, P. A., and Brindley, D. N., 1994, Phosphatidate phosphorylation, in: *Signal-Activated Phospholipases* (M. Liscovitch, ed.), pp. 139–164, R. G. Landes Company, Austin.

Martin, T. W., and Michaelis, K., 1989, P$_2$-purinergic agonists stimulate phosphodiesteratic cleavage of phosphatidylcholine in endothelial cells. Evidence for activation of phospholipase D, *J. Biol. Chem.* **264:**8847.

Martin, T. W., Feldman, D. R., Goldstein, K. E., and Wagner, J. R., 1989, Long-term phorbol ester treatment dissociates phospholipase D activation from phosphoinositide hydrolysis and prostacyclin synthesis in endothelial cells stimulated with bradykinin, *Biochem. Biophys. Res. Commun.* **165:**319.

Martinson, E. A., Goldstein, D., and Brown, J. H., 1989, Muscarinic receptor activation of phosphatidylcholine hydrolysis. Relationship to phosphoinositide hydrolysis and diacylglycerol metabolism, *J. Biol. Chem.* **264:**14748.

Martinson, E. A., Trilivas, I., and Brown, J. H., 1990, Rapid protein kinase C-dependent activation of phospholipase D leads to delayed 1,2-diglyceride accumulation, *J. Biol. Chem.* **265:**22282.

Martinson, E., Scheible, S., and Presek, P., 1994, Inhibition of phospholipase D of human platelets by protein tyrosine kinase inhibitors, *Cell. Mol. Biol.* **40:**627.

Massenburg, D., Han, J.-S., Liyanage, M., Patton, W. A., Rhee, S. G., Moss, J., and Vaughan, M., 1994, Activation of rat brain phospholipase D by ADP-ribosylation factors 1, 5, and 6: Separation of ADP-ribosylation factor-dependent and oleate-dependent enzymes, *Proc. Natl. Acad. Sci. USA* **91:**11718.

Metz, C. N., Zhang, Y., Guo, Y., Tsang, T. C., Kochan, J. P., Altszuler, N., and Davitz, M. A., 1991, Production of the glycosylphosphatidylinositol-specific phospholipase D by the islets of Langerhans, *J. Biol. Chem.* **266:**17733.

Metz, C. N., Thomas, P., and Davitz, M. A., 1992, Immunolocalization of a glycosylphosphatidylinositol-specific phospholipase D in mast cells found in normal tissue and neurofibromatosis lesions, *Am. J. Pathol.* **140:**1275.

Metz, C. N., Brunner, G., Choi-Muira, N. H., Nguyen, H., Gabrilove, J., Caras, I. W., Altszuler, N., Rifkin, D. B., Wilson, E. L., and Davitz, M. A., 1994, Release of GPI-anchored membrane proteins by a cell associated GPI-specific phospholipase D, *EMBO J.* **13:**1741.

Mitsuyama, T., Takeshige, K., and Minakami, S., 1993, Phosphatidic acid induces the respiratory burst of electropermeabilized human neutrophils by acting on a downstream step of protein kinase C, *FEBS Lett.* **328:**67.

Moehren, G., Gustavsson, L., and Hoek, J. B., 1994, Activation and desensitization of phospholipase D in intact rat hepatocytes, *J. Biol. Chem.* **269:**838.

Möhn, H., Chalifa, V., and Liscovitch, M., 1992, Substrate specificity of neutral phospholipase D from rat brain studied by selective labeling of endogenous synaptic membrane phospholipids *in vitro*, *J. Biol. Chem.* **267:**11131.

Moolenaar, W. H., Kruijer, W., Tilly, B. C., Verlaan, I., Bierman, A. J., and de Laat, S. W., 1986, Growth factor-like action of phosphatidic acid, *Nature* **323:**171.

Moolenaar, W. H., Jalink, K., and van Corven, E. J., 1992, Lysophosphatidic acid: A bioactive phospholipid with growth factor-like properties, *Rev. Physiol. Biochem. Pharmacol.* **119:**48.

Moolenaar, W. H., Jalink, K., Eichholtz, T., Hordijk, P. L., van der Bend, R., van Blitterswijk, W. J., and van Corven, E., 1994, Lysophosphatidic acid as a novel lipid mediator, *Curr. Top. Membr. Transp.* **40:**439–450.

Moritz, A., De Graan, P. N. E., Gispen, W. H., and Wirtz, K. W. A., 1992, Phosphatidic acid is a specific activator of phosphatidylinositol-4-phosphate kinase, *J. Biol. Chem.* **267:**7207.

Moss, J., and Vaughan, M., 1993, ADP-ribosylation factors, 20,000 M, guanine nucleotide-binding protein activators of cholera toxin and components of intracellular vesicular transport system, *Cell. Signal.* **5:**367.

Muir, J. G., and Murray, A. W., 1987, Bombesin and phorbol ester stimulate phosphatidylcholine hydrolysis by phospholipase C: Evidence for a role of protein kinase C, *J. Cell. Physiol.* **130:**382.

Mullmann, T. J., Siegel, M. I., Egan, R. W., and Billah, M. M., 1991, Sphingosine inhibits phosphatidate phosphohydrolase in human neutrophils by a protein kinase C-independent mechanism, *J. Biol. Chem.* **266:**2013.

Murayama, T., and Ui, M., 1987, Phosphatidic acid may stimulate membrane receptors mediating adenylate cyclase inhibition and phospholipid breakdown in 3T3 fibroblasts, *J. Biol. Chem.* **262:**5522.

Nakashima, S., Suganuma, A., Matsui, A., and Nozawa, Y., 1991, Thrombin induces a biphasic 1,2-diacylglycerol production in human platelets, *Biochem. J.* **275:**355.

Negami, A. I., Sasaki, H., and Yamamura, H., 1986, Activation of phosphorylase kinase through autophosphorylation by membrane component phospholipids, *Eur. J. Biochem.* **157:**597.

Nishimura, R., Li, W., Kashishian, A., Mondino, A., Zhou, M., Cooper, J., and Schlessinger, J., 1993, Two signaling molecules share a phosphotyrosine-containing binding site in the platelet-derived growth factor receptor, *Mol. Cell. Biol.* **13:**6889.

Ohguchi, K., Banno, Y., Nakashima, S., and Nozawa, Y., 1995, Activation of membrane-bound phospholipase D by protein kinase C in HL60 cells: synergistic action of a small GTP-binding protein RhoA, *Biochem. Biophys. Res. Commun.* **211:**306.

Ohsako, S., and Deguchi, T., 1981, Stimulation by phosphatidic acid of calcium influx and cyclic GMP synthesis in neuroblastoma cells, *J. Biol. Chem.* **256:**10945.

Ohtsuka, T., Ozawa, M., Okamura, N., and Ishibashi, S., 1989, Stimulatory effects of a short chain phosphatidate on superoxide anion production in guinea pig polymorphonuclear leukocytes, *J. Biochem.* **106:**259.

Okamura, S.-L., and Yamashita, S., 1994, Purification and characterization of phosphatidylcholine phospholpase D from pig lung, *J. Biol. Chem.* **269:**31207.

Olson, S. C., Bowman, E. P., and Lambeth, J. D., 1991, Phospholipase D activation in a cell-free system from human neutrophils by phorbol 12-myristate 13-acetate and guanosine 5′-O-(3-thiotriphosphate). Activation is calcium dependent and requires protein factors in both the plasma membrane and cytosol, *J. Biol. Chem.* **266:**17236.

Pachter, J. A., Pai, J.-K., Mayer-Ezell, R., Petrin, J. M., Dobek, E., and Bishop, W. R., 1992, Differential regulation of phosphoinositide and phosphatidylcholine hydrolysis by protein kinase C-β1 overexpression. Effects on stimulation by α- thrombin, guanosine 5′-O-(thiotriphosphate), and calcium, *J. Biol. Chem.* **267:**9826.

Pai, J.-K., Siegel, M. I., Egan, R. W., and Billah, M. M., 1988, Activation of phospholipase D by chemotactic peptide in HL-60 granulocytes, *Biochem. Biophys. Res. Commun.* **150:**355.

Pai, J.-K., Dobek, E. A., and Bishop, W. R., 1991, Endothelin-1 activates phospholipase D and thymidine incorporation in fibroblasts overexpressing protein kinase C_β1, *Cell Regul.* **2:**897.

Panagia, V., Ou, C., Taira, Y., Dai, J., and Dhalla, N. S., 1991, Phospholipase D activity in subcellular membranes of rat ventricular myocardium, *Biochim. Biophys. Acta* **1064:**242.

Perry, D. K., Stevens, V. L., Widlanski, T. S., and Lambeth, J. D., 1993, A novel *ecto*-phosphatidic acid phosphohydrolase activity mediates activation of neutrophil superoxide generation by exogenous phosphatidic acid, *J. Biol. Chem.* **268:**25302.

Pessin, M. S., Baldassare, J. J., and Raben, D. M., 1990, Molecular species analysis of mitogen-stimulated 1,2-diglycerides in fibroblasts. Comparison of α-thrombin, epidermal growth factor, and platelet-derived growth factor, *J. Biol. Chem.* **265:**7959.

Pete, M. J., and Exton, J. H., 1995, Phospholipid interactions affect substrate hydrolysis by bovine brain phospholipase A_1, *Biochim. Biophys. Acta* **1256:**367.

Pete, M. J., Ross, A. H., and Exton, J. H., 1994, Purification and properties of phospholipase A_1 from bovine brain, *J. Biol. Chem.* **269:**19494.

Peterson, M. W., and Walter, M. E., 1992, Calcium-activated phosphatidylcholine-specific phospholipase C and D in MDCK epithelial cells, *Am. J. Physiol.* **263:**C1216.

Pettit, T. R., Zaqqa, M., and Wakelam, M. J. O., 1994, Epidermal growth factor stimulate distinct diradylglycerol species generation in Swiss 3T3 fibroblasts: Evidence for a potential phosphatidylcholine-specific phospholipase C-catalysed pathway, *Biochem. J.* **298:**655.

Pfeffer, L. M., Strulovici, B., and Saltiel, A. R., 1990, Interferon-α selectively activates the β isoform of protein kinase C through phosphatidylcholine hydrolysis, *Proc. Natl. Acad. Sci. USA* **87**:6537.

Pfeilschifter, J., and Huwiler, A., 1993, A role for protein kinase C-ε in angiotensin II stimulation of phospholipase D in rat renal mesangial cells, *FEBS Lett.* **331**:267.

Plevin, R., and Wakelam, M. J. O., 1992, Rapid desensitization of vasopressin-stimulated phosphatidylinositol 4,5-bisphosphate and phosphatidylcholine hydrolysis questions the role of these pathways in sustained diacylglycerol formation in A10 vascular-smooth-muscle cells, *Biochem. J.* **285**:759.

Plevin, R., Cook, S. J., Palmer, S., and Wakelam, M. J. O., 1991a, Multiple sources of *sn*-1,2-diacylglycerol in platelet-derived-growth-factor-stimulated Swiss 3T3 fibroblasts, *Biochem. J.* **279**:559.

Plevin, R., MacNulty, E. E., Palmer, S., and Wakelam, M. J. O., 1991b, Differences in the regulation of endothelin-l- and lysophosphatidic-acid-stimulated Ins(1,4,5)P$_3$ formation in Rat-1 fibroblasts, *Biochem. J.* **280**:609.

Polverino, A. J., and Barritt, G. J., 1988, On the source of the vasopressin-induced increases in diacylglycerol in hepatocytes, *Biochim. Biophys. Acta* **970**:75.

Powis, G., Lowry, S., Forrai, L., Secrist, P., and Abraham, R., 1991, Inhibition of phosphoinositide phospholipase C by compounds U-73122 and D-609, *J. Cell Pharmacol.* **2**:257.

Powis, G., Seewald, M. J., Melder, D., Hoke, M., Gratas, C., Christensen, T. A., and Chapman, D. E., 1992, Inhibition of growth factor binding, Ca^{2+} signaling and cell growth by polysulfonated azo dies related to the antitumor agent suramin, *Cancer Chemother. Pharmacol.* **31**:223.

Price, B. D., Morris, J. D. H., and Hall, A., 1989, Stimulation of phosphatidylcholine breakdown and diacylglycerol production by growth factors in Swiss-3T3 cells, *Biochem. J.* **264**:509.

Proll, M. A., Clark, R. B., and Butcher, R. W., 1985, Phosphatidate and monoleylphosphatidate inhibition of fibroblast adenylate cyclase is mediated by the inhibitory coupling protein N$_i$, *Mol. Pharmacol.* **28**:331.

Purkiss, J. R., and Boarder, M. R., 1992, Stimulation of phosphatidate synthesis in endothelial cells in response to P$_2$-receptor activation, *Biochem. J.* **287**:31.

Putney, J. W., Jr., Weiss, S. J., Van De Walle, C. M., and Haddas, R. A., 1980, Is phosphatidic acid a calcium ionophore under neurohumoral control? *Nature* **284**:345.

Qian, Z., and Drewes, L. R., 1989, Muscarinic acetylcholine receptor regulates phosphatidylcholine phospholipase D in canine brain, *J. Biol. Chem.* **264**:21720.

Qualliotine-Mann, D., Agwu, D. E., Ellenburg, M. D., McCall, C. E., and McPhail, L. C., 1993, Phosphatidic acid and diacylglycerol synergize in a cell-free system for activation of NADPH oxidase from human neutrophils, *J. Biol. Chem.* **268**:23843.

Rakhimov, M. M., Gorbataia, O. N., and Almatox, K. T., 1989, Properties of phospholipase D from rat liver mitochondria, *Biokhimiya* **54**:1066.

Randall, R. W., Bonser, R. W., Thompson, N. T., and Garland, L. G., 1990, A novel and sensitive assay for phospholipase D in intact cells, *FEBS Lett.* **264**:87.

Reinhold, S. L., Prescott, S. M., Zimmerman, G. A., and McIntyre, T. M., 1990, Activation of human neutrophil phospholipase D by three separable mechanisms, *FASEB J.* **4**:208.

Rider, M. H., and Baquet, A., 1988, Activation of rat liver plasma-membrane diacylglycerol kinase by vasopressin and phenylephrine, *Biochem. J.* **255**:923.

Ridley, A. J., and Hall, A., 1992, The small GTP-binding protein rho regulates the assembly of focal adhesions and actin stress fibers in response to growth factors, *Cell* **70**:389.

Ridley, A. J., Paterson, H. F., Johnston, C. L., Dickmann, D., and Hall, A., 1992, The small GTP-binding protein rho regulates the assembly of focal adhesions and actin stress fibres in response to growth factors, *Cell* **70**:401.

Rosoff, P. M., Savage, N., and Dinarello, C. A., 1988, Interleukin-1 stimulates diacylglycerol production in T lymphocytes by a novel mechanism, *Cell* **54**:73.

Rossi, F., Grzeskowiak, M., Della Bianca, V., Calzetti, F., and Gandini, G., 1990, Phosphatidic acid and not diacylglycerol generated by phospholipase D is functionally linked to the activation of the NADPH oxidase by FMLP in human neutrophils, *Biochem. Biophys. Res. Commun.* **168**:320.

Saito, M., and Kanfer, J., 1975, Phosphatidohydrolase activity in a solubilized preparation from rat brain particulate fraction, *Arch. Biochem. Biophys.* **169**:318.

Sakane, F., Yamada, K., Kanoh, H., Yokoyama, C., and Tanabe, T., 1990, Porcine diacylglycerol kinase sequence has zinc finger and E-F hand motifs, *Nature* **344**:345.

Sakane, F., Yamada, K., Imai, S.-I., and Kanoh, H., 1991, Porcine 80-kDa diacylglycerol kinase is a calcium-binding and calcium/phospholipid-dependent enzyme and undergoes calcium-dependent translocation, *J. Biol. Chem.* **266**:7096.

Salari, H., Low, M., Howard, S., Edin, G., and Bittman, R., 1993, 1-O-hexadecyl-2-0-methyl-sn-glycero-3-phosphocholine inhibits diacylglycerol kinase in WEHI-3B cells, *Biochem. Cell. Biol.* **71**:36.

Salmon, D. M., and Honeyman, T. W., 1980, Proposed mechanism of cholingeric action in smooth muscle, *Nature* **284**:344.

Scallon, B. J., Fung, W.-J. C., Tsang, T. C., Li, S., Kado-Fong, H., Huang, K.-S., and Kochan, J. P., 1991, Primary structure and functional activity of a phosphatidylinositol-glycan-specific phospholipase D, *Science* **252**:446.

Schaap, D., de Widt, J., van der Wal, J., Vandekerckhove, J., van Damme, J., Gussow, D., Ploegh, H. L., van Blitterswijk, W. J., and van der Bend, R. L., 1990, Purification, cDNA cloning and expression of human diacylglycerol kinase, *FEBS Lett.* **275**:151.

Schaap, D., van der Wal, J., van Blitterswijk, W. J., van der Bend, R. L., and Ploegh, H. L., 1993, Diacylglycerol kinase is phosphorylated *in vivo* upon stimulation of the epidermal growth factor receptor and serine-threonine kinases, including protein kinase C-ε, *Biochem. J.* **289**:875.

Sebaldt, R. J., Adams, D. O., and Uhing, R. J., 1992, Quantification of contributions of phospholipid precursors to diradylglycerols in stimulated mononuclear phagocytes, *Biochem. J.* **284**:367.

Seewald, M. J., Olsen, R. A., and Powis, G., 1989, Suramin blocks intracellular Ca^{2+} release and growth factor induced increases in cytoplasmic free Ca^{2+} concentration, *Cancer Lett.* **49**:107.

Serhan, C., Anderson, P., Goodman, E., Dunham, P., and Weissmann, G., 1981, Phosphatidate and oxidized fatty acids are calcium iononnonphores. Studies employing arsenazo III in liposomes, *J. Biol. Chem.* **256**:2736.

Seufferlein, T., and Rozengurt, E., 1994, Lysophosphatidic acid stimulates tyrosine phosphorylation of focal adhesion kinase, paxillin, and p130. Signaling pathways and cross-talk with platelet-derived growth factor, *J. Biol. Chem.* **269**:9345.

Severson, D. L., and Hee-Cheong, M., 1989, Diacylglycerol metabolism in isolated aortic smooth muscle cells, *Am. J. Physiol.* **256**:C11.

Siddiqui, R. A., and Exton, J. H., 1992, Phospholipid base exchange activity in rat liver plasma membranes. Evidence for regulation by G-protein and $P_{2\gamma}$-purinergic receptor, *J. Biol. Chem.* **267**:5755.

Siddiqi, A. R., Smith, J. L., Ross, A. H., Qiu, R.-G., Symons, M., and Exton, J. H., 1995, Regulation of phospholipase D in HL60 cells. Evidence for a cytosolic/phospholipase D, *J. Biol. Chem.* **270**:8466.

Singer, W. D., Brown, H. A., Bokoch, G. M., and Sternweis, P. C., 1995, Resolved phospholipase D activity is modulated by cytosolic factors other than Arf, *J. Biol. Chem.* **270**:14944.

Slivka, S. R., Meier, K. E., and Insel, P. A., 1988, α_1-Adrenergic receptors promote phosphatidylcholine hydrolysis in MDCK-D1 cells. A mechanism for rapid activation of protein kinase C, *J. Biol. Chem.* **263**:12242.

Söling, H.-D., Fest, W., Schmidt, T., Esselmann, H., and Backmann, V., 1989, Signal transmission in exocrine cells is associated with rapid activity changes of acyltransferases and diacylglycerol kinase due to reversible protein phosphorylation, *J. Biol. Chem.* **264**:10643.

Song, J., Pfeffer, L. M., and Foster, D. A., 1991, v-Src increases diacylglycerol levels via a type D phospholipase-mediated hydrolysis of phosphatidylcholine, *Mol. Cell. Biol.* **11**:4903.

Songyang, Z., Shoelson, S. E., Chaudhuri, M., Gish, G., Pawson, T., Haser, W. G., King, F., Roberts, T., Ratnofsky, S., Lechleider, R. J., Neel, B. G., Birge, R. B., Fajardo, J. E., Chou, M. M., Hanafusa, H., Schaffhausen, B., and Cantley, L. W., 1993, SH2 domains recognize specific phosphopeptide sequences, *Cell* **72**:767.

Sozzani, S., Agwu, D. E., McCall, C. E., OFlaherty, J. T., Schmitt, J. D., Kent, J. D., and McPhail, L. C., 1992, Propranolol, a phosphatidate phosphohydrolase inhibitor, also inhibits protein kinase C, *J. Biol. Chem.* **267**:20481.

Stewart, S. J., Cunningham, G. R., Strupp, J. A., House, F. S., Kelley, L. L., Henderson, G. S., Exton, J. H., and Bocckino, S. B., 1991, Activation of phospholipase D: A signaling system set in motion by perturbation of the T lymphocyte antigen receptor/CD3 complex, *Cell Regul.* **2:**841.

Sugiyama, T., Sakai, T., Nozawa, Y., and Oka, N., 1994, Prostaglandin $F_{2\alpha}$-stimulated phospholipase D activation in osteoblast-like MC3T3-E1 cells: Involvement in sustained 1,2-diacylglycerol production, *Biochem. J.* **298:**479.

Taki, T., and Kanfer, J. N., 1979, Partial purification and properties of a rat brain phospholipase D, *J. Biol. Chem.* **254:**9761.

Thomson, F. J., and Clark, M. A., 1995, Purification of phosphatidic acid-hydrolysing phospholipase A_2 from rat brain, *Biochem. J.* **306:**305.

Tigyi, G., Dyer, D. L., and Miledi, R., 1994, Lysophosphatidic acid possesses dual action in cell proliferation, *Proc. Natl. Acad. Sci. USA* **91:**1908.

Tou, J.-s., Jeter, J. R., Jr., Dola, C. P., and Venkatesh, S., 1991, Accumulation of phosphatidic acid mass and increased *de novo* synthesis of glycerolipids in platelet-activating-factor-activated human neutrophils, *Biochem. J.* **280:**625.

Truett, A. P., III, Verghese, M. W., Dillon, S. B., and Snyderman, R., 1988, Calcium influx stimulates a second pathway for sustained diacylglycerol production in leukocytes activated by chemoattractants, *Proc. Natl. Acad. Sci. USA* **85:**1549.

Tsai, M.-H., Yu, C.-L., Wei, F.-S., and Stacey, D. W., 1988, The effect of GTPase activating protein upon Ras is inhibited by mitogenically responsive lipids, *Science* **243:**522.

Tyagi, S. R., Olson, S. C., Burnham, D. N., and Lambeth, J. D., 1991, Cyclic AMP-elevating agents block chemoattractant activation of diradylglycerol generation by inhibiting phospholipase D activation, *J. Biol. Chem.* **266:**3498.

Uhing, R. J., Prpic, V., Hollenbach, P. W., and Adams, D. O., 1989, Involvement of protein kinase C in platelet-activating factor-stimulated diacylglycerol accumulation in murine peritoneal macrophages, *J. Biol. Chem.* **264:**9224.

Uings, I. J., Thompson, N. T., Randall, R. W., Spacey, G. D., Bonser, R. W., Hudson, A. T., and Garland, L. G., 1992, Tyrosine phosphorylation is involved in receptor coupling to phospholipase D but not phospholipase C in the human neutrophil, *Biochem. J.* **281:**597.

Valius, M., and Kazlauskas, A., 1993, Phospholipase C-γ1 and phosphatidylinositol 3 kinase are the downstream mediators of the PDGF receptor's mitogenic signal, *Cell* **73:**321.

Valius, M., Bazenet, C., and Kazlauskas, A., 1993, Tyrosines 1021 and 1009 are phosphorylation sites in the carboxy terminus of the platelet-derived growth factor receptor β subunit and are required for binding of phospholipase Cγ and a 64-kilodalton protein, respectively, *Mol. Cell. Biol.* **13:**133.

van Blitterswijk, W. J., Hilkmann, H., de Widt, J., and van der Bend, R. L., 1991, Phospholipid metabolism in bradykinin-stimulated human fibroblasts. II. Phosphatidylcholine breakdown by phospholipases C and D; involvement of protein kinase C, *J. Biol. Chem.* **266:**10344.

van Corven, E. J., Groenink, A., Jalink, K., Eichholtz, T., and Moolenaar, W. H., 1989, Lysophosphatidate-induced cell proliferation: Identification and dissection of signaling pathways mediated by G proteins, *Cell* **59:**45.

van Corven, E. J., van Rijswijk, A., Jalink, K., van der Bend, R. L., van Blitterswijk, W. J., and Moolenaar, W. H., 1992, Mitogenic action of lysophosphatidic acid and phosphatidic acid on fibroblasts, *Biochem. J.* **281:**163.

van Corven, E. J., Hordijk, P. L., Medema, R. H., Bos, J. L., and Moolenaar, W. H., 1993, Pertussis toxin-sensitive activation of p21^{ras} by G protein-coupled receptor agonists in fibroblasts, *Proc. Natl. Acad. Sci. USA* **90:**1257.

van der Bend, R. L., de Widt, J., van Corven, E. J., Moolenaar, W. H., and van Blitterswijk, W. J., 1992, The biologically active phospholipid, lysophosphatidic acid, induces phosphatidylcholine breakdown in fibroblasts via activation of phospholipase D. Comparison with the response to endothelin, *Biochem. J.* **285:**235.

Vingaard, A. M., and Hansen, H. S., 1991, Phorbol ester and vasopressin activate phospholipase D in Leydig cells, *Mol. Cell. Endocrinol.* **79:**157.

Wahl, M., and Carpenter, G., 1988, Regulation of epidermal growth factor-stimulated formation of inositol phosphates in A-431 cells by calcium and protein kinase C, *J. Biol. Chem.* **263:**7581.

Wang, P., Anthes, J. C., Siegel, M. I., Egan, R. W., and Billah, M. M., 1991, Existence of a cytosolic phospholipase D. Identification and comparison with membrane-bound enzyme, *J. Biol. Chem.* **266**:14877.

Weiss, A., 1993, T cell antigen receptor signal transduction: A tale of tails and cytoplasmic protein-tyrosine kinases, *Cell* **73**:209.

Welsh, C. J., and Cabot, M. C., 1987, *sn*-1,2-diacylglycerols and phorbol diesters: Uptake, metabolism, and subsequent assimilation of the diacylglycerol metabolites into complex lipids of cultured cells, *J. Cell. Biochem.* **35**:231.

Whatmore, J., Cronin, P., and Cockcroft, S., 1994, ARF1-regulated phospholipase D in human neutrophils is enhanced by PMA and MgATP, *FEBS Lett.* **352**:113.

Wright, T. M., Rangan, L. A., Shin, H. S., and Raben, D. M., 1988, Kinetic analysis of 1,2-diacylglycerol mass levels in cultured fibroblasts. Comparison of stimulation by α-thrombin and epidermal growth factor, *J. Biol. Chem.* **263**:9374.

Wright, T. M., Shin, H. S., and Raben, D. M., 1990, Sustained increase in 1,2-diacylglycerol precedes DNA synthesis in epidermal-growth-factor-stimulated fibroblasts. Evidence for stimulated phosphatidylcholine hydrolysis, *Biochem. J.* **267**:501.

Wu, H., James-Kracke, M. R., and Halenda, S. P., 1992, Direct relationship between intracellular calcium mobilization and phospholipase D activation in prostaglandin E-stimulated human erythroleukemia cells, *Biochemistry* **31**:3370.

Wyke, A. W., Cook, S. J., MacNulty, E. E., and Wakelam, M. J. O., 1992, v-Src induces elevated levels of diglyceride by stimulation of phosphatidylcholine hydrolysis, *Cell. Signal.* **4**:267.

Xie, M., and Dubyak, G. R., 1991, Guanine-nucleotide- and adenine-nucleotide-dependent regulation of phospholipase D in electropermeabilized HL-60 granulocytes, *Biochem. J.* **278**:81.

Xie, M., and Low, M. G., 1994, Expression and secretion of glycosylphosphatidylinositol-specific phospholipase D by myeloid cell lines, *Biochem. J.* **297**:547.

Xie, M., Jacobs, L. S., and Dubyak, G. R., 1991, Regulation of phospholipase D and primary granule secretion by P_2-purinergic- and chemotactic peptide-receptor agonists is induced during granulocytic differentiation of HL-60 cells, *J. Clin. Invest.* **88**:45.

Xie, M., Sesko, A. M., and Low, M. G., 1993, Glycosylphosphatidylinositol-specific phospholipase D is localized in keratinocytes, *Am. J. Physiol.* **265**:C1156.

Yamashita, S., Hosaka, K., Miki, Y., and Numa, S., 1981, Glycerolipid acyltransferases from rat liver: 1-Acylglycerophosphate acyltransferase, 1-acylglycerophosphorylcholine acyltransferase, and diacylglycerol acyltransferase, *Methods Enzymol.* **71**:528.

Yeo, E.-J., and Exton, J. H., 1995, Stimulation of phospholipase D by epidermal growth factor requires protein kinase C activation in Swiss 3T3 cells, *J. Biol. Chem.* **270**:3980.

Yeo, E.-J., Kazlauskas, A., and Exton, J. H., 1994, Activation of phospholipase C-γ is necessary for stimulation of phospholipase D by platelet-derived growth factor, *J. Biol. Chem.* **269**:27823.

Younes, A., Kahn, D. W., Besterman, J. M., Bittman, R., Byun, H. S., and Kolesnick, R. N., 1992, Ceramide is a competitive inhibitor of diacylglycerol kinase in vitro and in intact human leukemia (HL-60) cells, *J. Biol. Chem.* **267**:842.

Yu, C.-L., Tsai, M.-H., and Stacey, D. W., 1988, Cellular *ras* activity and phospholipid metabolism, *Cell* **52**:63.

Zhao, Z., Shen, S.-H., and Fischer, E. H., 1993, Stimulation by phospholipids of a protein-tyrosine-phosphatase containing two *src* homology 2 domains, *Proc. Natl. Acad. Sci. USA* **90**:4251.

Chapter 4

PI 3-Kinase and Receptor-Linked Signal Transduction

Brian C. Duckworth and Lewis C. Cantley

4.1. Historical Background

4.1.1. Discovery of PI 3-Kinase

Classical phosphoinositide (PI) metabolism leading to the well-known second messengers diacylglycerol (DAG) and inositol-1,4,5-trisphosphate (InsP$_3$), was elucidated more than 10 years ago (Fig. 4-1A). Many mitogenic signals stimulate PI turnover and transformed cells have constitutively activated PI turnover. It was work on this classical pathway that eventually led to the discovery of the novel PI pathway. Let us first look briefly at the classical pathway, in which phosphatidylinositol (PtdIns) is phosphorylated by PtdIns 4-kinase to PtdIns-4-P, which is subsequently phosphorylated by PtdIns-4-P 5-kinase to form PtdIns-4,5-P$_2$. Much of the PtdIns-4,5-P$_2$ in the cell is found on the inner leaflet of the plasma membrane. This lipid can serve as a substrate for PI-specific phospholipase C (PLC), liberating DAG and IP$_3$. IP$_3$ is a water-soluble molecule which, when released into the cytosol, acts to liberate intracellular stores of Ca^{2+}, increasing the intracellular concentration of Ca^{2+} from the resting level of ~110 nM to 400–1000 nM, which in turn can activate a number of Ca^{2+}-sensitive enzymes and channels. The DAG released from PtdIns-4,5-P$_2$ remains in the membrane and serves as a cofactor in activating many of the protein kinase C (PKC) isotypes.

Ca^{2+} has been implicated in cellular responses for more than 50 years. Since the discovery that changes in internal Ca^{2+} levels are involved in the response to fertilization and development in sea urchin eggs, modulation of Ca^{2+} levels in the cell has been seen in response to many cellular activators and mitogens. DAG

Brian C. Duckworth • Department of Physiology, Tufts University, Boston, Massachusetts 02111. *Lewis C. Cantley* • Division of Signal Transduction, Beth Israel Hospital, and Department of Cell Biology, Harvard Medical School, Boston, Massachusetts 02115.

Handbook of Lipid Research, Volume 8: Lipid Second Messengers, edited by Robert M. Bell *et al.* Plenum Press, New York, 1996.

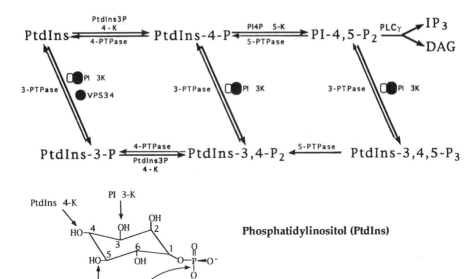

Figure 4-1. Metabolism of inositol lipids. Panel A depicts the classical and D-3 PI pathways. The classical pathway involves the sequential phosphorylation of PtdIns to PtdIns-4-P and then to PtdIns-4,5-P_2. PtdIns-4,5-P_2 is a substrate for PLCγ, which hydrolyzes the phosphodiester bond of PtdIns-4,5-P_2, releasing DAG and Ins-1,4,5-P_3. DAG can then activate PKCs and Ins-1,4,5-P_3 raises Ca^{2+} levels in the cell, which subsequently modulates a number of Ca^{2+}-sensitive proteins. Each of the lipids in the classical pathway can be phosphorylated at the D-3 position of the inositol ring by PI 3-kinase, resulting in the three "novel" or "D-3" lipid products shown. The only other D-3 kinase yet known, the yeast PtdIns 3-kinase VPS34, can utilize only PtdIns as a substrate, forming PtdIns-3-P. PtdIns-3,4-P_2 and PtdIns-3,4,5-P_3 are not detected in yeast. An alternate route for the synthesis for PtdIns-3,4-P_2 has been demonstrated in platelets (Graziani *et al.*, 1992; Sultan *et al.*, 1990; Yamamoto *et al.*, 1990). This reaction is catalyzed by a PtdIns-3-P 4-kinase, which is distinct from PtdIns 4-kinase (Graziani *et al.*, 1992). The D-3 lipids are not known to be substrates for any lipases (Serunian *et al.*, 1989), but are broken down by phosphatases specific for each site of phosphorylation as shown by the light arrows. Panel B shows the structure of PtdIns and the site of action of the lipid kinases that use it as a substrate and PLC.

became of central interest in signal transduction when it was shown that its cellular targets, the PKC isozymes, were the primary cellular receptors for the tumor promoters collectively known as phorbol esters. Thus, both of the products released during PtdIns-4,5-P_2 hydrolysis had been implicated as critical components of cellular mitogenesis by the early 1980s, but none of the enzymes involved in PtdIns metabolism had been purified or characterized.

Since PI turnover was thought to be constitutively activated in transformed cells, investigators began to characterize PI kinases in transformed cells, and found that pp60[v-src], mT/pp60[c-src], and pp68[v-ros] all had associated PtdIns kinase activities (Macara *et al.*, 1984; Sugimoto and Erikson, 1985; Whitman *et al.*, 1985).

Subsequently, two distinct PtdIns kinase activities (types I and II) were characterized in normal and mT-transformed fibroblasts (Whitman *et al.*, 1987). Type I was separated from type II and they were named based on the order of their elution from a monoS anion-exchange column. These two activities were further distinguished in that the type I enzyme was inhibited by detergent whereas the type II enzyme was activated by detergent and inhibited by adenosine. The type I enzyme was shown to be the same activity as that which associated with polyoma mT/pp60[c-src] and pp60[v-src] and was also shown to associate with partially purified PDGF receptor in a PDGF-dependent manner (Whitman *et al.*, 1987). This established a strong link between PtdIns kinase and both of the two main classes of molecules controlling cell growth, growth factor receptors and transforming oncogenes.

Soon it was shown that the type I PtdIns kinase produced a lipid that had not been previously known, that is, PtdIns-3-P. The first step in the "classical" PI metabolic pathway is the phosphorylation of PtdIns to PtdIns-4-P, and this was the only known PtdInsP known to exist. Thus, the expectation was that both type I and type II PtdIns kinases would catalyze the addition of a phosphate at the D-4 position of the inositol ring, the first step in the synthesis of PtdIns-4,5-P_2. However, it was noticed that the migration rate of the product of type I PtdIns kinase was slightly slower by TLC than the product of type II PtdIns kinase. This slower-migrating lipid was found to be chemically unique: the phosphate on the inositol ring was determined to be at the D-3 position, rather than the D-4 position (Whitman *et al.*, 1988). Thus, the type I PtdIns kinase was shown to be a PtdIns 3-kinase. In addition, the "unique" lipid was found *in vivo* as a minority component of PtdInsP in fibroblasts. Thus, the PtdIns kinase activity implicated in control of cell growth had a unique enzymatic activity that diverged from the classical PI pathway. A new subfield within signal transduction had been initiated.

4.1.2. Characterization of PI 3-Kinase

PtdIns 3-kinase had more surprises. Serunian *et al.* found that the lipid kinase activity associated with mT immunoprecipitates would not only use PtdIns as a substrate, but *in vitro* would also phosphorylate PtdIns-4-P and PtdIns-4,5-P_2 to form PtdIns-3,4-P_2 and PtdIns-3,4-5-P_3, respectively (Fig. 4-1) (Serunian *et al.*, 1990). Auger *et al.* also found that the lipid kinase activity that immunoprecipitated with anti-pTyr antibody on PDGF stimulation of vascular smooth muscle cells was also capable of using all three of the classical lipids as substrates to produce PtdIns-3-P, PtdIns-3,4-P_2, and PtdIns-3,4,5-P_3 (Auger *et al.*, 1989). Purification of rat liver PtdIns 3-kinase activity to homogeneity revealed that all three activities were carried out by the same enzyme (Carpenter *et al.*, 1990). Because of the broad specificity for phosphoinositides, this enzyme is properly called phosphoinositide 3-kinase (PI 3-kinase).

These three "novel" or "D-3" lipids were all found to exist *in vivo* and PtdIns-3,4-P_2 and PtdIns-3,4,5-P_3 were found to correlate with cell growth and transformation (Auger *et al.*, 1989; Serunian *et al.*, 1989). Vascular smooth muscle cells (SMCs) were labeled with [^{3}H]inositol, the lipids produced *in vivo* were isolated, and their chemically produced breakdown products were compared with known standards by anion-exchange HPLC. PtdIns-3-P was found in quiescent as well as PDGF-stimulated SMCs, but PtdIns-3,4-P$_2$ and PtdIns-3,4,5-P$_3$ were found only after stimulation with PDGF. These latter two lipids appeared within 1 min of PDGF stimulation (Auger *et al.*, 1989). Both normal and transformed cells contained PtdIns-3-P regardless of whether they were stimulated with PDGF or serum. However, in normal cells, PtdIns-3,4-P$_2$ and PtdIns-3,4,5-P$_3$ were found only after stimulation with PDGF or serum. Transformation of NIH3T3 cells with polyoma virus caused an elevation in PtdIns-3,4-P$_2$ and PtdIns-3,4,5-P$_3$ even in the absence of serum (Serunian *et al.*, 1990). The appearance of PtdIns-3,4-P$_2$ and PtdIns-3,4,5-P$_3$ in growing or transformed cells, the rapid production of these novel lipids in response to PDGF, and the association of PtdIns 3-kinase with anti-pTyr immunoprecipitates on PDGF stimulation and with mT immunoprecipitates from polyoma-transformed cells strongly implicated this new pathway in mitogenic signaling and transformation. The involvement of PI 3-kinase in transformation was further supported by the early evidence that mutants of polyoma mT that failed to associate with PI 3-kinase were compromised in their ability to transform cells (Courtneidge and Heber, 1987; Kaplan *et al.*, 1987; Whitman *et al.*, 1985). As will be discussed in detail below, there is now a great deal of correlative evidence linking PtdIns 3-kinase with cell growth and transformation.

4.1.3. Purification and Cloning of PI 3-Kinase

4.1.3.1. Purification of p85/p110 PI 3-Kinase

The ability to immunoprecipitate PI 3-kinase activity with anti-pTyr antibodies in cells stimulated with PDGF or transformed with mT gave investigators a handle on this enzyme. An 85-kDa phosphoprotein (pp85) was seen in anti-pTyr immunoprecipitates in cells that had been stimulated with PDGF or transformed with mT, v-fms, or v-sis, but not in unstimulated or nontransformed cells (Kaplan *et al.*, 1987). The appearance of this phosphoprotein correlated temporally and quantitatively with the appearance of PtdIns kinase activity in the immunoprecipitates: pp85 and type I PI kinase activity both appeared within 1 min of PDGF stimulation; pp85 appeared in the immunoprecipitates of cells transformed with wild-type mT but was reduced or absent in cells carrying transformation-defective mutants of mT. It was shown by peptide mapping that the 85-kDa protein immunoprecipitated from PDGF-stimulated cells and from mT-transformed cells were the same protein and that the 85-kDa protein and PtdIns kinase activity copurified from several sequential immunoprecipitation protocols (Kaplan *et al.*, 1987). This focused attention on the 85-kDa phosphoprotein as a component of the PI 3-kinase.

Two groups subsequently purified mammalian PI 3-kinases based on following its activity through several conventional column purification steps (Carpenter *et al.*, 1990; Morgan *et al.*, 1990). Both groups showed a correlation of an 85-kDa protein with PtdIns 3-kinase activity. However, Carpenter *et al.* purified the enzyme to near homogeneity (27,000-fold purification) from rat liver cytosol and found that the 85-kda protein copurified stoichiometrically with another protein of 110 kDa (p110). These two proteins were shown to be tightly associated into a 200-kDa heterodimer and to remain associated in gels containing up to 5 M urea (B.C.D. and L.C.C., unpublished data). Subsequent purification of PI 3-kinase to homogeneity from bovine brain and bovine thymus by other groups confirmed the presence of a p85/p110 heterodimer (Fry *et al.*, 1992; Shibasaki *et al.*, 1991).

Carpenter *et al.* (1990) showed that the 85-kDa subunit of purified PI 3-kinase was the same p85 protein that copurified with mT and the PDGF receptor. In fact, this protein could be blotted with tyrosine-phosphorylated mT by a "far-Western" procedure, indicating that the 85-kDa subunit mediates the association with mT. In the purified preparation from Carpenter *et al.* the 110-kDa protein could be resolved into an upper (p110$_{upper}$) and lower (p110$_{lower}$) band by SDS-PAGE. These two proteins were shown to be different but closely related by peptide mapping and partial protein sequencing. PI 3-kinase containing the upper and lower forms of p110 could be separated over the final column purification step. Both p85/p110$_{upper}$ and p85/p110$_{lower}$ were capable of phosphorylating all three lipid substrates (PtdIns, PtdIns-4-P, and PtdIns-4,5-P$_2$) and no significant difference in turnover rates was observed for the two enzymes. Thus, there were known to be at least two genes coding for similar 110-kDa subunits.

4.1.3.2. *Cloning of the 85- and 110-kDa subunits of PtdIns 3-Kinase*

Three groups cloned the p85 subunit of 3-kinase at about the same time, all using strategies that took advantage of p85's affinity for tyrosine-phosphorylated receptors. Escobedo and co-workers and Otsu and co-workers used tyrosine -phosphorylated PDGF receptor (Escobedo *et al.*, 1991), or a tyrosine-phosphorylated peptide derived from the PDGF receptor (Otsu *et al.*, 1991) known to be essential for PtdIns 3-kinase binding, to affinity-purify enough PtdIns 3-kinase to obtain protein sequence of the p85 subunit and subsequently clone its cDNA. Otsu *et al.* cloned two related bovine brain cDNAs for p85, termed α and β, with about 60% identity at the protein level. Escobedo *et al.* isolated a mouse fibroblast cDNA that turned out to be the mouse equivalent of bovine p85α. Skolnik *et al.* devised a general method for cloning receptor binding proteins (Skolnik *et al.*, 1991). The carboxy terminus of the EGF receptor, containing several tyrosine sites that are phosphorylated *in vivo*, was used as a probe to screen expression libraries. Several growth factor receptor binding (GRB) proteins were cloned and the first to be characterized (named GRB1) was found to be the human homologue of the p85α subunit. The sequences of these clones revealed regions of homology to several other genes for known proteins, as discussed below. None of these genes revealed sequences that indicated homology to any kinase catalytic region and the expressed proteins had no intrinsic PtdIns 3-kinase activity (Escobedo *et al.*, 1991; Otsu *et al.*, 1991).

The gene for the 110-kDa subunit of PI 3-kinase was cloned by Waterfield's group (Hiles *et al.*, 1992). The protein predicted by this clone (p110α) had 1068 amino acids and matched the sequences found in the rat p110$_{lower}$ subunit purified by Carpenter *et al.* (1990). When expressed alone in insect cells the 110-kDa protein possessed PtdIns 3-kinase activity. A clone for the 110-kDa subunit of a human PI 3-kinase was isolated later (Hu *et al.*, 1993). This clone was predicted to code for a 1070-amino-acid product with 42% identity to the bovine p110 protein. The sequence of the human p110 (named p110β) matched sequences found in the rat p110$_{upper}$ protein (Carpenter *et al.*, 1990). Thus, there appear to be at least two different p110 proteins and two different p85 proteins in several mammalian species.

4.2. Structure of p85/p110 PI 3-Kinase

4.2.1. Structure of p85

The p85 subunit of PI 3-kinase is composed almost entirely of modular domains that have been defined as regions of homology among signaling proteins (Fig. 4-2). It contains two src-homology 2 (SH2) domains, one src-homology 3 (SH3), domain two proline-rich motifs, a coiled-coil, and a region of homology to the breakpoint cluster region protein (bcr). This section will discuss general features of these domains as a preparation for a discussion of function in the next two sections.

4.2.1.1. SH2 Domains

SH2 domains are modular components of proteins that confer the ability to bind to phosphotyrosine-containing sequences, and thereby modulate binding to other proteins in response to tyrosine phosphorylation (Fig. 4-3). SH2 domains are shared among many proteins that are involved in protein-tyrosine kinase-mediated signaling. This 100-aa domain has a low degree of homology among the

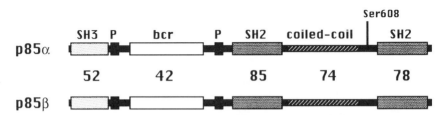

Figure 4-2. Structure of the p85 subunit of PI 3-kinase. This cartoon represents the structure of the p85 subunit of PI 3-kinase. Regions of homology between the p85 proteins and other signaling molecules are shown as larger shaded boxes. The percent identity between the different domains of p85α and p85β are indicated. The thick black boxes labeled with a "P" represent the polyproline sites. The major site of autophosphorylation, serine 608, is indicated.

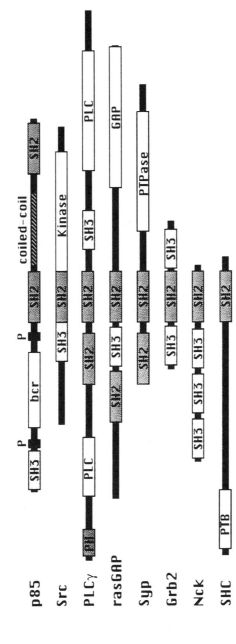

Figure 4-3. Comparison of p85 with other SH2- and SH3-containing signaling proteins. This cartoon shows the modular structure of proteins involved in receptor-mediated signaling. p85 is composed almost entirely of modular domains. Some of these SH2/SH3-containing proteins, such as Grb2, Nck, and SHC, have no catalytic domains and act only as adapter between tyrosine-phosphorylated proteins and their effectors. Although p85 has a putative region of homology to rho/rac/CDC42 GAP (the bcr domain), no enzymatic activity has been demonstrated for this domain, and p85's most important function may be to act as an adapter protein for p110. PTPase stands for protein tyrosine phosphatase. PTB stands for phosphotyrosine binding, a unique domain found in SHC that binds to phosphotyrosine residues found in an NPXpY motif (Blaikie et al., 1994; Kavanaugh and Williams, 1994).

many proteins that contain it but has a few invariant residues that make it relatively easy to distinguish. The structures of the SH2 domains from src (Waksman *et al.*, 1992), abl (Overduin *et al.*, 1992), PtdIns 3-kinase (Booker *et al.*, 1992), lck (Eck *et al.*, 1993), syp (Lee *et al.*, 1994), and PLCγ (Pascal *et al.*, 1994) have been determined. These proteins reveal a common structural motif shared by SH2 domains, the most invariant of the conserved residues making up the structural core. This independently folded motif has its amino- and carboxy-terminal ends arranged closely in space, underscoring its modular characteristic, i.e., that it can be inserted in a molecule without disrupting either's tertiary structure.

SH2 recognition of phosphotyrosine-containing peptides is dependent on the several residues immediately C-terminal to the phosphotyrosine residue as well as the phosphotyrosine itself. It was noted that the binding site for PI 3-kinase on mT shared the same primary sequence C-terminal to the phosphotyrosine (pY_{315}) as that found for the binding site on the PDGF receptor, and that this sequence could also be found in other molecules that were known to bind PI 3-kinase. From this comparison came a consensus sequence for binding of PI 3-kinase: pY-(M/V)-X-M (Cantley *et al.*, 1991). By synthesizing a group of phospho-peptides based on sites phosphorylated in the PDGF receptor, Fantl *et al.* confirmed that methionine (M) three residues C-terminal to phosphotyrosine is important for high-affinity binding to the C-SH2 domain of p85 (Fantl *et al.*, 1992). A general and rapid technique for determining the optimal peptide binding site of SH2 domains was developed by Songyang *et al.* (1993). A library of short degenerate peptides with a fixed phosphotyrosine at the same position in each peptide is passed over a column to which the SH2 domain of interest has been immobilized. The column is washed and peptides that bind to the SH2 domain column with high affinity are then eluted with phenylphosphate and sequenced. Amino acids that are important for binding to the SH2 domain are found to be enriched at specific positions C-terminal of the phosphotyrosine moiety. In this way the optimal binding sequence can be determined for any SH2 domain. The optimal sequence for p85's N-terminal SH2 domain was determined to be pY-(M/V/I/E)-X-M, which is in good agreement with the sequence PI 3-kinase has been shown to bind to on cellular receptors such as mT and PDGF. The optimal peptide sequence for the C-terminal SH2 domain was found to be similar but less selective at the +1 residue and more selective for methionine at the +4 residue. This sequence is useful in predicting potential binding sites for PI 3-kinase (Table 4-1).

The structure of SH2 domains complexed to their high-affinity phospho-tyrosine-peptides (which represents the tyrosine-phosphorylated cellular target) reveals a general pattern in which the phosphotyrosine residue is bound in a pocket lined by the most invariant residues of SH2 domains, while the residues immediately C-terminal to the phosphotyrosine contact surfaces which vary among the different SH2 domains (Eck *et al.*, 1993; Songyang *et al.*, 1993; Waksman *et al.*, 1993). In this way SH2 domains have evolved to bind to phosphotyrosine-containing targets but have adapted to distinguish different phosphotyrosine sites based on their context. Most of the energy of binding is contributed by the

Table 4-1. Known and Possible PI 3-Kinase Binding Sites[a]

Protein	Binding demonstrated	Binding site
DNA virus oncoproteins		
Hamster polyoma mT		$Y_{298}MPM$
Mouse polyoma mT	+	$Y_{315}MPM$
		$Y_{250}SVM$
Receptor protein-tyrosine kinases		
Human PDGFβ receptor	+	$Y_{740}MDM$
	+	$Y_{751}VPM$
Human PDGFα receptor		$Y_{731}MDM$
		$Y_{742}VPM$
Human HGF receptor (met)		$Y_{1331}EVM$
Mouse CSF-1 receptor	+	$Y_{721}VEM$
Human c-kit	+	$Y_{721}MDM$
Human ErbB4		$Y_{1056}TPM$
Human insulin receptor		$Y_{1334}THM$
Human FGF receptor		$Y_{728}MMM$
Tyrosine kinase substrates		
Rat IRS-1		$Y_{460}ICM$
		$Y_{608}MPM$
		$Y_{628}MPM$
		$Y_{658}MMM$
		$Y_{727}MNM$
		$Y_{939}MNM$
		$Y_{987}MTM$
Human ErbB3		$Y_{1054}PMP$
		$Y_{1197}EYM$
		$Y_{1222}EYM$
		$Y_{1260}EYM$
		$Y_{1289}EEM$
Cytosolic protein-tyrosine kinases		
Chicken FAK		$Y_{926}VPM$
src	+	SH3
abl	+	SH3
fyn	+	SH3
lck	+	SH3
lyn	+	SH3

[a]Proteins that are known to be involved in the activation of PI 3-kinase are shown and the known or possible sites of interaction with PI 3-kinase are given. A "+" indicates that the site or domain given has been shown to mediate an interaction with PI 3-kinase.

phosphotyrosine residue, but the specificity of binding is determined by the C-terminal flanking sequence, as the difference in binding between a low-affinity and an optimal phosphotyrosyl peptide can be three orders of magnitude (Piccione *et al.*, 1993; Songyang *et al.*, 1993). The specificity of binding is thus determined by the residues adjacent to the tyrosine, while phosphorylation of the tyrosine acts as a switch for binding of the appropriate SH2.

4.2.1.2. SH3 Domain

The 85-kDa subunit also contains an SH3 domain, which is a small modular domain that confers on a protein the ability to bind to certain proline-rich sequences. SH3 domains share some features in common with SH2 domains in that: (1) They are small (70 aa) modular domains that are found in many proteins involved in intracellular signal transduction (as well as some cytoskeletal proteins; Fig. 4-3). Structure studies show that the amino- and carboxy-terminal ends of the domain are positioned in space very closely and underscore the modular structure of the domain, i.e., through evolution these domains can be inserted into proteins with little disruption of the tertiary structure of the protein or the SH3 domain itself. (2) The small degree of sequence homology between different SH3 domains masks a high degree of structural conservation, the most invariant residues being involved in the structural core of the domain and in the binding of conserved target residues. (3) SH3 domains bind to short sequences that share a consensus sequence for binding but have variable positions that impart selectivity and specificity to the interaction.

That said, SH3 domains bear no sequence or structural homologies with SH2 domains. SH3 domains bind to proline-rich peptide with the consensus sequence Ψ-P-p-Ψ-P, where Ψ is an aliphatic amino acid and p is a weakly conserved proline. The structure of the SH3 domain of p85 (Booker *et al.*, 1993; Koyama *et al.*, 1993) and several other proteins have been determined and show the domain to consist primarily of β-sheets twisted into a barrel structure. Many of the highly conserved residues are clustered on one face of the SH3 domain, forming a hydrophobic platform. Structural studies with SH3–ligand complexes have shown that this hydrophobic platform forms the binding surface for ligand (Booker *et al.*, 1993; Feng *et al.*, 1994; Lim *et al.*, 1994; Wittekind *et al.*, 1994). This platform can be subdivided into three subdomains; the first two are hydrophobic sites which each bind a Ψ-P pair. The third site contains a well-conserved aspartic acid residue which frequently forms a salt bridge with arginine residues in the ligand. NMR chemical shifts on ligand binding to the p85 SH3 domain localize binding to this covered hydrophobic platform (Booker *et al.*, 1993). These same studies also show that the charged loops between β strands contribute to the binding pocket. The variability of these loops between SH3 domains as well as the variability of residues of the hydrophobic platform probably contribute to the binding specificity between various SH3 domains. The p85 protein has a unique 15-aa insert in one of its loops which probably influences binding specificity (Booker *et al.*, 1993).

The proline-rich peptides to which SH3 domains bind form left-handed (type II) helices with three residues per turn (Fig. 4-4). For each turn of the helix two of the three residues contact one subdomain of the SH3's binding platform. A surprising feature of SH3–ligand interaction is that different SH3 domains bind ligands in different orientations with respect to their amino- and carboxy-termini (Feng *et al.*, 1994; Lim *et al.*, 1994; Wittekind *et al.*, 1994). In fact, a given SH3 domain can bind different ligands in opposite orientations (Feng *et al.*, 1994). This flexibility expands the range of potential target proteins.

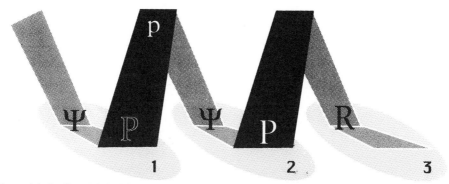

Figure 4-4. Proline-rich SH3-binding motif. This figure represents the structure of a proline-rich SH3-binding motif, Ψ-P-p-Ψ-P. The left-handed type II polyproline helix has three residues per turn; for each turn two residues can contact the surface of the SH3 domain. Ψ represents an aliphatic amino acid, and p a weakly conserved proline residue. The three binding pockets of the SH3 domain are represented as ovals.

The role of SH3 domains in signal transduction is less well understood than that of SH2 domains. While SH2 domain binding is triggered by a phosphorylation event, no such event is needed for SH3 binding to its target sequence. In contrast, some SH3-binding proteins have phosphorylation sites within their SH3-binding sequence which when phosphorylated could disrupt binding to their SH3-containing partner (Cherniack *et al.*, 1994). In this model, SH3-driven protein–protein interactions would be found under basal conditions and a kinase could be activated which would disrupt this interaction. Conversely, a phosphatase could act as a positive regulator by taking off a phosphate to allow the interaction of a protein with an SH3 domain. Alternatively the SH3 domain may provide an additional site of contact to add specificity to SH2 domain-regulated interactions.

The *in vivo* ligand for p85's SH3 domain is not yet clear, but several clues exist. Waterfield and co-workers have used p85's SH3 domain as an affinity matrix to identify binding proteins. One of the proteins that they identified is dynamin, a GTPase thought to be involved in pinching off coated pits (Booker *et al.*, 1993). Binding of dynamin to the p85 SH3 could be completed by a proline-rich peptide derived from dynamin (PAVPPARPGSRGPAPGPPAG), but the significance of this interaction is yet to be determined since the binding of p85 to dynamin has not been demonstrated *in vivo*. In a study using phage display libraries the p85 SH3 domain selected a set of phages with the consensus sequence RXXRPLPPLPP (Rickles *et al.*, 1994). Another study suggests the possibility that the p85 SH3 domain recognizes proline-rich sequences that are found within p85 itself (Kapeller *et al.*, 1994). These proline-rich sequences in p85 also mediate binding to SH3 domains of src-family tyrosine kinases (see below). An intramolecular association between the p85 SH3 domain and one of its proline-rich sequences could act as a means of regulating association with other targets.

4.2.1.3.   Proline-Rich Sequences

p85 contains two polyproline sites which fit the consensus sequence for SH3 binding. As just mentioned, one of the SH3 domains that we have reason to believe binds to these polyproline sites is the SH3 domain of p85 itself (Kapeller *et al.*, 1994). However, it is clear that these sites can also mediate binding to the SH3 domains of src-family protein-tyrosine kinases (Kapeller *et al.*, 1994; Karnitz *et al.*, 1994; Liu *et al.*, 1993; Pleiman *et al.*, 1993, 1994; Prasad *et al.*, 1993a; Vogel and Fujita, 1993).

4.2.1.4.   Bcr Homology Domain

p85 contains a region often referred to as the Bcr (breakpoint cluster region) domain. This region of p85 is homologous to the carboxy-terminal region of the Bcr protein, which is the product of the breakpoint cluster region gene, the translocation breakpoint in Philadelphia chromosome-positive chronic myeloid leukemias. The carboxy-terminus of the Bcr protein is the prototype for a family of homologous GTPase-activating proteins (GAPs), which interact with the rac/rho/CDC42 family of ras-related small GTP-binding proteins (for a review see Boguski and McCormick, 1993). GAPs accelerate the intrinsic rate of GTP hydrolysis of small GTP-binding proteins, which leads to a downregulation of the activated, GTP-bound form. Although GAPs downregulate small GTP-binding proteins, this does not exclude the possibility that a GAP is an effector for a G protein. Thus, there is the possibility that a small G protein may be regulated by the bcr domain of p85, or vice versa. Therefore, PI 3-kinase could be upstream or downstream of members of the small GTP-binding family.

The rac/rho/CDC42 family of small GTP-binding proteins has been shown to be involved in cytoskeletal organization (Chant and Stowers, 1995; Nobes and Hall, 1995). The bcr domain of p85 suggests a structural basis for the involvement of PI 3-kinase in cytoskeletal organization, which will be discussed below (Section 4.4.3.2.). Although p85 fits into this family of GAPs by sequence homology, no GAP activity has been demonstrated thus far. However, PI 3-kinase has been found to associate with CDC42 (Zheng *et al.*, 1994) and rac (Tolias *et al.*, 1995) in a GTP-dependent manner.

4.2.1.5.   Inter-SH2 Domain: Lipid Binding Site

The 190-aa region of p85 between the two SH2 domains is referred to as the inter-SH2 domain. This region is highly conserved between p85α and β. The inter-SH2 region fits in with the modular theme for p85; it is predicted to be an independently folded coiled-coil of two long antiparallel helices (Dhand *et al.*, 1994a). This prediction is based on amino acid sequence homology to members of the myosin family, which are known to be in a coiled-coil structure. A common feature of the underlying sequence of a coiled-coil are heptad repeats $(abcdefg)_n$, where the residues a and d are hydrophobic. These residues end up juxtaposed in an α-helix and form a hydrophobic surface, which zips together with a complementary surface on an antiparallel helix to form the coiled-coil. Although the sequence identity between the myosin family members and the inter-SH2 domain was low, the features of the heptad repeats were found to be conserved, predicting a coiled-coil structure (Fig. 4-5). This structure is predicted for all but the last 30 aa of the inter-SH2 domain.

Two features have been localized to the inter-SH2 domain: binding of acidic phospholipids and binding to the p110 catalytic subunit of PI 3-kinase. The p85 subunit of PI 3-kinase binds to lipid vesicles of phosphatidylserine (PS), PtdIns, PtdIns-4-P, or PtdIns-4,5-P_2. This binding is specific in that the lipid phosphatidylcholine (PC) bound only weakly and phosphatidylethanolamine (PE) and gangliosides did not bind at all (End *et al.*, 1993). Also, an increased phosphorylation state of the phosphoinositides corresponds to increased binding. The region of binding is defined by competition studies with two monoclonal antibodies that block binding of the lipids to p85. Each of these antibodies binds to the inter-SH2 region. The antibodies bind poorly to an inter-SH2 region constructed with a small deletion near the amino-terminal SH2 domain (Δ 478–513), suggesting that the epitope for the antibodies is within this deletion and the lipid binding site is at or near the site of this deletion (End *et al.*, 1993). This deletion contains several basic residues, as do sequences that have been shown to bind to PtdIns-4-P and PtdIns-4,5-P_2 (reviewed in Janmey, 1994), but it would be premature to say that the site of this deletion is directly involved in lipid binding. Because of the nature of a coiled-coil structure, a small deletion in one α-helix could have a profound effect on the tertiary structure of the entire inter-SH2 region and thereby affect the binding of an antibody whose epitope is located anywhere within the coiled-coil. Also, there are several other sites within this region that are rich in basic residues that could be involved in lipid binding.

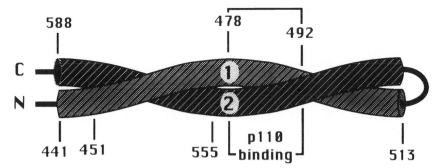

Figure 4-5. The inter-SH2 domain of p85. The inter-SH2 domain of p85 is thought to form a coiled-coil structure. A coiled-coil is formed by two α-helixes that wrap around each other; the amino-terminal α-helix is labeled 1, and the carboxy-terminal α-helix is labeled 2. This region has been implicated in lipid binding and interactions with the p110 catalytic subunit. Because of the nature of the coiled-coil, a small deletion in one α-helix can cause broader disruption of the tertiary structure. Residues indicating the range of different deletions which are discussed in the text are shown.

Why the noncatalytic subunit of PI 3-kinase would have a lipid binding site is not clear. Since it binds lipids that are substrates for PI 3-kinase, it is possible that the 85-kDa regulatory subunit presents the lipid substrate to the 110-kDa catalytic subunit. This fits well with the observations that p85 shows higher affinity for PtdIns-4,5-P_2 than for PtdIns, as PtdIns-4,5-P_2 is thought to be the primary substrate for receptor-activated PI 3-kinase *in vivo* (Hawkins *et al.*, 1992; Stephens *et al.*, 1991, 1993a). Although the site in the inter-SH2 region for lipid binding is not precisely defined, this region has been shown to be critical for PI 3-kinase activity. A 107-aa fragment from the inter-SH2 domain complexed with p110 is catalytically active, whereas the p110 subunit alone had very little activity (Klippel *et al.*, 1994). This fragment was defined by its capacity to mitigate the binding of p85 to p110 and thus is in intimate contact with the p110 catalytic subunit. A simple explanation for the necessity of this fragment for catalytic activity would be that it is required for substrate binding. A better definition of the region for lipid binding and of the region necessary for catalytic activity will have to be determined to support or disprove this theory. If these two regions do coincide, then site-directed mutagenesis could tell us whether the residues involved in lipid binding also affect catalytic activity.

An alternative explanation for the existence of a lipid binding site in p85 is that it is a site for localization of the enzyme to the inner leaflet of the plasma membrane where most of the PtdIns-4,5-P_2 in the cell is found, and that p110 has another site for substrate binding. This site on p85 could also be regulatory, binding PtdIns-3,4,5-P_3 as a negative feedback regulator. Although it was not determined if p85 binds to PtdIns-3,4,5-P_3 (End *et al.*, 1993), the tendency for the degree of binding to increase with increased phosphorylation of PI suggests that it would bind.

4.2.1.6.   Inter-SH2 Domain: p110 Binding Site

As mentioned, the inter-SH2 region of p85 is the region that mediates binding to the p110 subunit (Holt *et al.*, 1994; Hu *et al.*, 1993; Klippel *et al.*, 1993). Deletion studies have attempted to define the region within the inter-SH2 domain that is necessary and sufficient for binding to the p110 subunit. These studies are easiest to interpret with the coiled-coil structure of the inter-SH2 region in mind. The smallest fragment that bound to recombinant p110 was a 104-aa fragment (451–555), which includes all but the first few residues of helix-1 and little more than the first half of helix-2 (Dhand *et al.*, 1994a). This fragment would be expected to maintain the secondary structure of the second half of the coiled-coil. Deletion mutants define a 14-aa stretch of helix-1 within the third quarter of the coiled-coil (478–492) which is necessary for binding of p110 (see Fig. 4-5). This region is common to two overlapping deletions that do not bind to p110, but deletions of the adjacent first half or fourth quarter of helix-1 only slightly disrupt binding (Dhand *et al.*, 1994a). A prediction of the coiled-coil structure would be that the part of helix-2 that couples to this essential region of helix-1 would also be necessary for p110 binding. This is supported in that a construct containing the entire helix-1 but in which helix-2 is deleted does not bind p110 (Klippel *et al.*, 1993). Also, a synthetic peptide (470–497) of the essential region of helix-1 failed to associate with p110 (Dhand *et al.*, 1994a). Thus, functional analysis is consistent with the inter-SH2 domain being in a coiled-coil structure, and the integrity of a portion of this coiled-coil is required for binding of p85 to the p110 catalytic subunit.

4.2.2. Structure of p110

Where sequence analysis of the p85 subunit of PI 3-kinase revealed a protein consisting almost entirely of already relatively well-defined modular domains, the first clones for p110 showed little that looked familiar (Hiles *et al.*, 1992; Hu *et al.*, 1993). It did show some homology (27% identify at the amino acid level) with VPS34, a yeast gene that was isolated in a screen for mutants defective in vacuolar protein sorting (Herman and Emr, 1990). VPS34 was subsequently shown to be a PtdIns 3-kinase (Schu *et al.*, 1993). More recently, two yeast PtdIns 4-kinases (Flanagan *et al.*, 1993; Yoshida *et al.*, 1994). and a mammalian PtdIns 4-kinase (Wong and Cantley, 1994) were cloned and shown to be homologous to the C-terminus of p110 and VPS34 (Fig. 4-6).

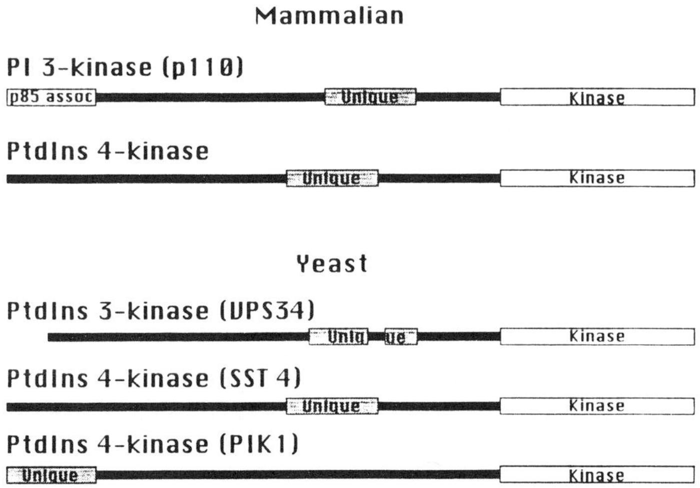

Figure 4-6. The structure of p110 and other known lipid kinases. The structure of the cloned lipid kinases or possible lipid kinases are shown. The list of clones for known lipid kinases now include: the p110α and β clones (Hu *et al.*, 1993); the clone for the human (type II) PtdIns 4-kinase (Wong and Cantley, 1994); the yeast PtdIns 3-kinase VPS34 (Herman and Emr, 1990); SST4, a *Saccharomyces cerevisiae* PtdIns 4-kinase (Yoshida *et al.*, 1994); PIK1, another yeast PtdIns 4-kinase (Flanagan *et al.*, 1993). All of the known lipid kinases share two regions of homology, a carboxy-terminal catalytic domain and a region termed the lipid kinase unique domain. The proteins encoded by the p110α and β genes also share a region of homology at the amino-terminal end that is involved in binding p85 (labeled "p85 assoc").

A comparison of the sequence of the two known mammalian p110 isoforms provides useful information. As discussed above, human p110β is only 42% identical (at the amino acid level) to bovine brain p110α, and so constitutes a separate isotype. p110α and β have three regions of high homology: (1) a short (~100 residue) amino-terminal region unique to these enzymes and not found in the PtdIns 4-kinases, (2) a lipid kinase unique sequence found in both PI 3-kinases and PtdIns 4-kinases, and (3) the carboxy-terminal lipid kinase domain that has distant homology to protein kinase domains.

4.2.2.1. The Kinase Domain

There are a few invariant residues common to almost all protein kinases which are involved in ATP binding and catalysis. Sequence comparison reveals that many of these invariant residues are conserved in lipid kinases. This homology with protein kinases indicates that the carboxy-terminal quarter of p110 (aa 799–1061 of bovine p110α) contains the catalytic domain. p110α and β share 57% identity in this region, and there is 28% identity between p110α, p110β, and the yeast PtdIns 3-kinase VPS34. Residues conserved between yeast and humans may indicate features that are important for function, and many of these residues are in fact homologous to the invariant ATP binding and catalytic residues of protein kinases. Individual point mutations of three invariant residues, which knock out catalytic activity in protein kinases, knock out lipid kinase activity when introduced into the VPS34 lipid kinase (Schu et al., 1993). The crystal structure of several protein kinases have been resolved, cyclic AMP-dependent protein kinase (PKA) being the prototype (Knighton et al., 1991). These crystal structures show that the position and orientation of the invariant residues in space are also highly conserved. Thus, the structure of protein kinases may serve as a model from which to predict the structure of lipid kinase catalytic domains.

There are other conserved sequences among the lipid kinase catalytic domain which are distinct from protein kinases. The importance and function of these will have to be determined from structural and mutagenic studies. Surprisingly, both mammalian PI 3-kinase and the yeast PtdIns 3-kinase have been shown to have a low level of protein kinase activity that is intrinsic to the same catalytic domain that phosphorylates lipids (Carpenter et al., 1993b; Dhand et al., 1994b; Stack and Emr, 1994). The relevance of this activity is discussed in Sections 4.3.5 and 4.4.2.

4.2.2.2. The Lipid Kinase Unique Domain

A domain of ~150 aa located N-terminal of the catalytic domain is conserved among all known PtdIns kinases. This domain stretches from amino acid 565 to 696 of p110α, has 39% identity between p110α and β, and is also conserved in VPS34, mammalian PtdIns 4-kinase, and the two yeast PtdIns 4-kinases. It is not found in conventional protein kinases and the function of this domain is not known.

4.2.2.3.   The Amino-Terminal Domain: p85 Binding Site

The N-terminal region of p110 interacts directly with the inter-SH2 domain of p85 (Holt *et al.*, 1994; Hu and Schlessinger, 1994; Klippel *et al.*, 1994). This interaction has been demonstrated both *in vitro* and *in vivo* and can be mediated by a fragment containing only the first 150 aa of p110α or β.

4.3.   Regulation of p85/p110 PI 3-Kinase

PtdIns-3,4-P_2 and PtdIns-3,4,5-P_3 are produced in response to the activation of a broad range of receptors, including receptors with intrinsic protein-tyrosine kinase activity, receptors that associate with cytosolic protein-tyrosine kinases, and receptors linked to heterotrimeric G proteins. Activation of PI 3-kinase is seen by agents that stimulate mitogenesis and by agents that activate various responses in nondividing cells such as histamine secretion from mast cells, oxidative burst in neutrophils, and platelet activation. PI 3-kinase is also activated by a number of transforming oncogenes such as polyoma mT and pp60[v-src], whose modes of action mimic one of the receptor types. In this section activation of PI 3-kinase will be discussed without regard to the effect of that activation, which will be discussed in the next section.

In nonstimulated cells the majority of PI 3-kinase activity is found in the cytosolic fraction of cell lysates. On stimulation of receptors, a universal feature of PI 3-kinase activation is recruitment from the cytosol to membranes. This recruitment is probably a major mechanism of activation of PI 3-kinase leading to the production of PtdIns-3,4-P_2 and PtdIns-3,4,5-P_3, as this brings PI 3-kinase into contact with its substrates PtdIns-4-P and PtdIns-4,5-P_2 in the membrane. As the mechanism of activation of PI 3-kinase has begun to be worked out, it has become apparent that different receptor types affect this recruitment in different ways, but that in most cases for the p85/p110[PI3k], the interaction is mediated by the p85 subunit. Depending on the receptor involved, this interaction may be through the SH2 domains, the proline-rich domain, the bcr domain, or a combination thereof. We will examine the activation of PI 3-kinase by looking closely at well-characterized examples of each type of activation.

4.3.1.   Activation through the SH2 Domains of p85

When receptors with protein-tyrosine kinase activity (RTK) are activated, the cytosolic domains become autophoshorylated (or transphosphorylated) on a set of specific tyrosine residues. Other cellular proteins also become tyrosine-phosphorylated. Phosphotyrosine residues, in the context of a few adjacent residues, are docking sites for SH2-containing signaling molecules such as PI 3-kinase. Thus, phosphorylation of a tyrosine residue within the context of a PI 3-kinase binding site serves as a trigger for the recruitment of PI 3-kinase from the cytosol to certain tyrosine-phosphorylated proteins. This association of PI

3-kinase, usually with the membrane-bound receptor or an associated protein, is thought to activate PI 3-kinase by bringing it into proximity with its substrate in the plasma membrane. There is also an increase in the specific activity of PI 3-kinase on association with tyrosine-phosphorylated proteins (see below).

We will discuss these mechanisms involved in the activation of PI 3-kinase through SH2 domain interactions in this section by looking closely at several well-defined systems. Each of these systems has evolved a slightly different means of recruiting and activating PI 3-kinase: middle T complexes with members of the src family of cytosolic membrane-bound protein-tyrosine kinases, itself becoming tyrosine-phosphorylated and binding PI 3-kinase; the PDGF receptor binds PI · 3-kinase directly; the EGF receptor heterodimerizes with and phosphorylates a closely related receptor (erbB3) to which PI 3-kinase is recruited; insulin receptor phosphorylates a cytosolic protein (IRS-1) to which PI 3-kinase is recruited. These examples are the best-defined representatives of the different types of SH2-mediated interactions. For a more exhaustive listing of the interactions between many of the other receptors that activate PI 3-kinase and PI 3-kinase, see Stephens *et al.* (1993).

4.3.1.1.   Middle T/pp60$^{c\text{-}src}$

Association of PI 3-kinase with the transforming polyoma middle T oncoprotein (mT) was one of the interactions that led to the discovery of PI 3-kinase. Middle T oncoprotein is a 58-kDa cytosolic protein that associates with the plasma membrane via a carboxy-terminal hydrophobic sequence. This protein forms a 1:1 complex with several members of the src family of cytosolic protein-tyrosine kinases, including pp60$^{c\text{-}src}$, pp59$^{c\text{-}fyn}$, and pp62$^{c\text{-}yes}$. Middle T has no intrinsic tyrosine kinase activity, but activates the protein-tyrosine kinase to which it binds. The mT/pp60$^{c\text{-}src}$ complex binds to many of the same signaling molecules that are recruited to receptor protein-tyrosine kinases, including PI 3-kinase (Fig. 4-7). The recruitment of PI 3-kinase to the mT/pp60$^{c\text{-}src}$ complex shares many attributes with receptor protein-tyrosine kinases. That is, the membrane-anchored mT/pp60$^{c\text{-}src}$ complex is tyrosine-phosphorylated, and these tyrosine phosphorylation sites serve as signals for binding of SH2-containing signaling molecules such as PI 3-kinase. Mutants of mT serve as an excellent model to study the mechanism of activation of PI 3-kinase because a large number of well-characterized mutants of mT exist.

Cells transformed with middle T have elevated levels of PtdIns-3,4-P$_2$ and PtdIns-3,4,5-P$_3$ (Ling *et al.*, 1992; Serunian *et al.*, 1990). A major site for tyrosine phosphorylation of mT is tyrosine 315, which lies in a YMXM motif, and tyrosine phosphorylation of this site appears to be the trigger for PI 3-kinase binding, as shown by the following. Middle T mutants that fail to associate with pp60$^{c\text{-}src}$ are nontransforming, do not become tyrosine-phosphorylated, and bind no detectable PI 3-kinase. Wild-type mT bound to kinase dead pp60$^{c\text{-}src}$ fails to become tyrosine-phosphorylated and fails to bind PI 3-kinase (Auger *et al.*, 1992). An mT mutant in which Y$_{315}$ is converted to a phenylalanine complexes with pp60$^{c\text{-}src}$ and

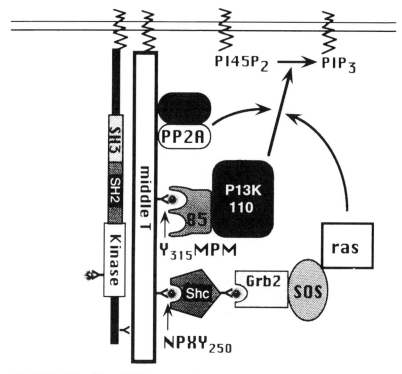

Figure 4-7. Middle T/pp60[c-src]. PI 3-kinase was discovered because of its association with middle T antigen. This cartoon illustrates that many features of middle T activation of PI 3-kinase mimic receptor activation. Middle T must complex with and be phosphorylated by pp60[c-src] at tyrosine 315 (YMPM) in order to activate PI 3-kinase. The binding site at tyrosine 250 is also necessary for activation of PI 3-kinase, and this site has been implied in the activation of ras.

activates its protein-tyrosine kinase activity, but its ability to associate with PI 3-kinase and its transforming ability are greatly reduced (Talmage *et al.*, 1989; Whitman *et al.*, 1985), and cells transfected with the mutant fail to show elevated levels of PtdIns-3,4-P_2 and PtdIns-3,4,5-P_3 *in vivo* (Serunian *et al.*, 1990). Lastly, tyrosine-phosphorylated peptides based on Y_{315}MXM are potent blockers of the association of PI 3-kinase with phosphorylated mT (Auger *et al.*, 1992). Thus, tyrosine 315 is phosphorylated by pp60[c-src], and this serves as a binding site for PI 3-kinase. This interaction is expectedly through the SH2 domains of p85, as both SH2 domains will bind to tyrosine-phosphorylated mT (Yoakim *et al.*, 1992).

Another mT mutant (mT$_{248m}$) reveals that binding of PI 3-kinase to mT/ pp60[c-src] is not sufficient for production of PtdIns-3,4-P_2 and PtdIns-3,4,5-P_3 or for transformation. A change of proline 248 to methionine causes a defect in transformation (Druker *et al.*, 1990). This proline lies in the sequence NPXY, which has been shown to bind to the adapter protein SHC (Campbell *et al.*, 1994; Dilworth *et al.*, 1994). [Shc binds to NPXpY motifs through a unique phosphotyrosine binding (PTB) domain distinct from SH2 domains (Blaikie *et al.*, 1994; Kavanaugh and Williams, 1994).] In this mutant, mT and pp60[c-src] complex and become phosphorylated normally, and PI 3-kinase binds to this complex, but transfection with this mutant middle T does not lead to an increase in PtdIns-3,4,-P$_2$ or PtdIns-3,4,5-P$_3$ (Ling *et al.*, 1992). This suggests that PI 3-kinase recruitment to the mT/pp60[c-src] complex at the plasma membrane is not sufficient for complete activation or coupling to its substrate. Since SHC is known to activate ras via the Grb2/SOS pathway, ras may be necessary for the activation of PI 3-kinase by mT/pp60[c-src]. Consistent with this idea, ras has been shown to interact with p110 *in vitro* (Rodriguez-Viciana *et al.*, 1994). Also, v-ras has been shown to activate p85/110[P13k] when both are expressed in yeast cells (Kodaki *et al.*, 1994) or COS cells (Rodriguez-Viciana *et al.*, 1994), and a dominant negative ras mutant (N17 ras) can disrupt activation of PI 3-kinase by growth factors (EGF and NGF) in PC12 cells (Rodriguez-Viciana *et al.*, 1994). These effects are dependent on an intact ras effector domain. In the latter study (Rodriguez-Viciana *et al.*, 1994), PI 3-kinase was still shown to be associated with tyrosine-phosphorylated proteins on stimulation with EGF or NGF, suggesting that, as for mT$_{248m}$/pp60[c-src], binding of PI 3-kinase to tyrosine-phosphorylated proteins is not sufficient to induce synthesis of the lipids. The role that ras plays in the activation of PI 3-kinase is discussed in more detail in Section 4.3.4.1.

4.3.1.2. Type III Receptors: PDGF Family

The prototype for this group of receptors is the PDGF receptor. This is also the first receptor shown to activate PI 3-kinase (Whitman *et al.*, 1987). PDGF is a dimeric ligand and its binding to PDGF receptor causes dimerization and cross-phosphorylation on several well-defined tyrosine residues, leading to the recruitment of several SH2-containing signaling molecules, including PI 3-kinase (Fig. 4-8). Two of these residues, pY740 and pY751 of the PDGFβ receptor, are found in consensus sequences for PI 3-kinase binding, Y(V/M)XM (Cantley *et al.*, 1991), and clearly mediate PI 3-kinase binding (Fantl *et al.*, 1992; Kazlauskas and Cooper, 1990; Kazlauskas *et al.*, 1992). What is not so clear is the individual contribution of each YXXM site and of each of the two SH2 domains in PI 3-kinase binding to the receptor.

To address the question as to the contribution of these sites, mutant receptors have been made and transfected into cells. These mutant receptors still retain protein-tyrosine kinase activity, become tyrosine-phosphorylated on other tyrosine residues, and bind other signaling molecules such as ras-GAP and PLCγ. Mutation of both Y740 and Y751 to phenylalanine completely abolishes binding of PI 3-kinase *in vivo* and *in vitro* (Fantl *et al.*, 1992; Kazlauskas *et al.*, 1992) and mutation of both of the +3 methionine residues to alanine also completely abolishes binding (Fantl *et al.*, 1992; Wennstrom *et al.*, 1994b).

Mutations of either one of these tyrosine residues alone reduces by 80–90% the amount of PI 3-kinase associated with the PDGF receptor, as seen by measuring the associated PI 3-kinase activity in PDGF receptor immunoprecipitates from

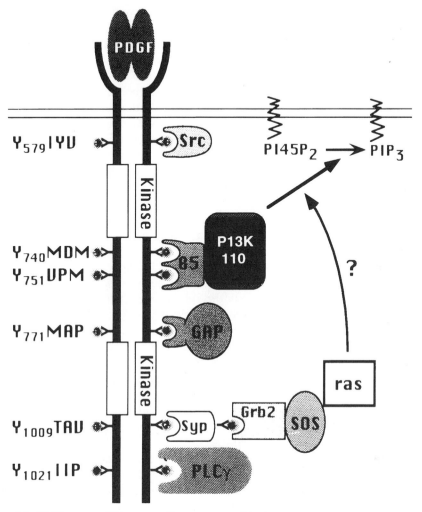

Figure 4-8. PDGF receptor. This cartoon illustrates most of the known interactions between the PDGF receptor and signaling molecules. The site of interaction is indicated on the left and a schematic of the proteins is indicated on the right. In addition to proteins that bind directly to the PDGF receptor, activation of ras is depicted through the binding of Grb2 to the protein tyrosine phosphatase Syp, and the ras activating nucleotide exchange factor SOS to Grb2. The arrow from ras to PI 3-kinase suggests that ras may play a role in activating PI 3-kinase.

cells (Fantl *et al.*, 1992; Kazlauskas *et al.*, 1992). Both residues would seem to be necessary for optimal binding of PI 3-kinase to the receptor, but the assay used requires an association that will survive the rigors of an immunoprecipitation. Neither of these studies measured the effect of these mutations on the *in vivo* synthesis of PtdIns-3,4-P_2 and PtdIns-3,4,5-P_3 in response to PDGF, so the physiological effect of these mutants on lipid synthesis is not yet known. However, in some cellular backgrounds where the double mutant was incapable of mediating an increase in DNA synthesis, each of the single mutants still retained some of this ability (Fantl *et al.*, 1992). Because of the correlation between PI 3-kinase activation and mitogenesis, this would suggest that in the single tyrosine mutants there is still enough PI 3-kinase binding to the remaining phosphotyrosine to transmit an essential signal *in vivo.*

The reciprocal question relates to the contribution the N-SH2 and C-SH2 domains make to the binding of PI 3-kinase to the PDGF receptor: are either alone sufficient or are both necessary? *In vitro* studies show that either SH2 domain alone can bind to phosphorylated PDGF receptor (Cooper and Kashishian, 1993; Hu *et al.*, 1992; Klippel *et al.*, 1992). *In vivo*, however, it appears that both SH2 domains are necessary for binding to the phosphorylated PDGF receptor and for being retained through immunoprecipitation protocols (Cooper and Kashishian, 1993). This difference points out the difficulty in assessing the physiological relevance of *in vitro* studies. For protein interaction, the high concentration of recombinant proteins used in *in vitro* experiments may push interactions that would not be seen at physiological concentration. This is also true for transfected cells expressing high concentrations of proteins in "*in vivo*" experiments. However, these experiments can give us the relative affinities and specificities for interactions between different proteins for a given target, which can be physiologically relevant.

Measurement of the binding affinity of the individual SH2 domains to phosphopeptides modeled around the Y740 and Y751 binding sites shows that the N-SH2 domain binds with higher affinity to the Y751 peptide (containing YVPM) than to the Y740 peptide (containing YMDM), whereas the C-SH2 domain binds with high affinity to both sites (Panayotou *et al.*, 1993). Also, *in vitro* the C-SH2 domain has a higher affinity for the phosphorylated PDGF receptor than the N-SH2 domain, matching that of the whole p85 subunit (Klippel *et al.*, 1992). These data agree with the peptide binding preferences defined for each SH2 domain by the peptide library technique. That is, the N-SH2 domain is more selective for residues at the +1 position, where the C-SH2 domain has no selectivity at this site; however, the C-SH2 domain has a higher selectivity for methionine at +3 than the N-SH2 domain (Songyang *et al.*, 1993). Thus, the C-SH2 domain may be less selective but have a higher affinity for sites with methionine at +3. This difference in the binding affinity could promote binding of the two SH2 domains of p85 to the Y740 and Y751 sites in a specific orientation, the N-SH2 binding to Y751 and the C-SH2 binding to Y740.

There is evidence that p85 bound to p110 is more selective in binding to receptors than the free p85 subunit or its SH2 domains. Purified PI 3-kinase will bind to the PDGF receptor but not to the EGF receptor, whereas the p85 subunit alone will bind to both (Otsu *et al.*, 1991). The p85/p110[PI3k] complex is also more selective in binding to phosphotyrosine-peptide columns (Fry *et al.*, 1992). p85 complexed to p110 could be under more stringent conformational constraints than free p85, and this could limit the range of substrates to which it will bind.

The activation of PI 3-kinase by the PDGF receptor may involve more than just recruitment to the membrane. Total cellular PI 3-kinase activity has been reported to increase almost immediately on PDGF activation, suggesting an increase in the specific activity of PI 3-kinase in the cell (Susa *et al.*, 1992). *In vitro*, phosphorylated peptides based on the PI 3-kinase binding site in the PDGF receptor activate PI 3-kinase severalfold (Carpenter *et al.*, 1993a). This activation is also seen with peptides based on other PI 3-kinase binding sites such as those found in IRS-1 and mT (Backer *et al.*, 1992a). Activation is much more potent by peptides with two phosphorylated YXXM sites than by singularly phosphorylated peptides (Carpenter *et al.*, 1993a). This phenomenon is best characterized for binding to IRS-1 and is discussed in detail below in the section on insulin stimulation. In short, it appears that doubly phosphorylated peptides are more potent because binding of both SH2 domains is required for complete activation, perhaps through a conformational change induced in (and transduced through) the p85 subunit.

4.3.1.3. *Type I Receptors: The EGF Receptor Family*

Type I protein-tyrosine kinase receptor family members include the EGF receptor, and the related proteins erbB2, erbB3, and erbB4. Members of this family are single-chain transmembrane receptors like the PDGF receptor, but they have no kinase insert domain. The kinase domain of erbB3 lacks some of the residues that are invariant in other kinases and when expressed in baculovirus shows negligible protein-tyrosine kinase activity compared with EGF receptor (Guy *et al.*, 1994). Each of these receptors has many potential tyrosine phosphorylation sites in a carboxy-terminal tail domain, but only erbB3 has YMXM PI 3-kinase binding motifs in the carboxy-terminal tail. [Most protein-tyrosine kinases, including the EGF receptor, have a YXXM motif in a conserved region of the kinase domain, but this site has not been shown to be phosphorylated in nontransformed cells (Cantley *et al.*, 1991).]

EGF stimulates production of PtdIns-3,4-P_2 and PtdIns-3,4,5-P_3 (Carter and Downes, 1992; Jackson *et al.*, 1992; Miller and Ascoli, 1990) and causes association of PI 3-kinase with anti-phosphotyrosine immunoprecipitates in several cell types (Carter and Downes, 1992; Jackson *et al.*, 1992; Miller and Ascoli, 1990; Raffioni and Bradshaw, 1992). In most cells where PI 3-kinase is activated by EGF stimulation, no PI 3-kinase activity is seen directly associated with activated EGF receptors (Soltoff *et al.*, 1994). In other cell types EGF fails to significantly activate PI 3-kinase in spite of high receptor numbers. Since p85/p110 PI 3-kinase is present in all cells investigated, the results suggested that a protein other than the EGF receptor is responsible for directly binding and activating PI 3-kinase, and that this protein is not present in all cell types.

How then does EGF receptor stimulation activate PI 3-kinase if not by direct interaction? The answer may be through erbB3. Members of the EGF/erbB family of receptors are able to heterodimerize in response to ligands that activate these proteins (reviewed in Carraway and Cantley, 1994). In COS7 cells expressing the erbB3 receptor, its ligand (heregulin) does not cause tyrosine phosphorylation of the receptor. But when erbB3 is coexpressed with erbB2, heregulin stimulates the tyrosine phosphorylation of both receptors (Sliwkowski *et al.*, 1994). Since erbB3 has little or no protein-tyrosine kinase activity, it is thought that erbB2 is responsible for phosphorylation of both receptors in the heterodimer. Once phosphorylated, erbB3 binds PI 3-kinase (Carraway *et al.*, 1995). The same model has been proposed for the heterodimerization of the EGF receptor and erbB3 (Kim *et al.*, 1994; Soltoff *et al.*, 1994). In cells expressing both of these receptors, EGF causes the tyrosine phosphorylation of both receptors and PI 3-kinase can be co-immunoprecipitated with erbB3 but not the EGF receptor. This effect can also be reproduced *in vitro*, where the EGF receptor is used to tyrosine-phosphorylate erbB3, which can then associate with p85 or PI 3-kinase activity (Kim *et al.*, 1994). In addition, studies using chimeric receptors with the erbB3 cytoplasmic domain also show association of PI 3-kinase with erbB3, and this association can be disrupted with phosphotyrosyl-peptides containing the PI 3-kinase binding motif (Fedi *et al.*, 1994; Prigent and Gullick, 1994). Thus, erbB3 can mediate both EGF and heregulin-dependent activation of PI 3-kinase (Fig. 4-9). When EGF binds to the EGF receptor, it causes heterodimer formation with erbB3 (in cells that express erbB3), which is then phosphorylated on one or more of its several YMXM motifs. This of course serves as a binding site for PI 3-kinase and causes its recruitment from the cytosol to the membrane, the site of its substrate, thereby contributing to the activation of PI 3-kinase. As if this were not complicated enough, there appears to be yet another mechanism of EGF-dependent activation of PI 3-kinase in cells lacking erbB3 (Soltoff *et al.*, 1994).

4.3.1.4. Type II Receptors: Insulin Family

A third group of receptors (Type II receptor protein-tyrosine kinases) include the insulin receptor (IR) and the insulinlike growth factor 1 (IGF-I) receptor. Receptors in this class have two identical α-chains and two identical β-chains; individual α- and β-chains are cleaved from the same peptide. The two extracellular α-chains are connected by cystine bonds to the β-chains and form the ligand binding domain; the two β-chains are connected via a cystine bond to each other. The cytoplasmic side of each β-chain contains the protein-tyrosine kinase domain and several tyrosine phosphorylation sites. This family of receptors activates PI 3-kinase by direct interaction with p85 and also by phosphorylation of an intermediate "adapter" protein. The adapter molecule, insulin receptor substrate-1 (IRS-1), is a cytosolic protein (Fig. 4-10).

Insulin has been shown to rapidly stimulate the production of PtdIns-3,4-P_2 and/or PtdIns-3,4,5-P_3 in Chinese hamster ovary cells transfected with the insulin receptor (CHO/IR) (Ruderman *et al.*, 1990) and Leydig tumor cells (Pignataro and Ascoli, 1990). Insulin-stimulated production of PtdIns-3,4-P_2 and PtdIns-3,4,5-P_3 in CHO/IR reaches a maximum within 1 min and is sustained for at least 60 min (Ruderman *et al.*, 1990). This increase in lipids correlates with an increase in PI 3-kinase activity in anti-pTyr immmunoprecipitates (Endemann *et al.*, 1990; Miller and Ascoli, 1990; Ruderman *et al.*, 1990). Anti-insulin receptor antibodies immunoprecipitated about 10% as much PI 3-kinase as could be precipitated with anti-

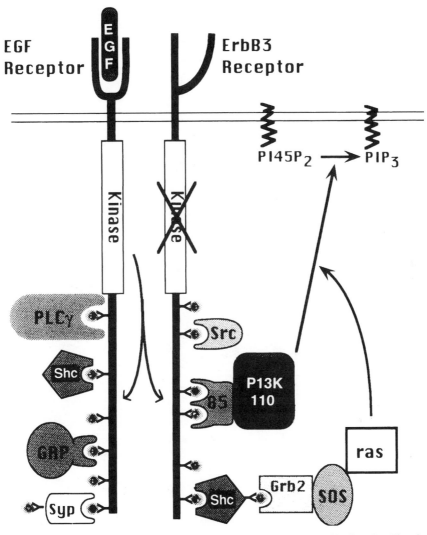

Figure 4-9. The EGF receptor. EGF receptor does not have a PI 3-kinase binding site. The closely related erbB3 receptor has an inactive kinase domain but several YMXM motifs. EGF receptor mediates activation of PI 3-kinase by heterodimerizing erbB3 and phosphorylating it at several sites, including PI 3-kinase binding sites.

pTyr antibodies (Ruderman *et al.*, 1990; Sjolander *et al.*, 1991; Sung and Goldfine, 1992; Yonezawa *et al.*, 1992b). This suggests that PI 3-kinase is either (1) phosphorylated directly by the insulin receptor without a stable association formed or (2) associated with another protein that is phosphorylated in response to insulin. Insulin receptor can phosphorylate p85 *in vitro* or when p85 and the insulin

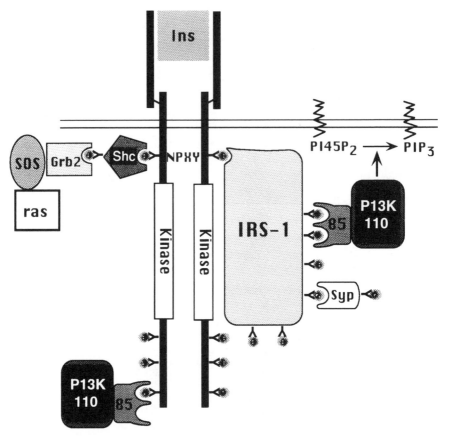

Figure 4-10. The insulin receptor. The insulin receptor activates PI 3-kinase through phosphorylation of insulin receptor substrate 1 (IRS-1). Some of the proteins known to associate with IRS-1 are also depicted. The exact residues involved in binding of these proteins to IRS-1 have not been defined.

receptor are overexpressed *in vivo* (Hayashi *et al.*, 1991, 1992, 1993), but under more physiological conditions others have been unable to detect tyrosine phosphorylation of PI 3-kinase (Folli *et al.*, 1992; Hadari *et al.*, 1992; Sung and Goldfine, 1992; Yonezawa *et al.*, 1992a). This result favors the second model, which became even more appealing with the cloning of IRS-1.

IRS-1 (previously called pp185) is a 185-kDa cytosolic protein which is one of the major tyrosine-phosphorylated proteins found in response to insulin (White *et al.*, 1985). The cDNA sequence shows that IRS-1 contains six YMXM consensus motifs for PI 3-kinase binding (Sun *et al.*, 1991). Many studies have now shown *in vivo* that PI 3-kinase associates with IRS-1 in an insulin-dependent manner, and that this correlates with the tyrosine phosphorylation of IRS-1, temporally and quantitatively (Folli *et al.*, 1992; Giorgetti *et al.*, 1992; Hadari *et al.*, 1992; Kelly and

Ruderman, 1993; Yonezawa *et al.*, 1992a). For example, insulin receptor mutants with reduced or no phosphorylation of IRS-1 show reduced or no activation of PI 3-kinase. One of the more interesting mutations of the insulin receptor is in the juxtamembrane region of the β-chain. This region is involved in coupling the receptor to IRS-1. Mutagenesis of this region does not affect the intrinsic protein-tyrosine kinase activity of the receptor nor its own autophosphorylation, but these receptor mutants do not phosphorylate IRS-1, and as a consequence do not activate PI 3-kinase (Backer *et al.*, 1992b; Kapeller *et al.*, 1991). Therefore, the ability to immunoprecipitate PI 3-kinase activity with antiphosphotyrosine antibodies after insulin stimulation is related in part to the association of PI 3-kinase with tyrosine-phosphorylated IRS-1.

In vitro, PI 3-kinase and its SH2 domains have been shown to bind to tyrosine-phosphorylated IRS-1 but not the unphosphorylated form, and this binding can be competed with phosphorylated YMXM-containing peptides (Backer *et al.*, 1992a; Myers *et al.*, 1992; Yonezawa *et al.*, 1992a). PI 3-kinase from lysates of unstimulated cells binds to phosphorylated IRS-1, so no modification of PI 3-kinase seems to be necessary for binding to IRS-1. Thus, the recruitment of PI 3-kinase to IRS-1 is triggered by tyrosine phosphorylation at one of the six YMXM (or three YXXM) motifs, and binding of PI 3-kinase through one or both of its SH2 domains (reviewed in White and Kahn, 1994).

Which sites on IRS-1 bind PI 3-kinase is not known, but Y(608)MPM and Y(739)MNM have been shown to be phosphorylated *in vivo*, and tryptic peptides from IRS-1 that contain these sites preferentially bind to the SH2 domains of p85 (Sun *et al.*, 1993). It is likely that both of the SH2 domains of p85 are involved in binding to IRS-1, based on the following experiments and analogy with the PDGF receptor. As for the PDGF receptor, the main mechanism of activation of PI 3-kinase may be its recruitment from the cytosol to membranes; but there is also a contribution made by an increase in its catalytic activity. Total cellular PI 3-kinase activity increases at least two- to threefold very rapidly (too fast to be accounted for by new translation) in insulin-stimulated cells (Folli *et al.*, 1992; Giorgetti *et al.*, 1993; Hayashi *et al.*, 1991, 1992; Herbst *et al.*, 1994). Fractionation of PI 3-kinase in insulin-stimulated cells shows that the specific activity of phosphotyrosine/IRS-1-associated PI 3-kinase is increased (Hayashi *et al.*, 1991; Kelly and Ruderman, 1993; Lamphere *et al.*, 1994). PI 3-kinase activity also can be increased *in vitro* by the addition of tyrosine-phosphorylated IRS-1 (Backer *et al.*, 1992a; Giorgetti *et al.*, 1993; Lamphere *et al.*, 1994). More directly, PI 3-kinase activity can be stimulated severalfold by peptides based on IRS-1 sequences containing tyrosine-phosphorylated YMXM motifs (Backer *et al.*, 1992a), and peptides containing two phosphorylated YMXM sites were far more potent at activating PI 3-kinase (Herbst *et al.*, 1994), in agreement with results previously reported using peptides based on the PI 3-kinase binding sites on the PDGF receptor (Carpenter *et al.*, 1993a). This suggests that, as for the PDGF receptor, PI 3-kinase binding to IRS-1 is mediated by both SH2 domains and that binding not only brings PI 3-kinase into proximity with its substrate but increases the specific activity of the enzyme, possibly via a conformational change mediated by p85.

4.3.2. Activation through the p85 Proline-Rich SH3-Binding Domains; Cytosolic Protein-Tyrosine Kinases

As we have seen in the previous section, an important aspect of PI 3-kinase activation is its recruitment from the cytosol to the membrane. Receptors without intrinsic protein-tyrosine kinases (as well as some RTKs) associate with cytosolic protein-tyrosine kinases of the src family and activate them on ligand binding. For these receptors, recruitment to the membrane is equally essential for activation of PI 3-kinase, but cytosolic protein-tyrosine kinases lack YMXM motifs for binding of PI 3-kinase through its SH2 domains. It appears that for receptor-linked cytosolic protein-tyrosine kinases this interaction is accomplished primarily through the association of the SH3 domain of the cytosolic protein-tyrosine kinase with one or both of the proline-rich domains of p85 (Fig. 4-11).

In vitro studies have shown association of p85 or PI 3-kinase activity with the SH3 domain of src (Liu *et al.*, 1993), abl (Kapeller *et al.*, 1994), lck (Kapeller *et al.*, 1994; Prasad *et al.*, 1993b; Vogel and Fujita, 1993), lyn (Pleiman *et al.*, 1994), fyn (Kapeller *et al.*, 1994; Karnitz *et al.*, 1994; Pleiman *et al.*, 1994; Prasad *et al.*, 1993a), and p85 itself (this will be discussed in the next section) (Kapeller *et al.*, 1994). These interactions can be mediated by constructs containing only the SH3 domain and the presence of an SH2 domain only slightly increases binding (Prasad *et al.*, 1993a,b; Vogel and Fujita, 1993). SH2 domains alone did not bind PI 3-kinase from resting cells. [In one study (Vogel and Fujita, 1993) the lck SH2 domain bound PI 3-kinase from v-src-transformed cells but not from normal cells; this interaction could be through tyrosine-phosphorylated proteins bound to PI 3-kinase.]

As we have discussed, SH3 domains interact with sequences that are generally proline-rich, and individual SH3 domains interact with distinct optimal peptides. The interaction between the fyn or lyn SH3 domain and p85 can be locked by polyprolyl peptides derived from the p85 sequence (Kapeller *et al.*, 1994; Pleiman *et al.*, 1994). In one study (Pleiman *et al.*, 1994) the amino-terminal polyproline region of P85 (K_{80}KISPPTPKPRPPRPTPVAPGSSKT$_{104}$) was far more potent in binding than the carboxy-terminal region (N_{299}ERQPAPATPPKPPKPTTVA$_{318}$), suggesting that the amino-terminal site is the relevant site for binding of the fyn and lyn SH3 domains. The amino-terminal peptide (but not the carboxy-terminal) also inhibited antigen receptor-mediated PI 3-kinase activation in permeabilized B cells, showing the significance of this interaction *in vivo*.

SH3 binding to PI 3-kinase is supported by mutational analysis and *in vivo* studies. A single point mutant of a conserved residue of the pp60[v-src] SH3 domain (lysine 106 to glutamate), which renders the mutant v-src only partially transforming but with normal protein-tyrosine kinase activity, resulted in a decreased association with PI 3-kinase (Wages *et al.*, 1992). Other src SH3 mutants support a correlation between a functional SH3 domain and PI 3-kinase binding (and transformation potential). Thus, it seems clear that the important interaction between src-family kinase and PI 3-kinase is between the SH3 domain of the src-family kinase and the proline-rich sites in PI 3-kinase.

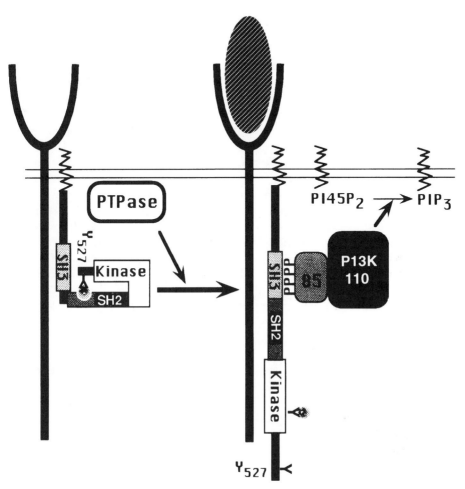

Figure 4-11. Activation of PI 3-kinase by src-family kinases. Association of PI 3-kinase with cytosolic protein-tyrosine kinases occurs through a polyproline site in p85 (depicted as PPPP) interacting with the SH3 domain of the protein-tyrosine kinase. This illustrates one means by which this interaction could be modulated. pp60[c-src] is known to be tyrosine-phosphorylated at its carboxy-terminus (tyrosine 527) when it is inactive, and this causes an intramolecular association between this phosphotyrosine and the pp60[c-src] SH2 domain. For this model we assume that in this conformation the pp60[c-src] SH3 domain would not be available for interaction with p85. Activation of pp60[c-src] occurs through dephosphorylation of tyrosine 527 and disruption of the intramolecular conformation, making the SH3 domain available for the recruitment and activation of PI 3-kinase.

The function of this interaction may be analogous to that for the SH2 interactions, that is, to bring PI 3-kinase from the cytosol to the membrane. src-family protein-tyrosine kinases are anchored to the plasma membrane by an amino-terminal myristoylation. Disruption of membrane localization by muta-genesis of the myristoylation site causes otherwise transforming variants of the kinases to be nontransforming. Transforming forms of v-abl cause an increase in PtdIns-3,4-P$_2$ and PtdIns-3,4,5-P$_3$ in cells; a nonmyristoylated v-abl mutant has normal tyrosine kinase activity and associates with PI 3-kinase, but does not cause an increase in the lipid products *in vivo* (Varticovski *et al.*, 1991). Thus, association of PI 3-kinase with abl is not sufficient to activate PI 3-kinase. It must be targeted to the plasma membrane. A similar conclusion was reached from investigation of myristoylation-defective mutants of pp60[c-src] (Fukui *et al.*, 1991).

Protein–protein interactions mediated through SH3 domains are not as well characterized as SH2-mediated interactions. What is the switch that triggers the association of an src-family SH3 domain with PI 3-kinase? Is it a conformational change in the src-like kinase or in PI 3-kinase? For src-family members there is a well-accepted model for activation which involves the intramolecular association of a carboxy-terminal tyrosine phosphate with the pp60[c-src] SH2 domain (Fig. 4-11; reviewed in Cantley *et al.*, 1991). In this conformation the pp60[c-src] kinase is inactive; disruption of this conformation by hydrolysis of the tyrosine phosphate or interaction of the SH2 domain with another tyrosine-phosphorylated protein activates pp60[c-src]. This conformational change could also affect the availability of the SH3 domain for binding to PI 3-kinase, as suggested by the crystal structure of the pp56[lck] SH2/SH3 phosphopeptide complex (Eck *et al.*, 1994).

4.3.3. Activation by G Protein-Linked Receptors

In addition to the proteins with intrinsic or associated protein-tyrosine kinase activity discussed above, PI 3-kinase is also activated by heterotrimeric G protein-linked receptors. An increase in PtdIns-3,4-P$_2$ and PtdIns-3,4,5-P$_3$ has been shown for several ligands that act through heterotrimeric G protein-linked receptors (reviewed in Stephens *et al.*, 1993a), but the mechanism of activation for these receptors is not well characterized. However, a PI 3-kinase activity has been partially purified from neutrophils and U937 myeloid cells and was shown to be activated by heterotrimeric G-protein βγ subunits (G$_{\beta\gamma}$) (Stephens *et al.*, 1994). This PI 3-kinase appears to be distinct from p85/p110[PI3k] by immunological and chromatographical criteria. Also, it is not activated by tyrosine-phosphrylated peptides that activate p85/p110[PI3k]. A G$_{\beta\gamma}$-sensitive PI 3-kinase activity also is found in platelets. Consistent with the G$_{\beta\gamma}$-sensitive PI 3-kinase activity found in neutrophils and U937 cells, the platelet enzyme does not appear to be a conventional p85/p110[PI3k] because it does not segregate with p85 in immunoprecipitation or cell fractionation protocols (Zhang *et al.*, 1995). A newly identified PI 3-kinase, called PI 3-K(γ), could be the G$_{\beta\gamma}$-sensitive enzyme discussed here. PI 3-K(γ) has been reported to be a 110-kDa protein with a conserved lipid kinase domain, but which does not possess the N-terminal p85-binding domain and does not bind to p85. PI 3-K(γ) has also been reported to be activated directly by G$_{\beta\gamma}$ *in vitro* (discussed in Zhang *et al.*, 1995). In addition, this enzyme segregates with the platelet G$_{\beta\gamma}$-sensitive PI 3-kinase (Zhang *et al.*, 1995). This makes PI 3-K(γ) a strong candidate for the G$_{\beta\gamma}$-sensitive PI 3-kinase in neutrophils, U937 cells, or platelets.

The possibility that a separate PI 3-kinase isozyme exists that is activated by heterotrimeric G proteins would be consistent with what is seen for PLC, another class of signaling molecules that are activated by both tyrosine kinases and heterotrimeric G proteins. That is, tyrosine kinase pathways and heterotrimeric G protein pathways activate different isotypes of PLC. The PLCγ isozyme is activated by tyrosine kinase pathways and contains two SH2 domains and an SH3 domain, while heterotrimeric G proteins activate the PLCβ isozyme which contains no SH2 or SH3 domains. Hopefully the cloning of this third isotype of PI 3-kinase will help with the further characterization of the activation of PI 3-kinase by heterotrimeric G protein-linked receptor.

4.3.4. Activation by ras and rho/rac/CDC42 Family Members

Because of the presence of the bcr domain in p85, it is not surprising that PI 3-kinase is found to interact with members of the rho/rac/CDC42 family of small GTP-binding proteins. PI 3-kinase has also been shown to interact with ras family members.

4.3.4.1. ras

Most factors that activate ras or PI 3-kinase activate both. This is consistent with recent findings suggesting that ras and PI 3-kinase are in the same pathway. Sjolander and Lapetina (1992) first demonstrated an association between ras and PI 3-kinase by showing that they could be co-immunoprecipitated with anti-ras antibodies after stimulation with IGF-1. Rodriguez-Viciana et al. (1994) showed that there is a direct interaction between ras and the p110α catalytic subunit of PI 3-kinase, and that this interaction is dependent on GTP binding to ras and an intact ras effector domain. The question now is which molecule is regulating which.

There is evidence to argue that ras is upstream of PI 3-kinase or vice versa. The initial connection between PI 3-kinase and ras came from PDGF receptor mutants, which showed a correlation between PI 3-kinase binding to the receptor and activation of ras (an increase in the percent of GTP-bound ras) (Satoh et al., 1993; Valius and Kazlauskas, 1993). This effect was dependent on the cell type used, but suggested that ras is downstream of PI 3-kinase. Other studies supporting this model involve insulin signaling (Jhun et al., 1994; Yamauchi et al., 1993). Insulin treatment induces transcription from the c-fos promoter in CHO cells expressing high levels of the insulin receptor (CHO-IR). This effect is mediated by ras and the MAP kinase pathway: transfection of cells with v-ras or v-raf will drive transcription from this promoter by activating the MAP kinase pathway, and dominant negative ras or raf will block induction by insulin. Transfection of the gene for p85 blocks c-fos induction by insulin, presumably by overproducing monomeric p85 and blocking the binding of an active p85/p110^{PI3k} to IRS-1, but does not affect induction by v-ras or v-raf (Yamauchi et al., 1993). The same results are obtained in microinjection experiments of insulin-responsive rat fibroblasts, where injection of the N-SH2 domain of p85 or antibody against p85 blocks insulin-induced c-fos induction, and this block can be overcome by injection of v-ras (Jhun *et al.*, 1994). These results are consistent with PI 3-kinase being upstream and activating ras, which in turn activates raf and the MAP kinase pathway to mediate transcription of fos.

But there is also compelling evidence that PI 3-kinase is downstream of ras (Rodriguez-Viciana *et al.*, 1994). Transfection of dominant negative ras into PC 12 cells blocks EGF- or NGF-dependent production of PtdIns-3,4,-P_2 and PtdIns-3, 4,5-P_3, and transfection of active ras increases the level of these lipids *in vivo*. This effect was specific in that transfection of a v-ras with a mutant effector domain, v-raf, rhoA, or rac1 had no effect on the production of the lipids. This study also indicates that there is a direct interaction between the ras effector domain and p110α. This unexpected mechanism of interaction could be used to explain the results presented above. That is, v-ras may activate PI 3-kinase directly through the p110 subunit, and overexpression of p85 would not act as a dominant negative for this means of activation. So in cells with high levels of v-ras, activation of PI 3-kinase may be strong enough through v-ras to overcome the inhibition of the activation of PI 3-kinase, through IRS-1, by dominant negative levels of p85. These results would be consistent with placing PI 3-kinase downstream of ras.

A model can be proposed that incorporates both of these sets of data if PI 3-kinase and ras are dependent on each other for their activation. PI 3-kinase would be considered upstream of ras if binding to its effector domain reduces the rate of GTP hydrolysis, increasing the amount of ras in a GTP-bound state. Other molecules that bind to the ras effector domain control GTP hydrolysis; rasGAP enhances intrinsic GTPase activity, and a monoclonal antibody to this region (Y13-259) inhibits GTPase activity. Binding of PI 3-kinase could block access of rasGAP, which also binds through the effector domain of ras, and/or binding itself could reduce the intrinsic GTPase activity of ras. Either or both ways, the effect would be for PI 3-kinase to activate ras.

In this same model, PI 3-kinase would also be considered downstream of ras if ras is necessary for the production of PtdIns-3,4,-P_2 and PtDIns-3,4,5-P_3. This could be accomplished by ras directly activating the catalytic activity of PI 3-kinase, or if ras presents PI 3-kinase to its substrate. There has been no evidence presented that ras increases the intrinsic catalytic activity of PI 3-kinase, but the importance of ras for PI 3-kinase function and subcellular localization is underlined by the mT mutant discussed in Section 4.3.1.1. This mutant mT is membrane-bound, associates with pp60[c-src], becomes phosphorylated, and associates with PI 3-kinase, but the increase in PtdIns-3,4,-P_2 or PtdIns-3,4,5-P_3 seen in wild-type mT- transformed cells is greatly reduced (Ling *et al.*, 1992). This mutant has a change in the NPXY sequence (proline$_{248}$ to methionine) which has been implicated in Shc binding and therefore ras activation (Campbell *et al.*, 1994; Dilworth *et al.*, 1994) and subcellular localization (Chen *et al.*, 1990). Thus, in two cases where ras activation is blocked, PtdIns-3,4-P_2 and PtdIns-3,4,5-P_3 synthesis is blocked, even though in both cases PI 3-kinase associates with tyrosine-phosphorylated proteins (Fig. 4-7).

Ras could activate PI 3-kinase by (1) presenting it to its substrate or (2) putting it in an environment where its lipid products have a longer half-life. The latter could be accomplished by producing PtdIns-3,4-P_2 and PtdIns-3,4,5-P_3 at the location of their targets, which would then bind and protect the lipids from phosphatases. This would be analogous to profilin binding to PtdIns-4,5-P_2, protecting it from hydrolysis by PLCγ (Goldschmidt-Clermont et al., 1990, 1991).

4.3.4.2. The rac/rho/CDC42 Family

rac, rho, and CDC42 are highly homologous small GTP-binding proteins that have been shown to be involved in the organization of the actin cytoskeleton. Several studies have now shown a connection between these proteins and PI 3-kinase. PI 3-kinase from whole-cell lysates has been shown to form a complex with GTP-rac in vitro, but not with GDP-rac (Tolias et al., 1995; Zheng et al., 1994). This association has also been demonstrated to form in vivo in Swiss 3T3 cells in a PDGF-dependent manner (Tolias et al., 1995). This interaction may be indirect as it was not seen with purified recombinant rac and p85α/p110^{PI3k} (Rodriguez-Viciana et al., 1994), or the interaction could be direct but with an isotype of PI 3-kinase other than p85α/p110α^{PI3k}. No effect of rac on PI 3-kinase activity could be demonstrated (Tolias et al., 1995), nor could p85 be shown to serve as a GAP for rac (Zhang et al., 1994). Rac has been shown to mediate cytoskeletal rearrangements, in this case membrane ruffling (Ridley et al., 1992), and the involvement of PI 3-kinase in this process is discussed below in Section 4.4.3.2.

CDC42Hs is the human homologue of a yeast protein (CDC42Sc) (Shinjo et al., 1990) that is involved in actin organization during bud site formation (Johnson and Pringle, 1990). CDC42Hs has been shown to complex with PI 3-kinase in vitro in a GTP-dependent manner, and this interaction can be reconstituted with recombinant p85 and CDC42, demonstrating a direct interaction (Zheng et al., 1994). This complex requires an intact CDC42 effector domain and can be disrupted by an excess of the bcr domain from p85. Thus, the interaction seems to be mediated by the effector domain of CDC42 and the bcr domain of p85. This complex has also been shown to exist in vivo by co-immunoprecipitation of PI 3-kinase with CDC42 in COS7 cells (Tolias et al., 1995). p85 does not appear to act as a GAP for CDC42 (Zheng et al., 1993), but CDC42 was shown to slightly stimulate the catalytic activity of PI 3-kinase (Zheng et al., 1994). This result suggests that CDC42 is upstream of PI 3-kinase, but is by no means conclusive. It is not known what event CDC42Hs mediates in mammalian cells, but its close relationship to rac and rho and its role in yeast budding point to an involvement in cytoskeletal organization.

Rho has been implicated in the activation of PI 3-kinase in platelets because toxins that are thought to be specific to rho partially block the activation of PI 3-kinase (Zhang et al., 1993, 1995). These results suggest that PI 3-kinase is downstream of rho, but the biochemical relationship between rho and PI 3-kinase is not well defined. In studies that have shown interactions between other small G proteins and PI 3-kinase, no interaction between PI 3-kinase and rho could be demonstrated (Rodriguez-Viciana *et al.*, 1994; Tolias *et al.*, 1995; Zheng *et al.*, 1994). It remains to be determined if rho acts directly on PI 3-kinase or through an intermediate.

It is interesting to speculate on the reason why many different G proteins are found to be involved in PI 3-kinase signaling. Since these G proteins mediate distinct cellular responses, it is unlikely that they are merely involved in a global activation of PI 3-kinase. Perhaps the relationship between ras and PI 3-kinase proposed above will apply to some of these other G proteins, where the G protein interacts with PI 3-kinase to direct it to a particular subcellular localization for a particular purpose. As discussed for ras, this type of cooperative interaction could be interpreted as the G protein being upstream of PI 3-kinase or vice versa, depending on the experimental approach. This complexity is discussed further in Section 4.4.3.

4.3.5. Regulation by Phosphorylation

There have been numerous studies suggesting that the p85 subunit of PI 3-kinase is tyrosine-phosphorylated, and an obvious implication would be that this plays a role in the regulation of PI 3-kinase. However, each of these studies shows phosphorylation under nonphysiological conditions, either *in vitro*, or in cells transfected and expressing high levels of receptors, transforming genes, and/or p85. Under more normal conditions, for many cellular backgrounds and growth factors, only a small fraction of the p85 subunit associated with p85/p110^{PI3k} is tyrosine-phosphorylated; this is true for total cellular PI 3-kinase or for phospho-tyrosine or receptor-associated PI 3-kinase. So as yet there is no strong indication that direct tyrosine phosphorylation of PI 3-kinase plays an important role in the regulation of PI 3-kinase activity; rather, it is association of PI 3-kinase with tyrosine-phosphrylated proteins that is central to its activation, as we have seen in the examples discussed above. There is some evidence that tyrosine phosphoryla-tion of PI 3-kinase results in accelerated release from mT/pp60$^{c\text{-}src}$, so this may be a shut-off mechanism (Cohen *et al.*, 1990).

Most of the phosphorylation of PI 3-kinase in PDGF-stimulated or mT-transformed cells is on serine (Kaplan *et al.*, 1987; Pallas *et al.*, 1988). As indicated above, p110 has been shown to be a dual-specificity kinase in that besides its lipid kinase activity it also possesses an unusual Mn^{2+}-dependent serine kinase activity (Carpenter *et al.*, 1993b; Dhand *et al.*, 1994b). Site directed mutagenesis has verified that both lipid kinase activity and protein kinase activity occurs in the C-terminal catalytic domain. In addition, wortmannin inhibits both activities with the same dose response (C. L. Carpenter, A. D. Couvillon, W.-M. Hou, B. Schaff-hausen, and L. C. Cantley, personal communication, 1995). The only known *in vivo* substrate of p110 is serine 608 on p85. This site is phosphorylated *in vivo* in transformed cells and is phosphorylated *in vitro* on purified p85/p110^{PI3k} (Car-penter *et al.*, 1993b; Dhand *et al.*, 1994b). Wortmannin blocks both *in vivo* and *in vitro* phosphorylation of serine 608 (Carpenter *et al.*, personal communication, 1995). Serine 608 is between the coiled-coil of the inter-SH2 domain and the carboxy-terminal SH2 domain. Phosphorylation at this site reduces the specific activity of PI 3-kinase severalfold, and this inhibition can be reversed by treatment with protein phosphatase 2A (Carpenter *et al.*, 1993b; Dhand *et al.*, 1994b). The obvious implication of this result is that the serine phosphorylation of p85 by the catalytic subunit is an autoregulatory feedback loop. The turnover number of the serine kinase is much slower than the lipid kinase [0.2 versus 160/min, respectively (Carpenter *et al.*, personal communication, 1995)]. Activation of PI 3-kinase would result in a burst of lipid synthesis which would be turned off over a slower time course through autophosphorylation by the serine kinase activity.

It is interesting to note that middle T associates with the phosphoserine phosphatase PP2A (Pallas *et al.*, 1990). This could be a means by which mT prevents downregulation of PI 3-kinase (e.g., by dephosphorylation of serine 608). Further work is needed to understand the role of the serine kinase in the regulation of PI 3-kinase *in vivo*. Mutants of p85 at serine 608 could help confirm the importance of this site in downregulation. It is also possible that the serine kinase has other targets *in vivo* and contributes to some of the positive signals attributed to the activation of PI 3-kinase.

4.4. Targets of or Responses to PI 3-Kinase

We will divide the discussion of the targets of PI 3-kinase into four categories: (1) proteins that directly bind and are modulated by the lipid products, (2) proteins that are phosphorylated by the protein kinase activity of p110, (3) proteins that directly bind to p85/p110 and are thought to be downstream, and (4) cellular responses downstream of PI 3-kinase with unknown biochemical links.

4.4.1. Proteins That Bind PtdIns-3,4-P_2 and PtdIns-3,4,5-P_3

PKCs are a growing family of serine/threonine kinases that are involved in signal transduction. They are divided into three groups based on their activation by Ca^{2+} and lipid cofactors. The conventional PKCs (cPKCs), α, βI, βII, and γ, are activated by a combination of Ca^{2+} and DAG (or the DAG analogues known as phorbol esters): these enzymes are also known as the Ca^{2+}-dependent PKCs. The nonconventional or novel PKCs (nPKCs), δ, ϵ, η, and θ, are Ca^{2+} independent but activated by DAG and phorbol esters. The third group is comprised of the atypical PKCs (aPKCs), ι/λ and ζ, which are not activated by Ca^{2+}, DAG, or phorbol esters. Nakanishi *et al.* (1993) reported the *in vitro* activation of purified bovine PKCζ by PtdIns-3,4-P_2 and PtdIns-3,4,5-P_3. PKCζ is fully activated (defined as the level of activation seen by calpain cleavage of the autoinhibitory peptide) by these lipids at 0.5 and 0.1 μM, respectively. PKCζ is also known to be fully activated by PS, and this activation is inhibited about 80% in the presence of PE. PtdIns-3,4,5-P_3 presented in PS/PE (25 μM each) micelles fully restored activation, but PtdIns-3,4,5-P_3 was not additive with PS alone. Toker *et al.* (1994) have shown that PtdIns-3,4-P_2 and PtdIns-3,4,5-P_3 also can activate nPKCs expressed and purified from baculovirus, especially PKCε (Toker *et al.*, 1994). (In this study, recombinant PKCζ was not activated by PtdIns-3,4-P_2 and PtdIns-3,4,5-P_3, perhaps because of constitutive activation of this enzyme in Sf9 cells.) For each of these lipids, activation was half-maximal at 0.1 μM, even in the presence of 10 μM ptdIns-4-P or PtdIns-4,5-P_2 and 10 μM PS. Pure synthetic short-chain (dioctanoyl) PtdIns-3,4,5-P_3 also activated PKCε, whereas Ins-1,3,4,5-P_4 had no effect. Neither of these studies showed any activation of the conventional PKCs, which in is disagreement with one other study (Singh *et al.*, 1993). However, this study used a mixture of brain PKCs consisting mainly of cPKC, and these apparently conflicting results could be explained by the presence of other PKC isoforms such as ε or ζ.

One of the striking things about these data is the specificity of activation by the lipids. PtdIns, PtdIns-4-P, and PtdIns-4,5-P_2 had little effect on PKCε or PKCζ activation, and conversion of 1% of PtdIns-4,5-P_2 to PtdIns-3,4,5-P_3 was sufficient to cause activation (Toker *et al.*, 1994). These results are consistent with these lipids acting as second messengers *in vivo*, as PtdIns-3,4,5-P_3 is about 1 to 3% as abundant as PtdIns-4,5-P_2 in stimulated cells. Thus, PtdIns-3,4,5-P_3 would have to have a much higher affinity for its target than PtdIns-4,5-P_2 in order to be effective. The ability of PtdIns-3,4-P_2 but not PtdIns-4,5-P_2 to activate PKCε indicates a stereospecific recognition rather than a nonspecific interaction based on charge. The concentration of lipids that activated these PKCs are consistent with the amount of these lipids produced in stimulated cells *in vivo* (Auger *et al.*, 1989). Evidence that PtdIns-3,4-P_2 or PtdIns-3,4,5-P_3 binds nPKCs *in vivo* is not yet available.

Many factors that activate PI 3-kinase lipid synthesis in cells also activate DAG production. This would seem to be redundant with respect to activation of PKCε and studies with mutant PDGF receptors would suggest it is. Receptors that bind neither PI 3-kinase nor PLCγ fail to transduce a mitogenic signal, whereas restoration of binding of either one restored the signaling capability of the receptor (Valius and Kazlauskas, 1993). However, this could be cell type specific as other studies have suggested that PI 3-kinase binding to the PDGF receptor is more critical for mitogenesis than is the activation of PLCγ (Fantl *et al.*, 1992). There are also several potential differences in signaling that could be accomplished by DAG versus PtdIns-3,4-P_2/PtdIns-3,4,5-P_3. These lipids would be expected to activate different sets of PKCs; DAG can activate both the cPKCs and nPKCs, but PtdIns-3,4-P_2/PtdIns-3,4,5-P_3 would only activate nPKC. Both may activate nPKCs *in vivo*, but may recruit them to different cellular location and thus to a different set of substrates that accomplish distinct functions. The temporal aspect of activation could also be different, as DAG and PtdIns-3,4-P_2/PtdIns-3,4,5-P_3 production may follow different time courses.

The *in vivo* substrates of PKCε or PKCζ have not been identified, but each kinase has been implicated in mitogenic signaling. Overexpression of these isotypes in mouse or rat fibroblasts causes a transformed phenotype (Berra *et al.*, 1993; Cacace *et al.*, 1993; Mischak *et al.*, 1993), and overexpression of a dominant negative PKCζ blocks mitogenic activation (Berra *et al.*, 1993). Thus, these PKCs, when overexpressed, are sufficient for transformation. It is too early to tell how significant the activation by PtdIns-3,4-P_2 and PtdIns-3,4,5-P_3 are for mitogenic signal transduction through PI 3-kinase. It is also unclear whether one can explain the plethora of cellular responses attributed to PI 3-kinase (see below) by activation of PKCs alone. Clearly much work is needed in this area.

4.4.2. Targets of the Serine/Threonine Kinase Activity of p110

The serine kinase activity of p110 could also be important in transducing a signal through PI 3-kinase. Besides p85, the only other substrate that has been implicated for p110's serine kinase activity is IRS-1 (Lam *et al.*, 1994). As discussed, stimulation of adipocytes causes p85/p110^{PI3k} to associate with IRS-1. This complex was co-immunoprecipitated and an immune complex kinase assay was performed, showing serine phosphorylation of IRS-1 as well as p85. This phosphorylation could be blocked with the PI 3-kinase inhibitor wortmannin. Thus, it appears that p110 can phosphorylate IRS-1, at least *in vitro*. This finding raises the possibility that the serine kinase activity of p110 is involved in insulin signaling. Also, other substrates for p110 could exist and it could be involved in signaling through other receptors.

4.4.3. Direct Interactions with p85/p110^{PI3k}

4.4.3.1. ras

As discussed in Section 4.3.4.1., there is a direct GTP-dependent interaction between the ras effector domain and p110α, but it is not clear if PI 3-kinase is an activator of ras or vice versa. As discussed in that section, it seems clear that ras can contribute to PI 3-kinase activation, but there is also the possibility that ras could be activated by (a target of) PI 3-kinase.

4.4.3.2. The rac/rho/CDC42 Family and Cytoskeletal Organization

These three highly homologous small GTP-binding proteins are each involved in distinct aspects of cytoskeletal organization. Rho mediates stress fiber and focal adhesion formation and rac mediates membrane ruffling. The function of CDC42 in mammalian cells is unknown, but the yeast homologue is involved in actin organization during bud site formation. Several studies have shown an interaction between PI 3-kinase and rac/rho/CDC42 family members (discussed in Section 4.3.4.2), and in at least one case (CDC42Hs) the interaction is direct. However, it is not clear if these small G proteins activate PI 3-kinase, which in turn mediates cytoskeletal rearrangements, or vice versa. In this section we will address this question.

Rac is known to mediate membrane ruffling, which is caused by a rearrangement of the cortical actin network located at the plasma membrane. Injection of recombinant activated rac into cells induces membrane ruffling, and injection of dominant negative rac will block growth factor-induced membrane ruffling

(Ridley *et al.*, 1992). PI 3-kinase has also been shown to be necessary for PDGF-induced membrane ruffling: PDGF receptor mutants that fail to bind PI 3-kinase fail to mediate membrane ruffling and wortmannin blocks PDGF-induced membrane ruffling (Wennstrom *et al.*, 1994a,b). [Insulin- and IGF-1-induced membrane ruffling also is blocked by the inhibition of PI 3-kinase signaling (Kotani *et al.*, 1994).] Consistent with these findings, rac has been shown to associate with PI 3-kinase *in vitro* in a GTP-dependent manner (Tolias *et al.*, 1995; Zheng *et al.*, 1994) or *in vivo* in a PDGF-dependent manner (Tolias *et al.*, 1995). PI 3-kinase appears to be upstream of rac in this pathway because wortmannin does not block membrane ruffling produced by injection of activated rac (discussed in Tolias *et al.*, 1995). Thus, a reasonable model is that the lipid product of PI 3-kinase stimulates GDP/GTP exchange on rac and that activated rac then binds to PI 3-kinase as a feedback response or to cluster the two proteins in the same compartment of the cell. It is significant that PtdIns-4-P 5-kinase was found to be constitutively associated with rac *in vivo* (Tolias *et al.*, 1995). The product of this enzyme, PtdIns-4,5-P_2, has been shown to bind to proteins that regulate actin polymerization such as profilin and gelsolin (reviewed in Janmey, 1994). Thus, the role of PI 3-kinase in membrane ruffling may be to localize the rac/PtdIns-4,5-P_2 complex.

Another indication of PI 3-kinase's involvement in cytoskeletal organization is its association with the focal adhesion kinase (FAK). Focal adhesions are distinct patches of organization formed at the cellular membrane in response to growth factors or cellular adhesion, where integrin receptors interact with the actin cytoskeleton through a complex of proteins. PI 3-kinase becomes associated with FAK *in vivo* in response to cell adhesion (Chen and Guan, 1994a) or PDGF stimulation (Chen and Guan, 1994b). This interaction can be reconstructed *in vitro* with recombinant p85. FAK has three YXXM sites, but it is not known if they are phosphorylated or if the association is mediated by p85's SH2 domains. However, FAK is tyrosine-phosphorylated in response to several factors, including PDGF and LPA. It would thus seem likely that PI 3-kinase is downstream of FAK.

4.4.4.   Downstream Responses to PI 3-Kinase

4.4.4.1.   p70S6 Kinase

P70 S6-kinase (p70^{S6k}) is a cytosolic kinase that phosphorylates the S6 ribosomal protein *in vivo*. Phosphorylation of S6 apparently acts to modulate preferential translation of a class of mRNAs containing polypyrimidine tracts at their 5' end (Jefferies *et al.*, 1994). p70^{S6k} may also regulate the activation of other targets besides ribosomes, such as modulation of glycogen synthase activity (Shepherd *et al.*, 1995), and is important for progression through G1 in the cell cycle. p70^{S6k} is activated in response to most growth factors by a series of serine and threonine phosphorylations. Activation appears to lie on a separate pathway from the ras/MAP kinase pathway, as dominant negative ras or raf, which block MAP kinase activation, do not interfere with EGF stimulation of p70^{S6k} (Ming *et al.*, 1994).

Two lines of evidence point to PI 3-kinase as an upstream activator of this kinase: receptor mutants and inhibitor studies. PDGF stimulation of p70[S6k] is largely dependent on an intact PI 3-kinase binding site. Stimulation of PLCγ also contributes to the activation of p70[S6k] (Chung *et al.*, 1994). PI 3-kinase-mediated activation is not sensitive to downregulation of PKCs by chronic PMA pretreatment, while PLCγ-mediated activation is sensitive to PKC downregulation. This finding indicates that if a PKC isoform mediates the PI 3-kinase activation of p70[S6k], then it must be resistant to PMA downregulation. This suggests parallel pathways: the PLCγ pathway utilizes a conventional PKC, and the PI 3-kinase pathway does not. This conclusion is also supported by inhibitor studies. Both PI 3-kinase inhibitors, wortmannin and LY294002, inhibit PDGF (Chung *et al.*, 1994), insulin (Cheatham *et al.*, 1994; Chung *et al.*, 1994; Sanchez *et al.*, 1994), or IL-2 (Monfar *et al.*, 1995)-induced p70[S6k] activation, but not activation by PMA. In contrast, the immunosuppressant rapamycin inhibits p70[S6k] activation by either growth factors or PMA. Since rapamycin has no effect on PI 3-kinase activity, it must act on a target downstream of PI 3-kinase or parallel to PI 3-kinase. The simplest model is that rapamycin acts on a common target of the PI 3-kinase and PLCγ pathways. Rapamycin does not affect p70[S6k] directly but rather binds to a protein with homology to lipid kinases (Heitman *et al.*, 1991). This protein could be the common target for these two pathways.

4.4.4.2. The MAP Kinase Pathway

Mitogen-activated protein (MAP) kinase (also known as ERK) has been a focal point in signal transduction because it is activated by a broad range of both mitogenic and nonmitogenic factors coupled to RTKs, cytosolic protein tyrosine kinases, and heterotrimeric G protein-linked receptors. Recently the linear pathway from some cell surface receptors to MAP kinase has been delineated (reviewed in Blumer and Johnson, 1994; Burgering and Bos, 1995; Marshall, 1994). Common features of this pathway are thought to be the activation of ras, which in turn binds to the serine kinase raf, bringing raf to the membrane. Raf then phosphorylates and activates MAPK/ERK kinase (MEK), a dual-specificity kinase that phosphorylates MAP kinase on threonine and tyrosine residues to activate it. Binding of ras to raf is not sufficient to activate raf and the additional component needed to activate raf is not known. Some studies have implicated PKC as the other component necessary to activate raf, its contribution being in addition to and independent of ras.

Recent studies have shown that PI 3-kinase may play an important role in the activation of the MAP kinase pathway. Wortmannin blocks the activation of MAP kinase by insulin or IGF-1 in a rat skeletal-muscle cell line (L6 cells) (Cross *et al.*, 1994) and by insulin or serum in CHO-IR cells (Welsh *et al.*, 1994). Wortmannin also partially blocks the activation of MAP kinase in neutrophils by platelet-activating factor (PAF), a factor that interacts with heterotrimeric G proteins and stimulates PI 3-kinase.

The contribution of PI 3-kinase to the activation of the MAP kinase pathway depends on the cell type and the factor used. In CHO-IR cells, wortmannin only partially inhibits TPA activation of MAP kinase, but has a much greater effect on insulin activation of MAP kinase (Welsh *et al.*, 1994). We have also seen that wortmannin almost completely blocks PDGF-dependent MAP kinase activation in CHO-IR cells, but has only about a 50% effect in a BALB/C fibroblast cell line, and no effect in Swiss 3T3 fibroblasts (B.C.D. and L.C.C., unpublished results). These results indicate that for a given cell, different factors may utilize different signals that feed into the MAP kinase pathway, and that the same factor (e.g., PDGF) may activate MAP kinase by different pathways depending on the cellular background.

Cross *et al.* (1994) showed that in L6 cells wortmannin also inhibited MEK and raf activation, but did not affect the activation of ras in response to IGF-1. This indicates that PI 3-kinase feeds into the MAP kinase pathway at raf. As mentioned above, PKC has also been implicated in the activation of raf. These results could be consistent with PI 3-kinase activating a PKC, as discussed above, which would then feed into the MAP kinase pathway at raf. In systems where wortmannin does not inhibit the MAP kinase pathway, activation of raf could be via another signaling pathway that activates PKC, such as PLCγ. It will be informative to find out if the isotypes of PKC that are activated by PI 3-kinase are able to activate raf or other members of the MAP kinase pathway.

Another type of experiment supports a role for PI 3-kinase in the activation of the MAP kinase pathway. As discussed in Section 4.3.4.1, overexpression of the p85 subunit (Yamauchi *et al.*, 1993) or injection of the N-SH2 domain of p85 or anti-p85 antibody (Jhun *et al.*, 1994) blocks c-fos induction, which is mediated by the MAP kinase pathway. v-raf can overcome this inhibition, consistent with raf being activated by PI 3-kinase (Yamauchi *et al.*, 1993).

4.4.4.3. *Mitogenesis*

Point mutations in a variety of growth factor receptors and oncoproteins that eliminate the binding sites for PtdIns 3-kinase without otherwise affecting intrinsic protein kinase activity or autophosphorylation have been shown to impair the ability of these proteins to mediate mitogenesis (Coughlin *et al.*, 1989; Choudhury *et al.*, 1991; Courtneidge and Heber, 1987; Fantl *et al.*, 1992; Kaplan *et al.*, 1987; Kazlauskas and Cooper, 1989; Reedijk *et al.*, 1990; Reedijk *et al.*, 1992; Talmage *et al.*, 1989; Valius and Kazlauskas, 1993; Whitman *et al.*, 1985; reviewed in Varticovski *et al.*, 1994, and Cantley *et al.*, 1994). This strong correlative evidence that PI 3-kinase is important for mitogenesis is now supported by inhibition of insulin- (Cheatham *et al.*, 1994; Sanchez *et al.*, 1994) or serum- (Cheatham *et al.*, 1994; Vemuri and Rittenhouse, 1994) stimulated DNA synthesis by wortmannin and/or LY294002. Also, injection of the p85 N-SH2 domain as a dominant negative inhibitor of PI 3-kinase blocked insulin-stimulated DNA synthesis (Jhun *et al.*, 1994). Injection of antibody to p110α blocked mitogenesis by PDGF and EGF (but not CSF-1, bombesin, or LPA) (Roche *et al.*, 1994). Consistent with a role for PI 3-kinase in mitogenesis, the PKC isoforms that are activated by PI 3-kinase can transform cells when overexposed (Berra *et al.*, 1993; Cacace *et al.*, 1993; Mischak *et al.*, 1993). It is clear that PI 3-kinase is necessary for mitogenesis in some cell lines, but there are a number of examples now where it is equally clear that PI 3-kinase is not necessary.

4.4.4.4.   Others

In addition to those already discussed, other cellular responses have been implicated as effectors of PI 3-kinase by their susceptibility to inhibition by wortmannin or LY294002. These include formyl peptide (fMLP)-stimulated superoxide production in neutrophils (Arcaro and Wymann, 1993; Okada *et al.*, 1994b), histamine secretion in basophils (Yano *et al.*, 1993), insulin-induced glucose uptake (Okada *et al.*, 1994a; Cheatham *et al.*, 1994; Hara *et al.*, 1994; Sanchez *et al.*, 1994; Clarke *et al.*, 1994; Berger *et al.*, 1994; Yeh *et al.*, 1995), and neurite outgrowth of PC12 cells (Kimura *et al.*, 1994).

4.5.   Closing Remarks

It is clear that the PI 3-kinase pathway can be used to control many different effects in the cell, from mitogenesis in dividing cells to metabolic functions of nondividing cells. Different cell types appear to be able to adapt a signaling mechanism for different purposes, and in this way signaling pathways such as the PI 3-kinase pathway can be thought of as modular units. Upstream, cells have adapted a number of different ways to activate PI 3-kinase, and downstream different effectors are linked to PI 3-kinase. So as we find out more about signaling through PI 3-kinase in a well-defined system such as PDGF-stimulated fibroblasts, we also find that we must be cautious in applying this knowledge to other systems. Now that the activation of some PKC isotypes by PtdIns-3,4-P_2 and PtdIns-3,4,5-P_3 has been demonstrated *in vitro*, it will be important to demonstrate this effect *in vivo*, and to find out if these PKCs are common components of PI 3-kinase signaling in its different roles in the cell. Another unanswered important question is what role the different domains of p85 and the serine kinase activity of PI 3-kinase play in signaling. Do they contribute to the propagation of the PI 3-kinase signal, or are they only involved in the regulation of the lipid kinase activity? These questions should keep us busy for some time to come.

References

Arcaro, A., and Wymann, M. P., 1993, Wortmannin is a potent phosphatidylinositol 3-kinase inhibitor: The role of phosphatidylinositol 3,4,5-trisphosphate in neutrophil responses, *Biochem. J.* **296:** 297–301.

Auger, K. R., Serunian, L. A., Soltoff, S. P., Libby, P., and Cantley, L. C., 1989, PDGF-dependent tyrosine phosphorylation stimulates production of novel polyphosphoinositides in intact cells, *Cell* **57:**167–175.

Auger, K. R., Carpenter, C. L., Shoelson, S. E., Piwnica, W. H., and Cantley, L. C., 1992, Polyoma virus middle T antigen–pp60c-src complex associates with purified phosphatidylinositol 3-kinase in vitro, *J. Biol. Chem.* **267:**5408–5415.

Backer, J. M., Myers, M. J., Shoelson, S. E., Chin, D. J., Sun, X. J., Miralpeix, M., Hu, P., Margolis, B., Skolnik, E. Y., Schlessinger, J., and White, M. F., 1992a, Phosphatidylinositol 3'-kinase is activated by association with IRS-1 during insulin stimulation, *EMBO J.* **11:**3469–3479.

Backer, J. M., Schroeder, G. G., Kahn, C. R., Myers, M. J., Wilden, P. A., Cahill, D. A., and White, M. F., 1992b, Insulin stimulation of phosphatidylinositol 3-kinase activity maps to insulin receptor regions required for endogenous substrate phosphorylation, *J. Biol. Chem.* **267:**1367–1374.

Berger, J., Hayes, N., Szalkowski, D. M., and Zhang, B., 1994, PI 3-kinase activation is required for insulin stimulation of glucose transport into L6 myotubes, *Biochem. Biophys. Res. Commun.* **205:**570–576.

Berra, E., Diaz, M. M., Dominguez, I., Municio, M. M., Sanz, L., Lozano, J., Chapkin, R. S., and Moscat, J., 1993, Protein kinase C zeta isoform is critical for mitogenic signal transduction, *Cell* **74:** 555–563.

Blaikie, P., Immanuel, D., Wu, J., Li, N., Yajnik, V., and Margolis, B., 1994, A region in Shc distinct from the SH2 domain can bind tyrosine-phosphorylated growth factor receptors, *J. Biol. Chem.* **269:**32031–32034.

Blumer, K. J., and Johnson, G. L., 1994, Diversity in function and regulation of MAP kinase pathways, *Trends Biochem. Sci.* **19:**236–240.

Boguski, M. S., and McCormick, F., 1993, Proteins regulating Ras and its relatives, *Nature* **366:**643–654.

Booker, G. W., Breeze, A. L., Downing, A. K., Panayotou, G., Gout, I., Waterfield, M. D., and Campbell, I. D., 1992, Structure of an SH2 domain of the p85 alpha subunit of phosphatidylinositol-3-OH kinase, *Nature* **358:**684–687.

Booker, G. W., Gout, I., Downing, A. K., Driscoll, P. C., Boyd, J., Waterfield, M. D., and Campbell, I. D., 1993, Solution structure and ligand-binding site of the SH3 domain of the p85 alpha subunit of phosphatidylinositol 3-kinase, *Cell* **73:**813–822.

Burgering, B. M. T., and Bos, J. L., 1995, Regulation of Ras-mediated signalling; more than one way to skin a cat, *Trends Biochem. Sci.* **20:**18–22.

Cacace, A. M., Guadagno, S. N., Krauss, R. S., Fabbro, D., and Weinstein, I. B., 1993, The epsilon isoform of protein kinase C is an oncogene when overexpressed in rat fibroblasts, *Oncogene* **8:**2095–2104.

Campbell, K. S., Ogris, E., Burke, B., Su, W., Auger, K. R., Druker, B. J., Schaffhausen, B. S., Roberts, T. M., and Pallas, D. C., 1994, Polyoma middle tumor antigen interacts with SHC protein via the NPTY (Asn-Pro-Thr-Tyr) motif in middle tumor antigen, *Proc. Natl. Acad. Sci. USA* **91:**6344–6348.

Cantley, L. C., Auger, K. R., Carpenter, C., Duckworth, B., Graziani, A., Kapeller, R., and Soltoff, S., 1991, Oncogenes and signal transduction, *Cell* **64:**281–302.

Carpenter, C. L., Duckworth, B. C., Auger, K. R., Cohen, B., Schaffhausen, B. S., and Cantley, L. C., 1990, Purification and characterization of phosphoinositide 3-kinase from rat liver, *J. Biol. Chem.* **265:**19704–19711.

Carpenter, C. L., Auger, K. R., Chanudhuri, M., Yoakim, M., Schaffhausen, B., Shoelson, S., and Cantley, L. C., 1993a, Phosphoinositide 3-kinase is activated by phosphopeptides that bind to the SH2 domains of the 85-kDa subunit, *J. Biol. Chem.* **268:**9478–9483.

Carpenter, C. L., Auger, K. R., Duckworth, B. C., Hou, W. M., Schaffhausen, B., and Cantley, L. C., 1993b, A tightly associated serine/threonine protein kinase regulates phosphoinositide 3-kinase activity, *Mol. Cell. Biol.* **13:**1657–1665.

Carraway, K. L., III, and Cantley, L. C., 1994, A neu acquaintance for erbB3 and erbB4: A role for receptor heterodimerization in growth signaling, *Cell* **78:**5–8.

Carraway, K. L., III, Soltoff, S. P., Diamonti, A. J., and Cantley, L. C., 1995, Heregulin stimulates mitogenesis and phosphatidylinositol 3-kinase in mouse fibroblasts transfected with erbB2/neu and erb3, *J. Biol. Chem.* **270:**7111–7116.

Carter, A. N., and Downes, C. P., 1992, Phosphatidylinositol 3-kinase is activated by nerve growth factor and epidermal growth factor in PC12 cells [published erratum appears in *J. Biol. Chem.* 1992, **267:**23434], *J. Biol. Chem.* **267:**14563–14567.

Chant, J., and Stowers, L., 1995, GTPase cascades choreographing cellular behavior: Movement, morphogenesis, and more, *Cell* **81:**1–4.

Cheatham, B., Vlahos, C. J., Cheatham, L., Wang, L., Blenis, J., and Kahn, C. R., 1994, Phosphatidyl-inositol 3-kinase activation is required for insulin stimulation of pp70 S6 kinase, DNA synthesis, and glucose transporter translocation, *Mol. Cell. Biol.* **14:**4902–4911.

Chen, H. C., and Guan, J. L., 1994a, Association of focal adhesion kinase with its potential substrate phosphatidylinositol 3-kinase, *Proc. Natl. Acad. Sci. USA* **91:**10148–10152.

Chen, H. C., and Guan, J. L., 1994b, Stimulation of phosphatidylinositol 3′-kinase association with focal adhesion kinase by platelet-derived growth factor, *J. Biol. Chem.* **269:**31229–31233.

Chen, W.-J., Goldstein, J. L., and Brown, M. S., 1990, NPXY, a sequence often found in cytoplasmic tails, is required for coated pit-mediated internalization of the low density lipoprotein receptor, *J. Biol. Chem.* **265:**3116–3128.

Cherniack, A. D., Klarlund, J. K., and Czech, M. P., 1994, Phosphorylation of the Ras nucleotide exchange factor son of sevenless by mitogen-activated protein kinase, *J. Biol. Chem.* **269:**4717–4720.

Choudhury, G. G., Wang, L. M., Pierce, J., Harvey, S. A., and Sakaguchi, A. Y., 1991, A mutational analysis of phosphatidylinositol-3-kinase activation by human colony-stimulating factor-1 receptor, *J. Biol. Chem.* **266:**8068–8072.

Chung, J., Grammer, T. C., Lemon, K. P., Kazlauskas, A., and Blenis, J., 1994, PDGF- and insulin-dependent pp70S6k activation mediated by phosphatidylinositol-3-OH kinase, *Nature* **370:**71–75.

Clarke, F. J., Young, P. W., Yonezawa, K., Kasuga, M., and Holman, G. D., 1994, Inhibition of the translocation of GLUT1 and GLUT4 in 3T3-L1 cells by the phosphatidylinositol 3-kinase inhibitor, wortmannin, *Biochem. J.* **306:**631–635.

Cohen, B., Yoakim, M., Piwnica, W. H., Roberts, T. M., and Schaffhausen, B. S., 1990, Tyrosine phosphorylation is a signal for the trafficking of pp85, an 85-kDa phosphorylated polypeptide associated with phosphatidylinositol kinase activity, *Proc. Natl. Acad. Sci. USA* **87:**4458–4462.

Cooper, J. A., and Kashishian, A., 1993, In vivo binding properties of SH2 domains from GTPase-activating protein and phosphatidylinositol 3-kinase, *Mol. Cell. Biol.* **13:**1737–1745.

Coughlin, S. R., Escobedo, J. A., and Williams, L. T., 1989, Role of phosphatidylinositol kinase in PDGF receptor signal transduction, *Science* **243:**1191–1194.

Courtneidge, S. A., and Heber, A., 1987, An 81 kd protein complexed with middle T antigen and pp60[c-src]: A possible phosphatidylinositol kinase, *Cell* **50:**1031–1037.

Cross, D. A., Alessi, D. R., Vandenheede, J. R., McDowell, H. E., Hundal, H. S., and Cohen, P., 1994, The inhibition of glycogen synthase kinase-3 by insulin or insulin-like growth factor 1 in the rat skeletal muscle cell line L6 is blocked by wortmannin, but not by rapamycin: Evidence that wortmannin blocks activation of the mitogen-activated protein kinase pathway in L6 cells between Ras and Raf, *Biochem. J.* **303:**21–26.

Dhand, R., Hara, K., Hiles, I., Bax, B., Gout, I., Panayotou, G., Fry, M. J., Yonezawa, K., Kasuga, M., and Waterfield, M. D., 1994a, PI 3-kinase: Structural and functional analysis of intersubunit interactions, *EMBO J.* **13:**511–521.

Dhand, R., Hiles, I., Panayotou, G., Roche, S., Fry, M. J., Gout, I., Totty, N. F., Truong, O., Vicendo, P., Yonezawa, K., Kasuga, M., Courtneidge, S. A., and Waterfield, M. D., 1994b, PI 3-kinase is a dual specificity enzyme: Autoregulation by an intrinsic protein-serine kinase activity, *EMBO J.* **13:**522–533.

Dilworth, S. M., Brewster, C. E., Jones, M. D., Lanfrancone, L., Pelicci, G., and Pelicci, P. G., 1994, Transformation by polyoma virus middle T-antigen involves the binding and tyrosine phosphorylation of Shc, *Nature* **367:**87–90.

Druker, B. J., Ling, L. E., Cohen, B., Roberts, T. M., and Schaffhausen, B. S., 1990, A completely transformation-defective point mutant of polyomavirus middle T antigen which retains full associated phosphatidylinositol kinase activity, *J. Virol.* **64:**4454–4461.

Eck, M. J., Shoelson, S. E., and Harrison, S. C., 1993, Recognition of a high-affinity phosphotyrosyl peptide by the Src homology-2 domain of p56[ck], *Nature* **362:**87–91.

Eck, M. J., Atwell, S. K., Shoelson, S. E., and Harrison, S. C., 1994, Structure of the regulatory domains of the Src-family tyrosine kinase Lck, *Nature* **368:**764–769.

End, P., Gout, I., Fry, M. J., Panayotou, G., Dhand, R., Yonezawa, K., Kasuga, M., and Waterfield, M. D., 1993, A biosensor approach to probe the structure and function of the p85 alpha subunit of the phosphatidylinositol 3-kinase complex, *J. Biol. Chem.* **268:**10066–10075.

Endemann, G., Yonezawa, K., and Roth, R. A., 1990, Phosphatidylinositol kinase or an associated protein is a substrate for the insulin receptor tyrosine kinase, *J. Biol. Chem.* **265:**396–400.

Escobedo, J. A., Navankasattusas, S., Kavanaugh, W. M., Milfay, D., Fried, V. A., and Williams, L. T., 1991, cDNA cloning of a novel 85 kd protein that has SH2 domains and regulates binding of PI3-kinase to the PDGF beta-receptor, *Cell* **65:**75–82.

Fantl, W. J., Escobedo, J. A., Martin, G. A., Turck, C. W., del Rosario, M., McCormick, F., and Williams, L. T., 1992, Distinct phosphotyrosines on a growth factor receptor bind to specific molecules that mediate different cellular signaling pathways, *Cell* **69:**413–423.

Fedi, P., Pierce, J. H., di Fiore, P., and Kraus, M. H., 1994, Efficient coupling with phosphatidylinositol 3-kinase, but not phospholipase C gamma or GTPase-activating protein, distinguishes ErbB-3 signaling from that of other ErbB/EGFR family members, *Mol. Cell. Biol.* **14:**492–500.

Feng, S., Chen, J. K., Yu, H., Simon, J. A., and Schreiber, S. L., 1994, Two binding orientations for peptides to the Src SH3 domain: Development of a general model for SH3–ligand interactions, *Science* **266:**1241–1247.

Flanagan, C. A., Schnieders, E. A., Emerick, A. W., Kunisawa, R., Admon, A., and Thorner, J., 1993, Phosphatidylinositol 4-kinase: Gene structure and requirement for yeast cell viability, *Science* **262:**1444–1448.

Folli, F., Saad, M. J., Backer, J. M., and Kahn, C. R., 1992, Insulin stimulation of phosphatidylinositol 3-kinase activity and association with insulin receptor substrate 1 in liver and muscle of the intact rat, *J. Biol. Chem.* **267:**22171–22177.

Fry, M. J., Panayotou, G., Dhand, R., Ruiz, L. F., Gout, I., Nguyen, O., Courtneidge, S. A., and Waterfield, M. D., 1992, Purification and characterization of a phosphatidylinositol 3-kinase complex from bovine brain by using phosphopeptide affinity columns, *Biochem. J.* **288:**383–393.

Fukui, Y., Saltiel, A. R., and Hanafusa, H., 1991, Phosphatidylinositol-3 kinase is activated in v-src, v-yes, and v-fps transformed chicken embryo fibroblasts, *Oncogene* **6:**407–411.

Giorgetti, S., Ballotti, R., Kowalski, C. A., Cormont, M., and Van Obberghen, E., 1992, Insulin stimulates phosphatidylinositol-3-kinase activity in rat adipocytes, *Eur. J. Biochem.* **207:**599–606.

Giorgetti, S., Ballotti, R., Kowalski, C. A., Tartare, S., and Van Obberghen, E., 1993, The insulin and insulin-like growth factor-I receptor substrate IRS-1 associates with and activates phosphatidylinositol 3-kinase in vitro, *J. Biol. Chem.* **268:**7358–7364.

Goldschmidt-Clermont, P. J., Machesky, L. M., Baldassare, J. J., and Pollard, T. D., 1990, The actin-binding protein profilin binds to PIP2 and inhibits its hydrolysis by phospholipase C, *Science* **247:**1575–1578.

Goldschmidt-Clermont, P. J., Kim, J. W., Machesky, L. M., Rhee, S. G., and Pollard, T. D., 1991, Regulation of phospholipase C-gamma 1 by profilin and tyrosine phosphorylation, *Science* **251:**1231–1233.

Graziani, A., Ling, L. E., Endemann, G., Carpenter, C. L., and Cantley, L. C., 1992, Purification and characterization of human erythrocyte phosphatidylinositol 4-kinase. Phosphatidylinositol 4-kinase and phosphatidylinositol 3-monophosphate 4-kinase are distinct enzymes, *Biochem. J.* **284:**39–45.

Guy, P. M., Platko, J. V., Cantley, L. C., Cerione, R. A., and Carraway, K. L., III, 1994, Insect cell-expressed p180erbB3 possesses an impaired tyrosine kinase activity, *Proc. Natl. Acad. Sci. USA* **91:**8132–8136.

Hadari, Y. R., Tzahar, E., Nadiv, O., Rotenberg, P., Roberts, C. J., LeRoith, D., Yarden, Y., and Zick, Y., 1992, Insulin and insulinomimetic agents induce activation of phosphatidylinositol 3′-kinase upon its association with pp185 (IRS-1) in intact rat livers [published erratum appears in *J. Biol. Chem.* 1993, **268:**9156], *J. Biol. Chem.* **267:**17483–17486.

Hara, K., Yonezawa, K., Sakaue, H., Ando, A., Kotani, K., Kitamura, T., Kitamura, Y., Ueda, H., Stephens, L., Jackson, T. R., Hawkins, P. T., Dhand, R., Clark, A. E., Holman, G. D., Waterfield, M. D., and Kasuga, M., 1994, 1-Phosphatidylinositol 3-kinase activity is required for insulin-stimulated glucose transport but not for RAS activation in CHO cells, *Proc. Natl. Acad. Sci. USA* **91:**7415–7419.

Hawkins, P. T., Jackson, T. R., and Stephens, L. R., 1992, Platelet-derived growth factor stimulates synthesis of PtdIns(3,4,5)P3 by activating a PtdIns(4,5)P2 3-OH kinase, *Nature* **358:**157–159.

Hayashi, H., Miyake, N., Kanai, F., Shibasaki, F., Takenawa, T., and Ebina, Y., 1991, Phosphorylation in vitro of the 85 kDa subunit of phosphatidylinositol 3-kinase and its possible activation by insulin receptor tyrosine kinase, *Biochem.* **280:**769–775.

Hayashi, H., Kamohara, S., Nishioka, Y., Kanai, F., Miyake, N., Fukui, Y., Shibasaki, F., Takenawa, T., and Ebina, Y., 1992, Insulin treatment stimulates the tyrosine phosphorylation of the alpha-type 85-kDa subunit of phosphatidylinositol 3-kinase in vivo, *J. Biol. Chem.* **267:**22575–22580.

Hayashi, H., Nishioka, Y., Kamohara, S., Kanai, F., Ishii, K., Fukui, Y., Shibasaki, F., Takenawa, T., Kido, H., Katsunuma, N., and Ebina, Y., 1993, The alpha-type 85-kDa subunit of phosphatidylinositol 3-kinase is phosphorylated at tyrosines 368, 580, and 607 by the insulin receptor, *J. Biol. Chem.* **268:**7107–7117.

Heitman, J., Movva, N. R., and Hall, M. N., 1991, Targets for cell cycle arrest by the immunosuppressant rapamycin in yeast, *Science* **253:**905–909.

Herbst, J. J., Andrews, G., Contillo, L., Lamphere, L., Gardner, J., Lienhard, G. E., and Gibbs, E. M., 1994, Potent activation of phosphatidylinositol 3′-kinase by simple phosphatyrosine peptides derived from insulin receptor substrate 1 containing two YMXM motifs for binding SH2 domains, *Biochemistry* **33:**9376–9381.

Herman, P. K., and Emr, S. D., 1990, Characterization of VPS34, a gene required for vacuolar protein sorting and vacuole segregation in Saccharomyces cerevisiae, *Mol. Cell. Biol.* **10:**6742–6754.

Hiles, I. D., Otsu, M., Volinia, S., Fry, M. J., Gout, I., Dhand, R., Panayotou, G., Ruiz, L. F., Thompson, A., Totty, N. F., Hsuan, J. J., Courtneidge, S. A., Parker, P. J., and Waterfield, M. D., 1992, Phosphatidylinositol 3-kinase: Structure and expression of the 110 kd catalytic subunit, *Cell* **70:**419–429.

Holt, K. H., Olson, L., Moye, R. W., and Pessin, J. E., 1994, Phosphatidylinositol 3-kinase activation is mediated by high-affinity interactions between distinct domains within the p110 and p85 subunits, *Mol. Cell. Biol.* **14:**42–49.

Hu, P., and Schlessinger, J., 1994, Direct association of p110 beta phosphatidylinositol 3-kinase with p85 is mediated by an N-terminal fragment of p110 beta, *Mol. Cell. Biol.* **14:**2577–2583.

Hu, P., Margolis, B., Skolnik, E. Y., Lammers, R., Ullrich, A., and Schlessinger, J., 1992, Interaction of phosphatidylinositol 3- kinase-associated p85 with epidermal growth factor and platelet-derived growth factor receptors, *Mol. Cell. Biol.* **12:**981–990.

Hu, P., Mondino, A., Skolnik, E. Y., and Schlessinger, J., 1993, Cloning of a novel, ubiquitously expressed human phosphatidylinositol 3-kinase and identification of its binding site on p85, *Mol. Cell. Biol.* **13:**7677–7688.

Jackson, T. R., Stephens, L. R., and Hawkins, P. T., 1992, Receptor specificity of growth factor-stimulated synthesis of 3-phosphorylated inositol lipids in Swiss 3T3 cells, *J. Biol. Chem.* **267:**16627–16636.

Janmey, P. A., 1994, Phosphoinositides and calcium as regulators of cellular actin assembly and disassembly, *Annu. Rev. Physiol.* **56:**169–191.

Jefferies, H. B., Reinhard, C., Kozma, S. C., and Thomas, G., 1994, Rapamycin selectively represses translocation of the "polypyrimidine tract" mRNA family, *Proc. Natl. Acad. Sci. USA* **91:**4441–4445.

Jhun, B. H., Rose, D. W., Seely, B. L., Rameh, L., Cantley, L., Saltiel, A. R., and Olefsky, J. M., 1994, Microinjection of the SH2 domain of the 85-kilodalton subunit of phosphatidylinositol 3-kinase inhibits insulin-induced DNA synthesis and c-fos expression, *Mol. Cell. Biol.* **14:**7466–7475.

Johnson, D. I., and Pringle, J. R., 1990, Molecular characterization of CDC42, a Saccharomyces cerevisiae gene involved in the development of cell polarity, *J. Cell Biol.* **111:**143–152.

Kapeller, R., Chen, K. S., Yoakim, M., Schaffhausen, B. S., Backer, J., White, M. F., Cantley, L. C., and Ruderman, N. B., 1991, Mutations in the juxtamembrane region of the insulin receptor impair activation of phosphatidylinositol 3-kinase by insulin, *Mol. Endocrinol.* **5:**769–777.

Kapeller, R., Prasad, K. V., Janssen, O., Hou, W., Schaffhausen, B. S., Rudd, C. E., and Cantley, L. C., 1994, Identification of two SH3-binding motifs in the regulatory subunit of phosphatidylinositol 3-kinase, *J. Biol. Chem.* **269:**1927–1933.

Kaplan, D. R., Whitman, M., Schaffhausen, B., Pallas, D. C., White, M., Cantley, L., and Roberts, T. M., 1987, Common elements in growth factor stimulation and oncogenic transformation: 85 kd phosphoprotein and phosphatidylinositol kinase activity, *Cell* **50:**1021–1029.

Karnitz, L. M., Sutor, S. L., and Abraham, R. T., 1994, The Src-family kinase, Fyn, regulates the activation of phosphatidylinositol 3-kinase in an interleukin 2-responsive T cell line, *J. Exp. Med.* **179:**1799–1808.

Kavanaugh, W. M., and Williams, L. T., 1994, An alternative to SH2 domains for binding tyrosine-phosphorylated proteins, *Science* **266:**1862–1865.

Kazlauskas, A., and Cooper, J. A., 1989, Autophosphorylation of the PDGF receptor in the kinase insert region regulates interactions with cell proteins, *Cell* **58:**1121–1133.

Kazlauskas, A., and Cooper, J. A., 1990, Phosphorylation of the PDGF receptor beta subunit creates a tight binding site for phosphatidylinositol 3 kinase, *EMBO J.* **9:**3279–3286.

Kazlauskas, A., Kashishian, A., Cooper, J. A., and Valius, M., 1992, GTPase-activating protein and phosphatidylinositol 3-kinase bind to distinct regions of the platelet-derived growth factor receptor beta subunit, *Mol. Cell. Biol.* **12:**2534–2544.

Kelly, K. L., and Ruderman, N. B., 1993, Insulin-stimulated phosphatidylinositol 3-kinase. Association with a 185-kDa tyrosine-phosphorylated protein (IRS-1) and localization in a low density membrane vesicle, *J. Biol. Chem.* **268:**4391–4398.

Kim, H. H., Sierke, S. L., and Koland, J. G., 1994, Epidermal growth factor-dependent association of phosphatidylinositol 3-kinase with the erbB3 gene product, *J. Biol. Chem.* **269:**24747–24755.

Kimura, K., Hattori, S., Kabuyama, Y., Shizawa, Y., Takayanagi, J., Nakamura, S., Toki, S., Matsuda, Y., Onodera, K., and Fukui, Y., 1994, Neurite outgrowth of PC12 cells is suppressed by wortmannin, a specific inhibitor of phosphatidylinositol 3-kinase, *J. Biol. Chem.* **269:**18961–18967.

Klippel, A., Escobedo, J. A., Fantl, W. J., and Williams, L. T., 1992, The C-terminal SH2 domain of p85 accounts for the high affinity and specificity of the binding of phosphatidylinositol 3-kinase to phosphorylated platelet-derived growth factor beta receptor, *Mol. Cell. Biol.* **12:**1451–1459.

Klippel, A., Escobedo, J. A., Hu, Q., and Williams, L. T., 1993, A region of the 85-kilodalton (kDa) subunit of phosphatidylinositol 3-kinase binds the 110-kDa catalytic subunit in vivo, *Mol. Cell. Biol.* **13:**5560–5566.

Klippel, A., Escobedo, J. A., Hirano, M., and Williams, L. T., 1994, The interaction of small domains between the subunits of phosphatidylinositol 3-kinase determines enzyme activity, *Mol. Cell. Biol.* **14:**2675–2685.

Knighton, D. R., Zheng, J. H., Ten, E. L., Ashford, V. A., Xuong, N. H., Taylor, S. S., and Sowadski, J. M., 1991, Crystal structure of the catalytic subunit of cyclic adenosine monophosphate-dependent protein kinase [see comments], *Science* **253:**407–414.

Kodaki, T., Woscholski, R., Hallberg, B., Rodriguez, V. P., Downward, J., and Parker, P. J., 1994, The activation of phosphatidylinositol 3-kinase by Ras, *Curr. Biol.* **4:**798–806.

Kotani, K., Yonezawa, K., Hara, K., Ueda, H., Kitamura, Y., Sakaue, H., Ando, A., Chavanieu, A., Calas, B., Grigorescu, F., Nishiyama, M., Waterfield, M. D., and Kasuga, M., 1994, Involvement of phosphoinositide 3-kinase in insulin- or IGF-l-induced membrane ruffling, *EMBO J.* **13:**2313–2321.

Koyama, S., Yu, H., Dalgarno, D. C., Shin, T. B., Zydowsky, L. D., and Schreiber, S. L., 1993, Structure of the PI3K SH3 domain and analysis of the SH3 family, *Cell* **72:**945–952.

Lam, K., Carpenter, C. L., Ruderman, N. B., Friel, J. C., and Kelly, K. L., 1994, The phosphatidylinositol 3-kinase serine kinase phosphorylates IRS-1. Stimulation by insulin and inhibition by wortmannin, *J. Biol. Chem.* **269:**20648–20652.

Lamphere, L., Carpenter, C. L., Sheng, Z. F., Kallen, R. G., and Lienhard, G. E., 1994, Activation of PI 3-kinase in 3T3-L1 adipocytes by association with insulin receptor substrate-1, *Am. J. Physiol.* **266:**E486–E494.

Lee, C. H., Kominos, D., Jacques, S., Margolis, B., Schlessinger, J., Shoelson, S. E., and Kuriyan, J., 1994, Crystal structures of peptide complexes of the amino-terminal SH2 domain of the Syp tyrosine phosphatase, *Structure* **2:**423–438.

Lim, W. A., Richards, F. M., and Fox, R. O., 1994, Structural determinants of peptide-binding orientation and of sequence specificity in SH3 domains, *Nature* **372:**375–379.

Ling, L. E., Druker, B. J., Cantley, L. C., and Roberts, T. M., 1992, Transformation-defective mutants of polyomavirus middle T antigen associate with phosphatidylinositol 3-kinase (PI 3-kinase) but are unable to maintain wild-type levels of PI 3-kinase products in intact cells, *J. Virol.* **66:**1702–1708.

Liu, X., Marengere, L. E., Koch, C. A., and Pawson, T., 1993, The v-Src SH3 domain binds phosphatidylinositol 3' kinase, *Mol. Cell. Biol.* **13:**5225–5232.

Macara, I. G., Marinetti, G. V., and Balduzzi, P. C., 1984, Transforming protein of avian sarcoma virus UR2 is associated with phosphatidylinositol kinase activity: Possible role in tumorigenesis, *Proc. Natl. Acad. Sci. USA* **81:**2728–2732.

Marshall, C. J., 1994, MAP kinase kinase kinase, MAP kinase kinase, and MAP kinase, *Curr Opin. Gen. Dev.* **4:**82–89.

Miller, E. S., and Ascoli, M., 1990, Anti-phosphotyrosine immunoprecipitation of phosphatidylinositol 3'-kinase activity in different cell types after exposure to epidermal growth factor, *Biochem. Biophys. Res. Commun.* **173:**289–295.

Ming, X. F., Burgering, B. M., Wennstrom, S., Claesson, W. L., Heldin, C. H., Bos, J. L., Kozma, S. C., and Thomas, G., 1994, Activation of p70/p85 S6 kinase by a pathway independent of p21ras, *Nature* **371:**426–429.

Mischak, H., Goodnight, J. A., Kolch, W., Martiny, B. G., Schaechtle, C., Kazanietz, M. G., Blumberg, P. M., Pierce, J. H., and Mushinski, J. F., 1993, Overexpression of protein kinase C-delta and -epsilon in NIH 3T3 cells induces opposite effects on growth, morphology, anchorage dependence, and tumorigenicity, *J. Biol. Chem.* **268:**6090–6096.

Monfar, M., Lemon, K. P., Grammer, T. C., Cheatham, L., Chung, J., Vlahos, C. J., and Blenis, J., 1995, Activation of pp70/85 S6 kinases in interleukin-2-responsive lymphoid cells is mediated by phosphatidylinositol 3-kinase and inhibited by cyclic AMP, *Mol. Cell. Biol.* **15:**326–337.

Morgan, S. J., Smith, A. D., and Parker, P. J., 1990, Purification and characterization of bovine brain type I phosphatidylinositol kinase, *Eur. J. Biochem.* **191:**761–767.

Myers, M. J., Backer, J. M., Sun, X. J., Shoelson, S., Hu, P., Schlessinger, J., Yoakim, M., Schaffhausen, B., and White, M. F., 1992, IRS-1 activates phosphatidylinositol 3'-kinase by associating with src homology 2 domains of p85, *Proc. Natl. Acad. Sci. USA* **89:**10350–10354.

Nakanishi, H., Brewer, K. A., and Exton, J. H., 1993, Activation of the zeta isozyme of protein kinase C by phosphatidylinositol 3,4,5-trisphosphate, *J. Biol. Chem.* **268:**13–16.

Nobes, C. D., and Hall, A., 1995, Rho, Rac, and CDC42 GTPases regulate the assembly of multimolecular focal complexes associated with actin stress fibers, lamellipodia, and filopodia, *Cell* **81:**53–62.

Okada, T., Kawano, Y., Sakakibara, T., Hazeki, O., and Ui, M., 1994a, Essential role of phosphatidylionositol 3-kinase in insulin-induced glucose transport and antilipolysis in rat adipocytes. Studies with a selective inhibitor wortmannin, *J. Biol. Chem.* **269:**3568–3573.

Okada, T., Sakuma, L., Fukui, Y., Hazeki, O., and Ui, M., 1994b, Blockage of chemotactic peptide-induced stimulation of neutrophils by wortmannin as a result of selective inhibition of phosphatidylinositol 3-kinase, *J. Biol. Chem.* **269:**3563–3567.

Otsu, M., Hiles, I., Gout, I., Fry, M. J., Ruiz, L. F., Panayotou, G., Thompson, A., Dhand, R., Hsuan, J., Totty, N., Smith, A. D., Morgan, S. J., Courtneidge, S. A., Parker, P. J., and Waterfield, M. D., 1991, Characterization of two 85 kd proteins that associate with receptor tyrosine kinases, middle-T/pp60c-src complexes, and PI3-kinase, *Cell* **65:**91–104.

Overduin, M., Rios, C. B., Mayer, B. J., Baltimore, D., and Cowburn, D., 1992, Three-dimensional solution structure of the src homology 2 domain of c-abl, *Cell* **70:**697–704.

Pallas, D. C., Cherington, V., Morgan, W., DeAnda, J., Kaplan, D., Schaffhausen, B., and Roberts, T. M., 1988, Cellular proteins that associate with the middle and small T antigens of polyomavirus, *J. Virol.* **62:**3934–3940.

Pallas, D. C., Shahrik, L. K., Martin, B. L., Jaspers, S., Miller, T. B., Brautigan, D. L., and Roberts, T. M., 1990, Polyoma small and middle T antigens and SV40 small t antigen form stable complexes with protein phosphatase 2A, *Cell* **60:**167–176.

Panayotou, G., Gish, G., End, P., Truong, O., Gout, I., Dhand, R., Fry, M. J., Hiles, I., Pawson, T., and Waterfield, M. D., 1993, Interactions between SH2 domains and tyrosine-phosphorylated platelet-derived growth factor beta-receptor sequences: Analysis of kinetic parameters by a novel biosensor-based approach, *Mol. Cell. Biol.* **13:**3567–3576.

Pascal, S. M., Singer, A. U., Gish, G., Yamazaki, T., Shoelson, S. E., Pawson, T., Kay, L. E., and Forman, K. J., 1994, Nuclear magnetic resonance structure of an SH2 domain of phospholipase C-gamma 1 complexed with a high affinity binding peptide, *Cell* **77**:461–472.

Piccione, E., Case, R. D., Domchek, S. M., Hu, P., Chaudhuri, M., Backer, J. M., Schlessinger, J., and Shoelson, S. E., 1993, Phosphatidylinositol 3-kinase p85 SH2 domain specificity defined by direct phosphospeptide/SH2 domain binding, *Biochemistry* **32**:3197–3202.

Pignataro, O. P., and Ascoli, M., 1990, Studies with insulin and insulin-like growth factor-I show that the increased labeling of phosphatidylinositol-3,4-bisphosphate is not sufficient to elicit the diverse actions of epidermal growth factor on MA-10 Leydig tumor cells, *Mol. Endocrinol.* **4**:758–765.

Pleiman, C. M., Clark, M. R., Gauen, L. K., Winitz, S., Coggeshall, K. M., Johnson, G. L., Shaw, A. S., and Cambier, J. C., 1993, Mapping of sites on the Src family protein tyrosine kinases p55blk, p59fyn, and p56lyn which interact with the effector molecules phospholipase C-gamma 2, microtubule-associated protein kinase, GTPase-activating protein, and phosphatidylinositol 3-kinase, *Mol. Cell. Biol.* **13**:5877–5887.

Pleiman, C. M., Hertz, W. M., and Cambier, J. C., 1994, Activation of phosphatidylinositol-3′ kinase by Src-family kinase SH3 binding to the p85 subunit, *Science* **263**:1609–1612.

Prasad, K. V., Janssen, O., Kapeller, R., Raab, M., Cantley, L. C., and Rudd, C. E., 1993a, Src-homology 3 domain of protein kinase p59fyn mediates binding to phosphatidylinositol 3-kinase in T cells, *Proc. Natl. Acad. Sci. USA* **90**:7366–7370.

Prasad, K. V., Kapeller, R., Janssen, O., Repke, H., Duke, C. J., Cantley, L. C., and Rudd, C. E., 1993b, Phosphatidylinositol (PI) 3-kinase and PI 4-kinase binding to the CD4–p56lack complex: The p56lck SH3 domain binds to PI 3-kinase but not PI 4-kinase, *Mol. Cell. Biol.* **13**:7708–7717.

Prigent, S. A., and Gullick, W. J., 1994, Identification of c-erbB-3 binding sites for phosphatidylinositol 3′-kinase and SHC using an EGF receptor/c-erbB-3 chimera, *EMBO J.* **13**:2831–2841.

Raffioni, S., and Bradshaw, R. A., 1992, Activation of phosphatidylinositol 3-kinase by epidermal growth factor, basic fibroblast growth factor, and nerve growth factor in PC12 pheochromocytoma cells, *Proc. Natl. Acad. Sci. USA* **89**:9121–9125.

Reedijk, M., Liu, X. Q., and Pawson, T., 1990, Interactions of phosphatidylinositol kinase, GTPase-activating protein (GAP), and GAP-associated proteins with the colony-stimulating factor 1 receptor, *Mol. Cell. Biol.* **10**:5601–5608.

Reedijk, M., Liu, X., van der Geer, P., Letwin, K., Waterfield, M. D., Hunter, T., and Pawson, T., 1992, Tyr721 regulates specific binding of the CSF-1 receptor kinase insert to PI 3′-kinase SH2 domains: A model for SH2-mediated receptor–target interactions, *EMBO J.* **11**:1365–1372.

Rickles, R. J., Botfield, M. C., Weng, Z., Taylor, J. A., Green, O. M., Brugge, J. S., and Zoller, M. J., 1994, Identification of Src, Fyn, Lyn, PI3K and Abl SH3 domain ligands using phage display libraries, *EMBO J.* **13**:5598–5604.

Ridley, A. J., Paterson, H. F., Johnston, C. L., Diekmann, D., and Hall, A., 1992, The small GTP-binding protein rac regulates growth factor-induced membrane ruffling, *Cell* **70**:401–410.

Roche, S., Koegl, M., and Courtneidge, S. A., 1994, The phosphatidylinositol 3-kinase alpha is required for DNA synthesis induced by some, but not all, growth factors, *Proc. Natl. Sci. USA* **91**:9185–9189.

Rodriguez-Viciana, P., Warne, P. H., Dhand, R., Vanhaesebroeck, B., Gout, I., Fry, M. J., Waterfield, M. D., and Downward, J., 1994, Phosphatidylinositol-3-OH kinase as a direct target of Ras, *Nature* **370**:527–532.

Ruderman, N. B., Kapeller, R., White, M. F., and Cantley, L. C., 1990, Activation of phosphatidylinositol 3-kinase by insulin, *Proc. Natl. Acad. Sci. USA* **87**:1411–1415.

Sanchez, M. V., Goldfine, I. D., Vlahos, C. J., and Sung, C. K., 1994, Role of phosphatidylinositol-3-kinase in insulin receptor signaling: Studies with inhibitor, LY294002, *Biochem. Biophys. Res. Commun.* **204**:446–452.

Satoh, T., Fantl, W. J., Escobedo, J. A., Williams, L. T., and Kaziro, Y., 1993, Platelet-derived growth factor receptor mediates activation of ras through different signaling pathways in different cell types, *Mol. Cell. Biol.* **13**:3706–3713.

Schu, P. V., Takegawa, K., Fry, M. J., Stack, J. H., Waterfield, M. D., and Emr, S. D., 1993, Phosphatidyl-inositol 3-kinase encoded by yeast VPS34 gene essential for protein sorting, *Science* **260**:88–91.

Serunian, L. A., Haber, M. T., Fukui, T., Kim, J. W., Rhee, S. G., Lowenstein, J. M., and Cantley, L. C., 1989, Polyphosphoinositides produced by phosphatidylinositol 3-kinase are poor substrates for phospholipases C from rat liver and bovine brain, *J. Biol. Chem.* **264:**17809–17815.

Serunian, L. A., Auger, K. R., Roberts, T. M., and Cantley, L. C., 1990, Production of novel poly-phosphoinositides in vivo is linked to cell transformation by polyomavirus middle T antigen, *J. Virol.* **64:**4718–4725.

Shepherd, P. R., Nave, B. T., and Siddle, K., 1995, Insulin stimulation of glycogen synthesis and glycogen synthase activity is blocked by wortmannin and rapamycin in 3T3-L1 adipocytes: Evidence for the involvement of phosphoinositide 3-kinase and p70 ribosomal protein-S6 kinase, *Biochem. J.* **305:**25–28.

Shibasaki, F., Homma, Y., and Takenawa, T., 1991, Two types of phosphatidylinositol 3-kinase from bovine thymus. Monomer and heterodimer form, *J. Biol. Chem.* **266:**8108–8114.

Shinjo, K., Koland, J. G., Hart, M. J., Narasimhan, V., Johnson, D. I., Evans, T., and Cerione, R. A., 1990, Molecular cloning of the gene for the human placental GTP-binding protein GP (G25K): Identification of this GTP-binding protein as the human homolog of the yeast cell-division-cycle protein CDC42, *Proc. Natl. Acad. Sci. USA* **87:**9853–9857.

Singh, S. S., Chauhan, A., Brockerhoff, H., and Chauhan, V. P., 1993, Activation of protein kinase C by phosphatidylinositol 3,4,5-trisphosphate, *Biochem. Biophys. Res. Commun.* **195:**104–112.

Sjolander, A., and Lapetina, E. G., 1992, Agonist-induced association of the p21ras GTPase-activating protein with phosphatidylinositol 3-kinase, *Biochem. Biophys. Res. Commun.* **189:**1503–1508.

Sjolander, A., Yamamoto, K., Huber, B. E., and Lapetina, E. G., 1991, Association of p21ras with phosphatidylinositol 3-kinase, *Proc. Natl. Acad. Sci. USA* **88:**7908–7912.

Skolnik, E. Y., Margolis, B., Mohammadi, M., Lowenstein, E., Fischer, R., Drepps, A., Ullrich, A., and Schlessinger, J., 1991, Cloning of PI3 kinase-associated p85 utilizing a novel method for expression/cloning of target proteins for receptor tyrosine kinases, *Cell* **65:**83–90.

Sliwkowski, M. X., Schaefer, G., Akita, R. W., Lofgren, J. A., Fitzpatrick, V. D., Nuijens, A., Fendly, B. M., Cerione, R. A., Vandlen, R. L., and Carraway, K. L., III, 1994, Coexpression of erbB2 and erbB3 proteins reconstitutes a high affinity receptor for heregulin, *J. Biol. Chem.* **269:**14661–14665.

Soltoff, S. P., Carraway, K. L., III, Prigent, S. A., Gullick, W. G., and Cantley, L. C., 1994, ErbB3 is involved in activation of phosphatidylinositol 3-kinase by epidermal growth factor, *Mol. Cell. Biol.* **14:**3550–3558.

Songyang, Z., Shoelson, S. E., Chaudhuri, M., Gish, G., Pawson, T., Haser, W. G., King, F., Roberts, T., Ratnofsky, S., Lechleider, R. J., Neel, B. G., Birge, R. B., Fajardo, J. E., Chou, M. M., Hanafusa, H., Schaffhausen, B., and Cantley, L. C., 1993, SH2 domains recognize specific phosphopeptide sequences, *Cell* **72:**767–778.

Stack, J. H., and Emr, S. D., 1994, Vps34p required for yeast vacuolar protein sorting is a multiple specificity kinase that exhibits both protein kinase and phosphatidylinositol-specific PI 3-kinase activities, *J. Biol. Chem.* **269:**31552–31562.

Stephens, L. R., Hughes, K. T., and Irvine, R. F., 1991, Pathway of phosphatidylinositol(3,4,5)-trisphosphate synthesis in activated neutrophils, *Nature* **351:**33–39.

Stephens, L., Jackson, T., and Hawkins, P. T., 1993a, Synthesis of phosphatidylinositol 3,4,5-trisphosphate in permeabilized neutrophils regulated by receptors and G-proteins, *J. Biol. Chem.* **268:**17162–17172.

Stephens, L. R., Jackson, T. R., and Hawkins, P. T., 1993b, Agonist-stimulated synthesis of phos-phatidylinositol(3,4,5)-trisphosphate: A new intracellular signalling system? *Biochim. Biophys. Acta* **1179:**27–75.

Stephens, L., Smrcka, A., Cooke, F. T., Jackson, T. R., Sternweis, P. C., and Hawkins, P. T., 1994, A novel phosphoinositide 3 kinase activity in myeloid-derived cells is activated by G protein beta gamma subunits, *Cell* **77:**83–93.

Sugimoto, Y., and Erikson, R. L., 1985, Phosphatidylinositol kinase activities in normal and Rous sarcoma virus-transformed cells, *Mol. Cell. Biol.* **5:**3194–3198.

Sultan, C., Breton, M., Mauco, G., Grondin, P., Plantavid, M., and Chap, H., 1990, The novel inositol lipid phosphatidylinositol 3,4-bisphosphate is produced in human blood platelets upon thrombin stimulation, *Biochem. J.* **269:**831–834.

Sun, X. J., Rothenberg, P., Kahn, C. R., Backer, J. M., Araki, E., Wilden, P. A., Cahill, D. A., Goldstein, B. J., and White, M. F., 1991, Structure of the insulin receptor substrate IRS-1 defines a unique signal transduction protein, *Nature* **352:**73–77.

Sun, X. J., Crimmins, D. L., Myers, M. J., Miralpeix, M., and White, M. F., 1993, Pleiotropic insulin signals are engaged by multisite phosphorylation of IRS-1, *Mol. Cell. Biol.* **13:**7418–7428.

Sung, C. K., and Goldfine, I. D., 1992, Phosphatidylinositol-3-kinase is a nontyrosine phosphorylated member of the insulin receptor signalling complex, *Biochem. Biophys. Res. Commun.* **189:**1024–1030.

Susa, M., Keeler, M., and Varticovski, L., 1992, Platelet-derived growth factor activates membrane-associated phosphatidylinositol 3-kinase and mediates its translocation from the cytosol. Detection of enzyme activity in detergent-solubilized cell extracts, *J. Biol. Chem.* **267:**22951–22956.

Talmage, D. A., Freund, R., Young, A. T., Dahl, J., Dawe, C. J., and Benjamin, T. L., 1989, Phosphorylation of middle T by pp60c-src: A switch for binding of phosphatidylinositol 3-kinase and optimal tumorigenesis, *Cell* **59:**55–65.

Toker, A., Meyer, M., Reddy, K. K., Falck, J. R., Aneja, R., Aneja, S., Parra, A., Burns, D. J., Ballas, L. M., and Cantley, L. C., 1994, Activation of protein kinase C family members by the novel polyphosphoinositides PtdIns-3,4-P2 and PtdIns-3,4,5-P3, *J. Biol. Chem.* **269:**32358–32367.

Tolias, K. F., Cantley, L. C., and Carpenter, C. L., 1995, Rho family GTPases bind to phosphoinositide kinases, *J. Biol. Chem.* **270:**17656–17659.

Valius, M., and Kazlauskas, A., 1993, Phospholipase C-gamma 1 and phosphatidylinositol 3 kinase are the downstream mediators of the PDGF receptor's mitogenic signal, *Cell* **73:**321–334.

Varticovski, L., Daley, G. Q., Jackson, P., Baltimore, D., and Cantley, L. C., 1991, Activation of phosphatidylinositol 3-kinase in cells expressing abl oncogene variants, *Mol. Cell. Biol.* **11:**1107–1113.

Varticovski, L., Harrison, F. D., Keeler, M. L., and Susa, M., 1994, Role of PI 3-kinase in mitogenesis, *Biochim. Biophys. Acta* **1226:**1–11.

Vemuri, G. S., and Rittenhouse, S. E., 1994, Wortmannin inhibits serum-induced activation of phosphoinositide 3-kinase and proliferation of CHRF-288 cells, *Biochem. Biophys. Res. Commun.* **202:**1619–1623.

Vogel, L. B., and Fujita, D. J., 1993, The SH3 domain of p56lck is involved in binding to phosphatidylinositol 3'-kinase from T lymphocytes, *Mol. Cell. Biol.* **13:**7408–7417.

Wages, D. S., Keefer, J., Rall, T. B., and Weber, M. J., 1992, Mutations in the SH3 domain of the src oncogene which decrease association of phosphatidylinositol 3'-kinase activity with pp60v-src and alter cellular morphology, *J. Virol.* **66:**1866–1874.

Waksman, G., Kominos, D., Robertson, S. C., Pant, N., Baltimore, D., Birge, R. B., Cowburn, D., Hanafusa, H., Mayer, B. J., Overduin, M., Resh, M. D., Rios, C. B., Silverman, L., and Kuriyan, J., 1992, Crystal structure of the phosphotyrosine recognition domain SH2 of v-src complexed with tyrosine-phosphorylated peptides [see comments], *Nature* **358:**646–653.

Waksman, G., Shoelson, S. E., Pant, N., Cowburn, D., and Kuriyan, J., 1993, Binding of a high affinity phosphotyrosyl peptide to the Src SH2 domain: Crystal structures of the complexed and peptide-free forms, *Cell* **72:**779–790.

Welsh, G. I., Foulstone, E. J., Young, W. S., Tavare, J. M., and Proud, C. G., 1994, Wortmannin inhibits the effects of insulin and serum on the activities of glycogen synthase kinase-3 and mitogen-activated protein kinase, *Biochem. J.* **303:**15–20.

Wennstrom, S., Hawkins, P., Cooke, F., Hara, K., Yonezawa, K., Kasuga, M., Jackson, T., Claesson, W. L., and Stephens, L., 1994a, Activation of phosphoinositide 3-kinase is required for PDGF-stimulated membrane ruffling, *Curr. Biol.* **4:**385–393.

Wennstrom, S., Siegbahn, A., Yokote, K., Arvidsson, A. K., Heldin, C. H., Mori, S., and Claesson, W. L., 1994b, Membrane ruffling and chemotaxis transduced by the PDGF beta-receptor require the binding site for phosphatidylinositol 3' kinase, *Oncogene* **9:**651–660.

White, M. F., and Kahn, C. R., 1994, The insulin signaling system, *J. Biol. Chem.* **269:**1–4.

White, M. F., Maron, R., and Kahn, C. R., 1985, Insulin rapidly stimulates tyrosine phosphorylation of a Mr-185,000 protein in intact cells, *Nature* **318:**183–186.

Whitman, M., Kaplan, D. R., Schaffhausen, B., Cantley, L., and Roberts, T. M., 1985, Association of phosphatidylinositol kinase activity with polyoma middle-T competent for transformation, *Nature* **315:**239–242.

Whitman, M., Kaplan, D., Roberts, T., and Cantley, L., 1987, Evidence for two distinct phosphatidyl-inositol kinases in fibroblasts. Implications for cellular regulation, *Biochem. J.* **247**:165–174.

Whitman, M., Downes, C. P., Keeler, M., Keller, T., and Cantley, L., 1988, Type I phosphatidylinositol kinase makes a novel inositol phospholipid, phosphatidylinositol-3-phosphate, *Nature* **332**:644–646.

Wittekind, M., Mapelli, C., Farmer, B., Suen, K. L., Goldfarb, V., Tsao, J., Lavoie, T., Barbacid, M., Meyers, C. A., and Mueller, L., 1994, Orientation of peptide fragments from Sos proteins bound to the N-terminal SH3 domain of Grb2 determined by NMR spectroscopy, *Biochemistry* **33**:13531–13539.

Wong, K., and Cantley, L. C., 1994, Cloning and characterization of a human phosphatidylinositol 4-kinase, *J. Biol. Chem.* **269**:28878–28884.

Yamamoto, K., Graziani, A., Carpenter, C., Cantley, L. C., and Lapetina, E. G., 1990, A novel pathway for the formation of phosphatidylinositol 3,4-bisphosphate. Phosphorylation of phosphatidylinositol 3-monophosphate by phosphatidylinositol-3-monophosphate 4-kinase, *J. Biol. Chem.* **265**:22086–22089.

Yamauchi, K., Holt, K., and Pessin, J. E., 1993, Phosphatidylinositol 3-kinase functions upstream of Ras and Raf in mediating insulin stimulation of c-fos transcription, *J. Biol. Chem.* **268**:14597–14600.

Yano, H., Nakanishi, S., Kimura, K., Hanai, N., Saitoh, Y., Fukui, Y., Nonomura, Y., and Matsuda, Y., 1993, Inhibition of histamine secretion by wortmannin through the blockade of phosphatidyl-inositol 3-kinase in RBL-2H3 cells, *J. Biol. Chem.* **268**:25846–25856.

Yeh, J. I., Gulve, E. A., Rameh, L., and Birnbaum, M. J., 1995, The effects of wortmannin on rat skeletal muscle. Dissociation of signaling pathways for insulin- and contraction-activated hexose transport, *J. Biol. Chem.* **270**:2107–2111.

Yoakim, M., Hou, W., Liu, Y., Carpenter, C. L., Kapeller, R., and Schaffhausen, B. S., 1992, Interactions of polyomavirus middle T with the SH2 domains of the pp85 subunit of phosphatidylinositol-3-kinase, *J. Virol.* **66**:5485–5491.

Yonezawa, K., Ueda, H., Hara, K., Nishida, K., Ando, A., Chavanieu, A., Matsuba, H., Shii, K., Yokono, K., Fukui, Y., Calas, B., Grigorescu, F., Dhand, R., Gout, I., Otsu, M., Waterfield, M. D., and Kasuga, M., 1992a, Insulin-dependent formation of a complex containing an 85-kDa subunit of phospha-tidylinositol 3-kinase and tyrosine-phosphorylated insulin receptor substrate 1, *J. Biol. Chem.* **267**:25958–25965.

Yonezawa, K., Yokono, K., Shii, K., Ogawa, W., Ando, A., Hara, K., Baba, S., Kaburagi, Y., Yamamoto, H. R., Momomura, K., Kadowaki, T., and Kasuga, M., 1992b, In vitro association of phosphatidyl-inositol 3-kinase activity with the activated insulin receptor tyrosine kinase, *J. Biol. Chem.* **267**:440–446.

Yoshida, S., Ohya, Y., Goebl, M., Nakano, A., and Anraku, Y., 1994, A novel gene, STT4, encodes a phosphatidylinositol 4-kinase in the PKC1 protein kinase pathway of Saccharomyces cerevisiae, *J. Biol. Chem.* **269**:1166–1172.

Zhang, J., King, W. G., Dillon, S., Hall, A., Feig, L., and Rittenhouse, S. E., 1993, Activation of platelet phosphatidylinositide 3-kinase requires the small GTP-binding protein Rho, *J. Biol. Chem.* **268**:22251–22254.

Zhang, J., Zhang, J., Benovic, J. L., Sugai, M., Wetzker, R., Gout, I., and Rittenhouse, S. E., 1995, Sequestration of a G-protein βγ subunit or ADP-ribosylation of rho can inhibit thrombin-induced activation of platelet phosphoinositide 3-kinases, *J. Biol. Chem.* **270**:6589–6594.

Zheng, Y., Hart, M. J., Shinjo, K., Evans, T., Bender, A., and Cerione, R. A., 1993, Biochemical comparisons of the Saccharomyces cerevisiae Bem2 and Bem3 proteins. Delineation of a limit Cdc42 GTPase-activating protein domain, *J. Biol. Chem.* **268**:24629–24634.

Zheng, Y., Bagrodia, S., and Cerione, R. A., 1994, Activation of phosphoinositide 3-kinase activity by Cdc42Hs binding to p85, *J. Biol. Chem.* **269**:18727–18730.

Chapter 5

Ceramide

A Novel Second Messenger and Lipid Mediator

Yusuf A. Hannun, Lina M. Obeid, and Ghassan S. Dbaibo

5.1.   Introduction

Phospholipids have long been known to be integral structural components of cell membranes. Their ability to spontaneously form a lipid bilayer which provides a permeability barrier between extracellular and intracellular compartments is essential for cell survival. Over the last several decades, another role became widely appreciated. Phospholipids, particularly glycerolipids, emerged as a rich reservoir of a broad variety of bioactive molecules generated in response to extracellular stimuli (Dennis *et al.*, 1991; Liscovitch and Cantley, 1994). These metabolites can function either as intracellular second messengers [such as diacylglycerol (DAG), inositol triphosphate, and arachidonic acid] or as intercellular mediators (such as platelet-activating factor and the eicosanoids) (Nishizuka, 1992; Rhee *et al.*, 1989; Majerus *et al.*, 1986; Berridge and Irvine, 1989; Nahorski and Potter, 1989; Hanahan, 1986; Exton, 1990). A number of signaling pathways have now been well established with increasingly complex cross-interactions. The common theme is that of an extracellular signal-activated phospholipase that has specificity to a particular phospholipid substrate present in the cell membrane. The subsequent hydrolysis of the phospholipid results in the generation of bioactive molecules that are either released extracellularly where they exert their functions by binding to specific receptors on target cells or diffuse intracellularly where they act on specific targets (e.g., the different protein kinase C isozymes or calcium channels) (Nishizuka, 1992; Rhee *et al.*, 1989). These pathways are discussed in detail elsewhere in this volume.

Yusuf A. Hannun and *Lina M. Obeid* • Department of Medicine and Cell Biology, Duke University Medical Center, Durham, North Carolina, 27710.   *Ghassan S. Dbaibo* • Department of Pediatrics, Duke University Medical Center, Durham, North Carolina 27710.

Handbook of Lipid Research, Volume 8: Lipid Second Messengers, edited by Robert M. Bell *et al.* Plenum Press, New York, 1996.

In contrast to the well-established role of glycerolipid-derived metabolites in cell signaling, a similar role for sphingolipids and their breakdown products has only recently been appreciated despite the ubiquity, diversity, and complexity of these membrane lipids (Hannun and Bell, 1989; Sweely, 1991). Much of the early interest in sphingolipids focused on the carbohydrate moiety present in the glycosphingolipids (Hakomori, 1990). These compounds emerged as important determinants of cell–cell interactions, specific targets of viral and bacterial products, and important tumor and host cell antigens (Hakomori, 1981, 1990). More recently, the discovery of inhibition of protein kinase C by sphingosine (Hannun *et al.*, 1986), a sphingolipid breakdown product, focused the attention on the lipid moiety of sphingolipids and the possibility that sphingolipids may serve as precursors to important biologic lipid mediators in a manner analogous to glycerolipids (Hannun and Bell, 1989). This possibility was further substantiated with the recent discovery of the sphingomyelin cycle (Okazaki *et al.*, 1989) in which several extracellular agents activate a sphingomyelinase resulting in cleavage of membrane sphingomyelin and the generation of ceramide. Intracellular ceramide levels are transiently elevated followed by regeneration of sphingomyelin. The elevation of intracellular ceramide levels or the addition of cell-permeable ceramide analogues appear to modulate a number of cell regulatory activities (Hannun, 1994; Hannun and Linardic, 1993; Kolesnick and Golde, 1994). These include inhibition of cell growth, induction of cell differentiation, dephosphorylation of the retinoblastoma protein (Rb), cell cycle arrest, apoptosis, and others. Ceramide has been shown to activate a membrane-bound serine/threonine protein kinase, a cytosolic serine/threonine protein phosphatase 2A, and protein kinase C ζ. These studies are beginning to elucidate a novel ceramide-dependent pathway of signal transduction. The predominantly growth-suppressive features of this pathway possibly provide a counterbalance to the generally mitogenic and tumor-promoting properties of the various glycerolipid pathways.

In this review, we summarize the current understanding of the regulation of the sphingomyelin cycle, the putative second messenger role of ceramide and its mechanism of action, and highlight recent developments implicating ceramide as a potential tumor suppressor lipid.

5.2. Overview of Sphingolipid Biology and Metabolism

Ceramide plays a central structural and functional role in sphingolipid metabolism analogous to the role played by diacylglycerol in glycerolipid metabolism (Bishop and Bell, 1988). It forms the basic structural backbone of sphingolipids (Fig. 5-1) and can be considered to be formed of two components: a long-chain sphingoid base and a fatty acyl group linked to it through an amide bond. D-*erythro*-(2S,3R)-sphingosine is the predominant sphingoid base in mammalian cells while fatty acyl groups vary in length, saturation, and hydroxylation. The substitution of an array of head groups to the C-1 position yields diverse and complex sphingolipids (Hannun and Bell, 1989). Substitution of phosphoryl-

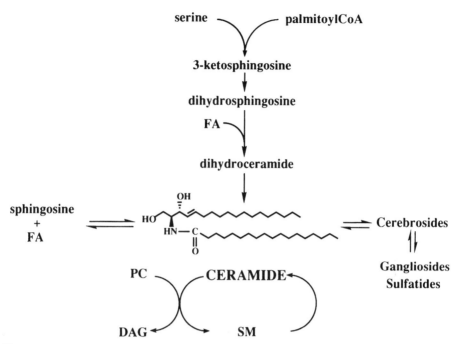

Figure 5-1. Metabolism of ceramide. This scheme illustrates the synthesis of ceramide described in the text. The central role of ceramide in sphingolipid synthesis and metabolism is shown. SM, sphingomyelin; PC, phosphatidylcholine; DAG, diacylglycerol; FA, fatty acid.

choline at C-1, for example, results in the formation of sphingomyelin (Fig. 5-1) while the addition of simple or complex glycosyl groups results in the formation of more than 300 different cerebrosides, sulfatides, and gangliosides (Sweely, 1991).

Ceramide synthesis starts with the condensation of serine and palmitoyl-CoA to form 3-ketosphinganine which in turn undergoes reduction to form sphinganine (dihydrosphingosine) (Fig. 5-1). A fatty acyl group is then added by an amide linkage to form dihydroceramide. Subsequent dehydrogenation results in the formation of a *trans* double bond between carbons 4 and 5 of the sphingoid base to yield ceramide (Merrill and Wang, 1992; Rother *et al.*, 1992). Introduction of the various head groups results in the formation of more complex sphingolipids (Van Echten and Sandhoff, 1993). The addition of the phosphorylcholine head group to ceramide to form sphingomyelin involves the transfer of choline phosphate from phosphatidylcholine through the action of phosphatidylcholine:ceramide choline phosphotransferase (Merrill and Jones, 1990).

Sphingolipids are catabolized by the action of a number of enzymes that have the ability to specifically cleave the different head groups usually in a stepwise fashion. In the case of sphingomyelin this is achieved through the action of sphingomyelinase, of which three different activities have been identified so far (Spence, 1993; Chatterjee, 1993; Okazaki *et al.*, 1994). Ceramide is generated as a result, and it is further catabolized by ceramidase to sphingosine and long-chain fatty acids. The sphingosine is then either recycled into new ceramide or further catabolized (Buehrer and Bell, 1992; Van Veldhoven and Mannaerts, 1993). Several rare inherited diseases of sphingolipid metabolism have been identified and found to be secondary to lack of specific lysosomal catabolic enzymes resulting in abnormal storage and buildup of various sphingolipids.

5.3. The Sphingomyelin Cycle

5.3.1. Components and Description

The discovery of protein kinase C inhibition by sphingosine and related compounds led to the hypothesis that membrane sphingolipids could function as reservoirs for bioactive molecules and that a "sphingomyelin cycle" may exist analogous to the "phosphatidylinositol (PI) cycle" which was generally recognized at the time (Hannun and Bell, 1989; Hannun, 1994). To test this hypothesis, sphingomyelin levels were measured in HL-60 human promyelocytic leukemia cells after treatment with $1\alpha,25$-dihydroxyvitamin D_3 (vitamin D_3), an inducer of monocytic differentiation (Okazaki et al., 1989). Time-dependent rapid hydrolysis of sphingomyelin was found to occur, peaking at 2 hr, and resulting in the generation of ceramide and phosphorylcholine. Sphingomyelin was subsequently regenerated, and its levels, as well as those of ceramide and phosphorylcholine, returned to baseline by 4 hr. Vitamin D_3 treatment was also found to result in a twofold increase in neutral sphingomyelinase activity in HL-60 cell extracts with the peak occurring at 1.5–2 hr after treatment (Okazaki et al., 1989). This finding provided preliminary evidence of the presence of a sphingomyelin cycle and its possible role in cell regulation.

Further evidence in support of the existence of the sphingomyelin cycle came from studies of the effects of other inducers of HL-60 differentiation (Kim et al., 1991). Both tumor necrosis factor-α (TNF-α) and γ-interferon (γ-IFN), inducers of monocytic differentiation of HL-60 cells similar to vitamin D_3, were found to cause significant sphingomyelin hydrolysis (Kim et al., 1991; Mathias et al., 1991; Dressler et al., 1992). In contrast, retinoic acid, dibutyryl cyclic AMP, and dimethyl sulfoxide (inducers of granulocytic differentiation), and phorbol esters (inducers of macrophagelike differentiation) failed to cause any sphingomyelin hydrolysis (Kim et al., 1991). Thus, the sphingomyelin cycle was associated specifically with monocytic differentiation of HL-60 cells. Furthermore, the addition of exogenous, cell-permeable ceramide resulted in monocytic differentiation of HL-60 cells within 2 hr implicating ceramide as the putative effector molecule (Kim et al., 1991).

A growing number of inducers of the sphingomyelin cycle in other cell types have since been described (Hannun, 1994). TNF-α has been shown to induce sphingomyelin hydrolysis in U937 monoblastic leukemia (Obeid et al., 1993) and in Jurkat T-cell leukemia cells (Dbaibo et al., 1993) with maximal effects occurring within 10 and 30 min, respectively, and baseline levels restored within 50–90 min. This activity appeared to be mediated through the 55-kDa TNF-α receptor (Wiegmann *et al.*, 1992). Nerve growth factor (NGF) also resulted in dose- and time-dependent sphingomyelin hydrolysis and ceramide generation in T9 glioblastoma cells (Dobrowsky *et al.*, 1994). NGF binds to two receptors: the high-affinity trkA receptor and a low-affinity p75 receptor. The effects of NGF on sphingomyelin were a result of interaction of NGF with the 75-kDa neurotrophin receptor (Dobrowsky *et al.*, 1994) which shares significant homology with the TNF-α receptor family (Smith *et al.*, 1994). Interleukin-1β (IL-1β) was shown to induce maximal sphingomyelin hydrolysis in dermal fibroblasts within 2 hr with sustained increase in ceramide levels lasting more than 24 hr (Ballou *et al.*, 1992). Similarly, IL-1β induced rapid sphingomyelin hydrolysis in EL4 murine thymoma cells in a dose- and time-dependent fashion (Mathias *et al.*, 1993). Additional inducers of the sphingomyelin cycle in other cell types include dexamethasone (Ramachandran *et al.*, 1990), complement (Niculescu *et al.*, 1993), ionizing radiation (Haimovitz-Friedman *et al.*, 1994), and ara C (Strum *et al.*, 1994). Brefeldin A, a macrolide capable of inhibiting intracellular protein transport and secretion, also induces sphingomyelin hydrolysis by undetermined mechanisms (Linardic *et al.*, 1992).

5.3.2. Regulation

An eventual common target of the action of extracellular activators of the sphingomyelin cycle appears to be a neutral sphingomyelinase (Okazaki *et al.*, 1989, 1990). This enzymatic activity was initially described in HL-60 cells in response to vitamin D_3 and later discovered to be related to a novel cytosolic magnesium-independent neutral sphingomyelinase which has been partially purified (Okazaki *et al.*, 1994). Evidence supportive of the tight coupling between receptor binding and the activation of sphingomyelinase came from the finding that TNF-α was capable of inducing sphingomyelin hydrolysis and ceramide generation in cell-free extracts at neutral pH within 1 min (Dressler, 1992). Later, IL-1β was similarly shown to induce rapid sphingomyelin hydrolysis in EL4 subcellular extracts at neutral pH within 2 min although magnesium dependence was observed (Mathias *et al.*, 1993). Additionally, an acidic, magnesium-independent sphingomyelinase was thought to be central to a pathway of NF-κB activation by TNF-α in permeabilized Jurkat cells (see below) (Schütze *et al.*, 1992). These observations raise the possibility that different sphingomyelinases are being activated by the different agonists.

The sequence of events that follows ligand–receptor binding and leads to activation of sphingomyelinase remains poorly understood. Recently, arachidonic acid emerged as a possible mediator of sphingomyelin hydrolysis in response to TNF-α in HL-60 cells (Jayadev *et al.*, 1994). Treatment of HL-60 cells with TNF-α resulted in rapid generation of arachidonic acid which shortly preceded the onset of sphingomyelin hydrolysis. Addition of exogenous arachidonic acid, a number of other fatty acids, or a nonmetabolizable arachidonic acid analogue resulted in prompt sphingomyelin hydrolysis. However, DAG, another second messenger generated by TNF-α, failed to induce sphingomyelin hydrolysis and actually caused a modest increase in baseline sphingomyelin levels similar to that produced by phorbol esters. Arachidonic acid was also found to activate neutral, but not acidic, sphingomyelinase *in vitro* (Jayadev *et al.*, 1994). Moreover, addition of melittin, a known activator of phospholipase A₂, resulted in similar sphingomyelin hydrolysis in intact cells but did not activate sphingomyelinase *in vitro*. These data suggested that arachidonic acid may play an important role in mediating early neutral sphingomyelinase activation in response to binding of TNF-α to its receptors.

In contrast, another mechanism for activating sphingomyelinase in response to TNF-α was suggested from experiments in Jurkat cell lysates where sphingomyelin hydrolysis was observed in response to DAG at acidic but not neutral pH (Schütze *et al.*, 1992). This effect was judged to be a result of activation of a phosphatidylcholine-specific phospholipase C (PC-PLC) by TNF-α since it was abrogated when a PC-PLC inhibitor was used. The authors postulated that activation of PC-PLC by TNF-α is the first step in a cascade that leads to generation of DAG in less than 1 min, subsequent activation of acidic sphiongomyelinase leading to generation of ceramide and sphingomyelin hydrolysis with peak effects within 2 min, activation of NF-κB, and regeneration of sphingomyelin by 10 min.

In summary, sphingomyelin cycle activation appears to be tightly regulated by receptor-mediated events and exhibits different kinetics depending on the agonist, cell type, and even cell batches within the same cell line (Jayadev *et al.*, 1994). Different sphingomyelinases could be involved whose activation is mediated by different pathways. However, pathways regulating the termination of the sphingomyelin cycle which allows regeneration of sphingomyelin remain unexplored.

5.3.3. The Second Messenger Function of Ceramide

The discovery of the sphingomyelin cycle gave birth to a whole new group of candidate sphingolipid second messengers the best studied of which is ceramide (Hannun, 1994; Hannun and Linardic, 1993; Kolesnick and Golde, 1994). Soon after the discovery of sphingomyelin turnover after vitamin D₃ treatment, the central question became whether ceramide was an effector molecule. Evidence supporting a second messenger role for ceramide has accumulated rapidly. First, cellular ceramide levels increase rapidly in response to a number of extracellular agents. This ceramide is generated from the hydrolysis of sphingomyelin, an abundant membrane lipid, and thus avoiding the need for *de novo* synthesis. Second, exogenous cell-permeable ceramide mediates a number of the biologic and biochemical effects of the pluripotent inducers of the sphingomyelin cycle, such as TNF-α, IL-1β, and vitamin D₃ with the naturally occurring D-*erythro*-ceramide stereoisomer being most effective. Third, specific biochemical targets for the action of ceramide such as ceramide-activated protein phosphatase (CAPP; discussed below) have been discovered and linked to several biologic effects mediated by this pathway. Fourth is the abrogation of certain effects mediated by ceramide when the phosphatase inhibitor, okadaic acid, is used at concentrations sufficient for inhibiting CAPP. The biochemical and biological events induced by ceramide as well as evidence supporting its role as a second messenger will be discussed below.

5.4. The Cell-Regulatory Functions of Ceramide

5.4.1. Role of Ceramide in Growth Suppression

The major emerging cellular activity of ceramide appears to be as a mediator of growth suppression through induction of differentiation, initiation of programmed cell death (apoptosis), activation of tumor suppressors (such as Rb), and induction of specific cell cycle arrest (Hannun and Linardic, 1993). These activities have been studied in different cell systems and will be summarized.

5.4.1.1. Cell Growth Inhibition

One of the earliest biologic effects attributed to ceramide was inhibition of HL-60 cell growth. This was studied after the discovery of sphingomyelin hydrolysis induced by vitamin D_3, a known growth inhibitor and inducer of differentiation of these cells (Okazaki *et al.*, 1990). Synthetic, cell-permeable C_2-ceramide (N-acetylsphingosine) was used to overcome the severe hydrophobicity of natural ceramides (Fig. 5-2). C_2-ceramide inhibited cell growth at low micromolar concentrations in a dose-dependent fashion (Okazaki *et al.*, 1990). In Jurkat (Dbaibo *et al.*, 1993), U937, Molt-4 (Dbaibo *et al.*, 1995), and T9 glioblastoma cells (Dobrowsky *et al.*, 1994), C_2-ceramide produced similar growth inhibition. Another cell-permeable ceramide analogue, C_6-ceramide (*N*-hexanoylsphingosine) (Fig. 5-2),

(2S, 3R) D-*erythro*-C$_{(n)}$-ceramide:

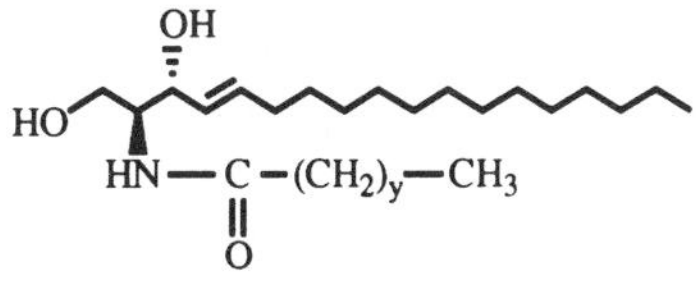

(2S, 3R) D-*erythro*-dihydro-C$_{(n)}$-ceramide:

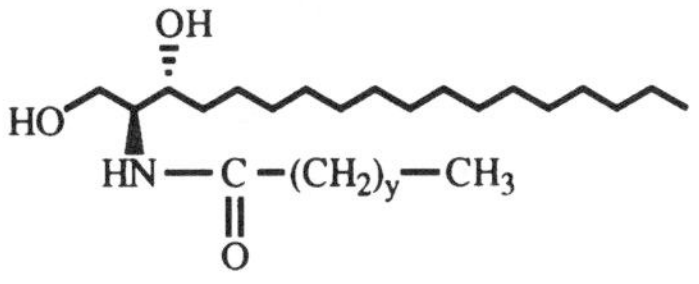

Figure 5-2. Structures of ceramide and dihydroceramide analogues. Shown are the structures of the D-*erythro* isomers of cell-permeable ceramide analogues where $n = 2, 6$, or 8. $y = n - 2$. Dihydroceramide lacks the double bond between carbons 4 and 5 on the sphingoid backbone.

produced growth inhibition of mink lung epithelial cells at low micromolar concentrations (Dbaibo *et al.*, 1995).

In another approach, endogenous ceramide levels were measured in human T lymphocytes that were induced to proliferate by IL-2 or a combination of phorbol ester and a calcium ionophore and compared to ceramide levels in nonproliferating T cells (Borchardt *et al.*, 1994). Ceramide levels were found to be approximately fourfold lower in the proliferating cells as compared to non-proliferating cells at 48 and 72 hr. Additionally, treatment of cells with D,L-*threo*-1-phenyl-2-decanoylamino-3-morpholino-1-propanol (PDMP), an inhibitor of UDP-glucosyl:ceramide transferase (Inokuchi *et al.*, 1989), resulted in a fourfold increase in endogenous ceramide levels accompanied by a tenfold decrease in thymidine incorporation into DNA by 72 hr.

In rat fibroblasts, cell-permeable ceramides were found to inhibit the mitogenic effects of phosphatidate (PA) and lysophosphatidate (lyso-PA) as measured by DNA synthesis (Gomez-Muñoz *et al.*, 1994). This inhibition was evident at low micromolar concentrations of C_2- or C_6-ceramide when combined with mitogenic concentrations of either PA or lyso-PA. The mechanism of this inhibition was found to involve decreased uptake and increased metabolism of both PA and lyso-PA as well as inhibition of phospholipase D (see below).

Detailed studies utilizing a number of synthetic ceramide analogues were performed to determine the specificity of ceramide's effects on growth inhibition and the relative importance of the different structural features of the ceramide molecule (Bielawska *et al.*, 1992a,b). Two important findings emerged from these studies. First, when the enantiomers of *N-acylphenylamino* alcohols, in which the hydrocarbon tail of sphingosine is replaced with a phenyl group, were compared, only D-*erythro*-2-(*N*-myristoyl-amino)-1-phenyl-1-propanol (D-*e*MAPP) was effective in inducing growth inhibition whereas L-*e*MAPP was not. This observation was particularly important since the D enantiomer of ceramide is the naturally occurring molecule. Second, the 4,5 *trans* double bond appeared to be critical in preserving the growth inhibitory activity of D-*erythro* ceramide, the naturally occurring molecule. Naturally occurring D-*erythro*-dihydroceramide, the immediate biosynthetic precursor of D-*erythro*-ceramide, which lacks this double bond was completely inert (Bielawska *et al.*, 1993; Obeid *et al.*, 1993; Dbaibo *et al.*, 1993). Thus, the growth-inhibitory effects of ceramide are dependent on the preservation of specific structural and stereochemical characteristics, which supports a second messenger function of ceramide.

5.4.1.2. Induction of Differentiation

Another early biologic effect ascribed to ceramide was its ability to induce monocytic differentiation of HL-60 cells similar to that induced by the sphingomyelin cycle activators vitamin D_3, TNF-α, and γ-IFN (Okazaki *et al.*, 1990; Kim *et al.*, 1991). In these experiments, C_2-ceramide induced monocytic differentiation of HL-60 cells as determined by the measurement of nitro blue tetrazolium-reducing ability and induction of nonspecific esterases which are induced specifically during monocytic differentiation. Concentrations of 1–6 µM were effective. Submicromolar concentrations were also effective when used with subthreshold concentrations of TNF-α or vitamin D_3. The selectivity of ceramide's effect to the monocytic pathway of differentiation was underscored by the lack of sphingomyelin hydrolysis and ceramide generation by inducers of granulocytic (retinoic acid, dibutyryl cAMP) or macrophagelike (phorbol ester) differentiation (Kim *et al.*, 1991).

Similar findings were observed in studies with rat T9 anaplastic glioblastoma cells (Dobrowsky *et al.*, 1994). These cells respond to treatment with NGF by growth inhibition and differentiation to an astrocytelike phenotype. As indicated above, NGF was found to initiate sphingomyelin hydrolysis and ceramide generation within minutes of binding to its p75 neurotrophin receptor. Low micromolar concentrations of C_2-ceramide mimicked NGF's effects during 4 days of treatment by inducing neuritic process formation in T9 cells indicative of astrocytic differentiation.

5.4.1.3. Role in Serum Deprivation-Induced Cell Cycle Arrest

Cultured cells are dependent on various serum growth factors for continued growth *in vitro*. In the absence of these factors, many cell types stop growing and may then die. In view of its observed growth-inhibitory activity, ceramide became a candidate mediator of these effects. Studies performed in serum dependent Molt-4 cells showed that cells deprived of serum quickly developed specific growth arrest in the G_0/G_1 phase of the cell cycle starting at 24 hr and becoming maximal at 48 hr with 96% of viable cells being in G_0/G_1 (Jayadev *et al.*, 1995). Cellular ceramide levels gradually increased in a time-dependent manner with prolonged serum starvation of these cells. Compared to control cells, ceramide levels increased to 3-, 8-, and 15-fold after 24, 48, and 96 hr of serum deprivation, respectively. Since the elevation of endogenous ceramide coincided with the development of cell cycle block, the next question to be examined was whether ceramide elevation played a role in mediating cell cycle arrest. To answer this question, Molt-4 cells were treated with C_6-ceramide and cell cycle analysis was performed. Indeed, C_6-ceramide produced rapid arrest in G_0/G_1 at concentrations as low as 15 µM and a time course mimicking the effects of serum deprivation. The specificity of these effects was confirmed when dihydroceramide failed to produce any inhibition of cell cycle progression. These findings were supportive of a possible role for ceramide as a mediator of cell cycle arrest. Interestingly, the sustained elevations in ceramide levels over several days and the accompanying cell cycle effects suggested, for the first time, a role for ceramide as a possible long-term modulator of cell function.

5.4.1.4. Tumor Suppressor Activation

The potent and specific effects of ceramide on growth inhibition, cellular differentiation, and cell cycle arrest prompted investigators to search for the molecular mechanisms of these effects. Tumor suppressor proteins emerged as possible targets of ceramide. Rb, the product of the retinoblastoma gene, belongs to this growing family of tumor suppressor proteins and is among the most extensively studied (Harlow, 1992; Weinberg, 1990). Loss of both copies of the *RB1* gene encoding this protein through deletion or mutation results in unrestrained cell growth and oncogenesis. Several human cancers have been associated with the loss of functional Rb including childhood retinoblastoma, osteosarcomas, bladder and breast carcinomas (Horowitz *et al.*, 1990). Rb phosphorylation appears to be tightly regulated in a cell cycle-dependent manner (DeCaprio *et al.*, 1992). The hypophosphorylated form is the molecularly active species which predominates in the G_0 and G_1 phases of the cell cycle (DeCaprio *et al.*, 1992; Ludlow *et al.*, 1990). It exerts its growth suppressive function, in part, through inhibitory interactions with several transcription factors important in the exit from G_1 and cell proliferation such as E2F (Nevins, 1992). This active form of Rb is targeted by several viral oncoproteins such as E1A of adenovirus, large T antigen of simian virus 40 (SV40), and E7 of human papilloma viruses (Nevins, 1992; Bagchi *et al.*, 1990; Ludlow, 1993; Vousden, 1993; Chellappan *et al.*, 1992). These viral oncoproteins, in concert with other viral products, form a well-orchestrated attack during viral infection in an attempt to relieve cell cycle block, release important transcription factors, and therefore promote cell proliferation and concomitant viral replication (Vousden, 1993; Moran, 1993). Sequential phosphorylation of Rb leads to its inactivation and usually proceeds in late G_1 leading to exit into S phase (Goodrich *et al.*, 1991). Rb remains in a phosphorylated/inactive form until it becomes dephosphorylated at the end of mitosis (DeCaprio *et al.*, 1992; Ludlow *et al.*, 1990).

In view of the results obtained in serum-deprived Molt-4 cells summarized above, it was imperative to examine the status of Rb phosphorylation after serum deprivation in order to determine whether Rb participates in the process of serum deprivation-induced cell cycle arrest. In serum-deprived Molt-4 cells, Rb dephosphorylation became evident at 24 hr, was more pronounced at 48 and 72 hr, and became almost complete by 96 hr (Dbaibo *et al.*, 1995). These results showed that Rb may be an important target for serum deprivation-induced factors during this process of cell cycle arrest.

Since one of the potential mediators produced during serum deprivation is ceramide, its effects on Rb phosphorylation were examined. Molt-4-cells treated with C_6-ceramide for 4 hr showed significant Rb dephosphorylation starting at concentrations of 10 μM (Dbaibo *et al.*, 1995). The degree of dephosphorylation increased with longer treatments. The specificity of this effect was confirmed by the use of several other lipids, including dihydro-C_6-ceramide, which failed to change the Rb phosphorylation status.

In earlier work, sphingosine had been similarly shown to induce rapid Rb dephosphorylation (Chao *et al.*, 1992). In this regard, determining the relative physiological significance of ceramide versus sphingosine became important. Two findings helped support ceramide as the relevant molecule in these experiments. First, unlike the dramatic elevation of endogenous ceramide levels in Molt-4 cells deprived of serum, sphingosine levels remained essentially unchanged (Dbaibo *et al.*, 1995). Second, exogenous C_6-ceramide was taken up rapidly by cells but was not significantly metabolized, particularly to sphingosine, for up to 24 hr (Dbaibo *et al.*, 1995). Therefore, although both ceramide and sphingosine are capable of potently inducing Rb dephosphorylation, only ceramide can claim a physiological role at this point.

Next, the importance of Rb in mediating ceramide's inhibitory effects on cell growth and cell cycle progression was examined (Dbaibo *et al.*, 1995). A series of experiments were performed in different cell lines either containing or lacking functional Rb. Growth of Molt-4 cells, known to contain functional Rb, was significantly inhibited at low micromolar concentrations of C_6-ceramide. In contrast, WERI-Rb-1 cells, from a retinoblastoma cell line known to lack functional Rb, were significantly more resistant to the growth-inhibitory effects of C_6-ceramide. Since these cells may differ in ways other than whether they have functional Rb or not, another approach was used. Mink lung epithelial cells transfected with either wild-type large T antigen from SV40 or the Kl mutant, which retains the transforming capability but lacks Rb binding, were compared. C_6-ceramide potently inhibited the growth of Mk-13 cells, transfected with the mutant large T antigen, which contain functional Rb. PVU-0 cells, transfected with wild-type large T antigen and therefore lacking functional Rb, were quite resistant to the growth-inhibitory effects of ceramide. To confirm these findings further, Molt-4 cells were infected with a retrovirus encoding the E1A gene of adenovirus or mock-infected. The two resulting cell lines differed only with respect to adenoviral E1A expression. Cell cycle analysis after treatment with C_6-ceramide was performed in these cells. Mock-infected cells behaved as expected; developing cell cycle arrest in G_0/G_1. The cells expressing E1A, therefore lacking functional Rb, showed no specific cell cycle arrest. Therefore, the presence of functional Rb is essential for mediating the effects of ceramide on growth inhibition and cell cycle arrest.

These findings showed conclusively that Rb is a downstream target of ceramide (Fig. 5-3). They also provided strong support for the hypothesis that the growth-inhibitory effects of the putative "tumor suppressor lipid," ceramide, are mediated, at least in part, through activation of Rb, an established tumor suppressor protein (Huang *et al.*, 1988).

5.4.1.5. *Induction of Apoptosis*

In early experiments with ceramide, antiproliferative effects were always accompanied by cytotoxicity especially at higher concentrations of the cell-permeable ceramides (Dbaibo *et al.*, 1993). These cytotoxic effects were only seen with ceramide analogues that were also effective in growth inhibition indicating that they were not related to nonspecific detergent effects leading to cell membrane injury (Bielawska *et al.*, 1993). Additionally, some of the inducers of ceramide generation, particularly TNF-α, are known for their potent cytotoxic effects (Wright *et al.*, 1992). These observations led to speculation as to whether ceramide was mediating these cytotoxic effects.

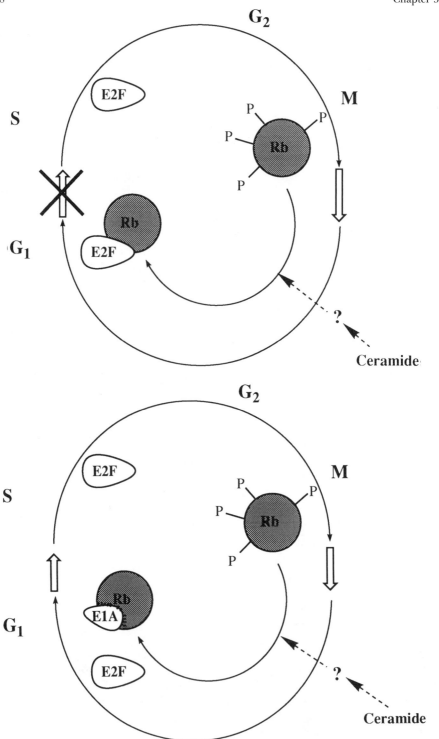

Inherent in most cell types is a program that initiates cell suicide also known as apoptosis (Gerschenson and Rotello, 1992). This process involves the orderly breakdown of cells into small, packaged, apoptotic bodies in an effort to protect the organism from the sudden release of potentially injurious intracellular molecules as occurs in necrosis. Endonuclease activation results in the characteristic fragmentation of DNA which gives the appearance of a DNA "ladder" on gel electrophoresis (Wyllie, 1980; Gerschenson and Rotello, 1992). Apoptosis appears to play an important role in development, differentiation, glandular atrophy following hormonal withdrawal, and maturation of the immune system (Song et al., 1992). Interfering with apoptosis appears to be one of the mechanisms leading to unrestrained growth and the development of cancer. Conversely, several chemotherapeutic agents as well as ionizing radiation impart their tumoricidal activities by initiating apoptosis (Hickman, 1992).

TNF-α, initially named for its ability to induce "necrosis" of certain tumors, was discovered to be a potent inducer of apoptosis. However, the intracellular mediators of these effects remained largely unknown. The observed effects of ceramide on growth suppression and cytotoxicity prompted investigators to examine its role in apoptosis (Obeid et al., 1993). An exciting discovery was that U937 cells treated with low micromolar concentrations of C_2-ceramide showed characteristic DNA fragmentation pathognomonic of apoptosis (Obeid et al., 1993). These effects occurred with 1–3 hr and were specific to ceramide: other amphiphilic lipids, particularly dihydro-C_2-ceramide, showed no effect. Interestingly, DNA fragmentation initiated by ceramide or TNF-α was abolished when cells were also treated with phorbol esters suggesting a protective role for PKC activation (Obeid et al., 1993). These results were later reproduced in HL-60, and murine fibrosarcoma cell lines L929/LM and WEHI-164/13 (Jarvis et al., 1994).

In serum deprivation studies with Molt-4 cells, apoptosis occurred in approximately 4–12% of cells in a time-dependent manner (Jayadev et al., 1995). When exogenous ceramide was used, more than 50% of cells underwent apoptosis, significantly exceeding the effects of serum deprivation. In view of the protective effects of PMA observed in previous experiments, the levels of DAG, its physiologic counterpart, were measured in serum-deprived cells. Surprisingly, DAG levels were three- to fourfold elevated after 72–96 hr of serum withdrawal. Exogenous, cell-permeable, dioctanoylglycerol (diC_8) was found to protect cells treated with ceramide from undergoing apoptosis by almost 50%. However, diC_8 did not significantly affect cell cycle arrest induced by ceramide in the same experiments. This provided a possible explanation for the lesser degree of apoptosis observed with serum starvation as opposed to that induced by exogenous ceramide.

Recent studies in bovine aortic endothelial cells demonstrated that these cells underwent apoptosis in response to ionizing radiation and TNF-α (Haimovitz-

Figure 5-3. Role of Rb in mediating cell cycle arrest by ceramide. (A) Scheme illustrating the activation of Rb by ceramide through dephosphorylation. The dephosphorylated Rb binds to transcription factors such as E2F and arrests cell cycle progression at the G_1/S boundary. (B) E1A binds to and inactivates dephosphorylated Rb. Ceramide in this situation has no effect on the cell cycle because its target, Rb, is permanently inactivated.

Friedman *et al.*, 1994). Ionizing radiation was subsequently found to induce sphingomyelin hydrolysis and generation of ceramide within 30 sec. Exogenous ceramide was capable of inducing apoptosis in these cells. Additionally, activation of PKC with phorbol esters 5 min prior to irradiation inhibited sphingomyelin hydrolysis, ceramide generation, and apoptosis. These findings again underscored the complex interplay between the two signaling pathways involving sphingolipids and glycerolipids.

This emerging role of ceramide in mediation of apoptosis is supported by other circumstantial evidence. Human immunodeficiency virus (HIV) infection has been demonstrated to induce apoptosis which was proposed as a possible mechanism for the dramatic loss of T4 cells and subsequent development of the acquired immunodeficiency syndrome (Meyaard *et al.*, 1992). T cells infected with HIV were also shown to have significantly increased levels of ceramide (Van Veldhoven *et al.*, 1992). Thus, ceramide may play a pathogenic role in HIV infection. Additionally, the induction of apoptosis by irradiation, tetrandrine, bistratene A, and cisplatin was abolished when cells were pretreated with okadaic acid, a known inhibitor of CAPP (Song *et al.*, 1992). In U937 cells the use of okadaic acid prior to treatment with ceramide attenuated the development of apoptosis, supporting a role for CAPP in mediating these events (L. M. Obeid, unpublished).

Insight into the mechanism of induction of apoptosis by ceramide was provided by recent studies implicating cysteine proteases belonging to the interleukin-1 β converting enzyme (ICE) family as triggers of apoptosis (Miura *et al.*, 1993). Proteolysis of specific targets by these proteases appears to be a central event during apoptosis. One of these targets is the enzyme poly(ADP-ribose) polymerase (PARP) which is involved in DNA repair. Disabling this enzyme by its cleavage to a signature 85 kDa fragment appears to accompany many forms of apoptosis (Kaufmann *et al.*, 1993; Lazebnik *et al.*, 1994). The protease responsible for targeting PARP has recently been identified as the ICE-like Yama/CPP32β/apopain (Tewari *et al.*, 1995; Fernandes-Alnemri *et al.*, 1994; Nicholson *et al.*, 1995). This protease emerged as a possible target for activation by ceramide. Studies in Molt-4 cells showed that ceramide treatment at concentrations of 10–20 μM resulted in specific cleavage of PARP to the apoptotic fragment within 3–4 hours (Smyth *et al.*, 1995). This activity was specific to ceramide and was not evident after treatment with dihydroceramide or diC_8. Moreover, inhibition of PARP proteolysis by overexpression of the antiapoptotic gene *bcl-2* rendered the cells resistant to ceramide-induced apoptosis. These studies suggested that ICE-like proteases are potential targets of activation by ceramide and may mediate its apoptotic effects.

5.4.1.6. Downregulation of c-myc

An important biochemical target of TNF-α is the protooncogene c-*myc* which plays a central role in cell proliferation and differentiation. TNF-α induces early downregulation of c-*myc* mRNA as a result of a block in elongation of the message rather than initiation of transcription (McCachren *et al.*, 1988). With the emergence of ceramide as a putative second messenger in several of TNF-α's biological and biochemical activities, ceramide's effects on the regulation of c-*myc* were investigated. In HL-60 cells, C_2-ceramide was found to induce prompt downregulation of c-*myc* mRNA beginning at 30 min (Kim *et al.*, 1991; Wolff *et al.*, 1994). Concentrations of ceramide used were similar to those that induced cell growth inhibition. Moreover, the mechanism by which ceramide exerted its effect was identical to that produced by TNF-α; namely, a block in elongation of the message at the exon I/intron I junction. The effects of both ceramide and TNF-α were abolished when cells were also treated with okadaic acid (Wolff *et al.*, 1994). In contrast, PMA-induced downregulation of c-*myc* was not inhibited by okadaic acid, indicating that its effects were specific to the ceramide pathway (Wolff *et al.*, 1994). The role of CAPP in this signaling pathway was further emphasized when four stereoisomers of C_2-ceramide, but neither of the stereoisomers of dihydro-C_2-ceramide, were able to activate CAPP *in vitro* or downregulate c-*myc* (Wolff *et al.*, 1994).

These results were supportive of an important role for ceramide and CAPP in transducing the effects of TNF-α on c-*myc* downregulation. They also emphasized the emerging role of ceramide as a mediator of growth-inhibitory signals.

5.4.1.7. Induction of Senescence

Recent work has suggested that ceramide may play an important role in inducing the senescent phenotype in human diploid fibroblasts (HDF) grown in culture. Cellular senescence has been defined as the loss of the ability to respond to mitogenic signals with DNA synthesis, growth, and proliferation (Hayflick, 1965). Biochemical changes in senescent cells include the inability to undergo AP-1 activation because of the inability to transcribe c-*fos* (Riabowol *et al.*, 1992; Seshadri and Campisi, 1990), and the inability to phosphorylate Rb in response to mitogenic signals, leaving Rb in a predominantly hypophosphorylated and growth-suppressive state (Stein *et al.*, 1990; Futreal and Barrett, 1991). In cultured HDF, the senescent phenotype is observed after approximately 50–60 population doublings when the cells stop dividing even in the absence of contact inhibition, but remain metabolically active (Goldstein, 1990). Measurement of ceramide in senescent HDF showed that ceramide levels were increased fourfold, in the absence of any changes in cellular levels of other lipids, compared to young cells (Venable *et al.*, 1995). Neutral, magnesium-dependent sphingomyelinase activity was also elevated eightfold in the senescent but not in the young cells. In order to test the ability of exogenous ceramide to induce the senescent phenotype in HDF, young cells were treated with exogenous ceramide for 1–3 days. Low concentrations of ceramide resulted in the cessation of DNA synthesis reminiscent of senescent cells. Additionally, when young HDF cells were treated with ceramide for 48 hr in the absence of serum, they failed to activate AP-1 or phosphorylate Rb when serum was reintroduced. In contrast, control cells treated with ethanol vehicle responded to the reintroduction of serum with strong AP-1 activation and

Rb phosphorylation. The lack of response to mitogenic signals in senescent cells prompted the evaluation of the DAG/PKC pathway which is also important in AP-1 activation (Venable *et al.*, 1994). Senescent cells were found to have an abundance of PKC that is easily activated, as determined by its translocation to the cell membrane, by PMA. However, PKC failed to translocate in response to serum stimulation in senescent, but not in young cells. Since DAG is the physiologic second messenger responsible for activating PKC, determination of DAG levels in response to serum stimulation was subsequently done. Senescent cells failed to produce DAG after stimulation with 10% serum whereas young cells produced a three- to fourfold increase in its levels. The defect in DAG production appeared to be secondary to ceramide-induced inhibition of phospholipase D in senescent cells.

5.4.2. Role of Ceramide in Inflammation and the Immune Response

5.4.2.1. Activation of NF-κB

TNF-α is now recognized as a mediator of inflammation and immune responses (Grunfeld and Palladino, 1990; Carswell *et al.*, 1975; Beutler and Cerami, 1988; Beutler and Van Huffel, 1994; Loetscher *et al.*, 1991). Induction of many genes by TNF-α is mediated, at least in part, by the activation of a family of transcription factors known collectively as NF-κB (Lowenthal *et al.*, 1989; Molitor *et al.*, 1990; Lenardo and Baltimore, 1989; Osborn *et al.*, 1989). These transcription factors are important in the expression of many immunoeffector genes, such as IL-2 and the α-subunit of its receptor (IL-2Rα), as well as a number of viruses, particularly HIV. NF-κB is present in most cells in the cytoplasm where it is bound to an inhibitory protein, I-κB, which prevents its translocation to the nucleus (Baeuerle and Baltimore, 1994). I-κB in turn is a family of proteins that are inactivated by either phosphorylation or dephosphorylation (Ghosh and Baltimore, 1990; Link *et al.*, 1992). On inactivation of I-κB, NF-κB is released and subsequently translocates to the nucleus where it binds to specific elements in responsive genes. Another recently described mechanism of NF-κB activation is through induction of proteolytic degradation of I-κB with release of NF-κB (Henkel *et al.*, 1993).

TNF-α activates NF-κB within minutes in most cell types (Osborn *et al.*, 1989). The mechanisms of this activation remain poorly understood despite being the subject of intense research. Although PKC, *in vitro*, can phosphorylate I-κB causing the release of NF-κB, several reports suggested its activity was not required in intact cells (Feuillard *et al.*, 1991; Meichle *et al.*, 1990). The search for other pathways led to the consideration of ceramide, generated in response to TNF-α, as a candidate mediator. Early experiments in Jurkat cells showed that while TNF-α induced strong activation of NF-κB within minutes, C_2-ceramide failed to induce NF-κB over a wide range of concentrations and durations (Dbaibo *et al.*, 1993). However, cotreatment of cells with TNF-α and ceramide showed enhancement of NF-κb activation as compared to TNF-α alone (Dbaibo *et al.*, 1993). The lack of activation of NF-κB by ceramide in intact cells was confirmed recently (Betts *et al.*, 1994).

In contrast, experiments performed in permeabilized Jurkat cells showed that using type III ceramide (C_{18}-ceramide), which is exceedingly hydrophobic NF-κB was activated at submicromolar concentrations within 30 min (Schütze *et al.*, 1992). Additionally, acidic and neutral sphingomyelinases added to cell homogenates at pH 5.0 and 7.4, respectively, activated NF-κB in a manner similar to TNF-α (Schütze *et al.*, 1992). Interestingly, in these experiments, both DAG and ceramide were generated in response to TNF-α starting within seconds and peaking at 1 and 2 min, respectively, although the adequacy of the metabolic labeling procedures used was subsequently questioned (Betts *et al.*, 1994). A mechanism of activation of sphingomyelin hydrolysis was proposed to occur through DAG generated by the action of PC-PLC which is in turn activated by TNF-α. An acidic sphingomyelinase was suspected to be the target since sphingomyelin hydrolysis in response to DAG occurred only at pH 5.0 in a cell-free system. In support of this hypothesis, a xanthogenate compound, D609, was found to inhibit NF-κB activation by TNF-α through inhibition of PC-PLC but not phospholipase A_2, phospholipase D, phosphatidylinositol-specific phospholipase C, or neutral sphingomyelinase (Schütze *et al.*, 1992). Activation of a serine-specific protease by ceramide and sphingomyelinase resulting in the degradation of I-κB and release of NF-κB was subsequently demonstrated (Machleidt *et al.*, 1994). Thus, a novel pathway for activation of NF-κB was proposed involving a cooperative relationship between two lipid second messengers.

More recently, PKCζ was shown to be involved in the activation of NF-κB by TNF-α through phosphorylation of I-κB (Diaz-Meco *et al.*, 1993, 1994; Dominguez *et al.*, 1993; Lozano *et al.*, 1994). Sphingomyelinase was subsequently shown to activate a promoter driven by three NF-κB sequences from the HIV enhancer similar to the effects of TNF-α (Lozano *et al.*, 1994). These effects were abolished when a PKCζ kinase-defective mutant was transfected. Ceramide activated recombinant PKCζ while sphingomyelinase and PC-PLC resulted in increased phosphorylation of I-κB (Lozano *et al.*, 1994). These results not only supported a role for ceramide in the activation of NF-κB, but also proposed PKCζ as a new molecular target for the action of ceramide. Intriguingly, little is known about the regulation of PKCζ which, unlike other PKC isoforms, is not activated by DAG or PMA (Nakanishi *et al.*, 1993; Berra *et al.*, 1993).

Therefore, the role of ceramide in the activation of NF-κB is far from clear. Based on the results summarized above, a necessary role for ceramide in a multistep process of NF-κB activation cannot be ruled out. However, in intact cells, cell-permeable ceramides are not sufficient for activating NF-κB.

5.4.2.2. Role in Eicosanoid and Cytokine Production

Recently, ceramide was proposed to play a role in enhancing prostaglandin E_2 production in response to IL-1β through enhanced cyclooxygenase gene expression in human fibroblasts (Ballou *et al.*, 1992). Other studies suggested a role for ceramide in augmenting the secretion of IL-2 induced by IL-1β in murine thymocytes (Mathias *et al.*, 1993). In HIV-infected HL-60 cells, sphingomyelinase, but not C_8-ceramide, treatment resulted in accumulation of soluble TNF-α (Rivas *et al.*, 1994).

5.4.2.3. Modulation of Superoxide Release by Neutrophils

TNF-α stimulates superoxide generation by neutrophils within 15–20 min (Yanaga and Watson, 1994). Ceramide was considered as a possible mediator of these effects. Neither C_2-ceramide nor sphingomyelinase resulted in any generation of superoxide (Yanaga and Watson, 1994). Contrarily, C_2-ceramide, as well as sphingosine, resulted in significant inhibition of H_2O_2 release as induced by formyl-Met-Leu-Phe (fMLP) (Nakamura *et al.*, 1994). Interestingly, fMLP also resulted in the generation of cellular ceramide starting at 90–120 min, a time coinciding with the termination of the respiratory burst. Since DAG formation in fMLP-stimulated neutrophils occurs through the action of phospholipase D and is necessary for superoxide production, ceramide's effects on phospholipase D were studied. Ceramide was found to inhibit phospholipase D by approximately 50% (Nakamura *et al.*, 1994). Additionally, okadaic acid, an inhibitor of CAPP, prevents the termination of oxidant release in PMA-stimulated neutrophils (Nakamura *et al.*, 1994). Therefore, a role for ceramide in terminating the oxidant release by neutrophils was proposed. Furthermore, this effect was thought to be secondary to phospholipase D inhibition and decreased DAG formation. Interestingly, these results suggest that ceramide may play a role in attenuating some of the effects that are stimulated by TNF-α.

5.4.3. Other Possible Functions of Ceramide

5.4.3.1. Inhibition of Protein Traffic

Recent studies have shown that C_6-ceramide significantly slowed the intracellular transport of vesicular stomatitis virus glycoprotein and eventual release of infectious virions from infected Chinese hamster ovary cells (Rosenwald and Pagano, 1993). Additionally, brefeldin A, a known inhibitor of protein secretion, induced sphingomyelin hydrolysis (Linardic *et al.*, 1992). Ceramide simulated the inhibitory effects of brefeldin A on protein secretion. These results not only implicate ceramide as a mediator of cellular defense against certain viral infections but also as a general regulator of cellular protein transport.

5.4.3.2. Role of Ceramide in Mitogenesis

Under certain conditions, ceramide appears to play a possible mitogenic role. In confluent fibroblasts and Swiss 3T3 cells, for example, ceramide results in an increase in thymidine uptake (Olivera *et al.*, 1992). This could represent new DNA synthesis or repair of damaged DNA. The mitogen-activated protein (MAP) kinase-related stress-activated protein kinases, essential for N-terminal phosphorylation of c-Jun as induced by TNF-α, can be activated by sphingomyelinase, indicating a role for this pathway in the mitogenic response (Kyriakis *et al.*, 1994; Westwick *et al.*, 1995). In HL-60 cells, ceramide and sphingomyelinase treatment resulted in phosphorylation, and activation, of the 42-kDa isoform of MAP kinase which is involved in mitogenic stimulation of fibroblasts by TNF-α (Raines *et al.*, 1993). Additionally, ceramide treatment has been shown to eventually result in the phosphorylation of the epidermal growth factor receptor (Goldkorn *et al.*, 1991). The biologic significance of these effects remains to be determined.

5.5. Mechanism of Action of Ceramide

The identification of ceramide as a putative second messenger directed research toward the identification of cellular targets that function as mediators of its actions. The starting point was to look at changes in cellular phosphorylation in response to ceramide. These studies led to the identification of CAPP and ceramide-activated protein kinase. More recently, protein kinase Cζ has emerged as a third possible target of ceramide action.

5.5.1. Ceramide-Activated Protein Phosphatase

The initial observation was that incubation of ceramide with crude cytosol from T9 glioma cells resulted in a dose-dependent hydrolysis of [^{32}P]phosphohistone (Dobrowsky and Hannun, 1992). CAPP activity was seen at concentrations of ceramide that parallel those resulting in growth inhibition and differentiation. The activation of CAPP was specific to ceramides (natural and synthetic). Other lipids, particularly sphingosine and dihydroceramide, did not activate CAPP. The activity was specific to the serine/threonine residues, was inhibited by low nanomolar concentrations of okadaic acid, and was cation-independent. These findings suggested that CAPP belonged to the protein phosphatase 2A (PP2A) family.

Subsequently, CAPP was partially purified from T9 and rat brain cells (Dobrowsky *et al.*, 1993). CAPP activity was also identified in HL-60 cells and in the yeast *Saccharomyces cerevisiae* whose growth is strongly inhibited by ceramide (Fishbein *et al.*, 1993). The latter suggested that the ceramide pathway is conserved in lower eukaryotes. CAPP was later hypothesized to be a subtype of heterotrimeric PP2A. Ceramide activated CAPP only when the B (regulatory) subunit was present but not when the catalytic subunit (C) was present with or without the structural subunit (A) (Dobrowsky *et al.*, 1993). These findings were interesting in view of the hypothesized role of the B subunits of PP2A in regulating the catalytic activity and substrate specificity. Moreover, these findings, for the first time, identified an *in vitro* molecular target for ceramide that could act as a proximal effector of its second messenger function.

The role of CAPP in mediating the biologic effects of ceramide was strengthened when okadaic acid treatment of cells reversed the effects of ceramide on c-*myc* and apoptosis as discussed above (Wolff *et al.*, 1994). The use of okadaic acid to study the role of CAPP in mediating ceramide's growth-inhibitory effects is hampered by its ability to independently suppress growth of cells in culture. Nevertheless, these observations support a role for CAPP in the ceramide signaling pathway (Dobrowsky and Hannun, 1994).

5.5.2. Ceramide-Activated Protein Kinase

The hypothesis that a protein kinase was activated by ceramide was proposed after sphingosine was found to stimulate phosphorylation of epidermal growth factor receptor (EGFR) in the human epidermoid carcinoma cell line, A431 (Faucher *et al.*, 1988). The suspicion that sphingosine's effects may have been mediated through conversion to ceramide led to the examination of ceramide's effect on EGFR phosphorylation. Although C_2-ceramide failed to induce this phosphorylation (Faucher *et al.*, 1988), C_8-ceramide induced identical phosphorylation of threonine-669 of EGFR in a time- and dose-dependent fashion (Goldkorn *et al.*, 1991). Metabolic measurements indicated that treatment of cells with either sphingosine or C_8-ceramide resulted in significant elevations of endogenous ceramide, but not sphingosine, levels, implicating ceramide as the active molecule (Goldkorn *et al.*, 1991).

The use of a 19-amino-acid synthetic peptide corresponding to the sequence around Thr-669 of EGFR as substrate allowed the identification of a ceramide-activated protein kinase (CAPK) activity in the membrane fraction of A431 and HL-60 cells (Mathias *et al.*, 1991). The activity of this kinase increased 2-fold after stimulation with C_8-ceramide and 1.5-fold after stimulation with TNF-α. CAPK was proposed to be a proline-directed serine/threonine kinase with specificity for -Leu-Thr-Pro-as a substrate (Joseph *et al.*, 1993). This kinase was tentatively identified as a 97-kDa autophosphorylating band and was renatured from SDS gels although, at that point, the kinase lost its responsiveness to ceramide (Liu *et al.*, 1994). The significance of this kinase in the ceramide-initiated signaling pathways awaits evidence of its involvement in biologic activities mediated by ceramide.

5.5.3. Protein Kinase C ζ Activation

The ζ isoform of PKC (PKCζ) is unique among the PKC isoforms in its lack of response to diacylglycerols and phorbol esters (Berra *et al.*, 1993; Nakanishi *et al.*, 1993). It is now known to be activated by phosphatidylserine and phosphatidylinositol 3,4,5-trisphosphate (Nakanishi *et al.*, 1993) and to be involved in mitogenic signaling (Berra *et al.*, 1993). PKCζ was recently shown to be involved in activation of NF-κB through activation of a putative I-κB kinase (Diaz-Meco *et al.*, 1994). Phosphorylation of I-κB results in its inactivation and the release of NF-κB as described above. The recent finding that ceramide may play a role in the activation of NF-κB prompted the examination of the ability of ceramide to activate PKCζ. Using I-κBα as a substrate and recombinant bacterially produced

PKCζ, ceramide was found to increase the activity of PKCζ almost threefold (Lozano, 1994). Moreover, treatment of NIH-3T3 cells with TNF-α, PC-PLC, or sphingomyelinase followed by immunoprecipitation of PKCζ and determination of I-κB phosphorylation showed that all three stimulated comparable I-κB phosphorylation (Lozano et al., 1994). Transfection of NIH-3T3 cells with a receptor plasmid driven by NF-κB resulted in activation after treatment with sphingomyelinase. Cotransfection of a defective PKCζ dominant negative mutant abolished this activation. These data suggested that PKCζ may be a target of ceramide activation that could play a role in some of ceramide's biologic activities. So far, the evidence available indicates a role for PKCζ in the activation of NF-κB by TNF-α (Dominguez et al., 1993; Diaz-Meco et al., 1993, 1994; Lozano et al., 1994). The proposed signaling pathway involves the activation of sphingomyelinase by TNF-α generating ceramide which activates PKCζ. In turn, PKCζ phosphorylates I-κB leading to the release of active NF-κB. The importance of PKCζ in mediating the various biologic effects of ceramide awaits further study.

5.6. Conclusions and Future Directions

The accumulating evidence suggests the operation of a ceramide-dependent signal transduction pathway activated in response to a number of extracellular agents (such as TNF-α) and perhaps additional extracellular and intracellular insults and injuries (such as chemotherapeutic agents and ionizing radiation). Ceramide, in turn, appears to act by modulating the activity of protein phosphatases and kinases resulting in modulation of c-*myc*, Rb, cyclooxygenase, and other intracellular targets. These effects appear to result in profound changes in cell growth behavior.

Indeed, a growing body of literature supports a role for ceramide as a pluripotent mediator of growth suppression. Thus, ceramide appears to participate in induction of terminal cell differentiation, cell cycle arrest, apoptosis, and senescence. The particular "end" biologic response appears to be determined by the particular cell type (e.g., differentiation in HL-60 cells and senescence in human fibroblasts) as well as by the operation of additional regulatory and counterregulatory signaling mechanisms. The latter is exemplified by the effect of activation of the PKC pathway by DAG or PMA which inhibit ceramide-induced apoptosis but do not modulate ceramide-induced cell cycle arrest. Therefore, in this case, the simultaneous operation of a ceramide-activated pathway and the DAG/PKC pathway results in steering of cells in the direction of cell cycle arrest away from apoptosis. One may, therefore, hypothesize that additional signal transduction pathways may modulate the particular responses of any given cell type to ceramide.

As a corollary to this hypothesis, we propose that ceramide functions as a signal of adversity/injury whereby the elevation in intracellular ceramide in response to cytokines such as TNF-α, extracellular injuries such as ionizing radiation, or intracellular injury (e.g., chemotherapeutic agents) provide the cell with an "index" of the degree of injury. This may then launch a series of biochemical reactions that will result in one of the end points of growth suppression (discussed above) depending on which additional signaling and regulatory mechanisms are launched.

As insight into the biologic regulation and function of ceramide accumulates, important questions arise as to the biochemical and molecular mechanisms operating in regulating ceramide production and transducing ceramide function. So far, a neutral sphingomyelinase has been implicated in the mechanism of action of vitamin D_3 and TNF-α and an acidic sphingomyelinase has also been suggested to mediate part of the effects of TNF-α. The mechanisms by which TNF-α and other stimuli regulate these sphingomyelinases and the role of sphingomyelinases (as well as other ceramide-generating enzymes) in regulating physiologic levels of ceramide are yet to be determined. Equally as important, the proximal targets for the action of ceramide (such as CAPP or protein kinases) need to be verified and their role in ceramide signaling needs to be determined. The investigation of this pathway at the biochemical and molecular level promises important insight into novel components of cell growth regulation with important consequences for the understanding of mechanisms of cell cycle arrest, apoptosis, terminal differentiation, and cell senescence.

References

Baeuerle, P. A., and Baltimore, D., 1994, I-κB: A specific inhibitor of the NF-κB transcription factor, *Science* **242**:540–546.

Bagchi, S., Raychaudhuri, P., and Nevins, J. R., 1990, Adenovirus E1A proteins can dissociate heteromeric complexes involving the E2F transcription factor: A novel mechanism for E1A *trans*-activation, *Cell* **62**:659–669.

Ballou, L. R., Chao, C. P., Holness, M. A., Barker, S. C., and Raghow, R., 1992, Interleukin-1-mediated PGE_2 production and sphingomyelin metabolism. Evidence for the regulation of cyclooxygenase gene expression by sphingosine and ceramide, *J. Biol. Chem.* **267**:20044–20050.

Berra, E., Diaz-Meco, M. T., Dominguez, I., Municio, M. M., Sanz, L., Lozano, J., Chapkin, R. S., and Moscat, J., 1993, Protein kinase C ζ isoform is critical for mitogenic signal transduction, *Cell* **74**:555–563.

Berridge, M. J., and Irvine, R. F., 1989, Inositol phosphates and cell signaling, *Nature* **341**:197–205.

Betts, J. C., Agranoff, A. B., Nabel, G. J., and Shayman, J. A., 1994, Dissociation of endogenous cellular ceramide from NF-κB activation, *J. Biol. Chem.* **269**:8455–8458.

Beutler, B., and Cerami, A., 1988, The history, properties, and biological effects of cachectin, *Biochemistry* **27**:7575–7582.

Beutler, B., and Van Huffel, C., 1994, Unraveling function in the TNF ligand and receptor families, *Science* **264**:667–668.

Bielawska, A., Linardic, C. M., and Hannun, Y. A., 1992a, Modulation of cell growth and differentiation by ceramide, *FEBS Lett.* **307**:211–214.

Bielawska, A., Linardic, C. M., and Hannun, Y. A., 1992b, Ceramide-mediated biology: Determination of structural and stereospecific requirements through the use of N-acyl-phenylaminoalcohol analogs, *J. Biol. Chem.* **267**:18493–18497.

Bielawska, A., Crane, H. M., Liotta, D., Obeid, L. M., and Hannun, Y. A., 1993, Selectivity of ceramide-mediated biology: Lack of activity of *erythro*-dihydroceramide, *J. Biol. Chem.* **268**:26226–26232.

Bishop, W. R., and Bell, R. M., 1988, Functions of diacylglycerol in glycerolipid metabolism, signal transduction and cellular transformation, *Oncogene Res.* **2**:205–218.

Borchardt, R. A., Lee, W. T., Kalen, A., Buckley, R. H., Peters, C., Schiff, S., and Bell, R. M., 1994, Growth-dependent regulation of cellular ceramides in human T-cells, *Biochim. Biophys. Acta* **1212**:327–336.

Buehrer, B. M., and Bell, R. M., 1992, Inhibition of sphingosine kinase *in vitro* and in platelets. Implications for signal transduction pathways, *J. Biol. Chem.* **267**:3154–3159.

Carswell, E. A., Old, L. J., Kassel, R. L., Green, S., Fiore, N., and Williamson, B., 1975, An endotoxin induced serum factor that causes necrosis of tumors, *Proc. Natl. Acad. Sci. USA* **72**:3666–3670.

Chao, R., Khan, W., and Hannun, Y. A., 1992, Retinoblastoma protein dephosphorylation induced by D-*erythro*-sphingosine, *J. Biol. Chem.* **267**:23459–23462.

Chatterjee, S., 1993, Neutral sphingomyelinase, *Adv. Lipid Res.* **26**:25–47.

Chellappan, S., Kraus, V. B., Kroger, B., Munger, K., Howley, P. M., Phelps, W. C., and Nevins, J. R., 1992, Adenovirus E1A, simian virus 40 tumor antigen, and human papillomavirus E7 protein share the capacity to disrupt the interaction between transcription factor E2F and the retinoblastoma gene product, *Proc. Natl. Acad. Sci. USA* **89**:4549–4553.

Dbaibo, G., Obeid, L. M., and Hannun, Y. A., 1993, TNFα signal transduction through ceramide: Dissociation of growth inhibitory effects of TNFα from activation of NF-κB, *J. Biol. Chem.* **268**:17762–17766.

Dbaibo, G. S., Pushkareva, M. Y., Jayadev, S., Schwarz, J. K., Horowitz, J. M., Obeid, L. M., and Hannun, Y. A., 1995, Rb as a downstream target for ceramide-dependent pathway of growth arrest, *Proc. Natl. Acad. Sci. USA* **92**:1347–1351.

DeCaprio, J. A., Furukawa, Y., Ajchenbaum, F., Griffin, J. D., and Livingston, D. M., 1992, The retinoblastoma-susceptibility gene product becomes phosphorylated in multiple stages during cell cycle entry and progression, *Proc. Natl. Acad. Sci. USA* **89**:1795–1798.

Dennis, E. A., Rhee, S. G., Billah, M. M., and Hannun, Y. A., 1991, Role of phospholipases in generating lipid second messengers in signal transduction, *FASEB J.* **5**:2068–2077.

Diaz-Meco, M. T., Berra, E., Municio, M. M., Sanz, L., Lozano, J., Dominguez, I., Diaz-Golpe, V., De Lera, M. T. L., Alcamí, J., Payá, C. V., Arenzana-Seisdedos, F., Virelizier, J.-L., and Moscat, J., 1993, A dominant negative protein kinase C ζ subspecies blocks NF-κB activation, *Mol. Cell. Biol.* **13**:4770–4775.

Diaz-Meco, M. T., Dominguez, I., Sanz, L., Dent, P., Lozano, J., Municio, M., Berra, E., Hay, R. T., Sturgill, T. W., and Moscat, J., 1994, ζPKC induces phosphorylation and inactivation of IκB-α *in vitro*, *EMBO J.* **13**:2842–2848.

Dobrowsky, R. T., and Hannun, Y. A., 1992, Ceramide stimulates a cytosolic protein phosphatase, *J. Biol. Chem.* **267**:5048–5051.

Dobrowsky, R. T., and Hannun, Y. A., 1994, The sphingomyelin cycle and ceramide second messengers, *Signal-Activated Phosphatases* (M. Liscovitch, ed.), Landes Company, pp. 85–99.

Dobrowsky, R. T., Kamibayashi, C., Mumby, M. C., and Hannun, Y. A., 1993, Ceramide activates heterotrimeric protein phosphatase 2A, *J. Biol. Chem.* **268**:15523–15530.

Dobrowsky, R. T., Werner, M. H., Castellino, A. M., Chao, M. V., and Hannun, Y. A., 1994, Activation of the sphingomyelin cycle through the low-affinity neurotrophin receptor, *Science* **265**:1596–1599.

Dominguez, I., Sanz, L., Arenzana-Seisdedos, F., Diaz-Meco, M. T., Virelizier, J.-L., and Moscat, J., 1993, Inhibition of protein kinase C ζ subspecies blocks the activation of an NF-κB-like activity in *Xenopus laevis* oocytes, *Mol. Cell. Biol.* **13**:1290–1295.

Dressler, K. A., Mathias, S., and Kolesnick, R. N., 1992, Tumor necrosis factor-α activates the sphingomyelin signal transduction pathway in a cell-free system, *Science* **255**:1715–1718.

Exton, J. H., 1990, Signaling through phosphatidylcholine breakdown, *J. Biol. Chem.* **265**:1–4.

Faucher, M., Girones, N., Hannun, Y. A., Bell, R. M., and Davis, R., 1988, Regulation of the epidermal growth factor receptor phosphorylation state by sphingosine in A431 human epidermoid carcinoma cells, *J. Biol. Chem.* **263**:5319–5327.

Fernandes-Alnemri, T., Litwack, G., and Alnemri, E. S., 1994, CPP32, a novel human apoptotic protein with homology to Caenorhabditis elegans cell death protein Ced-3 and mammalian interleukin-1 beta-converting enzyme, *J. Biol. Chem.* **269**:30761–30764.

Feuillard, J., Gouy, H., Bismuth, G., Lee, L. M., Debre, P., and Korner, M., 1991, NF-κB activation by tumor necrosis factor α in the Jurkat T cell line is independent of protein kinase A, protein kinase C, and Ca^{2+}-regulated kinases, *Cytokine* **3:**257–265.

Fishbein, J. D., Dobrowsky, R. T., Bielawska, A., Garrett, S., and Hannun, Y. A., 1993, Ceramide-mediated biology and CAPP are conserved in *Saccharomyces cerevisiae, J. Biol. Chem.* **268:**9255–9261.

Futreal, P. A., and Barrett, J. C., 1991, Failure of senescent cells to phosphorylate the RB protein, *Oncogene* **6:**1109–1113.

Gerschenson, L. E., and Rotello, R. J., 1992, Apoptosis: A different type of cell death, *FASEB J.* **6:**2450–2455.

Ghosh, S., and Baltimore, D., 1990, Activation in vitro of NF-κB by phosphorylation of its inhibitor IκB, *Nature* **344:**678–682.

Goldkorn, T., Dressler, K. A., Muindi, J., Radin, N. S., Mendelsohn, J., Menaldino, D., Liotta, D., and Kolesnick, R. N., 1991, Ceramide stimulates epidermal growth factor receptor phosphorylation in A431 human epidermoid carcinoma cells. Evidence that ceramide may mediate sphingosine action, *J. Biol. Chem.* **266:**16092–16097.

Goldstein, S., 1990, Replicative senescence: The human fibroblast comes of age, *Science* **249:**1129–1133.

Gomez-Muñoz, A., Martin, A., O'Brien, L., and Brindley, D. N., 1994, Cell-permeable ceramides inhibit the stimulation of DNA synthesis and phospholipase D activity by phosphatidate and lysophosphatidate in rat fibroblasts, *J. Biol. Chem.* **269:**8937–8943.

Goodrich, D. W., Wang, N. P., Qian, Y.-W., Lee, E. Y.-H. P., and Lee, W.-H., 1991, The retinoblastoma gene product regulates progression through the G1 phase of the cell cycle, *Cell* **67:**293–302.

Grunfeld, C., and Palladino, M. A., 1990, Tumor necrosis factor: Immunologic, antitumor, metabolic, and cardiovascular activities, *Adv. Intern. Med.* **35:**45–72.

Haimovitz-Friedman, A., Kan, C. C., Ehleiter, D., Persaud, R. S., McLoughlin, M., Fuks, Z., and Kolesnick, R. N., 1994, Ionizing radiation acts on cellular membranes to generate ceramide and initiate apoptosis, *J. Exp. Med.* **180:**525–535.

Hakomori, S., 1981, Glycosphingolipids in cellular interaction, differentiation, and oncogenesis, *Annu. Rev. Biochem.* **50:**733–764.

Hakomori, S., 1990, Bifunctional role of glycosphingolipids, *J. Biol. Chem.* **265:**18713–18716.

Hanahan, D. J., 1986, Platelet activating factor: A biologically active phosphoglyceride, *Annu. Rev. Biochem.* **55:**483–509.

Hannun, Y. A., 1994, The sphingomyelin cycle and the second messenger function of ceramide, *J. Biol. Chem.* **269:**3125–3128.

Hannun, Y. A., and Bell, R. M., 1989, Functions of sphingolipids and sphingolipid breakdown products in cellular regulation, *Science* **243:**500–507.

Hannun, Y. A., and Linardic, C. M., 1993, Sphingolipid breakdown products: Anti-proliferative and tumor-suppressor lipids, *Biochim. Biophys. Acta Bio-Membr.* **1154:**223–236.

Hannun, Y. A., Loomis, C. R., Merrill, A. H., Jr., and Bell, R. M., 1986, Sphingosine inhibition of protein kinase C activity and of phorbol dibutyrate binding *in vitro* and human platelets, *J. Biol. Chem.* **261:**12604–12609.

Harlow, E., 1992, For our eyes only, *Nature* **359:**270–271.

Hayflick, L., 1965, The limited *in vitro* lifetime of human diploid cell strains, *Exp. Cell Res.* **37:**614–636.

Henkel, T., Machleidt, T., Alkalay, I., Kronke, M., Ben-Neriah, Y., and Baeuerle, P., 1993, Rapid proteolysis of IκB-α is necessary for activation of transcription factor NF-κB, *Nature* **365:**182–185.

Hickman, J. A., 1992, Apoptosis is induced by anticancer drugs, *Cancer Metastasis Rev.* **11:**121–129.

Horowitz, J. M., Park, S.-H., Bogenmann, E., Cheng, J.-C., Yandell, D. W., Kaye, F. J., Minna, J. D., Dryja, T. P., and Weinberg, R. A., 1990, Frequent inactivation of the retinoblastoma anti-oncogene is restricted to a subset of human tumor cells, *Proc. Natl. Acad. Sci. USA* **87:**2775–2779.

Huang, H.-J. S., Yee, J.-K., Shew, J.-Y., Chen, P.-L., Bookstein, R., Friedmann, T., Lee, E. Y.-H. P., and Lee, W.-H., 1988, Suppression of the neoplastic phenotype by replacement of the RB gene in human cancer cells, *Science* **242:**1563–1566.

Inokuchi, J., Momosaki, K., Shimeno, H., Nagamatsu, A., and Radin, N. S., 1989, Effects of D-threo-PDMP, an inhibitor of glucosylceramide synthetase, on expression of cell surface glycolipid antigen and binding to adhesive proteins by B16 melanoma cells, *J. Cell. Physiol.* **141:**573–583.

Jarvis, W. D., Kolesnick, R. N., Fornari, F. A., Traylor, R. S., Gewirtz, D. A., and Grant, S., 1994, Induction of apoptotic DNA damage and cell death by activation of the sphingomyelin pathway, *Proc. Natl. Acad. Sci. USA* **91**:73–77.

Jayadev, S., Linardic, C. M., and Hannun, Y. A., 1994, Identification of arachidonic acid as a mediator of sphingomyelin hydrolysis in response to tumor necrosis factor α, *J. Biol. Chem.* **269**:5757–5763.

Jayadev, S., Liu, B., Bielawska, A. E., Lee, J. Y., Nazaire, F., Pushkareva, M. Y. U., Obeid, L. M., and Hannun, Y. A., 1995, Role for ceramide in cell cycle arrest, *J. Biol. Chem.* **270**:2047–2052.

Joseph, C. K., Byun, H.-S., Bittman, R., and Kolesnick, R. N., 1993, Substrate recognition by ceramide-activated protein kinase. Evidence that kinase activity is proline-directed, *J. Biol. Chem.* **268**:20002–20006.

Kaufmann, S. H., Desnoyers, S., Ottaviano, Y., Davidson, N. E., and Poirier, G. G., 1993, Specific proteolytic cleavage of poly(ADP-ribose) polymerase: an early marker of chemotherapy-induced apoptosis, *Cancer Res.* **53**:3976–3985.

Kim, M.-Y., Linardic, C., Obeid, L., and Hannun, Y., 1991, Identification of sphingomyelin turnover as an effector mechanism for the action of tumor necrosis factor α and gamma-interferon. Specific role in cell differentiation, *J. Biol. Chem.* **266**:484–489.

Kolesnick, R., and Golde, D. W., 1994, The sphingomyelin pathway in tumor necrosis factor and interleukin-1 signaling, *Cell* **77**:325–328.

Kyriakis, J. M., Banerjee, P., Nikolakaki, E., Dai, T., Rubie, E. A., Ahmad, M. F., Avruch, J., and Woodgett, J. R., 1994, The stress-activated protein kinase subfamily of c-Jun kinases, *Nature* **369**:156–160.

Lazebnik, Y. A., Kaufmann, S. H., Desnoyers, S., Poirier, G. G., and Earnshaw, W. C., 1994, Cleavage of poly(ADP-ribose) polymerase by a proteinase with properties like ICE, *Nature* **371**:346–347.

Lenardo, M. J., and Baltimore, D., 1989, NF-κ B: A pleiotropic mediator of inducible and tissue-specific gene control, *Cell* **58**:227–229.

Linardic, C. M., Jayadev, S., and Hannun, Y. A., 1992, Brefeldin A promotes hydrolysis of sphingomyelin, *J. Biol. Chem.* **267**:14909–14911.

Link, E., Kerr, L. D., Schreck, R., Zabel, U., Verma, I., and Baeuerle, P. A., 1992, Purified IκB-β is inactivated upon dephosphorylation, *J. Biol. Chem.* **267**:239–246.

Liscovitch, M., and Cantley, L. C., 1994, Lipid second messengers, *Cell* **77**:329–334.

Liu, J., Mathias, S., Yang, Z., and Kolesnick, R. N., 1994, Renaturation and tumor necrosis factor-α stimulation of a 97-kDa ceramide-activated protein kinase, *J. Biol. Chem.* **269**:3047–3052.

Loetscher, H. R., Brockhaus, M., Dembic, Z., Gentz, R., Gubler, U., Hohmann, H.-P., Lahm, H.-W., Van Loon, A. P. G. M., Pan, Y.-C. E., Schlaeger, E.-J., Steinmetz, M., Tabuchi, H., and Lesslauer, W., 1991, Two distinct tumor necrosis factor receptors—members of a new cytokine receptor gene family, in: *Oxford Survey on Eucaryotic Genes*, Vol. 7 (N. Maclean, ed.), pp. 119–142, Oxford University Press, London.

Lowenthal, J. W., Ballard, D. W., Bogerd, H., Böhnlein, E., and Greene, W. C., 1989, Tumor necrosis factor-α activation of the IL-2 receptor-α gene involves the induction of κB-specific DNA binding proteins, *J. Immunol.* **142**:3121–3128.

Lozano, J., Berra, E., Municio, M. M., Diaz-Meco, M. T., Dominguez, I., Sanz, L., and Moscat, J., 1994, Protein kinase C ζ isoform is critical for κB-dependent promoter activation by sphingomyelinase, *J. Biol. Chem.* **269**:19200–19202.

Ludlow, J. W., 1993, Interactions between SV40 large-tumor antigen and the growth suppressor proteins pRB and p53, *FASEB J.* **7**:866–871.

Ludlow, J. W., Shon, J., Pipas, J. M., Livingston, D. M., and DeCaprio, J. A., 1990, The retinoblastoma susceptibility gene product undergoes cell cycle-dependent dephosphorylation and binding to and release from SV40 large T, *Cell* **60**:387–396.

McCachren, S. S., Salehi, Z., Weinberg, J. B., and Niedel, J. E., 1988, Transcription interruption may be a common mechanism of c-myc regulation during HL-60 differentiation, *Biochem. Biophys. Res. Commun.* **151**:574–582.

Machleidt, T., Wiegmann, K., Henkel, T., Schütze, S., Baeuerle, P., and Krönke, M., 1994, sphingomyelinase activates proteolytic IκB-α degradation in a cell-free system, *J. Biol. Chem.* **269**:13760–13765.

Majerus, P. W., Connolly, T. M., Deckmyn, H., Ross, T. S., Bross, T. E., Ishii, H., Bansal, V., and Wilson, D., 1986, The metabolism of phosphoinositide-derived messenger molecules, *Science* **234:**1519–1526.

Mathias, S., Dressler, K. A., and Kolesnick, R. N., 1991, Characterization of a ceramide-activated protein kinase: Stimulation by tumor necrosis factor α, *Proc. Natl. Acad. Sci. USA* **88:**10009–10013.

Mathias, S., Younes, A., Kan, C.-C., Orlow, I., Joseph, C., and Kolesnick, R. N., 1993, Activation of the sphingomyelin signaling pathway in intact EL4 cells and in a cell-free system by IL-1β, *Science* **259:**519–522.

Meichle, A., Schütze, S., Hensel, G., Brunsing, D., and Krönke, M., 1990, Protein kinase C-independent activation of nuclear factor κB by tumor necrosis factor, *J. Biol. Chem.* **265:**8339–8343.

Merrill, A. H., Jr., and Jones, D. D., 1990, An update of the enzymology and regulation of sphingomyelin metabolism, *Biochim. Biophys. Acta Lipids Lipid Metab.* **1044:**1–12.

Merrill, A. H., Jr., and Wang, E., 1992, Enzymes of ceramide biosynthesis, *Methods Enzymol.* **209:**427–437.

Meyaard, L., Otto, S. A., Jonker, R. R., Mijnster, M. J., Keet, R. P. M., and Miedema, F., 1992, Programmed death of T cells of HIV-1 infection, *Science* **257:**217–219.

Miura, M., Zhu, H., Rotello, R., Hartwieg, E. A., and Yuan, J., 1993, Induction of apoptosis in fibroblasts by IL-1 beta-converting enzyme, a mammalian homolog of the C. elegans cell death gene ced-3, *Cell* **75:**653–660.

Molitor, J. A., Walker, W. H., Doerre, S., Ballard, D. W., and Greene, W. C., 1990, NF-κB: A family of inducible and differentially expressed enhancer-binding proteins in human T cell, *Proc. Natl. Acad. Sci. USA* **87:**10028–10032.

Moran, E., 1993, Interaction of adenoviral proteins with pRB and p53, *FASEB J.* **7:**880–885.

Nahorski, S. R., and Potter, B. V. L., 1989, Molecular recognition of inositol polyphosphates by intracellular receptors and metabolic enzymes, *Trends Pharm. Sci.* **10:**139–144.

Nakamura, T., Abe, A., Balazovich, K. J., Wu, D., Suchard, S. J., Boxer, L. A., and Shayman, J. A., 1994, Ceramide regulates oxidant release in adherent human neutrophils, *J. Biol. Chem.* **269:**18384–18389.

Nakanishi, H., Brewer, K. A., and Exton, J. H., 1993, Activation of the zeta isozyme of protein kinase C by phosphatidylinositol 3,4,5-trisphosphate, *J. Biol. Chem.* **268:**13–16.

Nevins, J. R., 1992, E2F: A link between the Rb tumor suppressor protein and viral oncoproteins, *Science* **258:**424–429.

Nicholson, D. W., Ali, A., Thornberry, N. A., Vaillancourt, J. P., Ding, C. K., Gallant, M., Gareau, Y., Griffin, P. R., Labelle, M., Lazebnik, Y. A., Munday, N. A., Raju, S. M., Smulson, M. E., Yamin, T., Yu, V. L., and Miller, D. K., 1995, Identification and inhibition of the ICE/CED-3 protease necessary for mammalian apoptosis, *Nature* **376:**37–43.

Niculescu, F., Rus, H., Shin, S., Lang, T., and Shin, M. L., 1993, Generation of diacylglycerol and ceramide during homologous complement activation, *J. Immunol.* **150:**214–224.

Nishizuka, Y., 1992, Intracellular signaling by hydrolysis of phospholipids and activation of protein kinase C, *Science* **258:**607–614.

Obeid, L. M., Linardic, C. M., Karolak, L. A., and Hannun, Y. A., 1993, Programmed cell death induced by ceramide, *Science* **259:**1769–1771.

Okazaki, T., Bell, R. M., and Hannun, Y. A., 1989, Sphingomyelin turnover induced by vitamin D_3 in HL-60 cells. Role in cell differentiation, *J. Biol. Chem.* **264:**19076–19080.

Okazaki, T., Bielawska, A., Bell, R. M., and Hannun, Y. A., 1990, Role of ceramide as a lipid mediator of 1α,25-dihydroxyvitamin D_3-induced HL-60 cell differentiation, *J. Biol. Chem.* **265:**15823–15831.

Okazaki, T., Bielawska, A., Domae, N., Bell, R. M., and Hannun, Y. A., 1994, Characteristics and partial purification of a novel cytosolic, magnesium-independent, neutral sphingomyelinase activated in the early signal transduction of 1α,25- dihydroxyvitamin D_3-induced HL-60 cell differentiation, *J. Biol. Chem.* **269:**4070–4077.

Olivera, A., Buckley, N. E., and Spiegel, S., 1992, Sphingomyelinase and cell-permeable ceramide analogs stimulate cellular proliferation in quiescent Swiss 3T3 fibroblasts, *J. Biol. Chem.* **267:**26121–26127.

Osborn, L., Kunkel, W., and Nabel, G., 1989, Tumor necrosis factor α and interleukin 1 stimulate the human immunodeficiency virus enhancer by activation of the nuclear factor κB, *Proc. Natl. Acad. Sci. USA* **86:**2336–2340.

Raines, M. A., Kolesnick, R. N., and Golde, D. W., 1993, Sphingomyelinase and ceramide activate mitogen-activated protein kinase in myeloid HL-60 cells, *J. Biol. Chem.* **268:**14572–14575.

Ramachandran, C. K., Murray, D. K., and Nelson, D. H., 1990, Dexamethasone increases neutral sphingomyelinase activity and sphingosine levels in 3T3-L1 fibroblasts, *Biochem. Biophys. Res. Commun.* **167:**607–613.

Rhee, S. G., Suh, P. G., Ryu, S. H., and Lee, S. Y., 1989, Studies of inositol phospholipid-specific phospholipase C, *Science* **244:**546–550.

Riabowol, K., Schiff, J., and Gilman, M. Z., 1992, Transcription factor AP-1 activity is required for initiation of DNA synthesis and is lost during cellular aging, *Proc. Natl. Acad. Sci. USA* **89:**157–161.

Rivas, C. I., Golde, D. W., Vera, J. C., and Kolesnick, R. N., 1994, Involvement of the sphingomyelin pathway in autocrine tumor necrosis factor signaling for human immunodeficiency virus production in chronically infected HL-60 cells, *Blood* **83:**2191–2197.

Rosenwald, A. G., and Pagano, R. E., 1993, Inhibition of glycoprotein traffic through the secretory pathway by ceramide, *J. Biol. Chem.* **268:**4577–4579.

Rother, J., Van Echten, G., Schwarzmann, G., and Sandhoff, K., 1992, Biosynthesis of sphingolipids: Dihydroceramide and not sphinganine is desaturated by cultured cells, *Biochem. Biophys. Res. Commun.* **189:**14–20.

Schütze, S., Potthoff, K., Machleidt, T., Berkovic, D., Wiegmann, K., and Krönke, M., 1992, TNF activates NF-κB by phosphatidylcholine-specific phospholipase C-induced "acidic" sphingomyelin breakdown, *Cell* **71:**765–776.

Seshadri, T., and Campisi, J., 1990, Repression of c-*fos* transcription and an altered genetic program in senescent human fibroblasts, *Science* **247:**205–209.

Smith, C. A., Farrah, T., and Goodwin, R. G., 1994, The TNF receptor superfamily of cellular and viral proteins: Activation, costimulation, and death, *Cell* **76:**959–962.

Song, Q., Baxter, G. D., Kovacs, E. M., Findik, D., and Lavin, M. F., 1992, Inhibition of apoptosis in human tumor cells by okadaic acid, *J. Cell. Physiol.* **153:**550–556.

Spence, M. W., 1993, Sphingomyelinases, *Adv. Lipid Res.* **26:**3–23.

Stein, G. H., Beeson, M., and Gordon, L., 1990, Failure to phosphorylate the retinoblastoma gene product in senescent human fibroblasts, *Science* **249:**666–669.

Strum, J. C., Small, G. W., Pauig, S. B., and Daniel, L. W., 1994, 1-β-D-arabinofuranosylcytosine stimulates ceramide and diglyceride formation in HL-60 cells, *J. Biol. Chem.* **269:**15493–15497.

Sweely, C. C., 1991, Sphingolipids, in: *Biochemistry of Lipids, Lipoproteins, and Membranes*, Vol. 20 (D. E. Vance and J. E. Vance, eds.), pp. 327–361, Elsevier, Amsterdam.

Tewari, M., Quan, L. T., O'Rourke, K., Desnoyers, S., Zeng, Z., Beidler, D. R., Poirier, G. G., Salvesen, G. S., and Dixit, V. M., 1995, Yama/CPP32 beta, a mammalian homolog of CED-3, is a CrmA-inhibitable protease that cleaves the death substrate poly(ADP-ribose) polymerase, *Cell* **81:**801–809.

Van Echten, G., and Sandhoff, K., 1993, Ganglioside metabolism, *J. Biol. Chem.* **268:**5341–5344.

Van Veldhoven, P. P., and Mannaerts, G. P., 1993, Sphingosine-phosphate lyase, *Adv. Lipids Res.* **26:**69–98.

Van Veldhoven, P. P., Matthews, T. J., Bolognesi, D. P., and Bell, R. M., 1992, Changes in bioactive lipids, alkylacylglycerol and ceramide, occur in HIV-infected cells, *Biochem. Biophys. Res. Commun.* **187:**209–216.

Venable, M. E., Blobe, G. C., and Obeid, L. M., 1994, Identification of a defect in the phospholipase D/diacylglycerol pathway in cellular senescence, *J. Biol. Chem.* **269:**26040–26044.

Venable, M. E., Lee, J. Y., Smyth, M. J., Bielawska, A., and Obeid, L. M., 1995, Role of ceramide in cellular senescence, *J. Biol. Chem.* **270** (in press).

Vousden, K., 1993, Interactions of human papillomavirus transforming proteins with the products of tumor suppressor genes, *FASEB J.* **7:**872–879.

Weinberg, R. A., 1990, The retinoblastoma gene and cell growth control, *Trends Biochem. Sci.* **15:**199–202.

Westwick, J. K., Bielawska, A. E., Dbaibo, G., Hannun, Y. A., and Brenner, D. A., 1995, Ceramide activates the stress-activated protein kinases, *J. Biol. Chem.* **270:**22689–22692.

Wiegmann, K., Schütze, S., Kampen, E., Himmler, A., Machleidt, T., and Krönke, M., 1992, Human 55-kDa receptor for tumor necrosis factor coupled to signal transduction cascades, *J. Biol. Chem.* **267:**17997–18001.

Wolff, R. A., Dobrowsky, R. T., Bielawska, A., Obeid, L. M., and Hannun, Y. A., 1994, Role of ceramide-activated protein phosphatase in ceramide-mediated signal transduction, *J. Biol. Chem.* **269:**19605–19609.

Wright, S. C., Kumar, P., Tam, A. W., Shen, N., Varma, M., and Larrick, J. W., 1992, Apoptosis and DNA fragmentation precede TNF-induced cytolysis in U937 cells, *J. Cell. Biochem.* **48:**344–355.

Wyllie, A. H., 1980, Glucocorticoid-induced thymocyte apoptosis is associated with endogenous endonuclease activation, *Nature* **284:**555–556.

Yanaga, F., and Watson, S. P., 1994, Ceramide does not mediate the effect of tumor necrosis factor α on superoxide generation in human neutrophils, *Biochem. J.* **298:**733–738.

Chapter 6

Bioactive Properties of Sphingosine and Structurally Related Compounds

Alfred H. Merrill, Jr., Dennis C. Liotta, and Ronald E. Riley

6.1. Introduction

Although sphingosine was discovered over 100 years ago (Thudichum, 1884), there was relatively little interest in the long-chain base backbones of sphingolipids until they were found to be potent inhibitors of protein kinase C (Hannun et al., 1986; Merrill et al., 1986; Wilson et al., 1986). This raised the possibility that cells utilize hydrolysis products of sphingolipids to regulate cell behavior, in analogy to the lipid second messengers (diacylglycerol, arachidonic acid and its metabolites, etc.) that are derived from phosphoglycerolipids (Hannun and Bell, 1989). Subsequent studies have uncovered other systems that are affected by sphingosine, and have found that N-acyl-sphingosines (ceramides) (Okazaki et al., 1989; Hannun, 1994; Kolesnik and Golde, 1994), sphingosine 1-phosphate (Spiegel, 1993), and probably other metabolites are involved in cell signaling. Furthermore, a number of potent mycotoxins have recently been shown to act via disruption of long-chain base metabolism (Merrill et al., 1993b), and long-chain bases have been directly linked to the etiology of diseases that range in symptoms from neurotoxicity, hepatotoxicity, nephrotoxicity, and immunotoxicity to cancer.

This review will focus on the current information about sphingosine and other "free" long-chain bases, but will address some issues involving related metabolites (e.g., sphingosine 1-phosphate and ceramides) because they are formed when sphingosine is provided to cells, and conditions that change the amounts of these compounds in vivo often also affect sphingosine.

Alfred H. Merrill, Jr. • Department of Biochemistry, Emory University School of Medicine, Atlanta, Georgia 30322-3050. *Dennis C. Liotta* • Department of Chemistry, Emory University, Atlanta, Georgia 30322-3050. *Ronald E. Riley* • Toxicology and Mycotoxins Research Unit, U.S. Department of Agriculture, Agriculture Research Service, Athens, Georgia 30613.

Handbook of Lipid Research, Volume 8: Lipid Second Messengers, edited by Robert M. Bell *et al.* Plenum Press, New York, 1996.

6.2. Properties of Sphingosine and Other Long-Chain Bases

6.2.1. Structures

Sphingosine is the common name for D-*erythro*-4-*trans*-sphingenine, the prevalent long-chain base of most mammalian sphingolipids. It has 18 carbon atoms and the stereochemistry shown in Fig. 6-1. There are a large number of other long-chain bases in mammalian sphingolipids (once estimated to exceed 70) (Karlsson, 1970) which differ in the presence or absence of: the 4,5-*trans*-double bond (the latter being named sphinganine or dihydrosphingosine); double bond(s) at other positions; a hydroxyl group at position 4 (D-4-hydroxysphinganine, also called "phytosphingosine") or elsewhere (Robson *et al.*, 1994); and methyl group(s) on the alkyl side chain or on the amino group. The 4-hydroxysphinganines are the major long-chain bases of yeast (Wells and Lester, 1983), plants (Lynch, 1993), and fungi (Merrill *et al.*, 1995a), but are also made by mammals (Crossman and Hirschberg, 1977). Other modifications of the long-chain base backbone include phosphorylation at carbon 1 (Buehrer and Bell, 1993), and acylation (Merrill and Wang, 1992) or methylation (Igarashi and Hakomori, 1989; Felding-Habermann *et al.*, 1990) of the amino group. Each of these compounds can be found in various alkyl chain lengths, with 18 carbons predominating in most sphingolipids, but other homologues can constitute a major portion of specific sphingolipids (as exemplified by the large amounts of C_{20}-sphingosine in brain gangliosides) (Valsecchi *et al.*, 1993) and in different sources (e.g., C_{16}-sphingosine is a substantial component of milk sphingomyelin) (Morrison, 1969). One difficulty in studying these compounds is that few are commericially available. For example, most of the sphinganine that can be purchased from various vendors is a mixture of the D and L enantiomers (therefore, commercially available dihydroceramides are also mixtures) and the metabolism, and some of the functions, of these compounds are sensitive to stereochemistry (Stoffel and Bister, 1973; Buehrer and Bell, 1992, 1993; Hauser *et al.*, 1994; Olivera *et al.*, 1994).

6.2.2. Chemistry

Sphingosines and other long-chain bases are cationic amphiphiles, which distinguishes them from most other naturally occurring lipids, which are neutral (including zwitterionic) or anionic. In the protonated form, they affect the phase behavior of both zwitterionic (Koiv *et al.*, 1993; López-García *et al.*, 1994) and acidic (Koiv *et al.*, 1993; López-García *et al.*, 1993) phospholipids. The pK_a for sphingosine is between 7 and 8 (Merrill *et al.*, 1989); therefore, both the neutral and protonated species are expected to be present at physiological pH. Sphingosine has a lower pK_a than is typical of aliphatic amines because there is a possibility for hydrogen bonding between the amino group and two vicinal hydroxyls (Fig. 6-2), and the proximity to a hydrophobic alkyl chain favors the uncharged species (Merrill *et al.*, 1989).

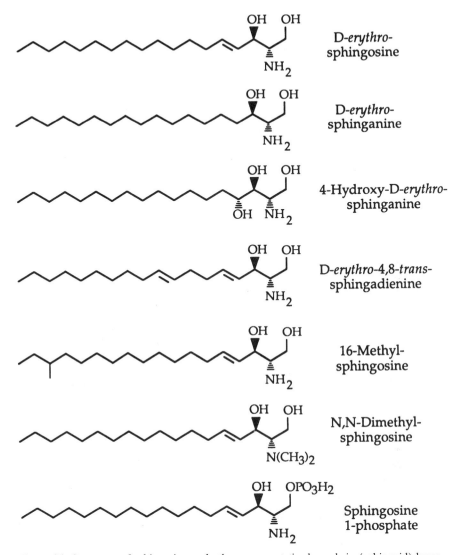

Figure 6-1. Structures of sphingosine and other representative long-chain (sphingoid) bases.

The hydroxyl groups at positions 1, 3, and sometimes 4 or 6 are also relevant to the behavior of these compounds. This has mostly been considered from the perspective of how hydrogen bonding in the interfacial region of the bilayer affects membrane structure (Thompson and Tillack, 1985). However, in a recent study of phosphatidic acid phosphatase purified from yeast (Wu *et al.*, 1993), inhibition of this enzyme by long-chain bases showed a considerable preference for phytosphingosine and sphinganine over sphingosine, which matches the types

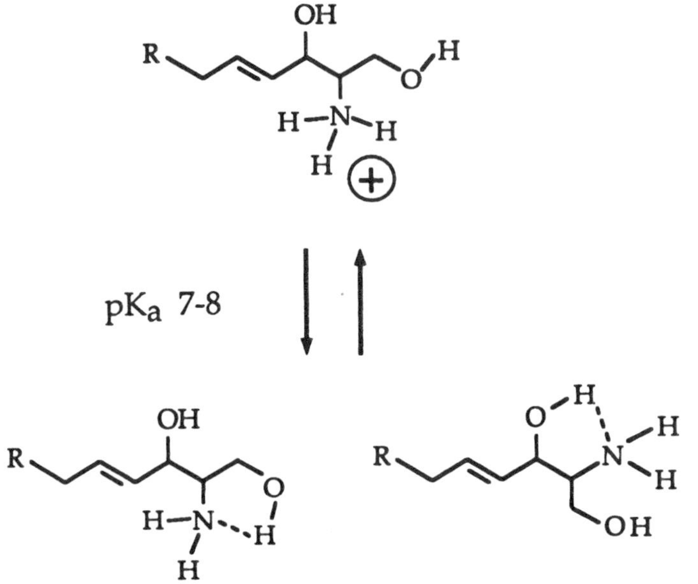

Figure 6-2. Ionization states of sphingosine, illustrating the opportunities for intramolecular hydrogen bonding.

of sphingoid bases found in yeast. Therefore, these functional groups appear to be present both for structural purposes and so that these compounds can optimally interact with cellular targets.

In strong acid, sphingosine can undergo loss of the 3-hydroxyl to form a resonance-stabilized carbonium ion that will react with the solvent as shown in Fig. 6-3. One can minimize this reaction during the hydrolysis of complex sphingolipids by using aqueous HCl in methanol (as recommended by Gaver and Sweeley); nonetheless, in our experience, it is critical to control both the temperature and the concentration of the acid, and to periodically check for decomposition by examining thin-layer chromatoplates (silica gel plates developed with chloroform:methanol: 2 N NH$_4$OH, 40:10:1, by volume) for higher-migrating, ninhydrin-positive compounds (Merrill and Wang, 1986). To improve quantitation, samples can be spiked with an internal standard before acid hydrolysis, such as N-acetyl-C$_{20}$-sphinganine (E. M. Schmelz, personal communication).

6.2.3. Properties

Sphingosine and other long-chain bases are relatively easy to handle because they are readily taken up by cells and move among membranes rapidly (Wilson et al., 1986; Slife et al., 1989; Hannun et al., 1991). The uptake depends on the alkyl chain length (Merrill et al., 1989), and 18-carbon homologues appear to be taken up

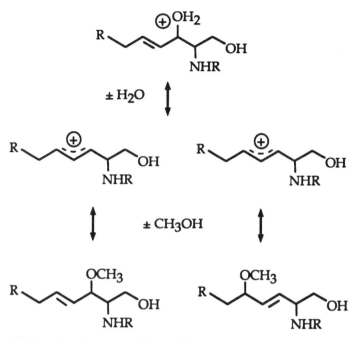

Figure 6-3. Examples of compounds formed from sphingosine under acidic conditions.

most rapidly. Movement across the membrane bilayer is also rapid unless the pH is low enough to trap the amino group in the protonated form (Hope and Cullis, 1987).

6.2.4. Handling

Long-chain bases, like many amines, tend to decompose to yellow products unless stored desiccated, at low temperature, and under a nitrogen or argon atmosphere. Stock solutions in organic solvents are generally stable when stored at low temperature unless the solutions are dilute or acidic. Phytosphingosines tend to be especially susceptible to decomposition (G. Schroepfer, personal communication). Decomposition can be particularly troublesome with [^{3}H]-labeled long-chain bases, which often require repurification before use (Merrill and Wang, 1992). This radiochemical instability may be related in part to contaminants because it is variable among different preparations.

Sphingosine and other long-chain bases can be dissolved in a variety of solvents (chloroform, alcohols, acetone, dimethyl sulfoxide); however, the sulfate salts are more difficult to dissolve. Long-chain bases can be dispersed by sonication and added to assays directly if low concentrations are used (ca. 1 µM). When higher concentrations are needed, or when slower delivery to cells is desired, long-chain bases can be prepared as complexes with albumin or as part of liposomes (Hannun *et al.*, 1991; Merrill and Wang, 1992). The binding to albumin is sufficiently tight for the complex to be dialyzed, if needed.

6.3. Sphingosine Metabolism

Elucidation of the details of the pathway(s) for the biosynthesis of sphingosine is still ongoing, but the first major steps in its characterization could be viewed as the identification of palmitoyl-CoA and serine as the precursors (Brady and Koval, 1957), followed by identification of the next enzymatic activities for long-chain base biosynthesis (Snell *et al.*, 1970) and turnover (Stoffel, 1970), and demonstration that the 4,5-*trans*-double bond is added after acylation of sphinganine (Ong and Brady, 1973; Merrill and Wang, 1986; Rother *et al.*, 1992). Although an enzymatic activity has been shown to desaturate dihydroceramides (Rother *et al.*, 1992), labeling experiments with cells in culture do not always show the precursor–product relationship that would be expected for ceramide as an obligatory intermediate. For example, when rat hepatocytes are incubated with [^{14}C]serine, there is rapid labeling of the ceramide pool (Merrill *et al.*, 1995b); however, experiments with L cells (Merrill and Wang, 1986), J774 macrophages (Smith and Merrill, 1995), or HT-29 cells (E. M. Schmelz, personal communication) yield sphingomyelins and glucosylceramides that contain dihydroceramides as the backbones, and ceramides only become the predominant backbones later. Therefore, it appears that desaturation can occur both at the level of dihydroceramides and after head groups have been added to dihydroceramides.

The initial and committed reaction of long-chain base biosynthesis is the condensation of serine and palmitoyl-CoA, which is catalyzed by serine palmitoyltransferase (EC 2.3.1.50), a pyridoxal 5'-phosphate-dependent enzyme that best utilizes linear, saturated fatty acyl-CoAs of 16 ± 1 carbon atoms (Williams *et al.*, 1984; Merrill and Williams, 1984). Serine palmitoyltransferase activity is reduced by adding long-chain bases to cells in culture (van Echten *et al.*, 1990; Mandon *et al.*, 1991), and the time course implies that the downregulation is at the level of transcription/translation rather than direct feedback inhibition.

Mutants in serine palmitoyltransferase have been isolated using Chinese hamster ovary cells (Hanada *et al.*, 1990, 1992) and yeast (Dickson *et al.*, 1990; Nagiec *et al.*, 1994). In the latter case, they fall into two complementation groups, LCB1 and LCB2, both of which have been sequenced and have considerable homology with δ-aminolevulinate synthase and other enzymes that catalyze reactions similar to serine palmitoyltransferase. LCB2 appears to have the requisite sequence for the active site of serine palmitoyltransferase; therefore, LCB1 may code for a regulatory subunit. Zhao *et al.* (1994) have isolated another series of mutants in *Saccharomyces cerevisiae* based on suppression of calcium sensitivity, and one of the genes that they have cloned (SCS1p) is identical to LCB2. Therefore, they hypothesize that sphingolipid metabolism either is regulated by calcium or is required for calcium homeostasis in yeast.

The product of serine palmitoyltransferase, 3-ketosphinganine, is reduced to sphinganine by a microsomal NADPH-dependent reductase (Snell *et al.*, 1970; Stoffel, 1970), which is apparently much more active than serine palmitoyl-transferase because none of the keto intermediate is found when microsomes are incubated with serine, palmitoyl-CoA, and NADPH (Williams *et al.*, 1984) or when cells are incubated with radiolabeled precursors (Merrill and Wang, 1986). The next step is the formation of the amide-linked fatty acid, which has been reported to occur by both reactions that involve fatty acyl-CoAs (Sribney, 1966; Morell and Radin, 1970; Ullman and Radin, 1972; Akanuma and Kishimoto, 1979) and a CoA-independent pathway (Singh, 1983; Mori *et al.*, 1985). There is also evidence for direct addition of head groups to long-chain bases (Yamaguchi *et al.*, 1994).

As already discussed, at some point(s) after the synthesis of dihydroceramides, the 4,5-*trans*-double bond is introduced into the backbone (Rother *et al.*, 1992). Therefore, to obtain free sphingosine, these complex sphingolipids must be hydrolyzed to ceramides, which are cleaved by ceramidases (*N*-acylsphingosine deacylases). There are at least three classes of ceramidase: a lysosomal enzyme with an acid pH optimum, a neutral enzyme, and an activity with an alkaline pH optimum (Spence *et al.*, 1986). The lysosomal turnover of ceramides is stimulated by the sphingolipid activator protein D (sap-D) (Klein *et al.*, 1994). Ceramides can also undergo phosphorylation by a ceramide kinase (Bajjalieh *et al.*, 1989; Kolesnick and Hemer, 1991). Ceramide phosphates are potentially potent mediators because they have been proposed to be responsible for damage caused by brown recluse spider venom (Rees *et al.*, 1984).

Once sphingosine has been released from more complex sphingolipids, it mainly undergoes reacylation or phosphorylation. Sphingosine (sphinganine) kinase (Buehrer and Bell, 1993) has been studied with extracts from platelets (Stoffel *et al.*, 1970; Buehrer and Bell, 1992), liver (Stoffel and Bister, 1973; Hirschberg *et al.*, 1970), and brain (Louie *et al.*, 1976), among others. The enzyme appears to be both cytosolic and membrane associated (Buehrer and Bell, 1992), and to interact with all four stereoisomers of the long-chain base, but in some cases as substrates and in others as inhibitors (Buehrer and Bell, 1992, 1993). Sphingosine kinase activity is induced by treatment of cells with PDGF (Olivera and Spiegel, 1993) and has also been reported to be affected by phorbol esters in some cell types (Mazurek *et al.*, 1994).

The next catabolic step is an aldolase-like reaction catalyzed by sphingosine 1-phosphate lyase, which yields ethanolamine phosphate and *trans*-2-hexadecenal (from sphingosine) or palmitaldehyde (from sphinganine) (Stoffel *et al.*, 1969; Stoffel, 1970; Van Veldhoven and Mannaerts, 1993). The lyase is most active in microsomal fractions (Van Veldhoven and Mannaerts, 1991), and is thought to require pyridoxal 5'-phosphate. Another metabolic fate for sphingosine 1-phosphate is hydrolysis to sphingosine by phosphatase(s) (Van Veldhoven and Mannaerts, 1994).

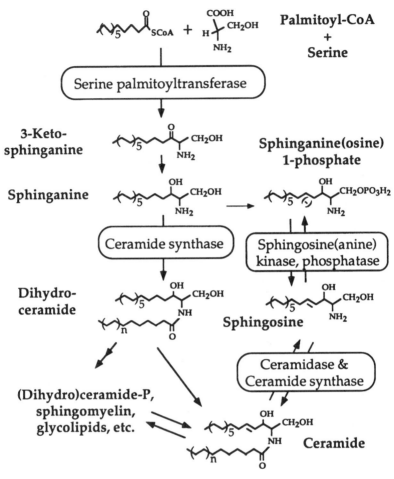

Figure 6-4. Sphingosine biosynthesis and metabolism. In addition to the reactions shown, the sphingosine- and sphinganine-1-phosphates can be cleaved to ethanolamine phosphate (which appears in phosphatidylethanolamine) plus hexadecenal or hexadecanal, respectively. There is also some evidence for introduction of the double bond after the head groups have been added; for direct addition of head groups to sphingosine; and for the methylation of sphingosine.

The enzymes of sphingolipid biosynthesis and turnover are found in multiple intracellular compartments, which has been generally indicated in Fig. 6-5. The early biosynthetic reactions occur on the cytosolic leaflet of the endoplasmic reticulum (Mandon *et al.*, 1992; Futerman, 1994), but activities for sphingolipid turnover are found in plasma membranes (Slife *et al.*, 1989) as well as in lysosomes. Sphingolipids also appear to undergo extensive recycling and modification (Trinchera *et al.*, 1990).

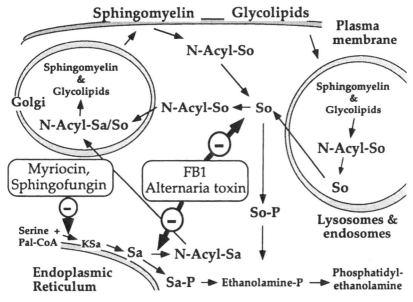

Figure 6-5. Sphingolipid metabolism and sites of inhibition by naturally occurring inhibitors. *De novo* sphingolipid biosynthesis begins in the endoplasmic reticulum with the condensation of serine and palmitoyl-CoA by serine palmitoyltransferase (inhibited by sphingofungins and myriocin); 3-keto-sphinganine is reduced to sphinganine, which is acylated to dihydroceramides by ceramide synthase (inhibited by fumonisins, FB_1, and alternaria toxin). Sphingolipid turnover is also presented to show the interrelationships between the biosynthetic and catabolic reactions.

6.4. Naturally Occurring Inhibitors of Sphingosine Metabolism

Fungi produce a variety of secondary metabolites (fumonisins, alternaria toxins, sphingofungins, and myriocins) that are structurally related to sphinganine (Fig. 6-6) and disrupt sphingolipid metabolism at the sites shown in Fig. 6-5. In addition, a number of novel long-chain base derivatives have been isolated from the sponge *Penares sp.* (Kobayashi *et al.*, 1991) that are potent activators of actomyosin ATPase and inhibitors of protein kinase C (Alvi *et al.*, 1992). These organisms have apparently recognized the advantages of mimicking long-chain bases to disrupt their metabolism and/or interact with their cellular targets.

One group (Caldas *et al.*, 1995) has suggested calling such compounds "*s*phinganine *a*nalogue *m*ycotoxins" (SAMs); however, this acronym may be too limiting because the compounds are not all "mycotoxins." An alternative acronym might be "SPHAM" for "*sph*ingolipid *a*nalogue bio*m*odulators." Considering how many of these compounds have been discovered in the last few years, it is likely that the list of SPHAMs will grow. Two classes are described in some detail below because they have been associated with disease, and are powerful tools for studies of sphingolipids.

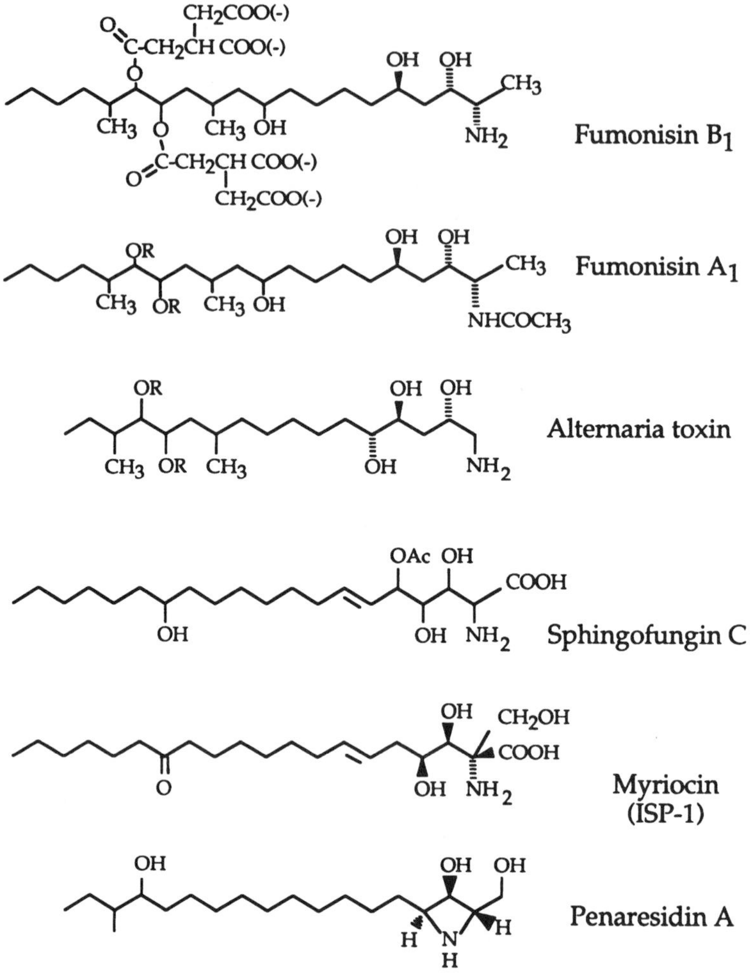

Figure 6-6. Structures of sphingolipid analogue biomodulators (SPHAM). The inhibitors of ceramide synthase (fumonisin B_1 and alternaria toxin) have a free amino group; fumonisin A_1 is not inhibitory. The "R" group of fumonisin A1 and alternaria toxin is the same as the side chain (termed a tricarballylic acid) for fumonisin B_1. Sphingofungin and myriocin (ISP-1) are inhibitors of serine palmitoyltransferase. Penaresidin A is an inhibitor of protein kinase C.

6.4.1.   Inhibitors of Ceramide Synthase

6.4.1.1.   Fumonisins

Fusarium moniliforme (Sheldon) is one of the most prevalent molds on corn, sorghum, and other grains throughout the world. Consumption of *F. moniliforme*-contaminated corn causes several diseases of agricultural concern (e.g., equine leukoencephalomalacia and porcine pulmonary edema) (Kriek *et al.*, 1981; Marasas *et al.*, 1988; Ross *et al.*, 1990); has been associated with human esophageal cancer (Yang, 1980; Lin *et al.*, 1980; Marasas, 1982; Riley *et al.*, 1994b); results in hepatotoxicity, nephrotoxicity, and liver cancer in rats (Marasas *et al.*, 1984; Gelderblom *et al.*, 1988, 1991; Riley *et al.*, 1994a); and alters immune function (Dombrink-Kurtzman *et al.*, 1993; Martinova and Merrill, 1995).

The agents thought to be responsible for these diseases are a group of mycotoxins termed *fumonisins*. The most prevalent compound, fumonisin B_1, is shown in Fig. 6-6 (Bezuidenhout *et al.*, 1988; ApSimon *et al.*, 1994; Harmange *et al.*, 1994; Hoye *et al.*, 1994; Poch *et al.*, 1994); the acetylated derivative, fumonisin A_1, is relatively nontoxic.

A primary target of the fumonisins is ceramide synthase (Wang *et al.*, 1991), as depicted in Fig. 6-4. The structural basis for this inhibition is unknown; however, one can speculate that similarities between the fumonisins and both long-chain (sphingoid) bases (in the vicinity of the amino group) and fatty acyl CoA (at the tricarballylic acids) may allow them to be recognized by both substrate binding sites of ceramide synthase (Merrill *et al.*, 1995a). In support of this hypothesis, fumonisin B_1 exhibits substrate-competitive inhibition against both sphinganine and fatty acyl CoA (Merrill *et al.*, 1993c), and removal of the tricarballylic acid side chains reduces the potency by about tenfold (Merrill *et al.*, 1993b).

With the inhibition of ceramide synthase, the mass of sphinganine increases as much as 110 fold (Wang *et al.*, 1991), so that it is within the levels that have been shown to affect cell behavior when added exogenously (such as inhibition of protein kinase C) (Wilson *et al.*, 1988; Merrill, 1991). In recent studies with J774 macrophages (E. R. Smith, unpublished observations), we have found that at least one isozyme of protein kinase C is affected when the levels of endogenous long-chain bases are elevated by fumonisin. Therefore, a plausible mechanism for the cellular effects of fumonisins is that they cause levels of long-chain bases to increase, and these bioactive compounds (including their metabolites, such as the 1-phosphates) trigger the subsequent changes that culminate in cell death (Norred *et al.*, 1992; Merrill *et al.*, 1995c). Studies using cells in culture are beginning to explore this hypothesis. Yoo *et al.* (1992, and personal communication) have related the accumulation of sphinganine to the inhibition of cell growth, and ultimately death, in LLC-PK1 cells. Schroeder *et al.* (1994) have shown that fumonisins can stimulate thymidine incorporation into DNA in Swiss 3T3 cells, and that this requires the formation of sphinganine; therefore, this may provide a mechanism that links fumonisins, sphingolipids, and carcinogenesis (Riley *et al.*, 1994b). Not all of the cellular effects of fumonisins are mediated through the accumulation of long-chain bases, however, Harel and Futerman (1993) have shown that fumonisin B$_1$ inhibits axonal outgrowth in cultured hippocampal neurons, and this can be bypassed by providing exogenous ceramides.

To determine if these *in vitro* observations are relevant *in vivo*, the amounts of sphingosine, sphinganine, and total sphingolipids have been measured using serum from ponies that had been given fumonisin-contaminated feed. There was a large increase in sphinganine and sphingosine, and a reduction in the amounts of complex sphingolipids in serum (Wang *et al.*, 1992). Similar findings have been obtained subsequently in feeding studies with pigs (Riley *et al.*, 1993), rats (Riley *et al.*, 1994a), and chickens (Weibking *et al.*, 1993). On this basis, the appearance of sphinganine in serum or urine has been proposed as a biomarker for exposure (Riley *et al.*, 1994c).

6.4.1.2. Alternaria Toxin

The presence of a sphinganine-like backbone is shared by a number of other mycotoxins, such as the host-specific phytotoxins produced by *Alternaria alternata f. sp. lycopersici*. This fungus produces the macroscopic symptoms of stem canker disease in tomatoes (Bottini *et al.*, 1981). Alternaria toxins ("AAL toxins") are similar to fumonisins (Bottini *et al.*, 1981; Boyle *et al.*, 1994), but lack the methyl group at position 1.

No animal diseases have been associated with consumption of AAL toxins. However, AAL toxin has been shown to be cytotoxic by *in vitro* assays of various mammalian cells in culture, with IC$_{50}$ values ranging from two- to sevenfold higher than fumonisin B1 (Shier *et al.*, 1991). We find that AAL toxin is also an inhibitor of sphingolipid biosynthesis in rat hepatocytes, but about tenfold less potent than fumonisin B$_1$. AAL toxin (and fumonisins) have been shown to induce accumulation of phytosphingosine and sphinganine in plants (Abbas *et al.*, 1994), which makes it likely that the fungi use these secondary metabolites against their host plants.

6.4.2. Inhibitors of Serine Palmitoyltransferase

6.4.2.1. Sphingofungins

Zweerink *et al.* (1992) have described a class of compounds produced by *Aspergillus fumigatus* that are potent and selective inhibitors of serine palmitoyltransferase. A representative "sphingofungin" is shown in Fig. 6-6. The IC$_{50}$ for inhibition of serine palmitoyltransferase *in vitro* is 100 nM for sphingofungin C (shown) and 20 nM for sphingofungin B (which lacks the *O*-acetyl group). Sphingofungins block the incorporation of [^{3}H]inositol into sphingolipids, and are toxic to *Saccharomyces cerevisiae*.

6.4.2.2. Myriocin (Thermozymocidin, ISP-1)

Another compound with long-chain base-like features was isolated from *Myriococcum albomyces* and *Mycelia sterilia* in the early 1970s (reviewed in Turner and

Aldridge, 1983) and named *myriocin* or *thermozymocidin*, respectively (Fig. 6-6). It received little further attention until Fujita *et al.* (1994) "rediscovered" this compound during the isolation of a potent immunosuppressive activity in the culture broth of *Isaria sinclairii* (ATCC 24400), which is described as the imperfect stage of *Cordyceps sinclairii*, a fungus that is parasitic on insects. They had given the activity the name "ISP-1." In the mouse allogeneic mixed lymphocyte reaction, ISP-1 was 5- to 10-fold more potent than cyclosporin A in inhibition of lymphocyte proliferation; when assayed for generation of alloreactive cytotoxic T lymphocytes, ISP-1 was 100-fold more potent than cyclosporin A. Miyake *et al.* (1994) have shown that ISP-1 inhibits serine palmitoyltransferase with a $K_i < 1$ nM. Based on the structure of this compound, one would predict that it resembles a transition-state intermediate of serine palmitoyltransferase.

6.4.2.3. Other Inhibitors of Sphingosine Metabolism

Serine palmitoyltransferase is sensitive to many compounds that act as active-site-directed (suicide) inhibitors of pyridoxal 5′-phosphate-dependent enzymes. This includes both natural products (cycloserine) (Sundaram and Lev, 1984; Holleran *et al.*, 1990) and synthetic compounds, such as the β-haloalanines (Medlock and Merrill, 1988). For the most part, these compounds have been used to manipulate sphingolipid biosynthesis by cells in culture (Medlock and Merrill, 1988; Holleran *et al.*, 1990; Merrill *et al.*, 1993c; Schroeder *et al.*, 1994); however, long-term feeding studies with L-cycloserine have been conducted. The administration of cycloserine to mice had little effect on the levels of sphingomyelin, sulfatides, or gangliosides, but reduced cerebroside levels in brain (Sundaram and Lev, 1989).

Compounds that have been used to inhibit other steps of sphingolipid metabolism include unnatural stereoisomers of sphinganine for sphingosine (anine) kinase (Buehrer and Bell, 1992), *N*-oleoyl-ethanolamine for ceramidase (Sugita *et al.*, 1975), PDMP for glucosylceramide synthase (Inokuchi and Radin, 1987; Shukla *et al.*, 1991; Abe *et al.*, 1995), epoxy-glucosylceramide for higher glycolipids (Zacharias *et al.*, 1994), and copper II for the neutral sphingomyelinase (Lister *et al.*, 1993).

6.5. Cellular Systems Affected by Sphingosine and Related Compounds

Since the initial discovery that sphingosine is an inhibitor of protein kinase C, a large number of other systems have been found to be affected by long-chain bases and their metabolites (for reviews see Merrill *et al.*, 1993a; Spiegel, 1993; Merrill, 1994; Hannun, 1994; Kolesnick and Golde, 1994). Some of these systems appear to be reasonable candidates for regulation by long-chain bases "physiologically" because they meet many of the following criteria: they are affected at the low levels of long-chain bases that are found in cells; effects can be seen both when long-chain bases are added to *in vitro* assays and when they are given to intact cells; the system exhibits stereoselectivity or other types of subtle structural specificity;

and there is a plausible association between the system of interest and changes in cellular sphingolipids. However, it is more difficult to prove that a particular long-chain base is a "second messenger" for a number of reasons:

- Few studies have linked the effects of exogenously added sphingolipids to changes in the mass of these molecules in cells.
- The intracellular localization of long-chain bases are not known with certainty; therefore, it is difficult to relate results with exogenously added compounds to the intracellular levels even when they are known.
- When long-chain bases are added to cells, they tend to be metabolized by at least three pathways (acylation, phosphorylation, and methylation), and these products also have biological activities. It is not known to what extent this diminishes the potency of the administered compound and/or leads to mixed responses.

6.5.1. Sphingosine and Other Long-Chain (Sphingoid) Bases

This review will summarize systems that have been reported to be affected by long-chain bases and, in only a few cases, their metabolites. Systems that have been reported to be affected by long-chain bases *in vitro*, or when added to cells, are listed in Table 6-1. From this listing, a few patterns are noteworthy.

The inhibition of protein kinase C has been studied extensively. Sphingosine inhibits most of the isozymes of protein kinase C, and exhibits competitive inhibition with diacylglycerol, phorbol dibutyrate, and Ca^{2+}, and blocks activation by unsaturated fatty acids and other lipids (Hannun *et al.*, 1986; Wilson *et al.*, 1986; Oishi *et al.*, 1988; El Touny *et al.*, 1990). The inhibition depends on the mole percentage of the sphingosine with respect to the other lipids or detergents. A positive charge on sphingosine is probably required for inhibition (Hannun *et al.*, 1986; Hannun and Bell, 1989; Merrill *et al.*, 1989; Bottega *et al.*, 1989). The inhibition of protein kinase C is not affected much by the stereochemistry of the long-chain base, but the most potent inhibitors have alkyl chain lengths of approximately 18 carbon atoms (Merrill *et al.*, 1989). This may indicate that a substantial portion of the inhibition occurs at the membrane surface. *N*-acyl-sphingosine has been reported also to affect protein kinase C-alpha translocation (Jones and Murray, 1995). In addition to inhibiting this enzyme *in vitro*, it has been possible to block numerous protein kinase C-dependent processes in cells (Table 6-1).

Sphingosine and other alkylamines affect glycerolipid signaling pathways at multiple steps. They are potent inhibitors of phosphatidic acid phosphohydrolase (Mullmann *et al.*, 1991; Jamal *et al.*, 1991; Aridor-Piterman *et al.*, 1992; Perry *et al.*, 1992), which forms diacylglycerols (and alkylacylglycerols) in glycerolipid bio-synthesis and turnover via the phospholipase D pathway. Studies with human neutrophils revealed that sphingosine is about tenfold more potent as an inhibitor of phosphatidic acid phosphohydrolase than as an inhibitor of protein kinase C (Perry *et al.*, 1992); therefore, this may be one of the most "physiological" targets for endogenous long-chain bases. Sphingosine also activates phosphatidylethanol-amine-specific phospholipase D (Kiss *et al.*, 1991) and phospholipase C delta (Pawelczyk and Lowenstein, 1992), and has been found to enhance phospha-tidylinositol turnover (Chao *et al.*, 1994) and prostaglandin E_2 production (Ballou *et al.*, 1990, 1992).

Long-chain bases affect several ion transport systems. Of particular interest is the finding that sphingosine (Table 6-1), sphingosine 1-phosphate, and sphin-gosylphosphorylcholine (Zhang *et al.*, 1991; Desai *et al.*, 1992; Spiegel, 1993; Mattie *et al.*, 1994; Ghosh *et al.*, 1994) induce calcium release from intracellular stores via an inositol triphosphate-independent mechanism (Mattie *et al.*, 1994). Sphin-gosine is also an inhibitor of sarcoplasmic reticulum calcium release in response to caffeine, doxorubicin, and other agents (Sabbadini *et al.*, 1992). Other ion trans-port systems that are affected by long-chain bases include the Na^+,K^+-ATPase (which is inhibited with approximately the same dose response as for protein kinase C inhibition) (Oishi *et al.*, 1990) and the K^+ channel in smooth muscle cells (Petrou *et al.*, 1994).

Long-chain bases have been found to stimulate and inhibit cell growth, and the possible mechanisms for these divergent effects are complex. Sphingosine increases phosphorylation of the EGF receptor (see references in Table 6-1) and increases AP1 (at least to some extent via metabolism to sphingosine 1-phosphate and ceramide) (Goldkorn *et al.*, 1991; Spiegel, 1993; Hauser *et al.*, 1994). However, long-chain bases *per se* are able to increase thymidine incorporation into DNA (Schroeder *et al.*, 1994). Growth inhibition by long-chain bases may involve the induction of dephosphorylation of the retinoblastoma gene product (Chao *et al.*, 1992; Pushkareva *et al.*, 1995). Sphingosine also appears to be able to induce apoptosis (Bai *et al.*, 1990; Ohta *et al.*, 1994, 1995). The mechanism for the induction of apoptosis is not known, but might involve inhibition of protein kinase C (Stevens *et al.*, 1990a).

Long-chain bases have various effects on other protein kinases (Table 6-1). Pushkareva *et al.* (1992) have reported that D-*erythro*-sphingosine induces the phosphorylation of a number of cytosolic proteins in Jurkat T cells, and have proposed that there are "sphingosine-activated protein kinases" based on the nature of the activities, including the finding that the *erythro*-sphinganines were less active than sphingosine, and *threo*-sphinganine was not active.

Based on the profile of these effects of long-chain bases (Table 6-1), it is possible that they would inhibit carcinogenesis.. Evidence for this has emerged from several studies, which include the inhibition of ornithine decarboxylase induction by phorbol esters (Gupta *et al.*, 1988; Enkvetchakul *et al.*, 1989); inhibi-tion of foci development in C3H10T1/2 cells treated with γ-irradiation and phorbol esters (Borek *et al.*, 1991); and inhibition of tumor growth and metastasis (Endo *et al.*, 1991; Okoshi *et al.*, 1991; Park *et al.*, 1994). Long-chain bases could also reduce neoplasia by promotion of differentiation (Stevens *et al.*, 1989, 1990b; Yung *et al.*, 1992; Yung, 1994) and/or induction of apoptosis.

The effects of long-chain bases are not limited to mammalian cell systems. It has been found that sphingosine can modulate ferricyanide reductase in response

Table 6-1. Systems That Have Been Found to Be Affected by Long-Chain (Sphingoid) Bases

System affected	References
Inhibition of protein kinase C	Hannun *et al.* (1986, 1991), Hannun and Bell (1987), Bazzi and Nelsestuen (1987), Bottega *et al.* (1989), Igarashi *et al.* (1989), Merrill *et al.* (1989), El Touny *et al.* (1990), Katoh (1993), Blobe *et al.* (1994)
Inhibition of phorbol ester-dependent differentiation	Merrill *et al.* (1986)
Inhibition of the neutrophil oxidative burst and other protein kinase C-dependent functions of leukocytes	Wilson *et al.* (1986, 1988), Stinavage and Spitznagel (1989), Kimura *et al.* (1992)
Inhibition of phosphatidic acid phosphohydrolase	Jamal *et al.* (1991), Mullmann *et al.* (1991), Aridor-Piterman *et al.* (1992), Gomez-Munoz *et al.* (1992), Perry *et al.* (1992), Wu *et al.* (1993)
Stimulation of phospholipase D	Kiss *et al.* (1991), Kiss (1994), Natarajan *et al.* (1994)
Regulation of phospholipase C delta	Pawelczyk and Lowenstein (1992)
Regulation of phosphoinositide hydrolysis	Ritchie *et al.* (1992)
Potentiates IL-1 and tumor necrosis factor α-induced prostaglandin E_2 production	Ballou *et al.* (1990, 1992), Candela *et al.* (1991)
Ca^{2+} release from intracellular stores	Ghosh *et al.* (1990), Sabbadini *et al.* (1992), Olivera *et al.* (1994), Tornquist and Ekokoski (1994), Breittmayer *et al.* (1994), Kim *et al.* (1995)
Enhancement of phosphatidylinositol turnover and calcium mobilization	Chao *et al.* (1994)
Inhibition of Ca^{2+}-ATPase and reloading of intracellular Ca^{2+} stores	Pandol *et al.* (1994)
Inhibition of cardiac sarcoplasmic reticulum ryanodine receptor	Dettbarn *et al.* (1994)
Inhibition of Ca^{2+} transients and L-type Ca^{2+} channel conductance	McDonough *et al.* (1994)
Inhibition of Na,K-ATPase and sodium pump	Oishi *et al.* (1990)
Stimulation of K^+ channel activity in smooth muscle cells	Petrou *et al.* (1994)
Induction of epidermal growth factor receptor phosphorylation	Faucher *et al.* (1988), Wedegaertner and Gill (1989), Goldkorn *et al.* (1991)
Enhancement of epidermal growth factor receptor autophosphorylation (by *N,N*-dimethylsphingosine)	Igarashi *et al.* (1990)
Induction of p125FAK and paxillin tyrosine phosphorylation, actin stress fiber formation, and focal contact assembly	Seufferlein and Rozengurt (1994)
Stimulation of Swiss 3T3 cell proliferation	Zhang *et al.* (1990), Spiegel (1993)
Stimulation of keratinocyte proliferation	Wakita *et al.* (1994)
Induction of retinoblastoma protein dephosphorylation	Chao *et al.* (1990), Pushkareva *et al.* (1995)

(continued)

Table 6-1. Systems That Have Been Found to Be Affected by Long-Chain (Sphingoid) Bases
(continued)

System affected	References
Reversal of growth inhibition caused by activation of protein kinase C in vascular smooth muscle cells	Weiss *et al.* (1991)
Growth inhibition and cytotoxicity	Stevens *et al.* (1990a)
Inhibition of neurite outgrowth	Uemura *et al.* (1993)
Induction of apoptosis	Bai *et al.* (1990), Ohta *et al.* (1994, 1995)
Activation of "sphingosine-activated" protein kinases	Pushkareva *et al.* (1992, 1993), Hudson *et al.* (1994)
Inhibition of insulin-stimulated hexose transport and glucose oxidation	Robertson *et al.* (1989)
Inhibition of the insulin receptor tyrosine kinase	Arnold and Newton (1991)
Inhibition of c-src and v-src tyrosine kinase	Igarashi *et al.* (1989)
Inhibition of myelin basic protein phosphorylation (by galactosylsphingosine)	Vartanian *et al.* (1989)
Downregulation of GMP-140 (CD62 or PADGEM) expression on platelets (by N,N-dimethyl and N,N,N-trimethyl-sphingosine)	Handa *et al.* (1991)
Platelet aggregation and ATP release	Knöfler *et al.* (1994)
Inhibition of fibronectin secretion	Scheidl *et al.* (1992), Scita and Wolf (1994)
Enhancement of differentiation of HL-60 cells	Stevens *et al.* (1989, 1990b), Yung *et al.* (1992, 1994)
Inhibition of sphingosine kinase (by threo-sphinganine)	Buehrer and Bell (1992)
Inhibition of sphingomyelinase-induced cholesteryl ester formation	Harmala *et al.* (1993)
Requirement for normal GPI-anchored protein function	Hanada *et al.* (1993, 1995), Horvath *et al.* (1994)
Inhibition of angiotensin-stimulated aldosterone synthesis	Elliott *et al.* (1991)
Inhibition of ornithine decarboxylase induction	Gupta *et al.* (1988), Enkvetchakul *et al.* (1989)
Inhibition of the induction of DNA polymerase and DNase in EBV-infected cells treated with phorbol esters and *n*-butyrate	Nutter *et al.* (1987)
Inhibition of synthesis of RNA primers by primase *in vitro*	Simbulan *et al.* (1994)
Alteration of the membrane association of DNA	Kinnunen *et al.* (1993)
Inhibition of multistage carcinogenesis in mouse C3H/10T1/2 cells	Borek *et al.* (1991)
Inhibition of tumor growth	Endo *et al.* (1991)
Inhibition of metastasis (by N,N,N-trimethylsphingosine)	Okoshi *et al.* (1991), Park *et al.* (1994)

to blue light in mesophyll cells (Dharmawardhane *et al.*, 1989) and pyrophosphatase activity in tonoplast vesicles and isolated vacuoles from *Chenopodium rubrum* (Bille *et al.*, 1992).

6.5.2. Long-Chain Base Metabolites

As already noted, when sphingosine is phosphorylated it is converted to another highly bioactive compound. Few investigators have determined whether the sphingosine (or other long-chain base) that they have added to cells has been converted to the 1-phosphate. Where this has been considered, the results have been somewhat surprising. Spiegel's group (Olivera *et al.*, 1944) has found that Swiss 3T3 cells phosphorylate both the D-*erythro* and L-*threo* stereoisomers of sphingosine; thus, phosphorylation is not restricted to the *erythro* forms as was predicted from previous work (Buehrer and Bell, 1992). Both long-chain base 1-phosphates caused phosphatidic acid to accumulate; however, only D-*erythro*-sphingosine 1-phosphate induced release of calcium from intracellular stores. Studies using sphingosine 1-phosphate should also recognize that cells have potent phosphatase(s) (Van Veldhoven and Mannaerts, 1994) that will release sphingosine.

A similar concern applies to the acylation of long-chain bases and the hydrolysis of ceramides. It is not clear if the *N*-methylation of sphingosine plays a major role in long-chain base metabolism because it has only been observed in a few systems (Felding-Habermann *et al.*, 1990).

6.5.3. The Natural Occurrence of Free Long-Chain Bases

Whereas it was once thought that free long-chain bases are not present in cells, many studies have now found that they are not only present in significant amounts, but the levels change when the cells are treated in a variety of ways. Most cells examined to date contain 1 to 10 nmole of free sphingosine per g tissue, wet weight (10 to 100 pmole/10^6 cells) (for examples, see Merrill *et al.*, 1986, 1988; Wilson *et al.*, 1988; Kobayashi *et al.*, 1988; Wertz and Downing, 1989; Van Veldhoven *et al.*, 1989; Paige *et al.*, 1993; Riboni *et al.*, 1994; Steen Law *et al.*, 1995). Neuronal cells seem to have higher levels of free long-chain bases (Merrill *et al.*, 1993c; Chigorno *et al.*, 1994), as do cells from patients with Niemann–Pick disease (Goldin *et al.*, 1992). Some of the factors that have been found to alter the levels of free long-chain bases are: lipoproteins (Wilson *et al.*, 1988); dexamethasone (Nelson and Murray, 1989; Miccheli *et al.*, 1994; Ricciolini *et al.*, 1994); phorbol dibutyrate (Wilson *et al.*, 1988; Miccheli *et al.*, 1991); changing the cell culture medium (Lavie *et al.*, 1994; Smith and Merrill, 1995); PDGF (Olivera and Spiegel, 1993); and fumonisins and alternaria toxins (Merrill *et al.*, 1995c). The levels of long-chain bases in cells fall within one order of magnitude of the amounts that are necessary to affect the most sensitive targets (such as phosphatidic acid phosphatase, protein kinase C, the Na^+/K^+-ATPase); therefore, it is plausible that they are involved in the regulation of some of these systems.

Fewer analyses have been conducted for the long-chain base 1-phosphates, but they are detectable (Van Veldhoven *et al.*, 1994) and the amounts are affected by growth factors (Olivera and Spiegel, 1993). The levels of cellular ceramides, on the other hand, are being studied extensively because they are involved in signaling pathways for cytokines (Kim *et al.*, 1991; Dressler *et al.*, 1992; Schütze *et al.*, 1994), 1α,25-dihydroxyvitamin D_3 (Okazaki *et al.*, 1989, 1990), retinoic acid (Kalen *et al.*, 1992), nerve growth factor (Dobrowsky *et al.*, 1994), HIV infection (Van Veldhoven *et al.*, 1992), and ionizing radiation (Haimovitz-Friedman *et al.*, 1994; Quintans *et al.*, 1994; Chen *et al.*, 1995). In these cases, the ceramide appears to be derived from the turnover of sphingomyelin by either (or in some cases both) neutral or acidic sphingomyelinase pathways (Schütze *et al.*, 1992; Hannun, 1994; Linardic and Hannun, 1994; Andrieu *et al.*, 1994; Wiegmann *et al.*, 1994).

Ceramide has been found to activate a serine/threonine protein phosphatase (Dobrowsky and Hannun, 1992; Wolff *et al.*, 1994). The ceramide-activated protein phosphatase (CAPP) is cation independent and sensitive to inhibition by okadaic acid (i.e., similar to the subgroup 2A protein phosphatases). CAPP is activated by a variety of ceramides with different hydrophobic moieties; however, studies with *N*-acyl-phenylaminoalcohol analogues indicate that it is stereospecific (Bielawska *et al.*, 1992) and that dihydroceramides are much less potent than ceramides. Kolesnick and co-workers have shown that ceramide also activates protein kinase(s) (Goldkorn *et al.*, 1991; Mathias *et al.*, 1991) that can be assayed using a polypeptide substrate representing amino acids 663–681 of the EGF receptor. Activation of this kinase activity has been demonstrated in a cell-free system (Dressler *et al.*, 1992) in which postnuclear supernatants from HL-60 cells were treated with TNF to induce sphingomyelin turnover to ceramide, and could be mimicked by adding exogenous sphingomyelinase to the preparation instead of TNF.

This sphingomyelin/ceramide cycle is thought to play a major role in some forms of growth arrest (Jayadev *et al.*, 1995; Dbaibo *et al.*, 1995), hypoxia (Kendler and Dawson, 1992), and apoptosis (Hannun and Obeid, 1995; Cifone *et al.*, 1994; Quintans *et al.*, 1994; Jarvis *et al.*, 1994; Chen *et al.*, 1995; Ji *et al.*, 1995).

6.6. Sphingolipids as Components of the Diet

Although it is not widely appreciated, long-chain bases are delivered to cells every day when they are released during the digestion of dietary sphingolipids (Nilsson, 1968, 1969; Schmelz *et al.*, 1994). A recent study (Schmelz *et al.*, 1994) has found that sphingomyelin metabolism occurs throughout the intestine (except the stomach) and yields choline, ceramide, sphingosine, and further degradation products (hexadecenal and ethanolamine phosphate). A portion of the ingested sphingomyelin proceeds through the small intestine to the cecum and colon (Nilsson, 1968; Schmelz *et al.*, 1994), where it is hydrolyzed, presumably by intestinal microflora. If long-chain bases and/or their metabolites are able to suppress carcinogenesis, dietary sphingolipids may inhibit the development of colon cancer.

To determine if dietary sphingolipids could affect colon carcinogenesis, female CF1 mice have been treated with 1,2-dimethylhydrazine (DMH) and then fed diets containing varying amounts of milk sphingomyelin and evaluated for colonic cell proliferation, aberrant colonic crypts, and colon tumors (Dillehay *et al.* 1994). None of the animals fed sphingomyelin without pretreatment with DMH exhibited abnormal histology or developed tumors. Mice treated with DMH and fed the diets containing the lowest amount of sphingomyelin (0.025%) for 3 to 7 weeks had a higher number of proliferating cells per crypt epithelium; however, sphingomyelin feeding decreased the number of aberrant colonic crypts. At the end of the study, the DMH-treated mice fed sphingomyelin had a 20% incidence of colon tumors compared with 47% in the control group. There was a significant reduction in the number of tumors per mouse. These results demonstrate that sphingomyelin inhibits the DMH-induced premalignant lesions (reflected as aberrant crypts) and appears to reduce DMH-induced colon carcinogenesis in CF1 mice. The mechanism(s) for these effects are not known, but probably involve modulation of some of the cellular systems that are sensitive to long-chain bases and ceramides (e.g., inhibition of protein kinase C, induction of differentiation, induction of apoptosis) (Merrill *et al.*, 1995d).

6.7.   Concluding Remarks

The identification of the cellular targets for long-chain bases and their metabolites is of considerable basic science interest, and is already leading to practical benefits—such as the discovery of biomarkers for exposure to mycotoxins, and the possibility that dietary sphingolipids may play a role in protection against colon cancer. As more is learned about these molecules, it should be possible to develop new drugs based on this knowledge.

ACKNOWLEDGMENTS.   The authors are grateful to many collaborators in various aspects of this work, namely Drs. H. Abbas, C. Alexander, C. W. Bacon, V. Beasley, R. M. Bell, G. Carman, H. M. Crane, D. L. Dillehay, V. Geisler, Y. A. Hannun, W. Haschek, B. Hennig, B. Lagu, R. LaRocque, E. Martinova, D. Menaldino, E. T. Morgan, W. P. Norred, R. D. Plattner, S. Ramasamy, R. Reddy, L. Rice, P. F. Ross, K. Sandhoff, E. M. Schmelz, J. J. Schroeder, E. R. Smith, P. Stancel, T. R. Vales, D. E. Vance, G. Van Echten-Deckert, K. A. Voss, E. Wang, T. Wilson, J. Xia, and H. Yoo. We also thank Ms. Winnie Scherer for help in preparing the text.
This work was supported by funds from the USDA (grant #91-37204-6684) and the NIH (GM46368).

References

Abbas, H. K., Tanaka, T., Duke, S. O., Porter, J. K., Wray, E. M., Hodges, L., Sessions, A. E., Wang, E., Merrill, A. H., Jr., and Riley, R. T., 1994, Fumonisin- and AAL-toxin-induced disruption of sphingolipid metabolism with accumulation of free sphingoid bases, *Plant Physiol.* **106**:1085–1093.

Abe, A., Radin, N. S., Shayman, J. A., Wotring, L. L., Zipkin, R. E., Sivakumar, R., Ruggieri, J. M., Carson, D. G., and Ganem, B., 1995, Structural and stereochemical studies of potent inhibitors of glucosylceramide synthase and tumor cell growth, *J. Lipid Res.* **36:**611–621.

Akanuma, H., and Kishimoto, Y., 1979, Synthesis of ceramides and cerebrosides containing both α-hydroxy and nonhydroxy fatty acids from lignoceroyl-CoA by rat brain microsomes, *J. Biol. Chem.* **254:**1050–1056.

Alvi, K. A., Palmer, W., and Crews, P., 1992, Protein kinase C inhibitory alkaloids from the marine sponge *Penares sollasi*, Abstracts of the 33rd Annual Meeting of the American Society of Pharmacognosy, Williamsburg, VA.

Andrieu, N., Salvayre, R., and Levade, T., 1994, Evidence against involvement of the acid lysosomal sphingomyelinase in the tumor-necrosis-factor- and interleukin-1-induced sphingomyelin cycle and cell proliferation in human fibroblasts, *Biochem. J.* **303:**341–345.

ApSimon, J. W., Blackwell, B. A., Edwards, O. E., and Fruchier, A., 1994, Relative configuration of the C-1 to C-5 fragment of fumonisin B_1, *Tetrahedron Lett.* **35:**7703–7706.

Aridor-Piterman, O., Lavie, Y., and Liscovitch, M., 1992, Bimodal distribution of phosphatidic acid phosphohydrolase in NG108-15 cells. Modulation by the amphiphilic lipids oleic acid and sphingosine, *Eur. J. Biochem.* **204:**561–568.

Arnold, R. S., and Newton, A. C., 1991, Inhibition of the insulin receptor tyrosine kinase by sphingosine, *Biochemistry* **30:**7747–7754.

Bai, C., Aw, T. Y., Wang, E., Merrill, A. H., Jr., and Jones, D. P., 1990, Effect of sphingosine, gangliosides, cyclic AMP, and interferons on programmed cell death, *FASEB J.* **4:**477.

Bajjalieh, S. M., Martin, T. F. J., and Floor, E., 1989, Synaptic vesicle ceramide kinase. A calcium-stimulated lipid kinase that co-purifies with brain synaptic vesicles, *J. Biol. Chem.* **264:**14354–14360.

Ballou, L. R., Barker, S. C., Postlethwaite, A. E., and Wang, A. K., 1990, Sphingosine potentiates IL-1-mediated prostaglandin E_2 production in human fibroblasts, *J. Immunol.* **145:**4245–4251.

Ballou, L. R., Chao, C. P., Holness, M. S., Barker, S. C., and Raghow, R., 1992, Interleukin-1-mediated prostaglandin E_2 production and sphingomyelin metabolism. Evidence for the regulation of cyclooxygenase gene expression by sphingosine and ceramide, *J. Biol. Chem.* **267:**20044–20050.

Bazzi, M. D., and Nelsestuen, G. L., 1987, Mechanism of protein kinase C inhibition by sphingosine, *Biochem. Biophys. Res. Commun.* **146:**203–207.

Bezuidenhout, C. S., Gelderblom, W. C. A., Gorstallman, C. P., Horak, R. M., Marasas, W. F. O., Spiteller, G., and Vleggaar, R., 1988, Structure elucidation of the fumonisins, mycotoxins from *Fusarium moniliforme*, *J. Chem. Soc. Commun.* **1988:**743–745.

Bielawska, A., Linardic, C. M., and Hannun, Y. A., 1992, Ceramide-mediated biology: Determination of structural and stereospecific requirements through the use of N-acyl-phenylaminoalcohol analogs, *J. Biol. Chem.* **267:**18493–18497.

Bille, J., Wkeiser, T., and Bentrup, F.-W., 1992, The lysolipid sphingosine modulates pyrophosphatase activity in tonoplast vesicles and isolated vacuoles from a heterotrophic cell suspension culture of *Chenopodium rubrum*, *Physiol. Plant* **84:**250.

Blobe, G. C., Obeid, L. M., and Hannun, Y. A., 1994, Regulation of protein kinase C and role in cancer biology, *Cancer Metastasis Rev.* **13:**411–431.

Borek, C., Ong, A., Stevens, V. L., Wang, E., and Merrill, A. H., Jr, 1991, Long-chain (sphingoid) bases inhibit multistage carcinogenesis in mouse C3H/10T1/2 cells treated with radiation and phorbol 12-myristate 13-acetate, *Proc. Natl. Acad. Sci. USA* **88:**1953–1957.

Bottega, R., Epand, R. M., and Ball, E. H., 1989, Inhibition of protein kinase C by sphingosine correlates with the presence of positive charge, *Biochem. Biophys. Res. Commun.* **164:**102–107.

Bottini, A. T., Bowen, J. R., and Gilchrist, D. G., 1981, Phytotoxins II. Characterization of a phytotoxic fraction from *Alternaria alternata f. sp. lycopersici*, *Tetrahedron Lett.* **22:**2723–2726.

Boyle, C. D., Harmange, J.-C., and Kishi, T., 1994, Novel structure elucidation of AAL toxin T_a backbone, *J. Am. Chem. Soc.* **116:**4995–4498.

Brady, R. O., and Koval, G. J., 1957, Biosynthesis of sphingosine in vitro, *J. Am. Chem. Soc.* **79:**2648–2649.

Breittmayer, J. P., Bernard, A., and Aussel, C., 1994, Regulation by sphingomyelinase and sphingosine of Ca^{2+} signals elicited by CD3 monoclonal antibody, thapsigargin, or ionomycin in the Jurkat T cell line, *J. Biol. Chem.* **269:**5054–5058.

Buehrer, B. M., and Bell, R. M., 1992, Inhibition of sphingosine kinase *in vitro* and in platelets. Implications for signal transduction pathways, *J. Biol. Chem.* **267**:3154–3159.

Buehrer, B. M., and Bell, R. M., 1993, Sphingosine kinase: Properties and cellular functions, *Adv. Lipid Res.* **26**:59–67.

Caldas, E. D., Jones, A. D., Winter, C. K., Ward, B., and Gilchrist, D. G., 1995, Electrospray ionization mass spectrometry of sphinganine analog mycotoxins, *Anal. Chem.* **67**:196–207.

Candela, M., Barker, S. C., and Ballow, L. R., 1991, Sphingosine synergistically stimulates tumor necrosis factor α-induced prostaglandin E2 production in human fibroblasts, *J. Exp. Med.* **174**:1363–1369.

Chao, C., Khan, W., and Hannun, Y. A., 1992, Retinoblastoma protein dephosphorylation induced by D-*erythro*-sphingosine, *J. Biol. Chem.* **267**:23459–23462.

Chao, C. P., L+au-lederkind, S. J., and Ballou, L. R., 1994, Sphingosine-mediated phosphatidylinositol metabolism and calcium mobilization, *J. Biol. Chem.* **269**:5849–5856.

Chen, M., Quintans, J., Fuks, Z., Thompson, C., Kufe, D. W., and Weichselbaum, R. R., 1995, Suppression of Bcl-2 messenger RNA production may mediate apoptosis after ionizing radiation, tumor necrosis factor alpha, and ceramide, *Cancer Res.* **55**:991–994.

Chigorno, V., Valsecchi, M., and Sonnino, S., 1994, Biosynthesis of gangliosides containing C18:1 and C20:1 [3-^{14}C]sphingosine after administrating [1-^{14}C]palmitic acid and [1-^{14}C]stearic acid to rat cerebellar granule cells in culture, *Eur. J. Biochem.* **221**:1095–1101.

Cifone, M. G., DeMaria, R., Roncaioli, P., Rippo, M. R., Azuma, M., Lanier, L. L., Santoni, A., and Testi, R., 1994, Apoptotic signaling through CD95 (Fas/Apo-1) activates an acidic sphingomyelinase, *J. Exp. Med.* **180**:1547–1552.

Crossman, M. W., and Hirschberg, D. B., 1977, Biosynthesis of phytosphingosine by the rat, *J. Biol. Chem.* **252**:5815–5819.

Dbaibo, G. S., Pushkareva, M. Y., Jayadev, S., Schwarz, J. K., Horowitz, J. M., Obeid, L. M., and Hannun, Y. A., 1995, Retinoblastoma gene product as a downstream target for a ceramide-dependent pathway of growth arrest, *Proc. Natl. Acad. Sci. USA* **92**:1347–1351.

Desai, N. N., Zhang, H., Olivera, A., Mattie, M. E., and Spiegel, S., 1992, Sphingosine-1-phosphate, a metabolite of sphingosine, increases phosphatidic acid levels by phospholipase D activation, *J. Biol. Chem.* **267**:23122–23128.

Dettbarn, C. A., Betto, R., Salviati, G., Palade, P., Jenkins, P. M., and Sabbadini, R. A., 1994, Modulation of cardiac sarcoplasmic reticulum ryanodine receptor by sphingosine, *J. Mol. Cell Cardiol.* **26**:229–242.

Dharmawardhane, S., Rubinstein, B., and Stern, A. I., 1989, Regulation of transplasmalemma electron transport in oat mesophyll cells by sphingoid bases and blue light, *Plant Physiol.* **89**:1345–1350.

Dickson, R. C., Wells, G. B., Schmidt, A., and Lester, R. L., 1990, Isolation of mutant *Saccharomyces cerevisiae* strains that survive without sphingolipids, *Mol. Cell. Biol.* **10**:2176–2181.

Dillehay, D. L., Webb, S. J., Schmelz, E.-M., and Merrill, A. H., Jr., 1994, Dietary sphingomyelin inhibits 1,2-dimethylhydrazine-induced colon cancer in CF1 mice, *J. Nutr.* **124**:615–620.

Dobrowsky, R. T., and Hannun, Y. A., 1992, Ceramide stimulates a cytosolic protein phosphatase, *J. Biol. Chem.* **267**:5048–5051.

Dobrowsky, R. T., Werner, M. H., Castellino, A. M., Chao, M. V., and Hannun, Y. A., 1994, Activation of the sphingomyelin cycle through the low-affinity neurotrophin receptor, *Science* **265**:1596–1599.

Dombrink-Kurtzman, M. A., Javed, T., Bennett, G. A., Richard, J. L., Cote, L. M., and Buck, W. B., 1993, Lymphocyte cytotoxicity and erythrocytic abnormalities induced in broiler chicks by fumonisins B$_1$ and B$_2$ and moniliformin from *Fusarium proliferatum*, *Mycopathologia* **124**:47–54.

Dressler, K. A., Mathias, S., and Kolesnick, R. N., 1992, Tumor necrosis factor-alpha activates the sphingomyelin signal transduction pathway in a cell-free system, *Science* **255**:1715–1718.

Elliott, M. E., Jones, H. M., Tomasko, S., and Goodfriend, T. L., 1991, Sphingosine inhibits angiotensin-stimulated aldosterone synthesis, *J. Steroid Biochem. Mol. Biol.* **38**:475–481.

El Touny, S., Khan, W., and Hannun, Y. A., 1990, Regulation of platelet protein kinase C by oleic acid. Kinetic analysis of allosteric regulation and effects on autophosphorylation, phorbol ester binding, and susceptibility to inhibition, *J. Biol. Chem.* **265**:16437–16443.

Endo, K., Igarashi, Y., Nisar, M., Zhou, Q. H., and Hakomori, S.-I., 1991, Cell membrane signaling as target in cancer therapy: Inhibitory effect of N,N-dimethyl and N,N,N-trimethyl sphingosine derivatives on *in vitro* and *in vivo* growth of human tumor cells in nude mice, *Cancer Res.* **51**:1613–1618.

Enkvetchakul, B., Merrill, A. H., Jr., and Birt, D. F., 1989, Inhibition of the induction of ornithine decarboxylase activity by 12-O-tetradecanoylphorbol-13-acetate in mouse skin by sphingosine sulfate, *Carcinogenesis* **10**:379–381.

Faucher, M., Girones, N., Hannun, Y. A., Bell, R. M., and Davis, R. A.,1988, Regulation of the epidermal growth factor receptor phosphorylation state by sphingosine in A431 human epidermoid carcinoma cells, *J. Biol. Chem.* **263**:5319–5327.

Felding-Habermann, B., Igarashi, Y., Fenderson, B. A., Park, L. S., Radin, N. S., Inokuchi, J.-I., Strassmann, G., Hanada, K., and Hakomori, S -I., 1990, A ceramide analogue inhibits T cell proliferative response through inhibition of glycosphingolipid synthesis and enhancement of N,N-dimethylsphingosine synthesis, *Biochemistry* **29**:6314–6322.

Fujita, T., Inoue, K., Yamamoto, S., Ikumoto, T., Sasaki, S., Toyama, R., Chiba, K., Hoshino, Y., and Okumoto, T., 1994, Fungal metabolites. Part 11. A potent immunosuppressive activity found in *Isaria sinclairii* metabolite, *J. Antibiot.* **47**:208–215.

Futerman, A. H., 1994, Ceramide metabolism is compartmentalized in the endoplasmic reticulum and Golgi apparatus, *Curr. Top. Membr.* **41**:93–110.

Gelderblom, W. C. A., Jaskiewicz, K., Marasas, W. F. O., Thiel, P. G., Dorak, R. M., Vleggaar, R., and Kriek, N. P. J., 1988, Cancer promoting potential of different strains of *Fusarium moniliforme* in a short-term cancer initiation/promotion assay, *Carcinogenesis* **9**:1405–1409.

Gelderblom, W. C. A., Kriek, N. P. J., Marasas, W. F. O., and Thiel, P.G., 1991, Toxicity and carcinogenicity of the *Fusarium moniliforme* metabolite, fumonisin B$_1$, in rats, *Carcinogenesis* **12**:1247–1251.

Ghosh, T. K., Bian, J., and Gill, D. L., 1990, Intracellular calcium release mediated by sphingosine derivatives generated in cells, *Science* **248**:1653–1656.

Ghosh, T. K., Bian, J., and Gill, D. L., 1994, Sphingosine 1-phosphate generated in the endoplasmic reticulum membrane activates release of stored calcium, *J. Biol. Chem.* **269**:22628–22635.

Goldin, E., Roff, C. F., Miller, S. P., Rodriguez-Lafrasse, C., Vanier, M. T., Brady, R. O., and Pentchev, P. G., 1992, Type C Niemann-Pick disease: A murine model of the lysosomal cholesterol lipidosis accumulates sphingosine and sphinganine in liver, *Biochim. Biophys. Acta* **1127**:303–311.

Goldkorn, T., Dressler, K. A., Muindi, J., Radin, N. S., Mendelsohn, J., Menaldino, D., Liotta, D., and Kolesnick, R. N., 1991, Ceramide stimulates epidermal growth factor receptor phosphorylation in A431 human epidermoid carcinoma cells: Evidence that ceramide may mediate sphingosine action, *J. Biol. Chem.* **266**:16092–16097.

Gomez-Munoz, A., Hamza, E. H., and Brindley, D. N., 1992, Effects of sphingosine, albumin and unsaturated fatty acids on the activation and translocation of phosphatidate phosphohydrolases in rat hepatocytes, *Biochim. Biophys. Acta* **1127**:49–56.

Gupta, A. K., Fischer, G. J., Elder, J. T., Nickoloff, B. J., and Voorhees, J. J., 1988, Sphingosine inhibits phorbol ester-induced inflammation, ornithine decarboxylase activity, and activation of protein kinase C in mouse skin, *J. Invest. Dermatol.* **91**:486–491.

Haimovitz-Friedman, A., Kan, C.-C., Ehleiter, D., Persaud, R. S., McLoughlin, M., Fuks, Z., and Kolesnick, R. N., 1994, Ionizing radiation acts on cellular membranes to generate ceramide and initiate apoptosis, *J. Exp. Med.* **180**:525–535.

Hanada, K., Mizawa, K., Nishijima, M., and Akamatsu, Y., 1993, Sphingolipid deficiency induces hypersensitivity of CD14, a glycosyl phosphatidylinositol-anchored protein, to phosphatidyl-inositol-specific phospholipase C, *J. Biol. Chem.* **268**:13820–13823.

Hanada, K., Nishijima, M., and Akamatsu, Y., 1990, A temperature-sensitive mammalian cell mutant with thermolabile serine palmitoyltransferase for the sphingolipid biosynthesis, *J. Biol. Chem.* **265**:22137–22142.

Hanada, K., Nishijima, M., Kiso, H., Hasegawa, A., Fujita, S., Ogawa, T., and Akamatsu, Y., 1992, Sphingolipids are essential for the growth of Chinese hamster ovary cells. Restoration of the growth of a mutant defective in sphingoid base biosynthesis with exogenous sphingolipids, *J. Biol. Chem.* **267**:23527–23533.

Hanada, K., Nishijima, M., Akamatsu, Y., and Pagano, R. E., 1995, Both sphingolipids and cholesterol participate in the detergent insolubility of alkaline phosphatase, a glycosylphosphatidylinositol-anchored protein, in mammalian membranes, *J. Biol. Chem.* **270:**6254–6260.

Handa, K., Igarashi, Y., Nisar, M., and Hakomori, S.-I., 1991, Downregulation of GMP-140 (CD62 or PADGEM) expression on platelets by N,N-dimethyl and N,N,N-trimethyl derivatives of sphingosine, *Biochemistry* **30:**11682–11686.

Hannun, Y. A., 1994, The sphingomyelin cycle and the second messenger function of ceramide, *J. Biol. Chem.* **269:**3125–3128.

Hannun, Y. A., and Bell, R. M., 1987, Lysosphingolipids inhibit protein kinase C: Implications for sphingolipidoses, *Science* **235:**670–674.

Hannun, Y. A., and Bell, R. M., 1989, Functions of sphingolipids and sphingolipid breakdown products in cellular regulation, *Science* **243:**500–507.

Hannun, Y. A., and Obeid, L. M., 1995, Ceramide: An intracellular signal for apoptosis, *Trends Biochem. Sci.* **20:**73–77.

Hannun, Y. A., Loomis, C. R., Merrill, A. H., Jr., and Bell, R. M., 1986, Sphingosine inhibition of protein kinase C activity and of phorbol bibutyrate binding in vitro and in human platelets, *J. Biol. Chem.* **261:**12604–12609.

Hannun, Y. A., Merrill, A. H., Jr., and Bell, R. M, 1991, Use of sphingosine as an inhibitor of protein kinase C, *Methods Enzymol.* **201:**316–328.

Harel, R., and Futerman, A. H., 1993, Inhibition of sphingolipid synthesis affects axonal outgrowth in cultured hippocampal neurons, *J. Biol. Chem.* **268:**14476–14482.

Harmala, A. S., Porn, M. I., and Slotte, J. P., 1993, Sphingosine inhibits sphingomyelinase-induced cholesteryl ester formation in cultured fibroblasts, *Biochim. Biophys. Acta* **1210:**97–104.

Harmange, J.-C., Boyle, C. D., and Kishi, Y., 1994, Relative and absolute stereochemistry of the fumonisin B2 backbone, *Tetrahedron Lett.* **35:**6819–6822.

Hauser, J. M. L., Buehrer, B. M., and Bell, R. M., 1994, Role of ceramide in mitogenesis induced by exogenous sphingoid bases, *J. Biol. Chem.* **269:**6803–6809.

Hirschberg, C. B., Kisic, A., and Schroepfer, G. J., 1970, Enzymatic formation of dihydrosphingosine-1-phosphate, *J. Biol. Chem.* **245:**3084–3090.

Holleran, W. M., Williams, M. L., Gao, W. N., and Elias, P. M., 1990, Serine palmitoyltransferase activity in cultured human keratinocytes, *J. Lipid Res.* **31:**1655–1661.

Hope, M. J., and Cullis, P. R., 1987, Lipid asymmetry induced by transmembrane pH gradients in large unilamellar vesicles, *J. Biol. Chem.* **262:**4360–4366.

Horvath, A., Sütterlin, C., Manning-Krieg, U., Movva, N. R., and Riezman, H., 1994, Ceramide synthesis enhances transport of GPI-anchored proteins to the Golgi apparatus in yeast, *EMBO J.* **13:**3687–3695.

Hoye, T. R., Jimenez, J. I., and Shier, W. T., 1994, Relative and absolute configuration of the fumonisin B1 backbone, *J. Am. Chem. Soc.* **116:**9409–9410.

Hudson, P. L., Pedersen, W. A., Saltsman, W. S., Liscovitch, M., MacLaughlin, D. T., Donahoe, P. K., and Blusztajn, J. K., 1994, Modulation by sphingolipids of calcium signals evoked by epidermal growth factor, *J. Biol. Chem.* **269:**21885–21890.

Igarashi, Y. and Hakomori, S.-I., 1989, Enzymatic synthesis of N,N-dimethyl-sphingosine: Demonstration of the sphingosine:N-methyltransferase in mouse brain, *Biochem. Biophys. Res. Commun.* **164:**1411–1416.

Igarashi, Y., Hakomori, S.-I., Toyokuni, T., Dean, B., Fujita, S., Sugimoto, M., Ogawa, T., El-Ghendy, K., and Racker, E., 1989, Effect of chemically well-defined sphingosine and its N-methyl derivatives on protein kinase C and src kinase activities, *Biochemistry* **28:**6796–6800.

Igarashi, Y., Kitamura, K., Toyokuni, T., Dean, B., Fenderson, B., Ogawa, T., and Hakomori, S.-I., 1990, A specific enhancing effect of N,N-dimethylsphingosine on epidermal growth factor receptor autophosphorylation, *J. Biol. Chem.* **265:**5385–5389.

Inokuchi, J., and Radin, N. S., 1987, Preparation of the active isomer of 1-phenyl-2-decanoylamino-3-morpholino-1-propanol, inhibitor of murine glucocerebroside synthetase, *J. Lipid Res.* **28:**565–571.

Jamal, H., Martin, A., Gomez-Munoz, A., and Brindley, D. N., 1991, Plasma membrane fractions from rat liver contain a phosphatidate phosphohydrolase distinct from that in the endoplasmic reticulum and cytosol, *J. Biol. Chem.* **266**:2988–2996.

Jarvis, W. D., Fornari, F. A., Jr., Browning, J. L., Gewirtz, D. A., Kolesnick, R. N., and Grant, S., 1994, Attenuation of ceramide-induced apoptosis by diglyceride in human myeloid leukemia cells, *J. Biol. Chem.* **269**:31685–31692.

Jayadev, S., Liu, B., Bielawska, A. E., Lee, J. Y., Nazaire, F., Pushkareva, M. Y., Obeid, L. M., and Hannun, Y. A., 1995, Role for ceramide in cell cycle arrest, *J. Biol. Chem.* **270**:2047–2052.

Ji, L., Zhang, G., Uematsu, S., Akahori, Y., and Hirabayashi, Y., 1995, Induction of apoptotic DNA fragmentation and cell death by natural ceramide, *FEBS Lett.* **358**:211–214.

Jones, M. J., and Murray, A. W., 1995, Evidence that ceramide selectively inhibits protein kinase C-alpha translocation and modulates bradykinin activation of phospholipase D, *J. Biol. Chem.* **270**:5007–5013.

Kalen, A., Borchardt, R. A., and Bell, R. M., 1992, Elevated ceramide levels in GH_4C_1 cells treated with retinoic acid, *Biochim. Biophys. Acta* **1125**:90–96.

Karlsson, K.-A., 1970, Sphingolipid long chain bases, *Lipids* **5**:878–891.

Katoh, N., 1993, Modulation by sphingosine of substrate phosphorylation by protein kinase C in bovine mammary gland, *Lipids* **28**:867–871.

Kendler, A., and Dawson, G., 1992, Hypoxic injury to oligodendrocytes: Reversible inhibition of ATP-dependent transport of ceramide from the endoplasmic reticulum to the Golgi, *J. Neurosci. Res.* **31**:205–211.

Kim, M.-Y., Linardic, C., Obeid, L., and Hannun, Y. A., 1991, Identification of sphingomyelin turnover as an effector mechanism for the action of tumor necrosis factor alpha and gamma-interferon, *J. Biol. Chem.* **266**:484–489.

Kim, S., Lakhani, V., Costa, D. J., Sharara, A. I., Fitz, J.G., Huang, L.-W., Peters, K. G., and Kindman, L. A., 1995, Sphingolipid-gated Ca^{2+} release from intracellular stores of endothelial cells is mediated by a novel Ca^{2+}-permeable channel, *J. Biol. Chem.* **270**:5266–5269.

Kimura, S., Kawa, S., Ruan, F., Nisar, M., Sadahira, Y., Hakomori, S.-I., and Igarashi, Y., 1992, Effect of sphingosine and its N-methyl derivatives on oxidative burst, phagokinetic activity, and trans-endothelial migration of human neutrophils, *Biochem. Pharmacol.* **44**:1585–1595.

Kinnunen, P. K., Rytomaa, M., Koiv, A., Lehtonen, J., Mustonen, P., and Aro, A., 1993, Sphingosine-mediated membrane association of DNA and its reversal by phosphatidic acid, *Chem. Phys. Lipids* **66**:75–85.

Kiss, Z., 1994, Sphingosine-like stimulatory effects of propranolol on phospholipase D activity in NIH 3T3 fibroblasts, *Biochem. Pharmacol.* **47**:1581–1586.

Kiss, Z., Crilly, K., and Chattopadhyay, J., 1991, Ethanol potentiates the stimulatory effects of phorbol ester, sphingosine and 4-hydroxynonenal on the hydrolysis of phosphatidylethanolamine in NIH 3T3 cells, *Eur. J. Biochem.* **197**:785–790.

Klein, A., Henseler, M., Klein, C., Suzuki, D., Harzer, K. and Sandhoff, K., 1994, Sphingolipid activator protein D (sap-D) stimulates the lysosomal degradation of ceramide in vivo, *Biochem. Biophys. Res. Commun.* **200**:1440–1448.

Knöfler, R., Urano, T., Takada, Y., and Takada, A., 1994, N,N,N-trimethylsphingosine modifies aggregatory response and ATP release from platelets in whole blood, *Thromb. Res.* **76**:323–332.

Kobayashi, J., Cheng, J.-F., Ishibashi, M., Walchli, M. R., Yamamura, S., and Ohizumi, Y., 1991, Penaresidin A and B, two novel azetidine alkaloids with potent actomyosin ATPase-activating activity from the Okinawan marine sponge *Penares sp.*, *J. Chem. Soc. Perkin Trans. 1* **1991**:1135–1137.

Kobayashi, T., Mitsuo, K., and Goto, I., 1988, Free sphingoid bases in normal murine tissues, *Eur. J. Biochem.* **171**:747–752.

Koiv, A., Mustonen, P., and Kinnunen, P. K., 1993, Influence of sphingosine on the thermal phase behaviour of neutral and acidic phospholipid liposomes, *Chem. Phys. Lipids* **66**:123–134.

Kolesnick, R N., and Golde, D. W., 1994, The sphingomyelin pathway in tumor necrosis factor and interleukin-1 signalling, *Cell* **77**:325–328.

Kolesnick, R. N., and Hemer, M. R., 1991, Characterization of a ceramide kinase activity from human leukemia (HL-60) cells. Separation from diacylglycerol kinase activity, *J. Biol. Chem.* **265**:18803–18808.

Kriek, N. P. J., Kellerman, T. S., and Marasas, W. F. O., 1981, A comparative study of the toxicity of *Fusarium verticillioides* (=*F. moniliforme*) to horses, primates, pigs, sheep and rats, *Onderstepoort J. Vet. Res.* **48**:129–131.

Lavie, Y., Blusztajn, J. K. and Liscovitch, M., 1994, Formation of endogenous free sphingoid bases in cells induced by changing medium conditions, *Biochim. Biophys. Acta* **1220**:323–328.

Lin, M., Lu, S., Ji, C., Wang, Y., Wang, M., Cheng, S., and Tian, G., 1980, Experimental studies on the carcinogenicity of fungus-contaminated food from Linxian County, in: *Genetic and Environmental Factors in Experimental and Human Cancer* (H. V. Gelboin, ed.), pp. 139–148, Japan Sci. Soc. Press, Tokyo.

Linardic, C. M., and Hannun, Y. A., 1994, Identification of a distinct pool of sphingomyelin involved in the sphingomyelin cycle, *J. Biol. Chem.* **269**:23530–23537.

Lister, M. D., Crawford-Redick, C. L., and Loomis, C. R., 1993, Characterization of the neutral pH-optimum sphingomyelinase from rat brain: Inhibition by copper II and ganglioside GM3, *Biochim. Biophys. Acta* **1165**:314–320.

López-García, F., Micol, V., Villalaín, J., and Gómez-Fernández, J. C., 1993, Interaction of sphingosine and stearylamine with phosphatidylserine as studied by DSC and NMR, *Biochim. Biophys. Acta* **1153**:1–8.

López-García, F., Villalaín, J., and Gómez-Fernández, J.C., 1994, A phase behavior study of mixtures of sphingosine with zwitterionic phospholipids, *Biochim. Biophys. Acta* **1194**:281–288.

Louie, D. D., Kisic, A., and Schroepfer, G. J., 1976, Sphingolipid base metabolism. Partial purification and properties of sphinganine kinase of brain, *J. Biol. Chem.* **251**:4557–4564.

Lynch, D. V., 1993, Sphingolipids, in: *Lipid Metabolism in Plants* (T. S. Moore, Jr., ed.), pp. 285–308, CRC Press, Baca Raton, FL.

McDonough, P. M., Yasui, K., Betto, R., Salviati, G., Glembotski, C. C., Palade, P. T., and Sabbadini, R. A., 1994, Control of cardiac Ca^{2+} levels: Inhibitory actions of sphingosine on Ca^{2+} transients and L-type Ca^{2+} channel conductance, *Circ. Res.* **75**:981–989.

Mandon, E. C., van Echten, G., Birk, R., Schmidt, R. R., and Sandhoff, K., 1991, Sphingolipid biosynthesis in cultured neurons. Down regulation of serine palmitoyltransferase by sphingoid bases, *Eur. J. Biochem.* **198**:667–674.

Mandon, E. C., Ehses, I., Rother, J., van Echten, G., and Sandhoff, K., 1992, Subcellular localization and membrane topology of serine palmitoyltransferase, 3-dehydrosphinganine reductase, and sphinganine N-acyltransferase in mouse liver, *J. Biol. Chem.* **267**:11144–11148.

Marasas, W.F.O., 1982, Mycotoxicological investigations on corn produced in oesophageal cancer areas in Transkei, in: *Cancer of the Oesophagus* (C. J. Pfeiffer, ed.), Vol. 1, pp. 29–40, CRC Press, Boca Raton, FL.

Marasas, W. F. O., Kriek, N. P. J., Fincham, J. E., and van Rensburg, S. J., 1984, Primary liver cancer and oesophageal basal cell hyperplasia in rats caused by *Fusarium moniliforme*, *Int. J. Cancer* **34**:383–387.

Marasas, W. F. O., Kellerman, T. S., Gelderblom, W. C. A., Coetzer, J. A. W., Thiel, P. G., and van der Lugt, J. J., 1988, Leukoencephalomalacia in a horse induced by fumonsin B_1 isolated from *Fusarium moniliforme*, *Onderstepoort J. Vet. Res.* **55**:197–203.

Martinova, E. A., and Merrill, A. H., Jr., 1995, Fumonisin B_1 alters sphingolipid metabolism and immune function in BALB/c mice, *Mycopathologia* **130**:163–170.

Mathias, S., Dressler, K. A., and Kolesnick, R. N., 1991, Characterization of a ceramide-activated protein kinase: Stimulation by tumor necrosis factor alpha, *Proc. Natl. Acad. Sci. USA* **88**:10009–10013.

Mattie, M., Brooker, G., and Spiegel, S., 1994, Sphingosine-1-phosphate, a putative second messenger, mobilizes calcium from internal stores via an inositol trisphosphate-independent pathway, *J. Biol. Chem.* **269**:3181–3188.

Mazurek, N., Megidish, T., Hakomori, S.-I., and Igarashi, Y., 1994, Regulatory effect of phorbol esters on sphingosine kinase in BALB/C 3T3 fibroblasts (variant A31): Demonstration of cell type-specific response—a preliminary note, *Biochem. Biophys. Res. Commun.* **198**:1–9.

Medlock, K. A., and Merrill, A. H., Jr., 1988, Inhibition of serine palmitoyltransferase in vitro and long-chain base biosynthesis in intact Chinese hamster ovary cells by β-chloroalanine, *Biochemistry* **27**:7079–7084.

Merrill, A. H., Jr., 1991, Cell regulation by sphingosine and more complex sphingolipids, *J. Bioeng. Biomembr.* **23**:83–104.

Merrill, A. H., Jr., 1994, Sphingosine and other long-chain bases that alter cell behavior, *Curr. Top. Membr.* **40**:361–386.

Merrill, A.H., Jr., and Wang, E., 1986, Biosynthesis of long-chain (sphingoid) bases from serine by LM-cells. Evidence for introduction of the 4-trans-double bond after *de novo* biosynthesis of N-acylsphinganine(s), *J. Biol. Chem.* **261**:3764–3769.

Merrill, A. H., Jr., and Wang, E., 1992, Enzymes of ceramide biosynthesis, *Methods Enzymol.* **209**:427–437.

Merrill, A. H., Jr., and Williams, R. D., 1984, Utilization of different fatty acyl-CoA thioesters by serine palmitoyl transferase from rat brain, *J. Lipid Res.* **25**:185–188.

Merrill, A. H., Jr., Sereni, A. M., Stevens, V. L., Hannun, Y. A., Bell, R. M., and Kinkade, J. M., Jr., 1986, Inhibition of phorbol ester-dependent differentiation of human promyelocytic leukemic (HL-60) cells by sphinganine and other long-chain bases, *J. Biol. Chem.* **261**:12610–12615.

Merrill, A. H., Jr., Wang, E., Mullins, R. E., Jamison, W. C. L., Nimkar, S., and Liotta, D. C., 1988, Quantitation of free sphingosine in liver by high-performance liquid chromatography, *Anal. Biochem.* **171**:373–381.

Merrill, A. H., Jr., Nimkar, S., Menaldino, D., Hannun, Y. A., Loomis, C., Bell, R. M., Tyagi, S. R., Lambeth, J. D., Stevens, V. L., Hunter, R., and Liotta, D. C., 1989, Structural requirements for long-chain (sphingoid) base inhibition of protein kinase C *in vitro* and for the cellular effects of these compounds, *Biochemistry* **28**:3138–3145.

Merrill, A. H., Jr., Hannun, Y. A., and Bell, R. M., 1993a, Sphingolipids and their metabolites in cell regulation, *Adv. Lipid Res.* **25**:1–24.

Merrill, A. H., Jr., Wang, E., Gilchrist, D. G., and Riley, R. T., 1993b, Fumonisins and other inhibitors of *de novo* sphingolipid biosynthesis, *Adv. Lipid Res.* **26**:215–234.

Merrill, A. H., Jr., van Echten, G., Wang, E., and Sandhoff, K., 1993c, Fumonisin B$_1$ inhibits sphingosine (sphinganine) N-acetyltransferase and *de novo* sphingolipid biosynthesis in cultured neurons in situ, *J. Biol. Chem.* **268**:27299–27306.

Merrill, A. H. Jr., Grant, A. M., Wang, E., and Bacon, C. W., 1995a, Lipids and lipid-like compounds of Fusarium, in: *Fungal Lipids* (R. Prasad and M. Ghanoum, eds.), CRC Press, Boca Raton, FL, in press.

Merrill, A. H., Jr., Lingrell, S., Wang, E., Nikolova-Karakashian, M., Vales, T. R., and Vance, D. E., 1995b, Sphingolipid biosynthesis *de novo* by rat hepatocytes in culture: Ceramide and sphingomyelin are associated with, but not required for, very-low density lipoprotein secretion, *J. Biol. Chem.* **270**:13834–13841.

Merrill, A. H., Jr., Wang, E. , Schroeder, J. J., Smith, E. R., Yoo, H. S., and Riley, R. T., 1995c, Disruption of sphingolipid metabolism in the toxicity and carcinogenicity of fumonisins, in: *Molecular Approaches to Food Safety Issues Involving Toxic Microorganisms* (M. Eklund, J. Richards, and K. Mise, eds.), Alaken Press, Fort Collins, CO, pp. 429–443.

Merrill, A. H., Schmelz, E. M., Wang, E., Schroeder, J. J., Dillehay, D. L., and Riley, R. T., 1995d, Role of dietary sphingolipids and inhibitors of sphingolipid metabolism in cancer and other diseases, *J. Nutr.* **125**:16775–16825.

Miccheli, A., Ricciolini, R., Lagana, A., Piccolella, E., and Conti, F., 1991, Modulation of the free sphingosine levels in Epstein–Barr virus transformed human B lymphocytes by phorbol dibutyrate, *Biochim. Biophys. Acta* **1095**:90–92.

Miccheli, A., Tomassini, A., Ricciolini, R., Di Cocco, M. E., Piccolella, E.,. Manetti, C., and Conti, F., 1994, Dexamethasone-dependent modulation of cholesterol levels in human lymphoblastoid B cell line through sphingosine production, *Biochim. Biophys. Acta* **1221**:171–177.

Miyake, Y., Kozutsumi, T., and Kawasaki, T., 1994, Action mechanism of sphingosine-like immunosuppressant, ISP-1, *Igaku No Ayumi* **171**:921–925.

Morell, P., and Radin, N. S., 1970, Specificity in ceramide biosynthesis from long chain bases and various fatty acyl coenzyme A's by brain microsomes, *J. Biol. Chem.* **245:**342–350.

Mori, M.-A., Shimeno, H., and Kishimoto, Y., 1985, Synthesis of ceramides and cerebrosides in rat brain: Comparison with synthesis of lignoceroyl-coenzyme A, *Neurochem. Int.* **7:**57–61.

Morrison, W. R., 1969, Polar lipids in bovine milk. I. Long-chain bases in sphingomyelin, *Biochim. Biophys. Acta* **176:**537–546.

Mullmann, T. J., Siegel, M. I., Egan, R. W., and Billah, M. M., 1991, Sphingosine inhibits phosphatidate phosphohydrolase in human neutrophils by a protein kinase C-independent mechanism, *J. Biol. Chem.* **266:**2013–2016.

Nagiec, M. M., Baltisberger, J. A., Wells, G. B., Lester, R. L., and Dickson, R. C., 1994, The LCB2 gene of *Saccharomyces* and the related LCB1 gene encode subunits of serine palmitoyltransferase, the initial enzyme in sphingolipid synthesis, *Proc. Natl. Acad. Sci. USA* **91:**7899–7902.

Natarajan, V., Jayaram, H. N., Scribner, W. M., and Garcia, J. G., 1994, Activation of endothelial cell phospholipase D by sphingosine and sphingosine-1-phosphate. *Am. J. Respir. Cell Mol. Biol.* **11:**221–229.

Nelson, D. H., and Murray, D. K., 1989, Dexamethasone and sphingolipids inhibit concanavalin A stimulated glucose uptake in 3T3-L1 fibroblasts, *Endocrine Res.* **14:**305–318. Dillehay, D. L., Webb, S. J., Schmelz, E.-M., and Merrill, A. H., Jr., 1994, Dietary sphingomyelin inhibits 1,2-dimethylhydrazine-induced colon cancer in CF1 mice, *J. Nutr.* **124:**615–620.

Nilsson, Å., 1968, Metabolism of sphingomyelin in the intestinal tract of the rat, *Biochim. Biophys. Acta* **164:**575–584.

Nilsson, Å., 1969, Metabolism of cerebrosides in the intestinal tract of the rat, *Biochim. Biophys. Acta* **187:**113–121.

Norred, W. P., Wang, E., Yoo, H., Riley, R. T., and Merrill, A. H., Jr., 1992, *In vitro* toxicology of fumonisins and the mechanistic implications, *Mycopathologia* **117:**73–78.

Nutter, L. M., Grill, S. P., Li, J. S., Tan, R. S., and Cheng, Y. C., 1987, Induction of virus enzymes by phorbol esters and n-butyrate in Epstein–Barr virus genome-carrying Raji cells, *Cancer Res.* **47:**4407–4412.

Ohta, H., Yatomi, Y., Sweeney, E. A., Hakomori, S.-I., and Igarashi, Y., 1994, A possible role of sphingosine in induction of apoptosis by tumor necrosis factor-alpha in human neutrophils, *FEBS Lett.* **355:**267–270.

Ohta, H., Sweeney, E. A., Masamune, A., Yatomi, Y., Hakomori, S.-I., and Igarashi, Y., 1995, Induction of apoptosis by sphingosine in human leukemic HL-60 cells: A possible endogenous modulator of apoptotic DNA fragmentation occurring during phorbol ester-induced differentiation, *Cancer Res.* **55:**691–697.

Oishi, K., Raynor, R. L., Charp, P. A., and Kuo, J. F., 1988, Regulation of protein kinase C by lysophospholipids. Potential role in signal transduction, *J. Biol. Chem.* **263:**6865–6871.

Oishi, K., Zheng, B., and Kuo, J. F., 1990, Inhibition of Na,K-ATPase and sodium pump by protein kinase C regulators sphingosine, lysophosphatidylcholine, and oleic acid, *J. Biol. Chem.* **265:**70–75.

Okazaki, T., Bell, R. M., and Hannun, Y. A., 1989, Sphingomyelin turnover induced by 1α 25-dihydroxyvitamin D_3 in HL-60 cells. Role in cell differentiation, *J. Biol. Chem.* **264:**19076–19080.

Okazaki, T., Bielawska, A., Bell, R. M., and Hannun, Y. A., 1990, Role of ceramide as a lipid mediator of 1α 25-dihydroxyvitamin D_3-induced HL-60 cell differentiation, *J. Biol. Chem.* **265:**15823–15831.

Okoshi, H., Hakomori, S.-I., Nisar, M., Zhou, Q. H., Kimura, S., Tashiro, K., and Igarashi, Y., 1991, Cell membrane signaling as target in cancer therapy. II: Inhibitory effect of N,N,N-trimethylsphingosine on metastatic potential of murine B16 melanoma cell line through blocking of tumor cell-dependent platelet aggregation, *Cancer Res.* **51:**6019–6024.

Olivera, A., and Spiegel, S., 1993, Sphingosine-1-phosphate as second messenger in cell proliferation induced by PDGF and FCS mitogens, *Nature* **365:**557–560.

Olivera, A., Zhang, H., Carlson, R. O., Mattie, M. E., Schmidt, R. R., and Spiegel, S., 1994, Stereospecificity of sphingosine-induced intracellular calcium mobilization and cellular proliferation, *J. Biol. Chem.* **269:**17924–17930.

Ong, D. E. and Brady, R. N., 1973, In vivo studies on the introduction of the 4-*trans*-double bond of the sphigenine moiety of rat brain ceramides, *J. Biol. Chem.* **248:**3884–3888.

Paige, D. G., Morse-Fisher, N., and Harper, J. I., 1993, The quantification of free sphingosine in the stratum corneum of patients with hereditary ichthyosis, *Br. J. Dermatol.* **129:**380–383.

Pandol, S. J., Schoeffield-Payne, M. S., Gukovskaya, A. S., and Rutherford, R. E., 1994, Sphingosine regulates Ca^{2+}-ATPase and reloading of intracellular Ca2+ stores in the pancreatic acinar cell, *Biochim. Biophys. Acta Bio-Membr.* **1195:**45–50.

Park, Y. S., Hakomori. S.-I., Kawa, S., Ruan, F., and Igarashi, Y., 1994, Liposomal N,N,N-tri-methylsphingosine (TMS) as an inhibitor of B16 melanoma cell growth and metastasis with reduced toxicity and enhanced drug efficacy compared to free TMS: Cell membrane signaling as a target for cancer therapy III, *Cancer Res.* **54:**2213–2217.

Pawelczyk, T., and Lowenstein, J. M., 1992, Regulation of phospholipase C delta activity by sphin-gomyelin and sphingosine, *Arch. Biochem. Biophys.* **297:**328–333.

Perry, D. K., Hand, W. L., Edmundson, D. E., and Lambeth, J. D., 1992, Role of phospholipase D-derived diradylglycerol in the activation of the human neutrophil respiratory burst oxidase. Inhibition by phosphatidic acid phosphohydrolase inhibitors, *J. Immunol.* **149:**2749–2758.

Petrou, S., Ordway, R. W., Hamilton, J. A., Walsh, J. V., Jr., and Singer, J. J., 1994, Structural require-ments for charged lipid molecules to directly increase or suppress K+ channel activity in smooth muscle cells. Effects of fatty acids, lysophosphatidate, acyl coenzyme A and sphingosine, *J. Gen. Physiol.* **103:**471–486.

Poch, G. K., Powell, R. G., Plattner, R. D., and Weisleder, D., 1994, Relative stereochemistry of fumonisin B1 at C-2 and C-3, *Tetrahedron Lett.* **35:**7707–7712.

Pushkareva, M. Y., Khan, W. A., Alessenko, A. V., Sahyoun, N., and Hannun, Y.A., 1992, Sphingosine activation of protein kinases in Jurkat T cells. *In vitro* phosphorylation of endogenous protein substrates and specificity of action, *J. Biol. Chem.* **267:**15246–15251.

Pushkareva, M. Y., Bielawska, A., Menaldino, D. C., and Hannun, Y. A., 1993, Regulation of sphingosine-activated protein kinases: Selectivity of activation by sphingoid bases and inhibition by non-esterified fatty acids, *Biochem. J.* **294:**699–703.

Pushkareva, M., Chao, R., Bielawska, A., Merrill, A. H., Jr., Crane, H. M., Lagu, B., Liotta, D. C., and Hannun, Y. A., 1995, Stereoselectivity of induction of the retinoblastoma gene product (pRb) dephosphorylation by D-*erythro*-sphingosine supports a role for pRb in growth suppression by sphingosine, *Biochemistry* **34:**1885–1892.

Quintans, J., Kilkus, J., McShan, C. L., Gottschalk, A. R., and Dawson, G., 1994, Ceramide mediates the apoptotic response of WEHI 231 cells to anti-immunoglobulin, corticosteroids and irradiation, *Biochem. Biophys. Res. Commun.* **202:**710–714.

Rees, R. S., Nanney, L. B., Yates, R. A., and King, L. J., 1984, Interaction of brown recluse spider venom on cell membranes: The inciting mechanism? *J. Invest. Dermatol.* **83:**270–275.

Riboni, L., Prinetti, A., Bassi, R., and Tettamanti, G., 1994, Formation of bioactive sphingoid molecules from exogenous sphingomyelin in primary cultures of neurons and astrocytes, *FEBS Lett.* **352:**323–326.

Ricciolini, R., Miccheli, A., Di Cocco, M. E., Piccolella, E., Marino, A., Sammartino, M. P., and Conti, F., 1994, Dexamethasone-dependent modulation of human lymphoblastoid B cell line through sphingosine production, *Biochim. Biophys. Acta* **1221:**103–108.

Riley, R. T., An, N. H., Showker, J. L., Yoo, H.-S., Norred, W. P., Chamberlain, W. J., Wang, E., Merrill, A. H., Jr., Motelin, G., Beasley, V. R., and Haschek, W. M., 1993, Alteration of tissue and serum sphinganine to sphingosine ratio: An early biomarker of exposure to fumonisin-containing feeds in pigs, *Toxicol. Appl. Pharmacol.* **118:**105–112.

Riley, R. T., Hinton, D. M., Chamberlain, W. J., Bacon, C. W., Wang, E., Merrill, A. H., Jr., and Voss, K. A., 1994a, Dietary fumonisin B_1 induces disruption of sphingolipid metabolism in Sprague–Dawley rats: A new mechanism of nephrotoxicity, *J. Nutr.* **124:**594–603.

Riley, R. T., Voss, K. A., Yoo, H.-S., Gelderblom, W. C. A., and Merrill, A. H., Jr., 1994b, Mechanism of fumonisin toxicity and carcinogenicity, *J. Food Protect.* **57:**638–645.

Riley, R. T., Wang, E., and Merrill, A. H., Jr., 1994c, Liquid chromatographic determination of sphinganine and sphingosine: Use of the free sphinganine-to sphingosine ratio as a biomarker for consumption of fumonisins, *J. Assoc. Off. Anal. Chem.* **77:**533–540.

Ritchie, T., Rosenberg, A., and Noble, E. P., 1992, Regulation of phosphoinositide hydrolysis in cultured astrocytes by sphingosine and psychosine, *Biochem. Biophys. Res. Commun.* **186**:790–795.

Robertson, D. G., DiGirolamo, M., Merrill, A. H., Jr,. and Lambeth, J.D., 1989, Insulin-stimulated hexose transport and glucose oxidation in rat adipocytes is inhibited by sphingosine at a step after insulin binding, *J. Biol. Chem.* **264**:6773–6779.

Robson, K. J., Stewart, M. E., Michelsen, S., Lazo, N. D., and Downing, D. T., 1994, 6-hydroxy-4-sphingenine in human epidermal ceramides, *J. Lipid Res.* **35**:2060–2068.

Ross, P. F., Nelson, P. E., Richard, J. L., Osweiler, G. D., Rice, L. G., Plattner, R. D., and Wilson, T. M., 1990, Production of fumonisins by *Fusarium moniliforme* and *Fusarium proliferatum* isolates associated with equine leukoencephalomalacia and a pulmonary edema syndrome in swine, *Appl. Environ. Microbiol.* **56**:3225–3226.

Rother, J., van Echten, G., Schwarzmann, G., and Sandhoff, K., 1992, Biosynthesis of sphingolipids: Dihydroceramide and not sphinganine is desaturated by cultured cells, *Biochem. Biophys. Res. Commun.* **189**:14–20.

Sabbadini, R. A., Betto, R., Teresi, A., Fachechi-Cassano, G., and Salviati, G., 1992, The effects of sphingosine on sarcoplasmic reticulum membrane calcium release, *J. Biol. Chem.* **267**:15475–15484.

Scheidl, H., Scita, G., Sampson, P. H., Park, H. Y., and Wolf, G., 1992, The effect of sphingosine and phorbol ester on the signal transduction enzymes and fibronectin release in cell culture, *Biochim. Biophys. Acta* **1135**:295–300.

Schmelz, E. M., Crall, K. J., LaRocque, R., Dillehay, D. L., and Merrill, A. H., Jr., 1994, Uptake and metabolism of sphingolipids in isolated intestinal loops of mice, *J. Nutr.* **124**:702–712.

Schroeder, J. J. Crane, H. M., Xia, J., Liotta, D. C., and Merrill, A. H., Jr., 1994, Disruption of sphingolipid metabolism and stimulation of DNA synthesis by fumonsin B_1: A molecular mechanism for carcinogenesis associated with *Fusarium moniliforme, J. Biol. Chem.* **269**:3475–3481.

Schütze, S., Potthoff, K., Machleidt, T., Berkovic, D., Wiegmann, K., and Kronke, M., 1992, TNF activates NF-kB by phosphatidylcholine-specific phospholipase C-induced "acidic" sphingomyelin breakdown, *Cell* **71**:765–776.

Schütze, S., Machleidt, T., and Krönke, M., 1994, The role of diacylglycerol and ceramide in tumor necrosis factor and interleukin-1 signal transduction, *J. Leukocyte Biol.* **56**:533–541.

Scita, G., and Wolf, G., 1994, The effect of sphingosine on the release of fibronectin from human lung fibroblasts, *Biochim. Biophys. Acta Mol. Cell Res.* **1223**:29–35.

Seufferlein, T., and Rozengurt, E., 1994, Sphingosine induces p125FAK and paxillin tyrosine phosphorylation, actin stress fiber formation, and focal contact assembly in Swiss 3T3 cells, *J. Biol. Chem.* **269**:27610–27617.

Shier, W. T., Abbas, H. K., and Mirocha, C. J., 1991, Toxicity of the mycotoxins fumonisins B_1 and B_2 and *Alternaria alternata f. sp. lycopersici* toxin (AAL) in cultured mammalian cells, *Mycopathologia* **116**:97–104.

Shukla, G., Shukla, A., Inokuchi, J.-I., and Radin, N. S., 1991, Rapid kidney changes resulting from glucosphingolipid depletion by treatment with a glucosyltransferase inhibitor, *Biochim. Biophys. Acta* **1083**:101–108.

Simbulan, C. M., Tamiya-Koizumi, K., Suzuki, M., Shoji, M.,Taki, T., and Yoshida, S., 1994, Sphingosine inhibits the synthesis of RNA primers by primase *in vitro*, *Biochemistry* **33**:9007–9012.

Singh, I., 1983, Ceramide synthesis from free fatty acids in rat brain: Function of NADPH and substrate specificity, *J. Neurochem.* **40**:1565–1570.

Slife, C. W., Wang, E., Hunter, R., Wang, S., Burgess, C., Liotta, D.C., and Merrill, A.H., Jr., 1989, Free sphingosine formation from endogenous substrates by a liver plasma membrane system with a divalent cation dependence and a neutral pH optimum, *J. Biol. Chem.* **264**:10371–10377.

Smith, E. R., and Merrill, A. H., Jr., 1995, Differential roles of *de novo* sphingolipid biosynthesis and turnover in the "burst" of free sphingosine and sphinganine, and their 1-phosphates and N-acylderivatives, that occurs upon changing the medium of cells in culture, *J. Biol. Chem.* **270**:18749–18758.

Snell, E. E., DiMari, S. J,. and Brady, R. N., 1970, Biosynthesis of sphingosine and dihydrosphingosine by cell-free systems from *Hansenula ciferri*, *Chem. Phys. Lipids* **55**:116–138.

Spence, M. W., Beed, S., and Cook, H. W., 1986, Acid and alkaline ceramidases of rat tissues, *Biochem. Cell Biol.* **64**:400–404.

Spiegel, S., 1993, Sphingosine and sphingosine 1-phosphate in cellular proliferation: Relationship with protein kinase C and phosphatidic acid, *J. Lipid Mediators* **8**:169–175.

Sribney, M., 1966, Enzymatic synthesis of ceramide, *Biochim. Biophys. Acta* **125**:542–547.

Steen Law, S. L., Squier, C. A., and Wertz, P. W., 1995, Free sphingosines in oral epithelium, *Comp. Biochem. Physiol. B* **110B**:511–513.

Stevens, V. L., Winton, E. F., Smith, E. E., Owens, N. E., Kinkade, J. M., Jr., and Merrill, A. H., Jr., 1989, Differential effects of long-chain (sphingoid) bases on the monocytic differentiation of human leukemia (HL-60) cells induced by phorbol esters, 1α, 25-dihydroxyvitamin D_3, or ganglioside G_{M3}, *Cancer Res.* **49**:3229–3234.

Stevens, V. L., Nimkar, S., Jamison, W. C., Liotta, D. C., and Merrill, A. H., Jr., 1990a, Characteristics of the growth inhibition and cytotoxicity of long-chain (sphingoid) bases for Chinese hamster ovary cells: Evidence for an involvement of protein kinase C, *Biochim. Biophys. Acta* **1051**:37–45.

Stevens, V. L., Owens, N. E., Winton, E. F., Kinkade, J. M., Jr., and Merrill, A. H., Jr., 1990b, Modulation of retinoic acid-induced differentiation of human leukemia (HL-60) cells by serum factors and sphinganine, *Cancer Res.* **50**:222–226.

Stinavage, P., and Spitznagel, J. K., 1989, Oxygen-independent and antimicrobial action in sphingosine-treated neutrophils, *J. Immunol. Methods* **124**:267–275.

Stoffel, W., 1970, Studies on the biosynthesis and degradation of sphingosine bases, *Chem. Phys. Lipids* **55**:139–158.

Stoffel, W., and Bister, K., 1973, Stereospecificities in the metabolic reactions of the four isomeric sphinganines (dihydrosphingosines) in rat liver, *Hoppe-Seyler's Z. Physiol. Chem.* **354**:169–181.

Stoffel, W., LeKim, D., and Sticht, G., 1969, Metabolism of sphingosine bases. XI. Distribution and properties of dihydrosphingosine-1-phosphate aldolase (sphinganine-1-phosphate alkanal-lyase), *Hoppe-Seyler's Z. Physiol. Chem.* **350**: 1233–1241.

Stoffel, W., Assmann, G., and Binczek, E., 1970, Metabolism of sphingosine bases. XIII. Enzymatic synthesis of 1-phosphate esters of 4t-sphingenine (sphingosine), sphinganine (dihydrosphingosine), 4-hydroxysphinganine (phytosphingosine), and 3-dehydrosphingosine by erythrocytes, *Hoppe-Seyler's Z. Physiol. Chem.* **351**:635–642.

Sugita, M., Williams, M., and Dulaney, J. T., 1975, Ceramidase and ceramide synthesis in human kidney and cerebellum. Description of a new alkaline ceramidase, *Biochim. Biophys. Acta* **398**:125–131.

Sundaram, K. S., and Lev, M., 1984, Inhibition of sphingolipid synthesis by cycloserine *in vitro* and *in vivo*, *J. Neurochem.* **42**:577–581.

Sundaram, K. S., and Lev, M., 1989, The long-term administration of L-cycloserine to mice: Specific reduction of cerebroside level, *Neurochem. Res.* **14**:245–248.

Thompson, T. E., and Tillack, T. W., 1985, Organization of glycosphingolipids in bilayers and plasma membranes of mammalian cells, *Annu. Rev. Biophys. Biophys. Chem.* **14**:361–386.

Thudichum, J. L. W., 1884, *A Treatise on the Chemical Constitution of Brain*, Baillière, Tindall, & Cox, London.

Tornquist, K., and Ekokoski, E., 1994, Effect of sphingosine derivatives on calcium fluxes in thyroid FRTL-5 cells, *Biochem. J.* **299**:213–218.

Trinchera, M., Ghidoni, R., Sonnino, S., and Tettamanti, G., 1990, Recycling of glucosylceramide and sphingosine for the biosynthesis of gangliosides and sphingomyelin in rat liver, *Biochem. J.* **270**:815–820.

Turner, W. B., and Aldridge, D. C., 1983, *Fungal Metabolites II*, p. 173, Academic Press, New York.

Uemura, K., Hara, A., and Taketomi, T., 1993, Inhibition of neurite outgrowth in neuroblastoma cells by sphingosine, *J. Biochem.* **114**:610–614.

Ullman, M. D., and Radin, N. S., 1972, Enzymatic formation of hydroxy ceramides and comparison with enzymes forming nonhydroxyceramides, *Arch. Biochem. Biophys.* **152**:767–777.

Valsecchi, M., Palestini, P., Chigorno, V., Sonnino, S., and Tettamanti, G., 1993, Changes in the ganglioside long-chain base composition of rat cerebellar granule cells during differentiation and aging in culture, *J. Neurochem.* **60**:193–196.

van Echten, G., Birk, R., Brenner-Weiss, G., Schmidt, R. R., and Sandhoff, K., 1990, Modulation of sphingolipid biosynthesis in primary cultured neurons by long chain bases, *J. Biol. Chem.* **265:**9333–9339.

Van Veldhoven, P. P., and Mannaerts, G. P., 1991, Subcellular localization and membrane topology of sphingosine-1-phosphate lyase in rat liver, *J. Biol. Chem.* **266:**12502–12507.

Van Veldhoven, P. P., and Mannaerts, G. P., 1993, Sphingosine-phosphate lyase, *Adv. Lipid Res.* **26:** 69–98.

Van Veldhoven, P. P., and Mannaerts, G. P., 1994, Sphinganine 1-phosphate metabolism in cultured skin fibroblasts: Evidence for the existence of a sphingosine phosphatase, *Biochem. J.* **299:**597–601.

Van Veldhoven, P. P., Bishop, W. R., and Bell, R. M., 1989, Enzymatic quantification of sphingosine in the picomole range in cultured cells, *Anal. Biochem.* **183:**177–189.

Van Veldhoven, P. P., Matthews, T., Bolognesi, D. P., and Bell, R. M., 1992, Changes in bioactive lipids, alkylacylglycerol and ceramide, occur in HIV-infected cells, *Biochem. Biophys. Res. Commun.* **187:**209–216.

Van Veldhoven, P. P., De Ceuster, P., Rozenberg, R., Mannaerts, G. P., and De Hoffmann, E., 1994, On the presence of phosphorylated sphingoid bases in rat tissues: A mass-spectrometric approach, *FEBS Lett.* **350:**91–95.

Vartanian, T., Dawson, G., Soliven, B., Nelson, D. J., and Szuchet, S., 1989, Phosphorylation of myelin basic protein in intact oligodendrocytes: Inhibition by galactosylsphingosine and cyclic AMP, *Glia* **2:**370–379.

Wakita, H., Tokure, Y., Yagi, H., Nishimure, K., Furukawa, F., and Takigawa, M., 1994, Keratinocyte differentiation is induced by cell-permeant ceramides and its proliferation is promoted by sphingosine, *Arch. Dermatol. Res.* **286:**350–354.

Wang, E., Norred, W. P., Bacon, C. W., Riley, R. T., and Merrill, A. H., Jr., 1991, Inhibition of sphingolipid biosynthesis by fumonisins. Implications for diseases associated with Fusarium moniliforme, *J. Biol. Chem.* **266:**14486–14490.

Wang, E., Ross, P. F., Wilson, T. M., Riley, R. T., and Merrill, A. H., Jr., 1992, Increases in serum sphingosine and sphinganine and decreases in complex sphingolipids in ponies given feed containing fumonisins, mycotoxins produced by Fusarium moniliforme, *J. Nutr.* **122:**1706–1716.

Wedegaertner, P. B., and Gill, G. N., 1989, Activation of the purified protein tyrosine kinase domain of the epidermal growth factor receptor, *J. Biol. Chem.* **264:**11346–11353.

Weibking, T. S., Ledoux, D. R., Bermudez, A. J., Turk, J. R., Rottinghaus, G. E., Wang, E., and Merrill, A. H., Jr., 1993, Effects of feeding *Fusarium moniliforme* culture material, containing known levels of fumonisin B$_1$, on the young broiler chick, *Poult. Sci* **72:**456–466.

Weiss, R. H., Huang, C.-H., and Ives, H. E., 1991, Sphingosine reverses growth inhibition caused by activation of protein kinase C in vascular smooth muscle cells, *J. Cell. Physiol.* **149:**307–312.

Wells, G. B., and Lester, R. L., 1983, The isolation and characterization of a mutant strain of *Saccharomyces cerevisiae* that requires a long chain base for growth and synthesis of phospho-sphingolipids, *J. Biol. Chem.* **258:**10200–10203.

Wertz, P. W., and Downing, D. T., 1989, Free sphingosines in porcine epidermis, *Biochim. Biophys. Acta* **1002:**213–217.

Wiegmann, K., Schütze, S., Machleidt, T., Witte, D., and Krönke, M., 1994, Functional dichotomy of neutral and acidic sphingomyelinases in tumor necrosis factor signaling, *Cell* **78:**1005–1015.

Williams, R. D., Wang, E., and Merrill, A. H., Jr., 1984, Enzymology of long-chain base synthesis by rat liver. Characterization of serine palmitoyltransferase of rat liver microsomes, *Arch. Biochem. Biophys.* **228:**282–291.

Wilson, E., Olcott, M. C., Bell, R. M., Merrill, A. H., Jr., and Lambeth, J. D., 1986, Inhibition of the oxidative burst in human neutrophils by sphingoid long-chain bases, *J. Biol. Chem.* **261:**12616–12623.

Wilson, E., Wang, E., Mullins, R. E., Liotta, D. C., Lambeth, J. D., and Merrill, A. H., Jr., 1988, Modulation of the free sphingosine levels in human neutrophils by phorbol esters and other factors, *J. Biol. Chem.* **263:**9304–9309.

Wolff, R. A., Dobrowsky, R. T., Bielawska, A., Obeid, L. A., and Hannun, Y. A., 1994, Role of ceramide-activated protein phosphatase in ceramide-mediated signal transduction, *J. Biol. Chem.* **269:**19605–19609.

Wu, W.-I., Lin, Y.-P., Wang, E., Merrill, A. H., and Carman, G. M., 1993, Regulation of phosphatidate phosphatase activity from the yeast *Saccharomyces cerevisiae* by sphingoid bases, *J. Biol. Chem.* **268**:13830–13837.

Yamaguchi, Y., Sasagasako, A., Goto, I., and Kobayashi, T., 1994, The synthetic pathway for glucosylsphingosine in cultured fibroblasts, *J. Biochem.* **116**:704–710.

Yang, C. S., 1980, Research on esophageal cancer in China: A review, *Cancer Res.* **40**:2633–2644.

Yoo, H., Norred, W. P., Wang, E., Merrill, A. H., Jr., and Riley, R. T., 1992, Sphingosine inhibition of *de novo* sphingolipid biosynthesis and cytotoxicity are correlated in LLC-PK1 cells, *Toxicol. Appl. Pharmacol.* **114**:9–15.

Yung, B. Y., 1994, Sphinganine potentiation of cellular differentiation induced by various antileukemia drugs in human leukemia cell line HL-60, *Naunyn Schmiedebergs Arch. Pharmacol.* **350**:575–581.

Yung, B. Y., Luo, K. J., and Hui, E. K., 1992, Interaction of antileukemia agents adriamycin and daunomycin with sphinganine on the differentiation of human leukemia cell line HL-60, *Cancer Res.* **52**:3593–3597.

Zacharias, C., van Echten-Deckert, G., Plewe, M., Schmidt, R. R., and Sandhoff, K., 1994, A truncated epoxy-glucosylceramide uncouples glycosphingolipid biosynthesis by decreasing lactosylceramide synthase activity, *J. Biol. Chem.* **269**:13313–13317.

Zhang, H., Buckley, N. E., Gibson, K., and Spiegel, S., 1990, Sphingosine stimulates cellular proliferation via a protein kinase C-independent pathway, *J. Biol. Chem.* **265**:76–81.

Zhang, H., Desai, N. N., Olivera, A., Seki, T., Booker, G., and Spiegel, S., 1991, Sphingosine-1-phosphate, a novel lipid, involved in cellular proliferation, *J. Cell Biol.* **114**:155–167.

Zhao, C., Beeler, T., and Dunn, T., 1994, Suppressors of the Ca^{2+}-sensitive yeast mutant (csg2) identify genes involved in sphingolipid biosynthesis. Cloning and characterization of SCS1, a gene required for serine palmitoyltransferase activity, *J. Biol. Chem.* **269**:21480–21488.

Zweerink, M. M., Edison, A. M., Wells, G. B., Pinto, W., and Lester, R. L., 1992, Characterization of a novel, potent, and specific inhibitor of serine palmitoyltransferase, *J. Biol. Chem.* **267**:25032–25038.

Chapter 7

Platelet-Activating Factor and PAF-Like Mimetics

Ralph E. Whatley, Guy A. Zimmerman, Stephen M. Prescott, and Thomas M. McIntyre

7.1. Introduction and Overview of PAF and Oxidized Phospholipids

Platelet-activating factor (PAF; 1-O-alkyl-2-acetyl-sn-glycero-3-phosphocholine; Fig. 7-1) is a potent bioactive lipid, with a diverse array of biologic effects in isolated systems, and which has as many roles in inflammatory states *in vivo*. Its name comes from observations in 1972 that a lipid extracted from the blood of rabbits undergoing anaphylaxis caused *ex vivo* activation of platelets (Benveniste *et al.*, 1972). At about the same time, other investigators found that a lipid extracted from kidneys lowered blood pressure in an animal model (Blank *et al.*, 1979). Subsequent elucidation of the structure of these compounds showed both to be 1-O-alkyl-2-acetyl-sn-glycero-3-phosphocholine. Although the name derives from its original description as an activator of platelets, it belies the vast array of other actions known to be mediated by this potent lipid. In fact, activation of platelets may be its least important role: rats, and many of their isolated cells, respond to PAF, but their platelets are resistant as they lack the high-affinity receptor for PAF.

PAF is now known to activate a variety of cell types: The activities of this lipid range from activation of inflammatory cells (e.g., platelets, neutrophils, mono-

Ralph E. Whatley • Department of Medicine, East Carolina University School of Medicine, Greenville, North Carolina 27858.   *Guy A. Zimmerman* • Department of Medicine, University of Utah School of Medicine, Salt Lake City, Utah 84112.   *Stephen M. Prescott* • Departments of Medicine and Biochemistry, the Nora Eccles Harrison Cardiovascular Research and Training Institute, and the Eccles Institute of Human Genetics, University of Utah School of Medicine, Salt Lake City, Utah 84112.   *Thomas M. McIntyre* • Departments of Medicine and Biochemistry, University of Utah School of Medicine, Salt Lake City, Utah 84112.

Handbook of Lipid Research, Volume 8: Lipid Second Messengers, edited by Robert M. Bell *et al.* Plenum Press, New York, 1996.

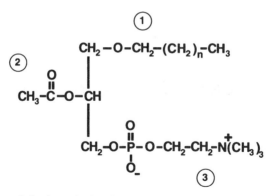

Figure 7-1. Structure of platelet-activating factor (PAF). PAF (1-O-alkyl-2-acetyl-sn-glycero-3-phos-phocholine) is a structural analogue of phosphatidylcholine and has the same stereochemistry as naturally occurring phospholipids. Three structural features are essential for full biological activity. These include: (1) a long-chain fatty alcohol in ether linkage at the sn-1 position, (2) an acetyl residue in ester linkage at the sn-2 position, and (3) the choline head group in phosphodiester linkage. Alterations of these features (e.g., replacement of the sn-1 alkyl group with an acyl group) result in analogues that may have bioactivity but are markedly less potent.

cytes, some activated T cells, eosinophils, and basophils) to vascular effects (hypotension, increased vascular permeability, anaphylaxis) to more steady-state physiologic effects (as a mediator of embryo implantation, parturition, and stimulation of hepatic glycogenolysis). Despite the increasing recognition of its diverse actions, the trivial name *platelet-activating factor* continues to be used. In order to impart more structural information, other names for the compound have been proposed including PAF acether and AGEPC (alkylglyceryl ether phosphorylcholine) although these are not widely used.

As will be seen in the following discussion, the synthesis and expression of PAF, its binding to a specific cell surface receptor, and its degradation are highly regulated events—speaking to its importance as a potent, and important, mediator. In addition, the production of PAF and its expression on the cell surface coincides with production and expression of other lipid and protein mediators of inflammation (e.g., eicosanoids and selectins). This coordinated production of inflammatory mediators and adhesion proteins, each with a specific and interactive role, comprises a first-line inflammatory defense. In some cases the interactions of PAF with other mediators is partially elucidated, and these will be discussed at some length below. However, it is clear that many aspects of these complex systems, and the variety of functional interactions of PAF with other signaling molecules remain to be discovered.

In addition to its role in the inflammatory response, there is increasing evidence that PAF serves as a signaling molecule in other physiologic processes.

The discussion below contains a description of its role in embryonic development, gestation, and parturition. Other evidence indicates that PAF (or its metabolites) may serve as intracellular second messengers. Hence, we can now ascribe many other roles to this biologically active compound beyond its role as simply an activator of platelets.

As the study of PAF has progressed, other structurally related compounds have been identified both as chemical entities and as biologically active mediators. Such newly defined structures range from those with head groups other than choline to those with different constituents at the sn-2 position to one with a different bond at the sn-1 position. Some of these mimic the effects of PAF but are less potent. Others have been found only in specific tissues; this suggests that these lipids may have a specialized role in specific sites that is unrelated to their rather modest PAF-like activity.

Perhaps the most exciting of these PAF-like lipids are certain oxidized phospholipids with sufficient structural similarity to PAF that they demonstrate biological activity similar, and in some cases identical, to that of PAF (Smiley et $al.$, 1991; McIntyre et $al.$, 1994, 1995). The importance of these compounds is that they arise from the unregulated oxidation of an abundant membrane and lipoprotein phospholipid, 1-acyl-2-arachidonoyl phosphatidylcholine. Arachidonate is highly susceptible to oxidation by virtue of its four olefinic bonds: oxidation of esterified arachidonate results in modification and carbon–carbon bond scission to yield a mixture of diacylglycerophosphocholines containing various sn-2 residues. Other long-chain polyunsaturated fatty acid residues are similarly susceptible to oxidation. The potential for unregulated production of these molecules via oxidation makes these compounds attractive candidates for inflammatory mediators in an ever broadening array of injuries where oxidants play a direct role in the development of the pathologic state: antioxidants have been found to suppress damage in animal models of atherogenesis and ischemia-reperfusion injury. Indeed, biologic effects of these oxidized PAF-like lipids have been shown in a number of isolated systems and in several whole animal models (McIntyre et $al.$, 1995; Lehr et $al.$, 1991, 1993). With the exception of the unique pathway to their formation, current evidence indicates that these molecules are functionally indistinguishable from the PAF derived from the highly regulated biosynthetic pathway: they bind to the PAF receptor, they are capable of mediating cell–cell interactions, and their degradation is catalyzed by the enzyme that degrades PAF. Consequently, understanding of the processes by which the prototype molecule, 1-O-alkyl-2-acetyl-sn-glycero-3-phosphocholine, exerts its effects and how it is metabolized to inactive products its critical to understanding the roles that these PAF-like molecules play in physiologic processes.

It is clear that the original description of PAF as an activator of platelets is incomplete and belies its role in an expanse of inflammatory processes. It is equally clear that PAF is but the prototype for a family of structurally related lipids with similar properties and biologic effects. The activity of PAF-like phospholipids in injury states requires a broadened perspective of these lipid autacoids. The following discussion is designed to provide an overview of these molecules and what is known about their role in pathophysiologic inflammation.

7.2. Platelet-Activating Factor

7.2.1. Synthesis and Regulation of Production

7.2.1.1. Cells of Origin

Production of PAF occurs in most nonlymphoid circulating inflammatory cells including neutrophils, monocytes, macrophages, basophils, and eosinophils (Benveniste *et al.*, 1972: Camussi *et al.*, 1981; Arnoux *et al.*, 1980; Clark *et al.*, 1980; Triggiani *et al.*, 1992; Lee *et al.*, 1984). PAF is also produced in endothelial cells derived from all parts of the vasculature including arteries, veins, and capillaries of both animals and humans (Whatley *et al.*, 1988). In fact, the production of PAF by endothelial cells appears to be ubiquitous and may be one of the defining functional characteristics of the endothelial cell phenotype. Production of PAF has been described in a variety of cells and tissues including the kidney (Blank *et al.*, 1979), nervous tissue (Goracci *et al.*, 1994; Kumar *et al.*, 1988; Marcheselli *et al.*, 1990), gastrointestinal epithelium (Sun and Hsueh, 1988), human amnion (Ban *et al.*, 1986; Billah *et al.*, 1985), and others (Bussolino *et al.*, 1986). In some cases, the precise cell of origin has been difficult to identify since most tissues are a mixed population of cells that includes inflammatory cells such as tissue macrophages. For example, a significant portion of PAF found in gastrointestinal tissues arises from intestinal mast cells (Hsueh *et al.*, 1987; Hogaboam and Wallace, 1994). In most cell types that have been examined, PAF is not present in significant amounts in the steady state (Zimmerman *et al.*, 1985b; Whatley *et al.*, 1990) (the possible exception being renal cells). However, with appropriate activation of the cell, PAF synthesis is rapidly initiated by activated synthetic enzymes. Although different cell types are activated by cell-type-specific agents, the activation schemes use several canonical pathways (such as G protein-coupled receptors) that will be described in detail below.

In addition to production of PAF by cells in mammalian tissues, some reports have described the presence of PAF and PAF-like lipids in nonmammalian tissues, including invertebrates. The precursor of PAF, alkylacyl glycerophosphocholine, can be found in surprisingly large amounts in several invertebrate species (the alkylacyl subclass comprises 80% of phosphatidylcholine in the sea sponge *Halochondria japonica*) (Suguira *et al.*, 1994). In addition, PAF and lipids that are structurally similar to PAF (PAF-like lipids) can be found in many lower animal forms (Suguira *et al.*, 1994). The presence of these lipids in the latter systems, while intriguing, is of uncertain significance and may represent signaling roles for PAF quite different from those in mammals. Alternatively, the presence of PAF, its precursor alkylacyl glycerophosphocholine and related lipids may simply have unique structural functions in nonmammalian cell membranes. Currently, the role that these lipids play in nonmammalian phylogeny is only speculative.

7.2.1.2. Structural Features of PAF

PAF (Fig. 7-1) has several key structural features, all of which contribute to high-affinity receptor recognition and corresponding bioactivity. It contains a glycerol backbone with a choline head group in phosphodiester linkage. The sn-1 position is occupied by a long-chain fatty alcohol in ether linkage while an acetyl residue is esterified at the sn-2 position. Exhaustive structure–activity studies have identified the structural components essential for bioactivity (Blank et al., 1982; O'Flaherty et al., 1983). The stereochemistry of naturally occurring PAF is the same as other mammalian phospholipids and this configuration is essential for its biologic effects. Another structural feature essential for high-affinity recognition by its receptor is the sn-2 acetyl residue: extending this residue by one methylene group to the propionyl homologue is benign, but for each further increase in chain length at the sn-2 position there is a corresponding order of magnitude decrease in potency. The hexanoyl homologue is inactive as is the underivatized sn-2 lyso homologue. This well-accepted paradigm is derived from an analysis of compounds with unadorned monocarboxy fatty acyl residues; the interesting and unexpected modifications to this are discussed below. Other structural features have definite effects on receptor activation. The ether linkage at the sn-1 position of the glycerol backbone, for instance, is critically important for biological efficacy. Substitution of the ether bond at the sn-1 position with an ester bond results in a compound that is biologically active, but whose potency is two to three orders of magnitude less than PAF itself (Blank et al., 1982; Smiley et al., 1991; Triggiani et al., 1991a). Substitution of the choline head group for other groups, particularly naturally occurring ethanolamine, also results in molecules with markedly decreased potency. The fatty alcohol at the sn-1 position is typically saturated and 16 carbons in length, although other chain lengths and fatty alcohols containing double bonds have been described (Sugiura and Waku, 1987). These differences in the fatty alcohol residue confer only modest differences in potency of the molecule.

7.2.1.3. Synthesis of PAF

PAF, like many other lipid mediators, is not constitutively present but is rapidly synthesized on appropriate activation of competent cells. Studies in a variety of cell types have demonstrated that there are two pathways for synthesis of this molecule (Fig. 7-2). The pathway that has been found in most cell types, termed the "remodeling" pathway (Fig. 7-2A), likely accounts for stimulated PAF synthesis. It consists of two enzymatic steps. The first of these is a phospholipase A_2-catalyzed hydrolysis of a long-chain fatty acid from the sn-2 position of a phospholipid precursor converting it to the respective lyso compound and a free fatty acid. Current evidence indicates that this step is catalyzed by the same PLA_2 that hydrolyzes fatty acids (e.g., arachidonic acid) from the sn-2 position of di-acylglycerophospholipids. Studies in intact cells, cell homogenates, and with the purified PLA_2 have shown that alkylphospholipid precursors that contain arach-

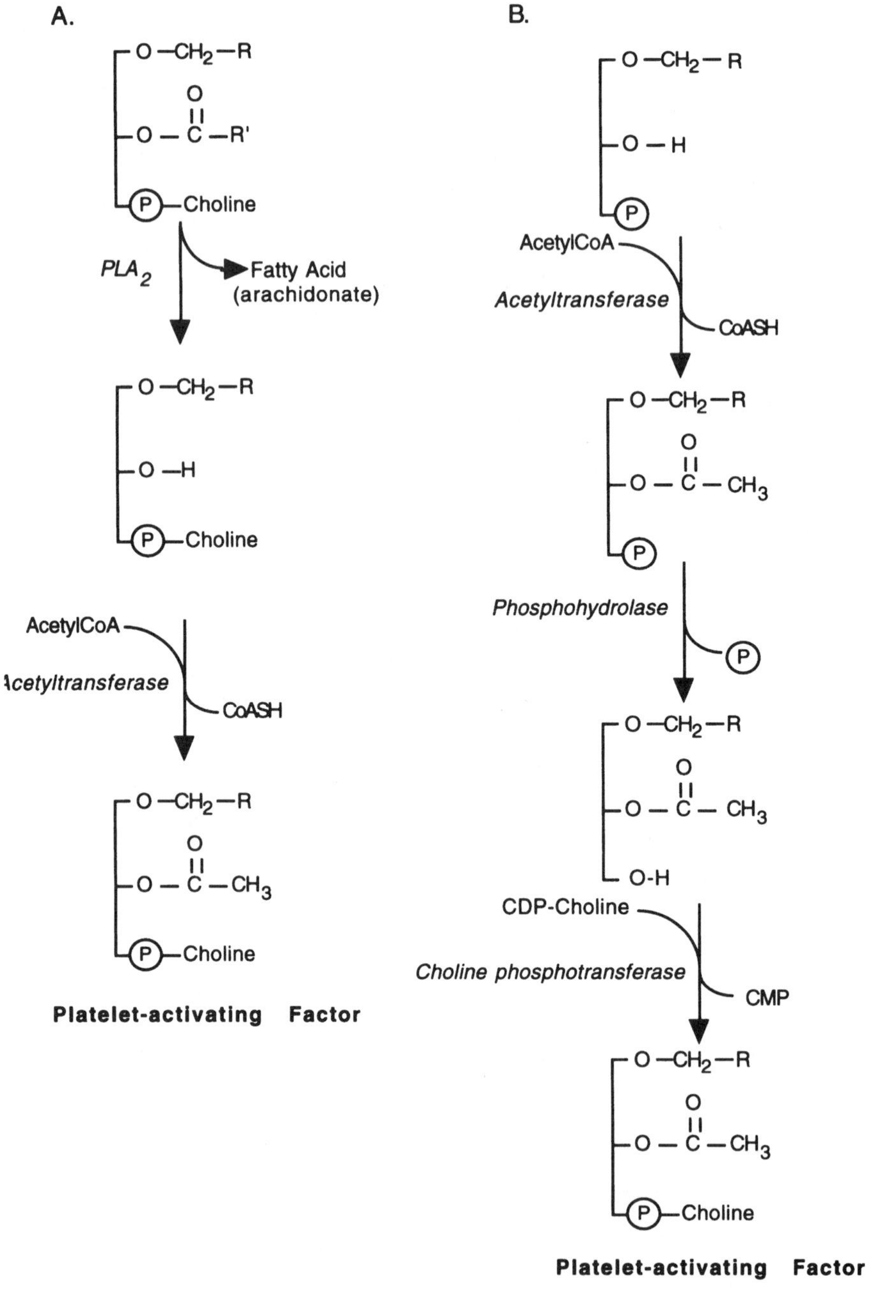

A.

$O-CH_2-R$

$O-\overset{\overset{\displaystyle O}{\|}}{C}-R'$

Ⓟ—Choline

PLA_2 → Fatty Acid (arachidonate)

$O-CH_2-R$

$O-H$

Ⓟ—Choline

AcetylCoA

Acetyltransferase → CoASH

$O-CH_2-R$

$O-\overset{\overset{\displaystyle O}{\|}}{C}-CH_3$

Ⓟ—Choline

Platelet-activating   Factor

B.

$O-CH_2-R$

$O-H$

Ⓟ

AcetylCoA

Acetyltransferase → CoASH

$O-CH_2-R$

$O-\overset{\overset{\displaystyle O}{\|}}{C}-CH_3$

Ⓟ

Phosphohydrolase → Ⓟ

$O-CH_2-R$

$O-\overset{\overset{\displaystyle O}{\|}}{C}-CH_3$

$O-H$

CDP-Choline

Choline phosphotransferase → CMP

$O-CH_2-R$

$O-\overset{\overset{\displaystyle O}{\|}}{C}-CH_3$

Ⓟ—Choline

Platelet-activating   Factor

idonic acid at the *sn*-2 position are equivalent, and possibly, preferential substrates for this enzyme (Chilton *et al.*, 1984; Suga *et al.*, 1990; Clark *et al.*, 1990; Winkler *et al.*, 1993; Qiu *et al.*, 1993). However, the first step in the remodeling pathway is somewhat more complex than the simple action of a PLA$_2$. Although formation of lysophospholipids must ultimately begin with a phospholipase A$_2$ the pathway leading to the formation of the specific PAF precursor also involves the action of a transacylase. There is abundant evidence that the immediate formation of the specific PAF precursor (1-*O*-alkyl-2-hydroxy-*sn*-glycero-3-phosphocholine) is catalyzed by a transacylase that transfers a fatty acid (e.g., arachidonate) from (1-*O*-alkyl-2-acyl-*sn*-glycero-3-phosphocholine to a lysophospholipid such as a lysoethanolamine plasmalogen (formed by action of the PLA$_2$ on an arachidonate-containing precursor ethanolamine plasmalogen) (Suguira *et al.*, 1990; Nieto *et al.*, 1991; Uemura *et al.*, 1991; Snyder, 1995a,b).

The net result of these first steps in the remodeling pathway is the hydrolysis of alkylacylglycerophosphocholine to lysoPAF and the release of arachidonic acid, thereby forming the precursors for both PAF and eicosanoid synthesis. This can be demonstrated experimentally as the coproduction of PAF and eicosanoids by stimulated cells (e.g., endothelial cell PAF and prostacyclin production; neutrophil PAF and leukotriene B$_4$ production). The lack of stringent specificity for the *sn*-1 bond in the phosphatidylcholine precursor results in the production of both alkyl and acyl lysophospholipids (collectively termed "radyl" lysophospholipids). The lysoPAF formed in this first step is then converted to the fully active molecule by the esterification of acetate; this reaction is catalyzed by a specific acetyl coenzyme A:lysoPAF acetyltransferase. Because the acetyltransferase fails to distinguish the nature of the *sn*-1 bond it acetylates both alkyl and acyl lysophospholipids; consequently, as will be discussed below, both PAF and its less potent acyl homologue can accumulate.

The second scheme for PAF biosynthesis is via the "de novo" pathway (Fig. 7-2B) which utilizes 1-*O*-alkyl-2-hydroxy-*sn*-glycerol as the precursor lipid. This is first converted by a specific acetyltransferase to 1-*O*-alkyl-2-acetyl-*sn*-glycerol, which is then converted by a unique CDP-choline:1-alkyl-2-acetyl-*sn*-glycerolcholinephosphotransferase (Lee *et al.*, 1986; Snyder, 1987) to PAF. While the intermediates of this reaction are somewhat unusual, this pathway is analogous to conventional phosphatidylcholine biosynthesis, although the enzymes employed are distinct.

←——

Figure 7-2. Biosynthesis of platelet-activating factor. (A) The remodeling pathway. PAF synthesis begins with the hydrolysis of a specific precursor, 1-*O*-alkyl-2-acyl-*sn*-glycero-3-phosphocholine, yielding lysoPAF and a free fatty acid. This process is the net result of the activities of a PLA$_2$ and one or more transacylases. Current evidence indicates that the initial PLA$_2$ reaction may hydrolyze another phospholipid (e.g., plasmenylethanolamine); this reaction is followed by a transacylase that transfers a fatty acid (arachidonate) to the resulting lysolipid from 1-*O*-alkyl-2-acyl-*sn*-glycero-3-phosphocholine. The net product of these reactions is a free fatty acid and lysoPAF. The lysoPAF is then converted to the fully active PAF by the action of an acetyl coenzyme A:lysoPAF acetyltransferase. (B) The *de novo* pathway. A precursor lipid, 1-*O*-alkyl-2-hydroxy-*sn*-glycerol, is converted to 1-*O*-alkyl-2-acetyl-*sn*-glycerol by a specific acetyltransferase. The product is converted to PAF by sequential actions of a phosphohydrolase and a CDP-choline:1-*O*-alkl-2-acetyl-*sn*-glycerol choline phosphotransferase.

The remodeling pathway is responsible for PAF accumulation following stimulation of many cell types, but the *de novo* pathway has been proposed to constitutively supply the kidney with low levels of PAF to maintain vascular tone; it is likely that this pathway is responsible for constitutive PAF formation in other tissues as well.

The purification and characterization of a key enzyme in the remodeling pathway, the PLA_2, remained elusive for some time. Mammalian cells had been known to contain several members of a large class of phospholipase A_2's referred to as "secretory" or "low-molecular-weight" PLA_2's that are analogous to the PLA_2 found in snake venom and digestive glands (Waite, 1985). These enzymes were not attractive candidates for the PLA_2 responsible for PAF formation or arachidonate release because of their extracellular location (or because they are secreted), their requirement for millimolar concentrations of calcium (a higher concentration than would occur in the cytosol of stimulated cells), and their lack of selectivity for fatty acid hydrolysis from the *sn*-2 position. In the past few years, several groups have purified a high-molecular-weight form of PLA_2 that is present in the cytosol of cells, catalyzes the selective release of arachidonate from phospholipids, and hydrolyzes fatty acids from the alkylglycerophosphocholine precursor of PAF to form lysoPAF (Clark *et al.*., 1991; Sharp *et al.*, 1991; Leslie *et al.*, 1988). This cytosolic PLA_2 is a 110-kDa protein that, in the presence of submicromolar concentrations of calcium, translocates to cell membranes where it is catalytically active (Leslie *et al.*, 1988; Kramer *et al.*, 1991; Clark *et al.*, 1991; Wijkander and Sundler, 1992a). This characteristic is particularly important since the calcium dependency of activation of the enzyme is in the same range as the changes in intracellular calcium concentrations that occur in response to activation of the cell (e.g., in response to the binding of a hormone to its cell surface receptor) (Channon and Leslie, 1990; Clark *et al.*, 1991; Kramer *et al.*, 1993). Moreover, other aspects of the regulation of this PLA_2 are consistent, as discussed later, with what had been inferred from a host of cell biologic studies prior to its molecular identification. As noted above, recent work has shown that the formation of lysoPAF (as shown in Fig. 7-2A) is somewhat more complex. In neutrophils and other cells, the formation of the lysoPAF intermediate arises from hydrolysis of an arachidonate-containing ethanolamine plasmalogen by PLA_2 followed by a transacylase reaction where an arachidonoyl residue is transferred from 1-*O*-alkyl-2-arachidonoyl-*sn*-glycero-3-phosphocholine to the newly formed lyso-ethanolamine plasmalogen (Suguira *et al.*, 1990; Nieto *et al.*, 1991; Uemura *et al.*, 1991; Snyder, 1995a,b). The net effect of these reactions is the formation of the lysoPAF intermediate and a free fatty acid (arachidonic acid). The interactions between the PLA_2 and the transacylases and the regulation of these enzymes are the objects of ongoing investigations. Of note, recent work has shown that the transacylase reaction may be catalyzed by a CoA-dependent enzyme (Blank *et al.*, 1995) as well as the CoA-independent enzyme originally described.

The final step in the remodeling pathway is the acetylation of lysoPAF via a specific acetyl-CoA:lysoPAF acetyltransferase. Although this enzyme has not yet been purified, its characteristics have been described by several groups (Wykle *et al.*, 1980; Ninio *et al.*, 1982, 1983; Lenihan and Lee, 1984; Lee, 1985; Gomez-

Cambronero *et al.*, 1986; Domenech *et al.*, 1987). These include specificity for the acetyl-CoA donor as opposed to a longer-chain acyl-CoA, activation by changes in intracellular calcium content and regulation by phosphorylation (Ninio *et al.*, 1983; Nieto *et al.*, 1988; Holland *et al.*, 1992; Gomez-Cambronero *et al.*, 1987; Domenech *et al.*, 1987; Lenihan and Lee, 1984).

7.2.1.4. Regulation

Production of PAF is a highly regulated event that is coordinated with other inflammatory processes, eicosanoid synthesis, and expression of vascular adhesion proteins, consistent with it being a multipotent inflammatory mediator. Characteristics of this regulation are presented schematically in Fig. 7-3. As shown,

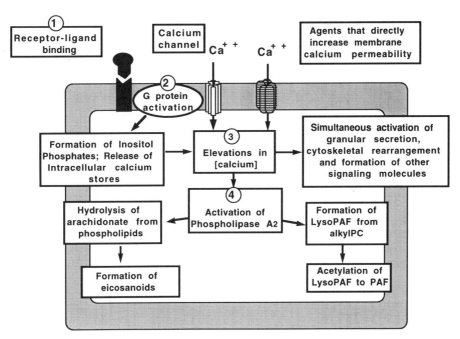

Figure 7-3. Synthesis of PAF occurs in concert with other cellular activation events. The initial stimulus for PAF formation is an elevation in intracellular calcium concentrations. This can be initiated by binding of a ligand to its receptor on the cell surface (1) with activation of a canonical G protein-coupled cascade (2). Alternatively, elevations in cell calcium that occur as a consequence of direct permeabilization of the cell membrane can also initiate PAF synthesis. The elevations in cell calcium concentration (3) result in activation of the cytosolic PLA_2 (4), the initial enzyme in the remodeling pathway. This results in formation of lysoPAF and release of free fatty acids (e.g., arachidonate). The lysoPAF is acetylated to PAF and the free fatty acids can be converted to a variety of bioactive products (e.g., eicosanoids). The key intracellular event is the elevation in cell calcium which also initiates a variety of other activation responses such as cytoskeletal rearrangement, granular secretion, and formation of nitric oxide. The array of activation pathways is diverse and cell type-dependent but typically occurs within the same time frame as PAF synthesis (seconds to minutes).

formation of PAF is intimately associated with release of arachidonic acid from phospholipid stores and its metabolism to eicosanoids. The proximal regulatory step in the synthesis of PAF is a change in intracellular calcium concentrations (Whatley *et al.*, 1989, 1990). Elevations in intracellular calcium concentrations can arise from both release of intracellular stores of calcium as well as entry of calcium from the extracellular space. The canonical scheme for such changes is the binding of an agonist to a specific membrane receptor (e.g., the serpentine family of receptors containing seven transmembrane-spanning domains) that activates one or more trimeric G protein(s) (Dohlman *et al.*, 1991). There is a consequent release of calcium from intracellular storage sites and entry of calcium from the extracellular space via specific membrane channels (Exton, 1988; Whatley *et al.*, 1989; Schilling *et al.*, 1992). This results in elevations in cell calcium to sub-micromolar concentrations from the steady-state level of 50–100 nM. Other mechanisms such as calcium ionophores or direct permeabilization of the plasma membrane (by pore-forming bacterial toxins or mellitin) result in similar changes in intracellular calcium (Whatley *et al.*, 1989, 1990; Suttorp and Habben, 1988). Regardless of the mechanism, elevations in cell calcium concentration are sufficient to initiate PAF biosynthesis (Whatley *et al.*, 1989, 1990). Although there are some modest differences depending on the cell type, robust formation of PAF requires a more sustained elevation in cell calcium concentrations that arises from the entry of calcium from the extracellular space.

With elevations in cytosolic calcium concentrations, the cytosolic PLA_2 ($cPLA_2$) translocates from the cytosol to a cell membrane where it hydrolyzes arachidonate-containing phospholipids, including the ether-linked PAF precursor. Activation of the acetyl-CoA:lysoPAF-acetyltransferase also occurs with these changes in cell calcium although the mechanism of that activation is less well characterized than in the case of the $cPLA_2$ (Nieto *et al.*, 1988; Holland *et al.*, 1992; Domenech *et al.*, 1987). Several characteristics of the $cPLA_2$ are intriguing and likely relevant to formation of other lipid hormones or autacoids. The $cPLA_2$ preferentially hydrolyzes phospholipids with long-chain polyunsaturated fatty acids at the *sn*-2 position (Clark *et al.*, 1991; Leslie *et al.*, 1988). Hence, its activation results in the selective, although not specific, release of arachidonic acid in addition to the formation of the immediate precursor for PAF (Fig. 7-4). This accounts for the observations that elevations in cell calcium concentrations regulated arachidonate release and eicosanoid formation in a manner that mirrors the regulation of PAF formation (Whatley *et al.*, 1989, 1990; Whorton *et al.*, 1984). Thus, an attractive hypothesis is that these changes in intracellular calcium activate an enzyme, the PLA_2, that is the rate-limiting enzyme in the pathways for both PAF formation and arachidonate release, consistent with a hypothesis originally proposed by Wykle and Chilton (Chilton *et al.*, 1984). At the biologic level such a close coordination of lipid mediator production is apparent in the activation of neutrophils adherent to activated endothelial cells. Transient expression of PAF by the endothelium activates the adherent neutrophils while simultaneous production of the arachidonate metabolite prostacyclin serves to dilate the vessel

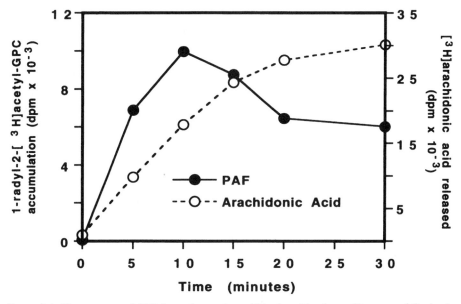

Figure 7-4. Time course of PAF formation and arachidonic acid release. Human umbilical vein endothelial cells in culture were exposed to thrombin (2 units/ml) in a buffer containing [^{3}H]acetate (25 μCi/ml). At the indicated times the reaction was terminated and cell lipids were extracted. The PAF fraction was separated chromatographically and the amount of label that accumulated in PAF was measured. In endothelial cells, label accumulates in PAF and in its 1-acyl analogue and are here expressed as 1-radyl-[^{3}H]acetyl glycerophosphocholine. The decrease in the amount of PAF that accumulates at longer time points results from the intracellular PAF acetylhydrolase that degrades PAF. In separate dishes, endothelial cells were labeled with [^{3}H]arachidonic acid for 1 hr. The labeling medium was then removed and the cells were washed with buffer. The cells were then stimulated in buffer containing thrombin (2 units/ml). At the indicated times the buffer was collected and the amount of label released from the cells was measured.

and thereby reduce blood flow and its attendant sheer stress: this sharpens the zone of neutrophil activation and suppresses neutrophil activation at distant sites.

It has also been reported that the cPLA$_2$ also has lysophospholipase activity, converting 1-acyl lysophospholipids to free fatty acids and the corresponding glycerophosphate (Leslie, 1991). Thus, the enzyme could further degrade the 1-acyl-linked lysophospholipid products of the initial hydrolysis of diacylglycero-phospholipids (e.g., lysophosphatidylcholine). This could be a mechanism for selective removal of lysolipids other than the precursor of PAF (the *sn*-1 ether bond of lysoPAF would not be hydrolyzed) and thereby alter the relative proportion of acyl and alkyl species of acetylated phosphoglycerides. This nuance of cPLA$_2$ activity has only been noted *in vitro* and its significance in whole cells is not known.

In addition to the key regulatory role played by calcium, PAF biosynthesis is also regulated by protein kinase activity. Prior to purification of the cPLA$_2$, several groups had reported that this step was modulated by protein kinases including protein kinase C (McIntyre *et al.*, 1987; Wijkander and Sundler, 1992b; Whatley *et al.*, 1989; Garcia *et al.*, 1991). Subsequent purification and cloning of the $cPLA_2$ has demonstrated that it is regulated by phosphorylation in a complicated manner. Activation of protein kinase C results in a specific phosphorylation of the enzyme and an increase in activity measured in *in vitro* assays; the effect of PKC activity is mediated by mitogen-activated protein kinase (MAP kinase) (Durstin *et al.*, 1994; Lin *et al.*, 1993; Kramer *et al.*, 1993). Correspondingly, we find that the phosphorylation state and ability to be activated in whole cells correlates with the growth state or, more accurately, the relative proportion of cells in $G_2 + M$ cell cycle (Whatley *et al.*, 1994). Given the well-known interaction of various protein kinases in a variety of systems, it is likely that the regulation of this enzyme by phosphorylation will seem to be even more complex with further scrutiny.

There is evidence that the acetyltransferase is also regulated by phosphorylation and dephosphorylation (Nieto *et al.*, 1988; Holland *et al.*, 1992; Gomez-Cambronero *et al.*, 1987; Domenech *et al.*, 1987), although the acetyltransferase has not been purified. Current evidence suggests that the enzyme is activated by phosphorylation with a time course and other characteristics that are similar to the $cPLA_2$ (Holland *et al.*, 1992); thus, it is tempting to speculate that the acetyltransferase and the $cPLA_2$ are activated by similar mechanisms.

Activation of these pathways results in rapid formation of PAF (Fig. 7-4). In this case, endothelial cells were exposed to thrombin which initiates PAF biosynthesis via interaction with a cell membrane receptor that is coupled, via G proteins, to elevations in cell calcium concentration (Khai *et al.*, 1991). With this stimulus, the synthesis of PAF is immediately initiated, accumulation peaks within 5 to 10 min and rapidly decreases; a process that is largely complete within 20–30 min. Decreases in the amount of PAF accumulation are the consequence of degradation by specific acetylhydrolases (Zimmerman *et al.*, 1990a; Stafforini *et al.*, 1987b). This pattern of formation is characteristic for PAF formation by a variety of cell types, although some qualitative and quantitative differences do exist (Elstad *et al.*, 1988; Sisson *et al.*, 1987).

The time course shown in Fig. 7-4 utilizes a standard assay of PAF formation, metabolic labeling with tritiated acetate. This technique utilizes the near-instantaneous labeling of the acetyl-CoA pool with exogenous labeled acetate. With an appropriate stimulus for PAF formation, the label is incorporated into PAF which can then be isolated and quantified (Zimmerman *et al.*, 1990b). The advantage of this assay is, in addition to its simplicity, its specificity. Since the relevant labeling period is only minutes, basal fatty acid and phospholipid synthesis contribute nothing to the metabolic labeling resulting in little, if any, nonspecific background. This is confirmed by the ability of phospholipase A_2 or PAF acetylhydrolase to hydrolyze all of the radiolabel from metabolically labeled PAF: only the acetyl residue was derived from acetyl-CoA during the few minutes of stimulation. This technique requires a qualification that illustrates an additional complexity to PAF biosynthesis. Several groups have shown that this technique does not distinguish between formation of 1-*O*-alkyl-2-acetyl-*sn*-glycero-3-phosphocholine and its analogue 1-acyl-2-acetyl-*sn*-glycero-3-phosphocholine, the PAF analogue formed by the PLA$_2$-mediated hydrolysis of a diacylphosphatidylcholine precursor (Garcia *et al.*, 1991; Triggiani *et al.*, 1991b; Whatley *et al.*, 1992). Although these *sn*-1 acyl analogues have biologic activity, their potency is at least two orders of magnitude less than that of authentic PAF (Triggiani *et al.*, 1991a; Blank *et al.*, 1982). Several studies using metabolic labeling with tritiated acetate and mass measurements by mass spectrometry have shown that the metabolic labeling technique does indeed measure the mix of 1-radyl-2-acetyl-*sn*-glycero-3-phosphocholines, and that the relative amount of each constituent appears to arise from the relative amounts of the specific precursors in the phospholipid pool (Whatley *et al.*, 1992; Garcia *et al.*, 1991). In some cells the precursor phosphatidylcholine pool as a relatively high alkyl choline phosphoglyceride content (e.g., neutrophils); this results in a correspondingly higher amount of PAF compared to its 1-acyl analogue. In other cells, the phosphatidylcholine pool has a smaller percentage of such precursor lipids (e.g., endothelial cells) and this results in a relatively greater amount of the 1-acyl PAF analogue compared to PAF. The significance of these differences, if any, is not known, but from a practical aspect the metabolic labeling assay safely overestimates the actual amount of PAF bioactivity. Mass spectrometry has confirmed that the amount of PAF in resting, unstimulated cells is negligible, consistent with metabolic labeling studies (Fig. 7-4) (Triggiani *et al.*, 1990; Whatley *et al.*, 1992). Also shown in Fig. 7-4 is the temporal relationship of arachidonic acid release to PAF formation. Consistent with the aforementioned concurrent release of arachidonate and *sn*-1 alkyl choline phosphoglyceride from the phospholipid pool, stimulated endothelial cells display a coordinated production of these potent lipid mediators.

7.2.2. Membrane Association and Secretion

Several of the early descriptions of PAF reported it as a circulating component in plasma (Pinckard *et al.*, 1979; Camussi *et al.*, 1983b), and some investigators have found PAF in biologic fluids, including saliva and urine (Billah and Johnston, 1983; Cox *et al.*, 1981); these findings indicate that PAF is released from some of the cells that synthesize it. *In vitro* studies with inflammatory cells of several types have shown that they are capable of releasing PAF into the surrounding media although the relative percentage of total PAF released is highly variable and dependent on several factors (Elstad *et al.*, 1988; Sisson *et al.*, 1987; Arnoux *et al.*, 1980; Clark *et al.*, 1980). For example, monocytes release a portion of the PAF synthesized into the surrounding medium; however, the percentage secretion increases with time and varies for unknown reasons depending on the agonist (Elstad *et al.*, 1988). Neutrophils also have been reported to release newly synthesized PAF, although this appears to depend on a serum releasing factor and/or suppression of PMN cross talk (Miwa *et al.*, 1992; Cluzel *et al.*, 1989). In contrast, none of the PAF produced by endothelial cells is released into the surrounding medium: it remains associated with the cell (including the extracellular surface) (McIntyre *et al.*, 1985). The presence of PAF in the fluid phase of blood and the ability of inflammatory cells to secrete a percentage of synthesized PAF suggest that it may play a role as a circulating intercellular signaling molecule. As will be discussed below, the half-life of PAF circulating in plasma is quite short (seconds to minutes). This is consistent with a role for PAF as a physiologic intercellular signaling molecule as it can be closely regulated and its effect transitory. Despite the short half-life, the biologic effects of PAF can occur at subnanomolar concentrations (Demopoulos *et al.*, 1979) so the levels detected in blood and body fluids are likely capable of evoking responses at sites distal to its production.

That portion of PAF that remains associated with the cell has a key role in cell–cell signaling: the expression of PAF on the surface of endothelial cells is part of a well-orchestrated series of events that results in activation of adherent neutrophils (Lorant *et al.*, 1991; Zimmerman *et al.*, 1990a). In this case, the PAF in the membrane microenvironment at the endothelial cell surface is a signaling molecule that activates inflammatory cells brought into close proximity by tethering molecules (Lorant *et al.*, 1993; Zimmerman *et al.*, 1992b). This is a cooperative interplay between PAF and a molecule, P-selectin (GMP-140, PADGEM, CD62P), that is expressed on the cell surface concomitant with PAF formation and that binds passing leukocytes and monocytes. The localized expression of an adhesion protein coupled with the restricted surface expression of PAF results in a narrow zone of leukocyte activation. The precise topographical localization of endothelial cell PAF is not known. However, studies suggest that a significant portion is located on the extracellular side of the plasma membrane: it is capable of activating other cells and it is accessible to hydrolysis by extracellular PAF acetylhydrolase (Zimmerman *et al.*, 1990a). The enzymes that catalyze PAF formation are present within cells; the initial PLA_2 is catalytically active on translocation from the cytosol to cell membranes—the presumed site of the ether-linked precursor lipids. Hence, it is likely that synthesis of PAF is initiated within intracellular membranes such as the endoplasmic reticulum or the membranes of other intracellular organelles. Consistent with this hypothesis, metabolic labeling studies have shown that PAF can first be found associated with the endoplasmic reticulum fraction and later in the plasma membrane (Vallari *et al.*, 1990; Record *et al.*, 1989). The mechanism responsible for the transport of PAF from its intracellular location to the plasma membrane is not known and may require transport proteins analogous to those that transport phospholipids to the outer membrane. Consistent with this is the report that, in some cells at least, PAF receptor antagonists block egress of newly synthesized PAF (Lachachi *et al.*, 1985). A potential role for the PAF receptor in what may be facilitated diffusion of PAF is strengthened by a recent report (Gerard and Gerard, 1994) that CHO cells expressing the cloned PAF receptor are capable of transporting and metabolizing PAF (in contrast to untransfected control cells). A specific transport mechanism is likely to exist as transbilayer movement of PAF is slow: in erythrocytes, which lack PAF receptors, the $t_{1/2}$ for flip-flop is approximately 17 hr (Schneider *et al.*, 1986). It is equally possible that PAF moves to the cell surface by fusion of intracellular membranes with the plasma membrane, a process used in other schemes of intracellular transport. Unfortunately, our overall understanding of intracellular localization and movement is hampered by the relatively small amounts of PAF present in subcellular fractions and the short half-life of the molecule (because of intracellular degradation).

7.2.3. Degradation

PAF degradation is catalyzed by specific PAF acetylhydrolases present in plasma and within many cells (Stafforini *et al.*, 1987b, 1991). These enzymes are Ca^{2+}-independent phospholipase A_2's that are exceptional by virtue of their marked specificity for short acyl chains at the *sn*-2 position (Stafforini *et al.*, 1987b). The distribution of these enzymes is widespread. One enzyme circulates in plasma; another is found intracellularly in a variety of cell types including, interestingly, cells that do not synthesize PAF (e.g., red blood cells) (Stafforini *et al.*, 1988) (Table 7-1). The intriguing implications of this distribution are developed below. The plasma form of the PAF acetylhydrolase has a molecular mass of 43 kDa and circulates in the plasma in association with lipoprotein particles that contain apo B or apo E (Stafforini *et al.*, 1987a). Thus, two-thirds of the plasma activity is associated with LDL, one-third with a specific HDL subfraction that contains apo E, and only a small amount of total activity is associated with VLDL or IDL because of their low relative abundance. The human plasma enzyme has been purified (and recently cloned) and its kinetic properties are well described (Stafforini *et al.*, 1987b). An acetylhydrolase from bovine brain, presumably intracellular in origin, has been purified and found to consist of three associated proteins (Hattori *et al.*, 1993): the catalytic 29-kDa subunit (Hattori *et al.*, 1994) has been cloned and one of the other associated proteins has been characterized. The plasma and the cellular PAF acetylhydrolases are not homologous to other lipases and they are not homologous to each other.

The intracellular PAF hydrolase isolated from red blood cells is approximately 25 kDa and, based on marked differences in inhibitor susceptibilities, is also likely to be unrelated to the two cloned enzymes. The distinguishing characteristic of PAF acetylhydrolases is the marked specificity for substrates with short

Table 7-1. Characteristics of PAF Acetylhydrolases

Location (size)	Associations	Catalytic properties and substrate specificity
Plasma	LDL/HDL	Preference for micellar substrate; hydrolyzes short-chain oxidized lipid moieties from the sn-2 position of phospholipids
		No dependence on divalent cations
		Inhibited by DFP; inactivated by oxidation
Intracellular (29 kDa)	Two protein cofactors	Similar kinetics and substrate specificity to the plasma form
		No dependence on divalent cations
		Requires a reducing environment; inactivated by oxidation

acyl chains at the *sn*-2 position: there is a progressive decrease in activity as the chain is lengthened beyond the acetyl group present in PAF (Table 7-1) (Stafforini *et al.*, 1987b). Other phospholipases do not have a preference for short *sn*-2 residues. A second distinguishing feature of PAF acetylhydrolases is their Ca^{2+}-independence. In contrast to a requirement for micromolar to millimolar levels of Ca^{2+} for cellular and secreted phospholipase A_2's, acetylhydrolases are active in the complete absence of calcium (in EDTA-containing buffers). This difference is consistent with the lack of structural homology among these lipases. Phospholipids with an acyl group at the *sn*-1 position (rather than an ether-linked fatty alcohol) are equally good substrates for PAF acetylhydrolases, as are phospholipids that contain head groups other than choline (Stafforini *et al.*, 1987b). Moreover, as will be noted later, these enzymes are capable of hydrolyzing phospholipids that contain fatty acids at the *sn*-2 position that have been shortened and modified by oxidation. Thus, the essential difference that identifies PAF acetylhydrolases is the requirement for a short or modified *sn*-2 residue.

The source of the plasma form of the enzyme is not completely known but both cultured macrophages and hepatocytes have the capacity to synthesize and secrete this enzyme (Stafforini *et al.*, 1990a; Elstad *et al.*, 1988; Tarbet *et al.*, 1991; Satoh *et al.*, 1993). Despite this capacity, Northern blot analysis of tissues revealed that human liver did not contain significant amounts of messenger RNA—this despite the presence of both hepatocytes and the tissue macrophages, Kupffer cells (Tjoelker *et al.*, 1995). The explanation for this observation is not yet in hand, and was unexpected on another count. Determination of plasma PAF acetylhydrolase activity from nearly 1000 individuals revealed that premenopausal women have a small, but significant, decrease in circulating levels compared to a matched male population, and that following menopause there was no gender-related difference. Correspondingly, we and others find (Satoh *et al.*, 1993; Tarbet *et al.*, 1991) that the synthesis and secretion of PAF acetylhydrolase by hepatocytes is decreased by half in the presence of estradiol, but not male steroids or mineralocorticoids. From this we anticipated that liver should be a significant, and regulated, source of circulating PAF acetylhydrolase activity. The half-life of circulating PAF acetylhydrolase is not known and it may be that both the rates of synthesis and turnover are slow. Elevations in the amounts of PAF acetylhydrolase activity in plasma occur in some diseases [e.g., essential hypertension and ischemic cerebrovascular disease (Satoh *et al.*, 1988, 1989)]. It is interesting that both conditions are associated with processes in which PAF has been postulated to play a role. Consistent with this association, PAF has been found (Satoh *et al.*, 1991) to induce the synthesis of PAF acetylhydrolase by a hepatoma cell line. The induction of the degradative enzyme by PAF itself suggests that the already rapid degradation of circulating PAF can be enhanced in the event that PAF is produced in quantities that exceed the basal degradative capacity. In most cases the amount of PAF acetylhydrolase present in plasma is such that the half-life of circulating PAF is limited to seconds to a few minutes (Stafforini *et al.*, 1990b). In the steady state the rapid constitutive degradation of PAF implies that changes in circulating PAF concentrations likely arise from changes in the synthetic rate.

Once PAF is degraded to the inactive lysoPAF compound and acetate, the lysoPAF may be reacylated with a long-chain fatty acid to form the PAF precursor. The fatty acids incorporated at the *sn*-2 position in stimulated cells, such as PMN (Reinhold *et al.*, 1989), are a random mix. Consequently, it is likely that a remodeling step is also present to subsequently enrich the *sn*-2 position with arachidonyl residues. In fact, there is considerable evidence in many cell types that trans-acylases that reacylate lysoPAF are highly selective for arachidonate (MacDonald and Sprecher, 1991; Snyder *et al.*, 1992). Alternatively, metabolic labeling studies performed in some cells have shown that the lysoPAF is selectively converted to an ethanolamine plasmalogen by sequential acylation, modification of the *sn*-1 linkage, and conversion of the head group from choline to ethanolamine (Tessner and Wykle, 1987). The significance of this pathway is not known at present.

7.2.4. PAF Receptor

The initial evidence for a receptor that recognizes PAF came from binding studies demonstrating that cells that responded to PAF possessed specific, high-affinity binding sites on the cell surface (Valone *et al.*, 1982; Valone, 1987). The most compelling evidence for a specific receptor is that only the naturally occurring stereoisomer of PAF is active (Wykle *et al.*, 1981): this proves that functional recognition of PAF is by a stereochemical element and is not a nonspecific perturbation of membrane–structure function. Binding studies, like functional studies, show constants for PAF binding at (or below) the nanomolar range. These studies are complicated by the difficulty in performing binding studies with lipid-soluble molecules and by the low number of these receptors on inflammatory cells (as few as several hundred per cell) (Kloprogge and Akkerman, 1984). Even so, some of the studies indicate that there are at least two receptors with differing affinities or one receptor that can achieve different affinity states (Hwang and Lam, 1991). Additionally, there is one report that human epithelial cells respond to PAF by both receptor-dependent and -independent mechanisms (Stoll *et al.*, 1994). The characterization of the PAF receptor initially progressed concurrently with the identification of PAF receptor antagonists. A few compounds structurally related to PAF were synthesized and found to be antagonists, but a number of chemically dissimilar compounds have been found to block the biologic effects of PAF in empiric studies (Shen *et al.*, 1987). A few compounds are relatively water-soluble and were important in confirming the presence of a specific receptor in addition to measuring receptor affinity. In general, these receptor antagonists have been used to confirm that a biologic effect is mediated by the PAF receptor or to identify what portion of a complex biologic response is mediated by PAF–receptor interaction (Fig. 7-5). It is worth noting that, for the most part, the PAF receptor antagonists have remained as specific inhibitors of the PAF receptor even though ample time has passed for nonspecific effects to become evident (in contrast to many other pharmacologic inhibitors). Perhaps this is a consequence of the rather unique nature and structure of PAF.

Recently, through the technique of expression cloning, the cDNA coding for

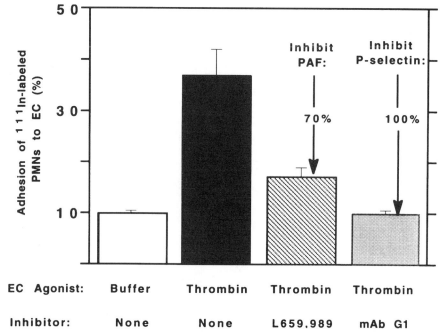

Figure 7-5. Effect of a PAF receptor antagonist on neutrophil binding to activated endothelium. Human umbilical vein endothelial cells in culture were exposed to buffer or to thrombin (2 units/ml) for 10 min. After the 10-min incubation, the solution was removed and ^{111}In-labeled neutrophils that had been pretreated with control buffer or the specific PAF receptor antagonist, L659,989 (100 µM), were added. In some wells, a monoclonal antibody to the PMN tethering or binding molecule, P-selectin (mAb G1; 10 µg/ml), was added to the endothelial cells immediately before the addition of thrombin and to the neutrophil suspension. After 5 min the neutrophil suspension was removed and the monolayers gently washed with buffer. The percent adhesion of neutrophils was determined by comparison of the amount of labeled neutrophils adherent to the monolayer to the total amount of labeled neutrophils (adherent + nonadherent). L659,989 antagonism of the binding of the neutrophil PAF receptor to endothelial cell PAF resulted in a 70% inhibition of neutrophil adhesion. This figure also illustrates the contribution of the adhesion molecule, P-selectin, in this process (as shown in the schematic Fig. 7-7).

the PAF receptor has been cloned. The predicted sequence indicates that it is a member of the superfamily of G protein-coupled receptors containing seven membrane-spanning domains (Honda *et al.*, 1991; Ye *et al.*, 1991; Kunz *et al.*, 1992; Dohlman *et al.*, 1991). Screening of a genomic library showed that it mapped to an intronless gene located on human chromosome 1 (Seyfried *et al.*, 1992). Despite functional evidence for more than one class of PAF receptors, a single gene codes for the identified receptor. The transcript is subject to alternative splicing, but this occurs in the 5′-untranslated region and only affects the promoter region (Mutoh *et al.*, 1993). This alternative splicing is interesting and suggests a rather complex scheme for PAF receptor expression. Transcript 1 contains consensus sequences for NF-κB and SP-1 transcription factors and is found in leukocytes; transcript 2 contains AP-1, AP-2, and SP-1 consensus elements and is expressed in all tissues examined, including leukocytes. This suggests that regulation of PAF receptor expression and sensitivity to PAF occurs at the transcriptional level. Such a process may be responsible for the changes in PAF responsiveness that occur as the promyelocytic cell line HL60 differentiates. With differentiation it progresses from a PAF-insensitive state to a responsive one concomitant with expression of PAF receptor message and protein (Ye *et al.*, 1991). The presence of potential AP-1 and NF-κB binding elements may well account for the induction of PAF receptors by PAF as PAF receptors activate the MAP kinase pathway (Honda *et al.*, 1994) which converges on transcription factors that bind these regulatory elements.

Subsequent studies using antibodies directed against predicted protein sequences or analysis of cell mRNA have demonstrated the presence of this receptor in cell types that are activated by PAF (Shimizu *et al.*, 1992). Transfection of the cDNA encoding the PAF receptor into cells that ordinarily do not respond to PAF changes their phenotype to a cell that responds to PAF with elevations of intracellular calcium and release of arachidonic acid (Fig. 7-6) (Ye *et al.*, 1991). Importantly, these effects are seen with PAF concentrations in the subnanomolar range, the same concentrations of PAF that stimulate inflammatory cells. The cellular responses that occur with PAF binding to its receptor are those mediated by a canonical G protein-coupled receptor. These include elevations in cellular calcium activation of phospholipases, protein kinases, and related activation responses (Shukla, 1992; Hwang, 1990; Bito *et al.*, 1992; Shimizu *et al.*, 1992; Honda *et al.*, 1994). Like other members of this class of receptors, the PAF receptor is desensitized by prolonged exposure to its agonist—a process mediated by phosphorylation of the receptor's cytoplasmic tail. In fact, the PAF receptor tail is a substrate for the β-adrenergic receptor kinase that desensitizes the β receptor (Takano *et al.*, 1994).

7.2.5. Biologic Effects

7.2.5.1. Activation of Inflammatory Cells

The original description of PAF was a substance circulating in animals in anaphylactic shock that caused platelets to aggregate and release histamine (Benveniste *et al.*, 1972). In addition to aggregation and degranulation, PAF causes platelets to undergo a dramatic shape change similar to that seen with other platelet activators (Benveniste *et al.*, 1972). PAF exerts similar effects on neutrophils and other circulating inflammatory cells. As in platelets, the responses are initiated by PAF binding to the PAF receptor; this initiates a characteristic G protein-coupled cascade resulting in increases in intracellular calcium concentrations, cytoskeletal rearrangement resulting in polarization of the neutrophil and ruffling of the neutrophil membrane (as seen on scanning electron microscopy), chemotaxis, and granular secretion (Shimizu *et al.*, 1992; Zimmerman *et al.*, 1992a). In addition, this stimulation causes functional upregulation of the CD11/CD18 integrin adhesion complex that causes the neutrophil to become

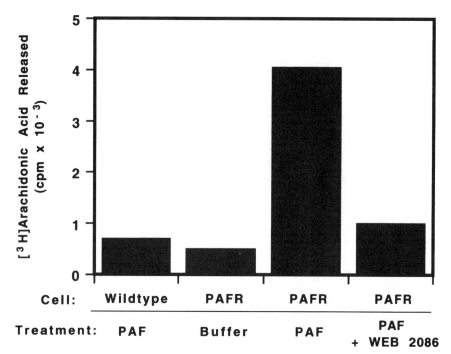

Figure 7-6. Expression of functional PAF receptors in CHO cells. Chinese hamster ovary (CHO) cells that were transfected with cDNA encoding the receptor for platelet-activating factor (PAFR) or wild-type CHO cells were labeled with [^{3}H]arachidonic acid for 1 hr. The labeling medium was then removed and the cells washed with buffer. The labeled cells were then exposed to PAF (10^{-9} M) in buffer or buffer alone for 20 min. Some of the cells were exposed to the PAF receptor antagonist WEB 2086 (4 μM) for 10 min prior to and during the exposure to PAF. Following the incubation the supernatant was removed and the amount of [^{3}H]arachidonic acid that had been released from the cells was measured. The presence of a functional PAF receptor in the transfected cells allows ligand (PAF) binding, resulting in cell activation, increases in intracellular calcium concentration (as measured by Indo-1 dye spectrofluorimetry; not shown), and arachidonic acid release.

more adherent to a wide variety of surfaces, including endothelial cell surfaces and components of the extracellular matrix (Rinder *et al.*, 1994; Zimmerman *et al.*, 1993). As in the case of platelets, these effects are blocked by PAF receptor antagonists. PAF has similar effects on other inflammatory cells although the spectrum of responses varies from cell to cell.

7.2.5.2. Cell–Cell Interactions

The secretion of PAF from some inflammatory cells, particularly monocytes (Camussi *et al.*, 1983a; Elstad *et al.*, 1988) but also neutrophils (Sisson *et al.*, 1987), may be a mechanism for intercellular signaling and amplification. Activated inflammatory cells that release PAF into the circulating fluid phase would activate other cells and so amplify the local inflammatory response. If so, it is difficult to determine the magnitude of this response in an experimental system or the physiologic significance in a whole animal. The relative percentage of PAF released from an inflammatory cell is small and it is rapidly degraded by the PAF acetylhydrolase present in plasma. In addition, it circulates bound to protein and lipoproteins, making its actual concentration difficult to determine. Nonetheless, infusion of PAF into experimental animals has dramatic physiologic effects. For instance, infusion of PAF into the aorta of a rat results in necrotizing enterocolitis (Hsueh *et al.*, 1987; Gonzalez-Crussi and Hsueh, 1983). Although it is not clear that such infusions are strictly relevant to more physiologic events, it is possible that PAF secreted from inflammatory cells has similar effects on other cells or tissues in the diverse inflammatory pathways that are active in pathologic states.

The role of PAF in cell–cell interactions has been well characterized in one model system, namely, the interaction of neutrophils, and other inflammatory cells, with activated endothelium. Expression of PAF on the surface of stimulated endothelial cells is the agonist generated by rapidly activated endothelial cells that transmits the activation signal to bound neutrophils (Zimmerman *et al.*, 1990a). PAF is rapidly synthesized by endothelial cells exposed to specific agonists (thrombin, histamine, LTD_4, bradykinin, and ATP). That production is coincident with increased adhesion of neutrophils and monocytes to the endothelial surface; this occurs in response to the translocation of the leukocyte adhesion molecule, P-selectin, from the specialized intracellular granules, the Weibel–Palade bodies, to the endothelial cell surface (Fig. 7-7) (Lorant *et al.*, 1991, 1993). The inflammatory cells are then activated by endothelial cell-associated PAF with increases in intracellular calcium concentrations, shape changes, and functional upregulation of their own adhesive integrins. Neutrophils express CD11/CD18 (the β2-integrin family) on their surface in their basal circulating state. Agonist stimulation, including that by free or endothelial-cell-associated PAF, alters these nonadhesive proteins to ones that bind many targets; the mechanism for this is undefined. PAF stimulation also causes the appearance of more CD11/CD18 on the neutrophil surface, but this, like the original CD11/CD18 complex before stimulation, is not functional (Hughes *et al.*, 1992). This sequence of events results in "bond trading" where the initial weak binding of endothelial cell P-selectin to its glycoprotein target on neutrophils is exchanged for a strong, sheer-resistant, multivalent binding of CD11/CD18 to its ligand on the endothelial cell (Tsuji *et al.*, 1994). The role of PAF in this process is confirmed by data demonstrating that activation, but not the initial tethering, can be blocked by PAF receptor antagonists or exposure of the stimulated endothelial cells to the PAF acethylhydrolase prior to exposing them to neutrophils (Zimmerman *et al.*, 1985a, 1990a). Thus, the role of PAF in this response is one part of a precisely orchestrated interaction (Fig. 7-7). The cooperativity of these various mediators in producing a full adhesion/activation response shows that their conjoint synthesis and expression is likely to be physiologically important and reaffirms the role of PAF (and related compounds) as important mediators of cell–cell interactions.

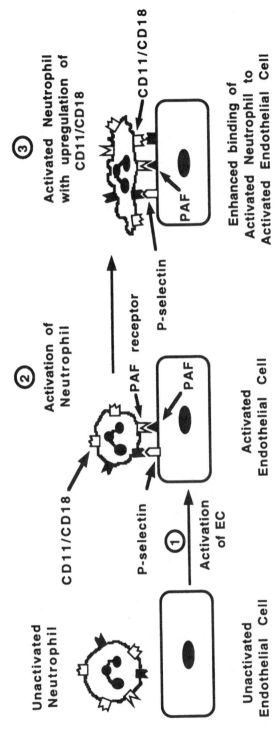

Figure 7-7. The role of PAF in binding and activation of neutrophils on activated endothelium. Activation of endothelial cells (1) results in synthesis of PAF and its expression on the endothelial cell surface. At the same time the adhesion molecule, P-selectin, is expressed on the endothelial cell surface and P-selectin binds to its ligand on the surface of the neutrophil. This brings the neutrophil in proximity to the endothelial cell and facilitates binding of endothelial PAF to its receptor, resulting in neutrophil activation (2). PAF-induced activation of the neutrophil further amplifies this process by inducing cytoskeletal rearrangement, shape change, and upregulation of the neutrophil integrin, CD11/CD18 (3), allowing their binding to counterreceptors on the endothelial cell and greatly enhancing the strength of the neutrophil–endothelial bond. (After Lorant *et al.*, 1991; with permission of the publisher.)

7.2.5.3. Intracellular Signaling

Most PAF remains associated with the cell and is not secreted; in addition the existence of intracellular PAF acetylhydrolases indicates that PAF accumulation within the cell is tightly regulated. These characteristics suggest the intriguing (although admittedly speculative) possibility that PAF may be an intracellular signaling molecule. The unusual structural features of PAF (compared to more conventional phospholipids) suggest that it might alter intracellular membrane characteristics such as fluidity (Bratton et al., 1988), and thus affect membrane function, fusion, or transport. Additionally, a potential role for PAF as an intracellular second messenger has been proposed: PAF receptor antagonists block eicosanoid and reactive oxygen species generation by leukocytes and bovine aortic endothelial cells (Stewart et al., 1989, 1990). Intracellular binding sites for PAF in stimulated neutrophils have been proposed as an essential signal-transducing element in Ca^{2+}-dependent activation of PLA_2 activity (Muller and Nigam, 1992). Another possible intercellular signaling role for PAF arises from observations in several cell types that PAF can be metabolized to 1-alkyl-2-acetyl-sn-glycerol (Stoll and Spector, 1989). This compound has been shown to have several interesting biologic effects that are not mediated via the PAF receptor (Stoll et al., 1994). In some cell types the alkylacetylglycerol activates PKC and in others it has been shown to interfere with diacylglycerol activation of PKC (Stoll and Spector, 1989; Daniel et al., 1988), perhaps reflecting differences in PKC isoforms. The mechanism by which alkylacetylglycerol exerts such effects and its role in intracellular signaling are not known. In addition, the mechanisms by which it is formed (presumably a phospholipase C-catalyzed hydrolysis of PAF) have not been demonstrated. As noted above, several groups have shown that PAF and lysoPAF can be metabolized by intact cells to other phospholipids, including plasmenylethanolamine (Frenkel and Johnston, 1992). The significance of this is intriguing; it may represent a mechanism to maintain the intracellular level of lysoPAF at low levels, to prevent its resynthesis to PAF, while defining a pathway that specifically recognizes lysoPAF. It is possible that the metabolism of lysoPAF is different from that of lysoPC (Daniel et al., 1986); such a scheme would serve as a mechanism of precisely controlling the accumulation of PAF and its precursors.

7.2.6. Physiologic Effects

One of the early descriptions of PAF was a compound extracted from renal tissue that lowered blood pressure in several animal models (Blank et al., 1979). The infusion of PAF into such animals results in hypotension and in some cases vascular collapse sufficient to cause death. This effect varies somewhat, depending on the animal model, and is sometimes associated with full-blown anaphylaxis making it difficult to define the effect of PAF on the vascular tone alone. Since these early descriptions, it has become clear that PAF causes diverse physiologic effects in whole animals and in isolated organ preparations. The diversity of those effects precludes a complete discussion here and a more expansive discussion can be found in reviews (Hanahan, 1986).

In some models, the infusion of PAF causes other, complex physiologic effects. For example, infusion of PAF, alone or in combination with lipopolysaccharide, into the mesenteric circulation of rats results in ischemic bowel necrosis (Gonzalez-Crussi and Hsueh, 1983). Intravenous administration of PAF into a rabbit results in bradycardia, hypotension, thrombocytopenia, leukopenia, alterations in the respiratory pattern (including apnea), and changes in ventricular filling pressures (Halonen *et al.*, 1980). Similar hemodynamic effects were observed during intravenous administration of PAF in dogs (Kenzora *et al.*, 1984). Infusion of PAF into the pulmonary circulation or installation into the airway can result in a variety of physiologic effects that are dependent on the method of installation and the model used (Mojarad *et al.*, 1985). These include changes in pulmonary vascular tone, airway resistance, lung lymphatic flow, and lung edema. Similar effects are found in more reduced systems. In an isolated lung preparation, infusion of PAF causes pulmonary artery vasoconstriction (Voelkel *et al.*, 1982). In similar preparations, infusion of PAF results in increases in lung water content (Mojarad *et al.*, 1985). While these observations comprise a very complicated spectrum of physiologic effects (that are quite species-dependent), they illustrate the variety of roles that PAF may play in vascular pathophysiology.

Perhaps some of the most intriguing physiologic effects of PAF are in reproduction and fetal development. PAF has been shown to have a role in ovulation (Abisogun *et al.*, 1989); although the precise mechanisms are not completely clear, a PAF receptor antagonist prevented follicle rupture in a fashion that could be overcome by the addition of synthetic PAF. Prior observations that early stage embryos produce PAF led to studies showing that this production was associated with an increased incidence of successful ovarian implantation (Ryan *et al.*, 1990). This is consistent with a number of studies that have now found PAF in multiple uterine and fetal tissues (Johnston and Miyaura, 1990). These observations have even been extended to clinical studies of *in vitro* fertilization: embryos pretreated with PAF were associated with significantly higher pregnancy rates compared to a control group implanted with embryos that had not been exposed to PAF activating factor (O'Neill *et al.*, 1992). While obviously intriguing, the precise role that PAF plays in this process and the mechanism involved are not known. It does, however, extend the observations that inflammatory mediators also play key roles in pregnancy (Cooper *et al.*, 1994).

Equally intriguing is the potential role of PAF in parturition. PAF has been found in amniotic fluid obtained from women in active labor. Women at term, but not in active labor, had much lower concentrations of PAF (Billah and Johnston, 1983). The PAF in amniotic fluid was associated with the lamellar body fraction suggesting that the PAF may be synthesized by fetal lung, although fetal kidneys may also be a source of this lipid. Perhaps related to these observations, the activity of the PAF acetylhydrolase in the maternal plasma of rabbits changes during pregnancy with a dramatic fall immediately prior to parturition (Maki *et al.*, 1988). Following delivery, levels returned to that of the prepregnancy state. This is thought to be related to the suppression of PAF acetylhydrolase production by the hyperestrogenic state, as this then would depress hepatic synthesis, with a return to the basal rate of synthesis following parturition. In contrast, in fetal plasma the levels of the PAF acetylhydrolase increased during the latter stages of gestation resulting in levels that were several times greater than that found in the corresponding maternal plasma (Maki *et al.*, 1988). These observations are consistent with the possibility that the decrease in maternal PAF acetylhydrolase activity allows access of PAF in the amniotic fluid to the uterine wall and thereby effects parturition. Several of the reports have shown that PAF causes contraction of myometrium and will induce uterine contractions in both experimental animals and humans (Tetta *et al.*, 1986). In addition, the administration of PAF receptor antagonists to rats greatly lengthens the time of parturition and will block delivery altogether in a significant percentage of the animals (Johnston and Miyaura, 1990). As a whole these data suggest that alterations in PAF concentrations in the uterine environment as a result of changes in PAF acetylhydrolase activity are a mechanism for initiating parturition. A direct pathological correlate of interference with this system seemingly is in hand. Smoking during pregnancy, in addition to other well-established consequences, increases the risk for preterm labor, and it increases blood levels of PAF (or biologically active mimetics) (Imaizumi *et al.*, 1991). One contributing mechanism for the accumulation of such lipids is the direct inhibition of PAF acetylhydrolase by both cigarette smoke extract component (Miyaura *et al.*, 1992) and oxygen radicals and by the inhibition of secretion of PAF acetylhydrolase by smoke extract-treated macrophages (Narahara and Johnston, 1993). Decreased PAF degradation, increased levels of PAF bioactivity, and preterm labor are smoking related and may represent cause and effect.

7.3. Oxidized Phospholipids

7.3.1. Production of Biologic Activity in Oxidant-Exposed Cells

Exposure of endothelial cells to millimolar concentrations of hydrogen peroxide results in permeabilization of the plasma membrane to calcium, and a rapid formation of PAF and increased adherence of neutrophils to the endothelial cells (Lewis *et al.*, 1988). The characteristics of this production are similar, but not identical, to the production of PAF by more conventional agonists (e.g., thrombin). As with conventional receptor-mediated agonists, the synthesis of PAF is rapid and the PAF remains cell-associated (Lewis *et al.*, 1988). However, compared to receptor-mediated agonists, the time course of PAF accumulation in hydrogen peroxide-treated endothelial cells is more prolonged with significant PAF levels present in the cells as long as 1 hr following the initial exposure (Lewis *et al.*, 1988). This expression of PAF is accompanied by an increased adherence of neutrophils similar to that seen with endothelial cells treated with conventional agonists, including the blockade of this enhanced adhesion by PAF receptor antagonists.

The precise mechanism by which hydrogen peroxide initiates this response is not completely clear; the effect is blocked by catalase but not superoxide dismutase, demonstrating that the active agent is hydrogen peroxide and not superoxide anion (Lewis *et al.*, 1988). It is likely that the effect of hydrogen peroxide is to directly permeabilize the plasma membrane to calcium, allowing a calcium entry that initiates the PAF synthetic pathway. Similar effects have been found in isolated vascular preparations exposed to hydrogen peroxide (Gasic *et al.*, 1991). Isolated vessels respond to hydrogen peroxide by synthesizing PAF, and this, in turn, results in enhanced leukocyte adhesion to the treated vessel (Hughes *et al.*, 1994). In another study, superperfusion of exposed mesentery with hydrogen peroxide resulted in increased adherence of neutrophils to venular endothelium when examined by intravital microscopy (Suzuki *et al.*, 1991). The role of PAF in that process was confirmed by the reduction in neutrophil adhesion following pretreatment of the animal with a PAF receptor antagonist. These studies confirm that the effect of hydrogen peroxide on cultured cells has relevance in whole animals and demonstrates that oxidant injury may play an important role in vascular pathophysiology.

In studies examining the effects of other oxidants, it was found that exposure of endothelial cells to lower concentrations of hydrogen peroxide, intracellular radical generators, or the lipid-soluble peroxide *tert*-butylhydroperoxide for longer times (> 30 min) also cause increased adherence to neutrophils to the endothelial cell surface (Patel *et al.*, 1991). In addition to the time course, this process differed in other ways from the increased adhesion caused by millimolar concentrations of hydrogen peroxide. Importantly, the increased adhesion could not be blocked by PAF receptor antagonists. The reason for these differences became apparent when it was determined that increased neutrophil adhesion to cells exposed to these oxidants was mediated by expression of P-selectin on the surface of the endothelial cells (Patel *et al.*, 1991). However, coincident with the increased adhesion for neutrophils, the exposed endothelial cells also had dramatic morphologic changes, including the formation of large membranous blebs that were shed as vesicles over the several hours of oxidant exposure (Patel *et al.*, 1992). In addition, a lipid extract of the vesicles (and to a lesser extent, the cells themselves) had the capacity to activate neutrophils *in vitro*, an effect blocked by PAF receptor antagonists or by treatment with the PAF acetylhydrolase (Patel *et al.*, 1992), implicating the PAF receptor as the target of the lipid extract. This suggested that PAF might be produced in this form of oxidant exposure similar to that seen with exposure to high concentrations of hydrogen peroxide. However, no PAF could be found in the cells of origin or in the vesicles themselves. Further characterization showed that while a lipid extract of the vesicles lost bioactivity following treatment with phospholipase C (as would occur in the case of PAF), the bioactivity was also lost when the extract was treated with phospholipase A_1 (which does not hydrolyze the ether linkage at the *sn*-1 position of PAF) (Patel *et al.*, 1992). Moreover, high-resolution chromatographic separation of the biologically active compounds showed that the product arising from this kind of oxidant exposure did not cochromatograph with authentic PAF (Patel *et al.*, 1992). Although the precise chemical structure of the molecule derived from oxidant exposure has not been elucidated, it is clear that it is a glycerophosphocholine containing a long-chain fatty acid at the *sn*-1 position and likely a short-chain constituent that is an oxidation product of a long-chain polyunsaturated fatty acyl residue at the *sn*-2 position. Two lines of evidence are consistent with this hypothesis.

First, exposure to synthetic 1-acyl-2-arachidonoyl-glycerophosphocholine to oxidants (lipoxygenase-initiated oxidation, ozonolysis) results in a mixture of 1-acyl-glycerophosphocholines with short-chain oxidation products at the *sn*-2 position (oxidized fatty acyl fragments containing aldehydic, carboxylic, and other functions) (Smiley *et al.*, 1991). This mixture of oxidized phospholipids has biologic activity that is indistinguishable from PAF: it will activate the PAF receptor (Fig. 7-8), is blocked by PAF receptor antagonists, and is degraded by exposure to the PAF acethylhydrolase (Stremler *et al.*, 1991). The ability to be hydrolyzed by the PAF acetylhydrolase not only means this enzyme is the likely route of degradation,

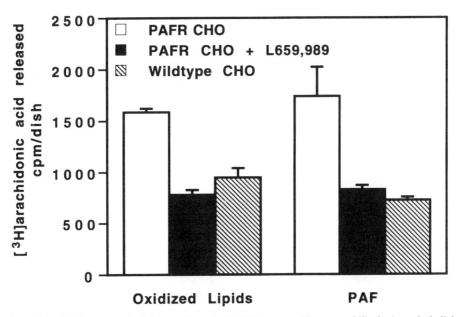

Figure 7-8. Oxidized phospholipids act through the PAF receptor. Human umbilical vein endothelial cells in culture were exposed to *tert*-butylhydroperoxide (250 μM) for 4 hr. Cell lipids were isolated and separated by HPLC. The fractions containing oxidized phospholipids were pooled and assayed for their ability to activate cells via the PAF reeptor. Chinese hamster ovary (CHO) cells that were transfected with cDNA encoding the receptor for platelet-activating factor (PAFR) or wild-type CHO cells were labeled with [^{3}H]arachidonic acid for 1 hr. The labeling medium was then removed and the cells washed with buffer. The cells were then exposed to buffer (Hanks' balanced salt solution + 1 mg/ml bovine serum albumin) containing oxidized phospholipids or PAF (10 nM). Some of the dishes were preincubated for 10 min in buffer containing the PAF receptor antagonist L659,989 (100 μM). Control incubations (containing buffer only) released 836 ± 91 cpm (PAFR CHO) or 991 ± 80 cpm (wild-type CHO). Data are expressed as the mean ± S.D. of measurements in three separate dishes.

it also provides confirmation that the *sn*-2 residue has been oxidatively modified. The acetylhydrolase catalyzes only phospholipids with short or oxidized *sn*-2 residues; it cannot hydrolyze the native long-chain diacyl phospholipid (Stremler *et al.*, 1991). In addition, these compounds cochromatograph on HPLC with the biologically active lipids that are produced with oxidant treatment of cells (Patel *et al.*, 1992).

The second line of evidence implicating these compounds is that chemical synthesis of several candidate phosphatidylcholines containing short-chain residues at the *sn*-2 position results in compounds that possess biologic activity like that of PAF (Goetzl *et al.*, 1980). From these data we can predict that the biologically active lipid that is produced with exposure to lipid hydroperoxides or similar oxidants is a PAF-like glycerophosphocholine that possesses a short-chain oxidized constituent at the *sn*-2 position. In most respects this molecule (or, more precisely, mixture of compounds) cannot be distinguished from PAF itself; in fact, it is recognized by the cloned human PAF receptor expressed in a heterologous cell (Fig. 7-8). Similar compounds have been identified in lipid extracts of bovine brain (Tokumura *et al.*, 1988, 1989) and in other lipids that have been subjected to oxidative attack (Itabe *et al.*, 1988; Tanaka *et al.*, 1993; Kayganich-Harrison and Murphy, 1994). In fact, it is likely that such compounds occur in many other situations that involve oxidative injury. For instance, oxidation of low-density lipoprotein (LDL) leads to the formation of PAF-like bioactivity that is destroyed by purified PAF acetylhydrolase. In addition, oxidized LDL contains PAF-like bioactivity that induces leukocyte rolling along the vascular wall, the first observable event in inflammation (Lehr *et al.*, 1991). Moreover, oxidative modification of LDL is thought to play an early, key role in atherosclerosis; the PAF acetylhydrolase blocks this process (Stafforini *et al.*, 1992). In addition, PAF receptor antagonists specifically block vascular cholesterol deposition and foam cell formation, a process thought to involve oxidized LDL (Feliste *et al.*, 1989).

7.3.2. Biologic Effects of Oxidized PAF-Like Lipids

Oxidative injury has been proposed as an important mediator in a variety of pathophysiologic states. Notable among these are lung injury with exposure to high concentrations of oxygen and ischemia-reperfusion injury. In addition, it has been proposed that oxidant-induced modification of LDL is a process that leads to atherosclerotic vascular disease (Witztum and Steinberg, 1991; Yagi, 1984). Oxidant injury may play a role in a host of other injuries where activation of neutrophils (or other radical-generating systems) plays a role. In a number of models of such injury, data support a role for PAF, either because the effect is blocked by a PAF receptor antagonist or because PAF-like bioactivity can be recovered from biologic material (McIntyre *et al.*, 1995). As can be seen from the preceding discussion, these effects could arise from a PAF-like oxidized phospholipid since they are not separable by most biologic or chemical techniques. PAF receptor antagonists are effective in blocking the effects of both PAF and PAF-like lipids. Oxidized PAF-like lipids are not separated from authentic PAF in most separation systems used to isolate PAF (thin-layer chromatography; most high-performance chromatographic systems). The essential difference between PAF and oxidatively fragmented phospholipids with PAF-like bioactivity is that only the synthesis of PAF is highly regulated. When chemical oxidation leads to the creation of bioactive PAF-like lipids, there is no regulation of synthesis. Only the PAF acetyl-hydrolase remains to regulate the accumulation of these compounds and the acetylhydrolase is exquisitely sensitive to oxidative inactivation itself. This important degradative responsibility may explain why the PAF acetylhydrolase has such a widespread distribution—including intracellular localization in cells that do not make authentic PAF. At present, such a scheme is admittedly speculative; however, cloning and expression of recombinant acetylhydrolase will allow elucidation of the precise role of the PAF acetylhydrolase and testing of the hypothesis that PAF and its mimetics inappropriately extend and prolong the inflammatory process in pathologic states.

7.4. Summary

As noted in a number of places throughout this review, the use of the term *platelet-activating factor* is inadequate for what we now know to be a multipotent activator of a variety of cellular, physiologic, and pathologic processes. It is a mediator of vascular inflammation, a regulator of ovulation and parturition, and an initiator of anaphylaxis. The actions of PAF in these and other processes will be better understood with further elucidation of the precise mechanisms by which it is synthesized and degraded and by further study of its targets, especially its cell-surface receptor. Of equal interest to PAF itself is the emerging role of PAF-like lipids. While their biologic activity is indistinguishable from PAF, they may represent a more significant biologic force. These lipids arise from the oxidation of lipid precursors that are in far greater abundance than the precursor for PAF; stated differently, the potential for their production is far greater than for PAF. In addition, their production occurs as a consequence of oxidation, a process implicated in a variety of pathologic states and a process that has none of the careful regulatory safeguards that modulate the synthesis of PAF. Given this, we are likely to find that these PAF-like lipids play exciting and important roles in pathologic states. Finally, it is well worth noting that our understanding of this area has arisen only as a result of careful and persistent dissection of the biology of PAF itself.

References

Abisogun, A. O., Braquet, P., and Tsafriri, A., 1989, The involvement of platelet activating factor in ovulation, *Science* **243**:381–382.

Arnoux, B., Duval, D., and Benveniste, J., 1980, Release of platelet-activating factor (PAF-acether) from alveolar macrophages by the calcium ionophore A23187 and phagocytosis, *Eur. J. Clin. Invest.* **10**:437–441.

Ban, C., Billah, M. M., Truong, C. T., and Johnston, J. M., 1986, Metabolism of platelet-activating factor (1-O-alkyl-2-sn-glycero-3-phosphocholine) in human fetal membranes and decidua vera, *Arch. Biochem. Biophys.* **246:**9–18.

Benveniste, J., Henson, P. M., and Cochrane, C., 1972, Leukocyte-dependent histamine release from rabbit platelets, *J. Exp. Med.* **136:**1356–1375.

Billah, M. M., and Johnston, J. M., 1983, Identification of phospholipid platelet-activating factor in human amniotic fluid and urine, *Biochem. Biophys. Res. Commun.* **113:**51–58.

Billah, M. M., Di Renzio, G. C., Ban, C., Truong, C. T., Hoffman, D. R., Anceschi, M. M., Bleasdale, J. E., and Johnston, J. M., 1985, Platelet-activating factor metabolism in human amnion and the responses of this tissue to extracellular platelet-activating factor, *Prostaglandins* **30:** 841–850.

Bito, H., Nakamura, M., Honda, Z., Izumi, T., Iwatsubo, T., Seyama, Y., Ogura, A., Kudo, Y., and Shimizu, T., 1992, Platelet-activating factor (PAF) receptor in rat brain: PAF mobilizes intracellular Ca^{2+} in hippocampal neurons, *Neuron* **9:**285–294.

Blank, M. L., Snyder, F., Byers, L. W., Brooks, B., and Muirhead, E. E., 1979, Antihypertensive activity of an alkyl ether analog of phosphatidylcholine, *Biochem. Biophys. Res. Commun.* **90:**1194–1200.

Blank, M. L., Cress, E. A., Lee, T.-c., Malone, B., Surles, J. R., Piantadosi, C., Hajdu, J., and Snyder, F., 1982, Structural features of platelet-activating factor (1-O-alkyl-2-acetyl-*sn*-glycero-3-phosphocholine) required for hypotensive and platelet serotonin responses. *Res. Commun. Chem. Pathol. Pharmacol.* **38:**3–20.

Blank, M. L., Fitzgerald, V., Smith, Z. L., and Snyder, F., 1995, Generation of the precursor (lysoPAF) of platelet-activating factor via a CoA-dependent transacylase, *Biochem. Biophys. Res. Commun.* **210:**1052–1058.

Bratton, D. L., Harris, R. A., Clay, K. L., and Henson, P. M., 1988, Effects of platelet activating factor on calcium–lipid interactions and lateral phase separations in phospholipid vesicles, *Biochim. Biophys. Acta* **943:**211–219.

Bussolino, F., Gremo, F., Tetta, C., Pescarmosa, P., and Camussi, G., 1986, Production of platelet-activating factor by chick retina, *J. Biol. Chem.* **261:**16502–16508.

Camussi, G., Aglietta, M., Coda, R., Bussolino, F., Piacibello, W., and Tetta, C., 1981, Release of platelet-activating factor (PAF) and histamine, *Immunology* **42:**191–199.

Camussi, G., Bussolino, F., Tetta, C., Piacibello, W., and Aglietta, M., 1983a, Biosynthesis and release of platelet-activating factor from human monocytes, *Int. Arch. Allergy Appl. Immunol.* **70:**245.

Camussi, G., Pawlowski, I., Bussolino, F., Caldwell, P. R. B., Brentjens, J., and Andres, G., 1983b, Release of platelet activating factor in rabbits with antibody-mediated injury of the lung: The role of leukocytes and of pulmonary endothelial cells, *J. Immunol.* **131:**1802–1807.

Channon, J. Y., and Leslie, C. C., 1990, A calcium-dependent mechanism for associating a soluble arachidonoyl-hydrolyzing phospholipase A_2 with membrane in the macrophage cell line RAW 264.7, *J. Biol. Chem.* **265:**5409–5413.

Chilton, F. H., Ellis, J. M., Olson, S. C., and Wykle, R. L., 1984, O-*O*-Alkyl-2-arachidonoyl-*sn*-glycero-3-phosphocholine: A common source of platelet-activating factor and arachidonate in human polymorphonuclear leukocytes, *J. Biol. Chem.* **259:**12014–12019.

Clark, J. D., Milona, N., and Knopf, J. L., 1990, Purification of a 110-kilodalton cytosolic phospholipase A_2 from the human monocytic cell line U937, *Proc. Natl. Acad. Sci. USA* **87:**7708–7712.

Clark, J. D., Lin, L. L., Kriz, R. W., Ramesha, C. S., Sultzman, L. A., Lin, A. Y., Milona, N., and Knopf, J. L., 1991, A novel arachidonic acid-selective cytosolic PLA_2 contains a Ca^{2+}-dependent translocation domain with homology to PKC and GAP, *Cell* **65:**1043–1051.

Clark, P. O., Hanahan, D. J., and Pinckard, R. N., 1980, Physical and chemical properties of platelet-activating factor obtained from human neutrophils and monocytes and rabbit neutrophils and basophils, *Biochim. Biophys. Acta* **628:**69–75.

Cluzel, M., Undem, B. J., and Chilton, F. H., 1989, Release of platelet-activating factor and the metabolism of leukotriene B_4 by the human neutrophil when studied in a cell superfusion model, *J. Immunol.* **143:**3659–3665.

Cooper, D., Butcher, C. M., Berndt, M. C., and Vadas, M. A., 1994, P-selectin interacts with a β2-integrin to enhance phagocytosis, *J. Immunol.* **153:**3199–3209.

Cox, C. P., Wardlow, M. L., Jorgenson, R., and Farr, R. S., 1981, The presence of platelet-activating factor (PAF) in normal human mixed saliva, *J. Immunol.* **127**:46–50.

Daniel, L. W., Waite, M., and Wykle, R. L., 1986, A novel mechanism of diglyceride formation, *J. Biol. Chem.* **261**:9128–9132.

Daniel, L. W., Small, G. W., and Schmitt, J. D., 1988, Alkyl-linked diglycerides inhibit protein kinase C activation by diacylglycerols, *Biochem. Biophys. Res. Commun.* **151**:291–297.

Demopoulos, C. A., Pinckard, R. N., and Hanahan, D. J., 1979, Platelet-activating factor: Evidence for 1-O-alkyl-2-acetyl-*sn*-glyceryl-3-phosphorylcholine as the active component (a new class of lipid chemical mediators), *J. Biol. Chem.* **254**:9355–9358.

Dohlman, H. G., Thorner, J., Caron, M. G., and Lefkowitz, R. J., 1991, Model systems for the study of seven-transmembrane-segment receptors, *Annu. Rev. Biochem.* **60**:653–658.

Domenech, C., Domenech, E. M.-d., and Soling, H.-D., 1987, Regulation of acetyl-CoA:1-alkyl-*sn*-glycero-3-phosphocholine acetyltransferase (lyso-PAF-acetyltransferase) in exocrine glands: Evidence for an activation via phosphorylation by calcium/calmodulin-dependent protein kinase, *J. Biol. Chem.* **262**:5671–5676.

Durstin, M., Durstin, S., Molski, T. F. P., Becker, E. L., and Sh'afi, R. I., 1994, Cytoplasmic phospholipase A_2 translocates to membrane fraction in human neutrophils activated by stimuli that phosphorylate mitogen-activated protein kinase, *Proc. Natl. Acad. Sci. UA* **91**:3142–3146.

Elstad, M. R., Prescott, S. M., McIntyre, T. M., and Zimmerman, G. A., 1988, Synthesis and release of platelet-activating factor by stimulated human mononuclear phagocytes, *J. Immunol.* **140**:1618–1624.

Exton, J. H., 1988, Mechanisms of action of calcium-mobilizing agonists: Some variations on a young theme, *FASEB J.* **2**:2670–2676.

Feliste, R., Perret, B., Braquet, P., and Chap, H., 1989, Protective effect of BN 52021, a specific antagonist of platelet-activating factor (PAF-acether) against diet-induced cholesteryl ester deposition in rabbit aorta, *Atherosclerosis* **78**:151–158.

Frenkel, R. A., and Johnston, J. M., 1992, Metabolic conversion of platelet-activating factor into ethanolamine plasmalogen in an amnion-derived cell line, *J. Biol. Chem.* **267**:19186–19191.

Garcia, M. C., Mueller, H. W., and Rosenthal, M. D., 1991, C20 polyunsaturated fatty acids and phorbol myristate acetate enhance agonist-stimulated synthesis of 1-radyl-2-acetyl-sn-glycero-3-phosphocholine in vascular endothelial cells, *Biochim. Biophys. Acta* **1083**:37–45.

Gasic, A. C., McGuire, G., Krater, S., Farhood, A. I., Goldstein, M. A., Smith, C. W., Entman, M. L., and Taylor, A. A., 1991, Hydrogen peroxide pretreatment of perfused canine vessels induces ICAM-1 and CD18-dependent neutrophil adherence, *Circulation* **84**:2154–2166.

Gerard, N. P., and Gerard, C., 1994, Receptor-dependent internalization of platelet-activating factor, *J. Immunol.* **152**:793–800.

Goetzl, E. J., Derian, C. K., Tauber, A. I., and Valone, F. H., 1980, Novel effects of 1-O-hexadecyl-2-acyl-*sn*-glycero-3-phosphorylcholine mediators on human leukocyte functions: Delineation of the specific roles of the acyl substituents, *Biochem. Biophys. Res. Commun.* **94**:881–888.

Gomez-Cambronero, J., Velasco, S., Sanchez-Crespo, M., Vivanco, F., and Mato, J. M., 1986, Partial purification and characterization of 1-O-alkyl-2-lyso-*sn*-glycero-3-phosphocholine:acetyl-CoA acetyltransferase from rat spleen, *Biochem. J.* **237**:439–445.

Gomez-Cambronero, J., Mato, J. M., Vivanco, F., and Sanchez-Crespo, M., 1987, Phosphorylation of partially purified 1-O-alkyl-2-lyso-*sn*-glycero-3-phosphocholine:acetyl-CoA acetyltransferase from rat spleen, *Biochem. J.* **246**:893–898.

Gonzalez-Crussi, F., and Hsueh, W., 1983, Experimental model of ischemic bowel necrosis. The role of platelet-activating factor and endotoxin, *Am. J. Pathol.* **112**:127–135.

Goracci, G., Francescangeli, E., Dreyfus, H., Boila, A., and Freysz, L., 1994, The synthesis of platelet-activating factor in brain and neural cells, *J. Lipid Mediat.* **10**:7–8.

Halonen, M., Palmer, J. D., Lohman, C., McManus, L. M., and Pinckard, R. N., 1980, Respiratory and circulatory alterations induced by acetyl glyceryl ether phosphorylcholine, a mediator of IgE anaphylaxis in the rabbit, *Am. Rev. Respir. Dis.* **122**:915–924.

Hanahan, D. J., 1986, Platelet activating factor: A biologically active phosphoglyceride, *Annu. Rev. Biochem.* **55**:483–509.

Hattori, M., Arai, H., and Inoue, K., 1993, Purification and characterization of bovine brain platelet-activating factor acetylhydrolase, *J. Biol. Chem.* **268**:18748–18753.

Hattori, M., Adachi, H., Tsujimoto, M., Arai, H., and Inoue, K., 1994, The catalytic subunit of bovine brain platelet-activating factor acetylhydrolase is a novel type of serine esterase, *J. Biol. Chem.* **269**:23150–23155.

Hogaboam, C. M., and Wallace, J. M., 1994, Intestinal PAF synthesis: The role of the mast cell, *J. Lipid Mediat.* **10**:103–105.

Holland, M. R., Venable, M. E., Whatley, R. E., Zimmerman, G. A., McIntyre, T. M., and Prescott, S. M., 1992, Activation of the acetyl-coenzyme A: lysoplatelet-activating factor acetyltransferase regulates platelet-activating factor synthesis in human endothelial cells, *J. Biol. Chem.* **267**:22883–22890.

Honda, Z., Nakamura, M., Miki, I., Minami, M., Watanabe, T., Seyama, Y., Okado, H., Toh, H., Ito, K., Miyamoto, T., and Shimizu, T., 1991, Cloning by functional expression of platelet-activating factor receptor from guinea-pig lung, *Nature* **349**:342–346.

Honda, Z.-i., Takano, T., Gotoh, Y., Nishida, E., Ito, K., and Shimizu, T., 1994, Transfected platelet-activating factor receptor activates mitogen-activated protein (MAP) kinase and MAP kinase in Chinese hamster ovary cells, *J. Biol. Chem.* **269**:2307–2315.

Hsueh, W., Gonzalez-Crussi, F., and Arroyave, J. L., 1987, Platelet-activating factor: An endogenous mediator for bowel necrosis in endotoxemia, *FASEB J.* **1**:403–405.

Hughes, B. J., Hollers, J. C., Crocket-Torabi, E., and Smith, C. W., 1992, Recruitment of CD11b/CD18 to the neutrophil surface and adherence-dependent cell locomotion, *J. Clin. Invest.* **90**:1687–1696.

Hughes, H., Sands, M. A., McGuire, G. M., and Taylor, A. A., 1994, PAF formation by H_2O_2-stimulated perfused canine carotid arteries, *Prostaglandins Leuk. Essent. Fatty Acids* **51**:323–328.

Hwang, S.-B., 1990, Specific receptors of platelet-activating factor, receptor heterogeneity, and signal transduction mechanisms, *J. Lipid Res.* **2**:123–158.

Hwang, S.-B., and Lam, M.-H., 1991, L-659,989: A useful probe in the detection of multiple conformational states of PAF receptors, *Lipids* **26**:1148–1153.

Imaizumi, T., Satoh, K., Yoshida, H., Kawamura, H., Hiramoto, M., and Takamatsu, S., 1991, Effect of cigarette smoking on the levels of platelet-activating factor-like lipid(s) in plasma lipoproteins, *Atherosclerosis* **87**:47–55.

Itabe, H., Kushi, Y., Handa, S., and Inoue, K., 1988, Identification of a 2-azelaoylphosphatidylcholine as one of the cytotoxic products generated during oxyhemoglobin-induced peroxidation of phosphatidylcholine, *Biochim. Biophys. Acta* **962**:8–15.

Johnston, J. M., and Miyaura, S., 1990, Platelet-activating factor: The alpha and omega of reproductive biology, in: *Advances in Applied Technology Series*, Vol. 9 (J. T. O'Flaherty and P. W. Ramwell, eds.), pp. 139–160, Portfolio Publishing, The Woodlands, TX.

Kayganich-Harrison, K. A., and Murphy, R. C., 1994, Characterization of chain-shortened oxidized glycerophosphocholine lipids using fast atom bombardment and tandem mass spectrometry, *Anal. Biochem.* **221**:16–24.

Kenzora, J. L., Perez, J. E., Bergmann, S. R., and Lange, L. G., 1984, Effects of acetyl glyceryl ether of phosphorylcholine (platelet-activating factor) on ventricular preload, afterload, and contractility in dogs, *J. Clin. Invest.* **74**:1193–1203.

Khai, T.-K., Hung, D. T., Wheaton, V. I., and Coughlin, S. R., 1991, Molecular cloning of a functional thrombin receptor reveals a novel proteolytic mechanism of receptor activation, *Cell* **64**:1057–1068.

Kloprogge, E., and Akkerman, J. W. N., 1984, Binding kinetics of PAF-acether to intact human platelets, *Biochem. J.* **223**:901–909.

Kramer, R. M., Roberts, E. F., Manetta, J., and Putnam, J. E., 1991, The Ca^{2+}-sensitive cytosolic phospholipase A_2 is a 100-kDa protein in human monoblast U937 cells, *J. Biol. Chem.* **266**:5268–5272.

Kramer, R. M., Roberts, E. F., Manetta, J. V., Hyslop, P. A., and Jakubowski, J. A., 1993, Thrombin-induced phosphorylation and activation of Ca^{++}-sensitive cytosolic phospholipase A_2 in human platelets, *J. Biol. Chem.* **268**:26796–26804.

Kumar, R., Harvey, S. A. K., Kester, M., Hanahan, D. J., and Olson, M. S., 1988, Production and effects of platelet-activating factor in the rat brain, *Biochim. Biophys. Acta* **963:**375–383.

Kunz, D., Gerard, N. P., and Gerard, C., 1992, The human leukocyte platelet-activating factor receptor, *J. Biol. Chem.* **267:**9101–9106.

Lachachi, H., Plantavid, M., Simon, M. F., Chap, H., Braquet, P., and Douste-Blazy, L., 1985, Inhibition of transmembrane movement and metabolism of platelet activating factor (PAF-acether) by a specific antagonist, BN 52021, *Biochem. Biophys. Res. Commun.* **132:**460–466.

Lee, T.-c., 1985, Biosynthesis of platelet-activating factor: Substrate specificity of 1-alkyl-2-lyso-*sn*-glycero-3-phosphocholine:acetyl-CoA acetyltransferase in rat spleen microsomes, *J. Biol. Chem.* **260:**10952–10955.

Lee, T.-c., Lenihan, D. J., Malone, B., Roddy, L. L., and Wasserman, S. I., 1984, Increased biosynthesis of platelet-activating factor in activated human eosinophils, *J. Biol. Chem.* **259:**5526–5530.

Lee, T.-c., Malone, B., and Snyder, F., 1986, A new *de novo* pathway for the formation of 1-alkyl-2-acetyl-*sn*-glycerols, precusors of platelet activating factor, *J. Biol. Chem.* **261:**5373–5377.

Lehr, H. A., Hubner, C., Nolte, D., Finckh, B., Beisiegel, U., Kohlschutter, A., and Mebmer, K., 1991, Oxidatively modified human low-density lipoprotein stimulates leukocyte adherence to the microvascular endothelium in vivo, *Res. Exp. Med.* **191:**85–90.

Lehr, H. A., Seemuller, J., Hubner, C., Menger, M. D., and Messmer, K., 1993, Oxidized LDL-induced leukocyte/endothelium interaction in vivo involves the receptor for platelet-activating factor, *Arterioscler. Thromb.* **13:**1013–1018.

Lenihan, D. J., and Lee, T.-c., 1984, Regulation of platelet-activating factor synthesis: Modulation of 1-alkyl-2-lyso-*sn*-glycero-3-phosphocholine: Acetyl-CoA acetyltransferase by phosphorylation and dephosphorylation in rat spleen microsomes, *Biochem. Biophys. Res. Commun.* **120:**834–839.

Leslie, C. C., 1991, Kinetic properties of a high molecular mass arachidonoyl-hydrolyzing phospholipase A_2 that exhibits lysophospholipase activity, *J. Biol. Chem.* **266:**11366–11371.

Leslie, C. C., Voelker, D. R., Channon, J. Y., Wall, M. M., and Zelarney, P. T., 1988, Properties and purification of an arachidonoyl-hydrolyzing phospholipase A_2 from a macrophage cell line, RAW 264.7, *Biochim. Biophys. Acta* **963:**476–492.

Lewis, M. S., Whatley, R. E., Cain, P., McIntyre, T. M., Prescott, S. M., and Zimmerman, G. A., 1988, Hydrogen peroxide stimulates the synthesis of platelet-activating factor by endothelium and induces endothelial cell-dependent neutrophil adhesion, *J. Clin. Invest.* **82:**2045–2055.

Lin, L.-L., Wartmann, M., Lin, A.-Y., Knopf, J. L., Seth, A., and Davis, R. J., 1993, cPLA$_2$ is phosphorylated and activated by MAP kinase, *Cell* **72:**269–278.

Lorant, D. E., Patel, K. D., McIntyre, T. M., McEver, R. P., Prescott, S. M., and Zimmerman, G. A., 1991, Coexpression of GMP-140 and PAF by endothelium stimulated by histamine or thrombin: A juxtacrine system for adhesion and activation of neutrophils, *J. Cell Biol.* **115:**223–234.

Lorant, D. E., Topham, M. K., Whatley, R. E., McEver, R. P., McIntyre, T. M., Prescott, S. M., and Zimmerman, G. A., 1993, Inflammatory roles of P-selectin, *J. Clin. Invest.* **92:**559–570.

MacDonald, J. I. S., and Sprecher, H., 1991, Phospholipid fatty acid remodeling in mammalian cells, *Biochim. Biophys. Acta* **1084:**105–121.

McIntyre, T. M., Zimmerman, G. A., Satoh, K., and Prescott, S. M., 1985, Cultured endothelial cells synthesize both platelet-activating factor and prostacyclin in response to histamine, bradykinin, and adenosine triphosphate, *J. Clin. Invest.* **76:**271–280.

McIntyre, T. M., Reinhold, S. L., Prescott, S. M., and Zimmerman, G. A., 1987, Protein kinase C activity appears to be required for the synthesis of platelet-activating factor and leukotriene B$_4$ by human neutrophils, *J. Biol. Chem.* **262:**15370–15376.

McIntyre, T., Patel, K. D., Smiley, P. L., Stafforini, D., Prescott, S. M., and Zimmerman, G. A., 1994, Oxidized phospholipids with PAF-like bioactivity, *J. Lipid Med.* **10:**37–40.

McIntyre, T. M., Patel, K. D., Zimmerman, G. A., and Prescott, S. M., 1995, Oxygen radical-mediated leukocyte adherence, in: *Physiology and Pathophysiology of Leukocyte Adhesion* (D. N. Granger and G. W. Schmid-Schonbein, eds.), pp. 261–277, Oxford University Press, London.

Maki, N., Hoffman, D. R., and Johnston, J. M., 1988, Platelet-activating factor acetylhydrolase activity in maternal, fetal, and newborn rabbit plasma during pregnancy and lactation *Proc. Natl. Acad. Sci. USA* **85**:728–732.

Marcheselli, V. L., Rossowska, M. J., Domingo, M.-T., Braquet, P., and Bazan, N. G., 1990, Distinct platelet-activating factor binding sites in synaptic endings and in intracellular membranes of rat cerebral cortex, *J. Biol. Chem.* **265**:9140–9145.

Miwa, M., Sugatani, J., Ikemura, T., Okamoto, Y., Ino, M., Saito, K., Suzuki, Y., and Matsumoto, M., 1992, Release of newly synthesized platelet-activating factor (PAF) from human polymorphonuclear leukocytes under in vivo conditions, *J. Immunol.* **148**:872–880.

Miyaura, S., Eguchi, H., and Johnston, J. M., 1992, Effect of a cigarette smoke extract on the metabolism of the proinflammatory autocoid, platelet-activating factor, *Circ. Res.* **70**:341–347.

Mojarad, M., Cox, C. P., and Said, S. I., 1985, Platelet-activating factor and acute lung injury, in: *The Pulmonary Circulation and Acute Lung Injury* (S. I. Said, ed.) pp. 375–386, Futura Publishing, Mount Kisco, NY.

Muller, S., and Nigam, S., 1992, Enhancement by staurosporine of platelet-activating factor formation in n-formyl peptide-challenged human neutrophils is mediated by intracellular platelet-activating factor binding sites, *Biochem. Biophys. Res. Commun.* **189**:771–776.

Mutoh, H., Bito, H., Minami, M., Nakamura, M., Honda, Z., Izumi, T., Nakata, R., Kurachi, Y., Terano, A., and Shimizu, T., 1993, Two different promoters direct expression of two distinct forms of mRNAs of human platelet-activating factor receptor, *FEBS* **322**:129–134.

Narahara, H., and Johnston, J. M., 1993, Smoking and preterm labor: Effect of a cigarette smoke extract on the secretion of platelet-activating factor-acetylhydrolase by human decidual macrophages, *Am. J. Obstet. Gynecol.* **169**:1321–1326.

Nieto, M. L., Velasco, S., and Sanchez-Crespo, M., 1988, Modulation of acetyl-CoA:1-alkyl-2-lyso-sn-glycero-3-phosphocholine (lyso-PAF) acetyltransferase in human polymorphonuclear leukocytes: The role of cyclic AMP-dependent and phospholipid sensitive, calcium-dependent protein kinases, *J. Biol. Chem.* **263**:4607–4611.

Nieto, M. L., Venable, M. E., Bauldry, S. A., Greene, D. G., Kennedy, M., Bass, D. A., and Wykle, R., 1991, Evidence that hydrolysis of ethanolamine plasmalogens triggers synthesis of platelet-activating factor via a transacylase reaction, *J. Biol. Chem.* **266**:18699–18706.

Ninio, E., Mencia-Huerta, J. M., Heymans, F., and Benveniste, J., 1982, Biosynthesis of platelet-activating factor: I. Evidence for an acetyltransferase activity in murine macrophages, *Biochim. Biophys. Acta* **710**:23–31.

Ninio, E., Mencia-Huerta, J. M., and Benveniste, J., 1983, Biosynthesis of platelet-activating factor: V. Enhancement of acetyltransferase activity in murine peritoneal cells by calcium ionophore A23187, *Biochim. Biophys. Acta* **751**:298–304.

O'Flaherty, J. T., Salzer, W. L., Cousart, S., McCall, C. E., Piantadosi, C., Surles, J. R., Hammett, M. J., and Wykle, R. L., 1983, Platelet-activating factor and analogues: Comparative studies with human neutrophils and rabbit platelets, *Res. Commun. Chem. Pathol. Pharmacol.* **39**:291–304.

O'Neill, C., Ryan, J. P., Collier, M., Saunders, D. M., Ammit, A. J., and Pike, I. L., 1992, Outcome of a trial of supplementing human IVF culture media with platelet-activating factor, *Reprod. Fertil. Dev.* **4**:109–112.

Patel, K. D., Zimmerman, G. A., Prescott, S. M., McEver, R. P., and McIntyre, T. M., 1991, Oxygen radicals induce human endothelial cells to express GMP-140 and bind neutrophils, *J. Cell Biol.* **112**:749–759.

Patel, K. D., Zimmerman, G. A., Prescott, S. M., and McIntyre, T. M., 1992, Novel leukocyte agonists are released by endothelial cells exposed to peroxide, *J. Biol. Chem.* **267**:15168–15175.

Pinckard, R. N., Farr, R. S., and Hanahan, D. J., 1979, Physicochemical and functional identity of rabbit platelet-activating factor (PAF) release in vivo during IgE anaphylaxis with PAF released in vitro from IgE-sensitized basophils, *J. Immunol.* **123**:1847–1855.

Qui, Z.-H., de Carvalho, M. S., and Leslie, C. C., 1993, Regulation of phospholipase A_2 activation by phosphorylation in mouse peritoneal macrophages, *J. Biol. Chem.* **268**:24506–24513.

Record, M., Ribbes, G., Terce, F., and Chap, H., 1989, Subcellular localization of phospholipids and enzymes involved in PAF-acether metabolism, *J. Cell Biochem.* **40**:353–359.

Reinhold, S. L., Zimmerman, G. A., Prescott, S. M., and McIntyre, T. M., 1989, Phospholipid remodeling in human neutrophils, *J. Biol. Chem.* **264**:21652–21659.

Rinder, H. M., Tracey, J. L., Rinder, C. S., Leitenberg, D., and Smith, B. R., 1994, Neutrophil but not monocyte activation inhibits P-selectin-mediated platelet adhesion, *Thromb. Haemost.* **71**:750–756.

Ryan, J. P., Spinks, N. R., O'Neill, C., and Wales, R. G., 1990, Implantation potential and fetal viability of mouse embryos cultured in media supplemented with platelet-activating factor, *J. Immunol.* **123**:1847–1855.

Satoh, K., Imaizumi, T., Kawamura, Y., Yoshida, H., Takamatsu, S., and Mizuno, S., 1988, Activity of platelet-activating factor (PAF) acetylhydrolase in plasma from patients with ischemic cerebrovascular disease, *Prostaglandins* **35**:685–698.

Satoh, K., Imaizumi, T.-A., Kawamura, Y., Yoshida, H., Takamatsu, S., and Takamatsu, M., 1989, Increased activity of the platelet-activating factor acetylhydrolase in plasma low density lipoprotein from patients with essential hypertension, *Prostaglandins* **37**:673–682.

Satoh, K., Imaizumi, T. A., Kawamura, Y., Yoshida, H., Hiramoto, M., Takamatsu, S., and Takamatsu, M., 1991, Platelet-activating factor (PAF) stimulates the production of PAF acetylhydrolase by the human hepatoma cell line, HepG2, *J. Clin. Invest.* **87**:476–481.

Satoh, K., Imaizumi, T.-a., Yoshida, H., and Takamatsu, S., 1993, Effect of 17β-estradiol on secretion of platelet-activating factor acetylhydrolase by HepG2 cells, *Metabolism* **42**:672–677.

Schilling, W. P., Cabello, O. A., and Rajan, L., 1992, Depletion of the inositol 1,4,5-trisphosphate-sensitive intracellular Ca^{2+} store in vascular endothelial cells activates the agonist-sensitive Ca^{2+}-influx pathway, *Biochem. J.* **284**:521–530.

Schneider, E., Haest, C. W. M., and Deuticke, B., 1986, Transbilayer reorientation of platelet-activating factor in the erythrocyte membrane, *FEBS* **198**:311–313.

Seyfried, C. E., Schweickart, V. L., Godiska, R., and Gray, P. W., 1992, The human platelet activating factor receptor gene (PAFR) contains no introns and maps to chromosome 1, *Genomics* **13**:832–834.

Sharp, J. D., White, D. L., Chiou, X. G., Goodson, T., Gamboa, G. C., McClure, D., Burgett, S., Hoskins, J., Skatrud, P. L., Sportsman, J. R., Becker, G. W., Kang, L. H., Roberts, E. F., and Kramer, R. M., 1991, Molecular cloning and expression of human Ca^{++}-sensitive cytosolic phospholipase A_2, *J. Biol. Chem.* **266**:14850–14853.

Shen, T. Y., Hwang, S.-B., Doebber, T. W., and Robbins, J. C., 1987, The chemical and biological properties of PAF agonists, antagonists, and biosynthetic inhibitors, in: *Platelet-Activating Factor and Related Lipid Mediators* (F. Snyder, ed.), pp. 153–190, Plenum Press, New York.

Shimizu, T., Honda, Z., Hakamura, M., Bito, H., and Izumi, T., 1992, Platelet-activating factor receptor and signal transduction, *Biochem. Pharmacol.* **44**:1001–1008.

Shukla, S. D., 1992, Platelet activating factor receptor and signal transduction mechanisms, *FASEB J.* **6**:2296–2301.

Sisson, J. H., Prescott, S. M., McIntyre, T. M., and Zimmerman, G. A., 1987, Production of platelet-activating factor by stimulated human polymorphonuclear leukocytes: Correlation of synthesis with release, functional events, and leukotriene B_4 metabolism, *J. Immunol.* **138**:3918–3926.

Smiley, P. L., Stremler, K. E., Prescott, S. M., Zimmerman, G. A., and McIntyre, T. M., 1991a, Oxidatively-fragmented phosphatidylcholines activate human neutrophils through the receptor for platelet-activating factor, *J. Biol. Chem.* **266**:11104–11110.

Snyder, F., 1987, Enzymatic pathways for platelet-activating factor, related alkyl glycerolipids, and their precursors, in: *Platelet-Activating Factor and Related Lipid Mediators* (F. Snyder, ed.), pp. 89–113, Plenum Press, New York.

Snyder, F., 1995a, Platelet-activating factor and its analogs: Metabolic pathways and related intracellular processes, *Biochim. Biophys. Acta* **1254**:231–249.

Snyder, F., 1995b, Platelet-activating factor: The biosynthetic and catabolic enzymes, *Biochem. J.* **305**:689–705.

Snyder, F., Lee, T.-c., and Blank, M. L., 1992, The role of transacylases in the metabolism of arachidonate and platelet-activating factor, *Prog. Lipid Res.* **31**:65–86.

Stafforini, D. M., McIntyre, T. M., Carter, M. E., and Prescott, S. M., 1987a, Human plasma platelet-

activating factor acetylhydrolase: Association with lipoprotein particles and role in the degradation of platelet-activating factor, *J. Biol. Chem.* **262**:4215–4222.

Stafforini, D. M., Prescott, S. M., and McIntyre, T. M., 1987b, Human plasma platelet-activating factor
acetylhydrolase: Purification and properties, *J. Biol. Chem.* **262**:4223–4230.

Stafforini, D. M., Prescott, S. M., and McIntyre, T. M., 1988, Platelet-activating factor acetylhydrolase in
human erythrocytes, *FASEB J.* **2**:A1375.

Stafforini, D. M., Elstad, M. E., McIntyre, T. M., Zimmerman, G. A., and Prescott, S. M., 1990a, Human
macrophages secrete platelet-activating factor acetylhydrolase, *J. Biol. Chem.* **265**:9682–9687.

Stafforini, D. M., McIntyre, T. M., and Prescott, S. M., 1990b, Platelet-activating factor acetylhydrolase
from human plasma, *Methods Enzymol.* **187**:344–357.

Stafforini, D. M., Prescott, S. M., Zimmerman, G. A., and McIntyre, T. M., 1991, Platelet-activating
factor acetylhydrolase activity in human tissues and blood cells, *Lipids* **26**:979–985.

Stafforini, D. M., Zimmerman, G. A., McIntyre, T. M., and Prescott, S. M., 1992, The platelet-activating
factor acetylhydrolase from human plasma prevents oxidative modification of low-density lipoprotein, *Trans. Assoc. Am. Physicians* **105**:44–63.

Stewart, A. G., Dubbin, P. N., Harris, T., and Dusting, G. J., 1989, Evidence for an intracellular action
of platelet-activating factor in bovine cultured aortic endothelial cells, *Br. J. Pharmacol.* **96**:
503–505.

Stewart, A. G., Dubbin, P. N., Harris, T., and Dusting, G. J., 1990, Platelet-activating factor may act as a
second messenger in the release of eicosanoids and superoxide anions from leukocytes and
endothelial cells, *Proc. Natl. Acad. Sci. USA* **87**:3215–3219.

Stoll, L. L., and Spector, A. A., 1989, Interaction of platelet-activating factor with endothelial and
vascular smooth muscle cells in coculture, *J. Cell. Physiol.* **139**:253–261.

Stoll, L. L., Denning, G. M., Kasner, N. A., and Hunninghake, G. W., 1994, Platelet-activating factor may
stimulate both receptor-dependent and receptor-independent increases in $[Ca^{2+}]$ in human
airway epithelial cells, *J. Biol. Chem.* **269**:4254–4259.

Stremler, K. E., Stafforini, D. M., Prescott, S. M., and McIntyre, T. M., 1991, Human plasma platelet-
activating factor acetylhydrolase: Oxidatively-fragmented phospholipids as substrates, *J. Biol.
Chem.* **266**:11095–11103.

Suga, K., Kawasaki, T., Blank, M. L., and Snyder, F., 1990, An arachidonoyl (polyenoic)-specific
phospholipase A2 activity regulates the synthesis of platelet-activating factor in granulocytic
HL-60 cells, *J. Biol. Chem.* **265**:12363–12371.

Sugiura, T., and Waku, K., 1987, Composition of alkyl ether-linked phospholipids in mammalian
tissues, in: *Platelet-Activating Factor and Related Lipid Mediators* (F. Snyder, ed.), pp. 55–85, Plenum
Press, New York.

Suguira, T., Fukuda, T., Masuzawa, Y., and Waku, K., 1990, Ether lysophospholipid-induced production
of platelet-activating factor in human polymorphonuclear leukocytes *Biochim. Biophys. Acta*
1047:223–232.

Suguira, T., Yamashita, A., Kudo, N., Kishimoto, S., and Waku, K., 1994, Platelet-activating factor and
related lipid molecules in invertebrates, *J. Lipid Med.* **10**:185–186.

Sun, X. M., and Hsueh, W., 1988, Bowel necrosis induced by tumor necrosis factor in rats is mediated by
platelet-activating factor, *J. Clin. Invest.* **81**:1328–1331.

Suttorp, N., and Habben, E., 1988, Effect of staphylococcal alpha-toxin on intracellular Ca^{2+} in
polymorphonuclear leukocytes, *Infect. Immun.* **56**:2228–2234.

Suzuki, M., Asako, H., Kubes, P., Jennings, S., Grisham, M. B., and Granger, D. N., 1991, Neutrophil-
derived oxidants promote leukocyte adherence in postcapillary venules, *Microvasc. Res.* **42**:
125–138.

Takano, T., Honda, Z.-i., Sakanaka, C., Izumi, T., Kameyama, K., Haga, K., Haga, T., Kurokawa, K., and
Shimizu, T., 1994, Role of cytoplasmic tail phosphorylation sites of platelet-activating factor
receptor in agonist-induced desensitization, *J. Biol. Chem.* **269**:22453–22458.

Tanaka, T., Minamino, H., Unezaki, S., Tsukatani, H., and Tokumura, A., 1993, Formation of platelet-
activating factor-like phospholipids by Fe^{2+}/ascorbate/EDTA-induced lipid peroxidation, *Biochim. Biophys. Acta* **1166**:264–274.

Tarbet, E. B., Stafforini, D. M., Elstad, M. R., Zimmerman, G. A., McIntyre, T. M., and Prescott, S. M., 1991, Liver cells secrete the plasma form of platelet-activating factor acetylhydrolase, *J. Biol. Chem.* **266:**16667–16673.

Tessner, T. G., and Wykle, R. L., 1987, Stimulated neutrophils produce an ethanolamine plasmalogen analog of platelet-activating factor, *J. Biol. Chem.* **262:**12660–12664.

Tetta, C., Montrucchio, G., Alloatti, G., Roffinello, C., Emanuelli, G., Benedetto, C., Camussi, G., and Massobrio, M., 1986, Platelet-activating factor contracts human myometrium in vitro, *Proc. Soc. Exp. Biol. Med.* **183:**376–381.

Tjoelker, L. W., Wilder, C., Eberhardt, C., Stafforini, D. M., Dietsch, G., Schimpf, B., Hooper, S., Trong, H. L., Cousens, L. S., Zimmerman, G. A., Yamada, Y., McIntyre, T. M., Prescott, S. M., and Gray, P. W., 1995, Cloning of human plasma platelet-activating factor acetylhydrolase: A lipase that inhibits PAF-mediated inflammation, *Nature* **374:**549–553.

Tokumura, A., Asai, T., Takauchi, K., Kamiyasu, K., Ogawa, T., and Tsukatani, H., 1988, Novel phospholipids with aliphatic dicarboxylic acid residues in a lipid extract from bovine brain, *Biochem. Biophys. Res. Commun.* **155:**863–869.

Tokumura, A., Takauchi, K., Asai, T., Kamiyasu, K., Ogawa, T., and Tsukatani, H., 1989, Novel molecular analogues of phosphatidylcholines in a lipid extract from bovine brain: 1-long-chain acyl-2-short-chain acyl-sn-glycero-3-phosphocholines, *J. Lipid Res.* **30:**219–224.

Triggiani, M., Hubbard, W. C., and Chilton, F. H., 1990, Synthesis of 1-acyl-2-acetyl-sn-glycero-3-phosphocholine by an enriched preparation of the human lung mast cell, *J. Immunol.* **144:**4773–4780.

Triggiani, M., Goldman, D. W., and Chilton, F. H., 1991a, Biological effects of 1-acyl-2-acetyl-sn-glycero-3-phosphocholine in the human neutrophil, *Biochim. Biophys. Acta* **1084:**41–47.

Triggiani, M. K., Schleimer, R. P., Warner, J. A., and Chilton, F. H., 1991b, Differential synthesis of 1-acyl-2-acetyl-*sn*-glycero-3-phosphocholine and platelet-activating factor by human inflammatory cells, *J. Immunol.* **147:**660–666.

Triggiani, M., Schleimer, R. P., Tomioka, K., Hubbard, W. C., and Chilton, F. H., 1992, Characterization of platelet-activating factor synthesized by normal and granulocyte-macrophage colony-stimulating factor-primed human eosinophils, *Immunology* **77:**500–504.

Tsuji, T., Nagata, K., Koike, J., Todoroki, N., and Irimura, T., 1994, Induction of superoxide anion production from monocytes and neutrophils by activated platelets through the P-selectin-sialyl Lewis X interaction, *J. Leukoc. Biol.* **56:**583–587.

Uemura, Y., Lee, T.-c., and Snyder, F., 1991, A coenzyme A-independent transacylase is linked to the formation of platelet-activating factor by generating the lyso-PAF intermediate in the remodeling pathway, *J. Biol. Chem.* **266:**8268–8272.

Vallari, D. S., Record, M., and Snyder, F., 1990, Conversion of alkylacetylglycerol to platelet-activating factor in HL-60 cells and subcellular localization of the mediator, *Arch. Biochem. Biophys.* **276:** 538–545.

Valone, F. H., 1987, Platelet-activating factor binding to specific cell membrane receptors, in: *Platelet-Activating Factor and Related Lipid Mediators* (F. Snyder, ed.), pp. 137–151, Plenum Press, New York.

Valone, F. H., Coles, C., Reinhold, V. R., and Goetzl, E. J., 1982, Specific binding of phospholipid platelet-activating factor by human platelets, *J. Immunol.* **129:**1637–1641.

Voelkel, N. F., Worthen, S., Reeves, J. T., and Henson, P. M., 1982, Nonimmunological production of leukotrienes induced by platelet activating factor, *Science* **218:**286–288.

Waite, M., 1985, Approaches to the study of mammalian cellular phospholipases, *J. Lipid Res.* **26:**1379–1388.

Whatley, R. E., Zimmerman, G. A., McIntyre, T. M., and Prescott, S. M., 1988, Endothelium from diverse vascular synthesizes platelet-activating factor, *Arteriosclerosis* **8:**321–331.

Whatley, R. E., Nelson, P., Zimmerman, G. A., Stevens, D. L., Parker, C. J., McIntyre, T. M., and Prescott, S. M., 1989, The regulation of platelet-activating factor production in endothelial cells, *J. Biol. Chem.* **264:**6325–6333.

Whatley, R. E., Zimmerman, G. A., McIntyre, T. M., and Prescott, S. M., 1990, Lipid metabolism and signal transduction in endothelial cells, *Prog. Lipid Res.* **29:**45–63.

Whatley, R. E., Clay, K. L., Chilton, F. H., Triggiani, M., Zimmerman, G. A., McIntyre, T. M., and Prescott, S. M., 1992, Relative amounts of 1-*O*-alkyl- and 1-acyl-2-acetyl-*sn*-glycero-3-phosphocholine in stimulated endothelial cells, *Prostaglandins* **43**:21–29.

Whatley, R. E., Satoh, K., Zimmerman, G. A., McIntyre, T. M., and Prescott, S. M., 1994, Proliferation-dependent changes in release of arachidonic acid from endothelial cells, *J. Clin. Invest.* **94**:1889–1900.

Whorton, A. R., Willis, C. E., Kent, R. S., and Young, S. L., 1984, The role of calcium in the regulation of prostacyclin synthesis by porcine aortic endothelial cells, *Lipids* **19**:17–24.

Wijkander, J., and Sundler, R., 1992a, Macrophage arachidonate-mobilizing phospholipase A$_2$: Role of Ca^{2+} for membrane binding but not for catalytic activity, *Biochem. Biophys. Res. Commun.* **184:** 118–124.

Wijkander, J., and Sundler, R., 1992b, Regulation of arachidonate-mobilizing phospholipase A$_2$ by phosphorylation via protein kinase C in macrophages, *FEBS* **311**:299–301.

Winkler, J. D., Sung, C.-M., Hubbard, W. C., and Chilton, F. H., 1993, Influence of arachidonic acid on indices of phospholipase A$_2$ activity in the human neutrophil, *Biochem. J.* **291**:825–831.

Witztum, J., and Steinberg, D., 1991, Role of oxidized low density lipoprotein in atherogenesis, *J. Clin. Invest.* **88**:1785–1792.

Wykle, R. L., Malone, B., and Snyder, F., 1980, Enzymatic synthesis of 1-alkyl-2-acetyl-*sn*-glycero-3-phosphocholine, a hypotensive and platelet aggregating lipid, *J. Biol. Chem.* **268**:10256–10260.

Wykle, R. L., Miller, C. H., Lewis, J. C., Schmitt, J. D., Smith, J. A., Surles, J. R., Piantadosi, C., and O'Flaherty, J. T., 1981, Stereospecific activity of 1-*O*-alkyl-2-*O*-acetyl-*sn*-glycero-3-phosphocholine and comparison of analogs in the degranulation of platelets and neutrophils, *Biochem. Biophys. Res. Commun.* **100**:1651–1658.

Yagi, K., 1984, Increased serum lipid peroxides initiate atherogenesis, *BioEssays* **1**:58–60.

Ye, R. D., Prossnitz, E. R., Zou, A., and Cochrane, C. G., 1991, Characterization of a human cDNA that encodes a functional receptor for platelet activating factor, *Biochem. Biophys. Res. Commun.* **180**:105–111.

Zimmerman, G. A., McIntyre, T. M., and Prescott, S. M., 1985a, Thrombin stimulates the adherence of neutrophils to human endothelial cells in vitro, *J. Clin. Invest.* **76**:2235–2246.

Zimmerman, G. A., McIntyre, T. M., and Prescott, S. M., 1985b. Production of platelet-activating factor by human vascular endothelial cells: Evidence for a requirement for specific agonists and modulation by prostacyclin, *Circulation* **76**:718–727.

Zimmerman, G. A., McIntyre, T. M., Mehra, M., and Prescott, S. M., 1990a, Endothelial cell-associated platelet-activating factor: A novel mechanism for signaling intercellular adhesion, *J. Cell Biol.* **110**:529–540.

Zimmerman, G. A., Whatley, R. E., McIntyre, T. M., Benson, D. M., and Prescott, S. M., 1990b, Endothelial cells for studies of platelet-activating factor and arachidonate metabolites, *Methods Enzymol.* **187**:520–535.

Zimmerman, G. A., Prescott, S. M., and McIntyre, T. M., 1992a, Platelet-activating factor: A fluid-phase and cell-mediator of inflammation, in: *Inflammation: Basic Principles and Clinical Correlates* (J. I. Gallin and I. M. Goldstein, eds.), pp. 149–176, Raven Press, New York.

Zimmerman, G. A., Prescott, S. M., and McIntyre, T. M., 1992b, Endothelial cell interactions with granulocytes: Tethering and signaling molecules, *Immunol. Today* **13**:93–100.

Zimmerman, G. A., Lorant, D. E., McIntyre, T. M., and Prescott, S. M., 1993, Juxtacrine intercellular signaling: Another way to do it, *Am. J. Respir. Cell Mol. Biol.* **9**:573–577.

Chapter 8

Lysophosphatidic Acid

Wouter H. Moolenaar and Kees Jalink

8.1. Introduction

Among the various phospholipid derived signaling molecules, relatively little attention has been paid to lysophospholipids as potential messengers. This is perhaps not too surprising in view of the detergentlike and lytic properties of many of these lipids. However, one notable exception is lysophosphatidic acid (LPA) or monoacyl-glycerol-3-phosphate, the smallest and structurally simplest of all (lyso)phospholipids. LPA evokes striking hormone- and growth factor-like effects when added exogenously to intact cells at submicromolar doses, far below the critical micelle concentration (Moolenaar, 1994; Jalink *et al.*, 1994a). LPA is rapidly produced and released by activated platelets and, as such, is a normal constituent of serum (but not plasma), where it is present in an albumin-bound form (Eichholtz *et al.*, 1993). Although its precise physiological and pathological functions *in vivo* remain to be explored, platelet-derived LPA has all the hallmarks of an important growth factor that may participate in wound healing and tissue remodeling. Thus, LPA not only stimulates the growth of fibroblasts (van Corven *et al.*, 1989; Tigyi *et al.*, 1994), vascular smooth muscle cells (Tokumura *et al.*, 1994), and keratinocytes (Piazza *et al.*, 1995), it also promotes cellular tension (Kolodney and Elson, 1993) and cell-surface fibronectin binding (Zhang *et al.*, 1994), which are important events in wound repair. Where tested, cellular responses to LPA show a striking overlap with those to whole serum; it therefore seems safe to conclude that LPA (albumin-bound) accounts for much of the biological activity of serum.

Although the concept of LPA acting on a cell-surface receptor was initially met with skepticism, it is now generally accepted that LPA activates its own G protein-coupled receptor(s) present in many diverse cell types. While the cDNA of the LPA receptor has not been cloned to date, much has been learned in recent

Wouter H. Moolenaar and *Kees Jalink* • Division of Cellular Biochemistry, The Netherlands Cancer Institute, 1066 CX Amsterdam, The Netherlands.

Handbook of Lipid Research, Volume 8: Lipid Second Messengers, edited by Robert M. Bell *et al.* Plenum Press, New York, 1996.

years about the LPA-induced signaling events. It appears that the LPA receptor couples to multiple independent effector pathways in a G protein-dependent manner. These pathways include not only "classic" second messenger pathways, such as stimulation of PLC and inhibition of adenylyl cyclase, but also novel routes, notably activation of the small GTP-binding proteins Ras and Rho. This chapter summarizes our current understanding of LPA as a bioactive phospholipid. For a more comprehensive introduction and background the reader is referred to recent reviews (Moolenaar, 1994, 1995; Jalink *et al.*, 1994a).

8.2. *Physicochemical Properties*

The physicochemical properties of LPA, such as its phase behavior and cation-binding properties, have not been studied in great detail. LPA has a free hydroxyl and phosphate moiety linked to the glycerol bacbone and hence is much more water-soluble than other phospholipids with similar acyl chain length (Jalink *et al.*, 1990). This hydrophilicity entails that LPA readily escapes detection in conventional lipid extraction procedures (Bjerve *et al.*, 1974). Unlike other lysophospholipids, LPA is not lytic to cells under physiological conditions. In nominally calcium-free solutions, however, LPA (100 nM) cay be lytic to cells (Jalink *et al.*, 1993b). Calcium ions tend to precipitate LPA from aqueous solutions, which can be prevented by addition of albumin (Jalink *et al.*, 1990). As a word of caution, the lytic potential of LPA (in calcium-free solutions) and its precipitation behavior (in the presence of calcium) could possibly cause nonspecific cellular responses under certain conditions and, furthermore, may disturb dose–response analyses. In other words, results obtained with high doses of LPA (or related lipids) and/or experiments carried out in calcium-free solutions must be interpreted with great care.

In this respect, we wish to emphasize a number of practical considerations. First, while LPA itself is quite stable, related lipids (e.g., other lysophospholipids, phosphatidic acid) are more prone to degradation and may contain significant amounts of breakdown products, including LPA (Jalink *et al.*, 1990). Second, in common with all natural lipids, LPA is subject to metabolic conversion resulting in generation of products that may be bioactive themselves, and in a decrease in effective concentration within minutes after its addition (van der Bend *et al.*, 1992b). Third, in diluted stock solutions, adsorption of (non-albumin-bound) LPA to plastic and glassware can be a serious source of variation. Thus, 75% of radioactivity of a 5 μM stock solution of [^{32}P]-LPA was found to stick tightly to the wall of an Eppendorf vial within 30 min. This radioactivity could not be dissolved with excess medium, but was completely recoverable with chloroform/methanol (K. Jalink, T. Hengeveld, and R. van der Bend, unpublished data). Therefore, diluted stock solutions of LPA and analogues are to be prepared immediately before use. Fourth, differences may arise from the method of administration of LPA to cells. For example, 1-stearoyl-LPA is only poorly soluble in water and yields a turbid suspension after sonication. In this physicochemical state the compound fails to induce calcium mobilization in responsive A431 cells (Jalink *et al.*, 1995).

Yet, in other cells, stearoyl-LPA elicits biological effects at nanomolar concentrations (e.g., Tigyi and Miledi, 1992). When we mixed stearoyl-LPA with fatty acid-free BSA (1:1 molar ratio), a clear stock solution was obtained that could readily induce calcium mobilization. In contrast, 1-oleoyl-LPA is equally potent in the presence and absence of albumin (Jalink *et al.*, 1995). Finally, when analyzing tissues or body fluids for LPA content, the relative hydrophilicity of the molecule may lead to severe underestimation as a result of selective loss in conventional lipid extraction procedures (e.g., Bjerve *et al.*, 1974).

8.3. Natural Occurrence

Besides serving a well-established precursor role in *de novo* lipid biosynthesis, LPA can be generated through the hydrolysis of preexisting phospholipids following cell activation. In particular, thrombin activated platelets rapidly produce and release LPA, apparently through phospholipase A_2 (PLA_2)-mediated deacylation of newly generated PA (Gerrard and Robinson, 1989). Distinct PA- specific PLA_2 activity has been identified in platelets (Ca^{2+}-dependent; Billah *et al.*, 1981) and in rat brain (Ca^{2+}-independent; Thompson and Clark, 1995). Preliminary evidence suggests that growth factor-stimulated fibroblasts can also produce LPA (Fukami and Takenawa, 1992). Obviously, many other cell systems remain to be examined for LPA production and release.

Being produced during blood clotting, LPA is present in freshly prepared mammalian serum (Tigyi and Miledi, 1992; Eichholtz *et al.*, 1993). Serum LPA levels are estimated to be in the 2–20 μM range (Eichholtz *et al.*, 1993); Tokumura *et al.*, 1994). LPA is not detectable in platelet-poor plasma, whole blood, or cerebrospinal fluid (Tigyi and Miledi, 1992). LPA behaves like long-chain fatty acids in that it binds to the primary high-affinity binding sites for long-chain fatty acids on serum albumin at a molar ratio of about 3:1 (Thumser *et al.*, 1994). In addition to LPA, serum albumin contains several other, as-yet-unidentified lipids (methanol-extractable) with LPA-like biological activity (Tigyi and Miledi, 1992). This would suggest that LPA belongs to a new family of phospholipid mediators showing overlapping biological activities and acting on distinct receptors.

As one would expect for a naturally occurring phospholipid, exogenous LPA is rapidly hydrolyzed by intact cells. Exogenous LPA is converted in part to monoacylglycerol and, to a much lesser extend, diacylglycerol and PA (van der Bend *et al.*, 1992b). Ecto-phosphohydrolase activity toward LPA has recently been identified in keratinocytes (Xie and Low, 1994) and an LPA-specific phospholipase has been purified from rat brain (Thompson and Clark, 1994).

8.4. Biological Actions

Table 8-1 lists the most important biological and cellular responses to LPA known to date. The responses are seen to be quite diverse, as they range from induction of cell proliferation to stimulation of neurite retraction (Tigyi and

Table 8-1. Biological Responses to LPA[a]

Platelet aggregation
Smooth muscle contraction
Cell proliferation
Epidermal cell proliferation in vivo (mouse skin)
Growth cone collapse and neurite retraction (neuronal cell lines)
Focal adhesion assembly and stress fiber formation (fibroblasts)
Isometric contraction (fibroblasts)
Cell-surface fibronectin binding (adherent cells)
Tumor cell invasion in vitro
Inhibition of gap-junctional communication (liver cells)
Neurotransmitter release (PC12 cells)
Chloride efflux (Xenopus oocytes)
Chemotaxis (Dictyostelium amoebae)

[a]For references see text and Moolenaar (1995).

Miledi, 1992; Jalink *et al.*, 1993a) and even slime mold chemotaxis (Jalink *et al.*, 1993b). Where studied, the effects of LPA are highly specific and fulfill the criteria of being receptor-mediated. But, as mentioned above, great caution must be exerted when interpreting results obtained with high doses of LPA.

The mitogenic action of LPA is almost completely inhibited by pertussis toxin (PTX), implicating the critical involvement of one or more G_i-type heterotrimeric G proteins (van Corven *et al.*, 1989). LPA also stimulates the growth of mouse preimplantation embryos in a PTX-sensitive manner (Kobayashi *et al.*, 1994). In keratinocytes, LPA-induced cell proliferation appears to be indirect as it involves the production of transforming growth factor alpha acting on endogenous EGF receptors in an autocrine manner (Piazza *et al.*, 1995). In confluent keratinocyte cultures, LPA stimulates terminal differentiation; its topical application to mouse skin induces thickening of the epidermis (Piazza *et al.*, 1995). Besides stimulating cell proliferation, LPA (but not other lipids) induces invasion of carcinoma and hepatoma cells into monolayers of mesothelial cells (Imamura *et al.*, 1993). The underlying mechanism is not clear, but may conceivably involve increased cell–matrix adhesion or enhanced cell motility, or both.

Many mitogens induce rapid and pronounced changes in the actin-based cytoskeleton, and so does LPA. In quiescent fibroblasts, LPA stimulates focal adhesion assembly followed by stress fiber formation (Ridley and Hall, 1992), which is expected to lead to increased cell–matrix adhesion. In murine neuroblastoma and PC 12 cells LPA triggers acute growth cone collapse followed by neurite retraction (Jalink *et al.*, 1993a; Tigyi and Miledi, 1992), a process driven by contractile force generation within the cortical actin cytoskeleton (Jalink *et al.*, 1994b). Thus, LPA suppresses and reverses the morphological differentiation of neuroblastoma cells (Jalink *et al.*, 1994a), an intriguing phenomenon the physiological (and perhaps even clinical) relevance of which remains to be elucidated. Finally, recent evidence indicates that LPA inhibits gap-junctional communication in rat liver cells as well as fibroblasts (Hii *et al.*, 1994; F. Postma and W. Moolenaar, unpublished observations), apparently via phosphorylation of the gap-junctional connexin-43 protein. The physiological relevance of this phenomenon remains to be addressed.

8.5. Receptor Identification and Receptor-Mediated Signal Transduction

It is now well established that LPA acts on its cognate G protein-coupled receptor(s), apparently present in many different cell types. Radioligand binding studies have revealed the presence of specific LPA binding sites in membranes from 3T3 cells and rat brain, with K_d values in the lower nanomolar range (Thompson *et al.*, 1994). Moreover, cross-linking experiments have identified a candidate high-affinity LPA receptor with an apparent molecular mass of 38–40 kDa in LPA-responsive cells and mammalian brain (van der Bend *et al.*, 1992a), but the putative receptor has not been purified or cloned.

Considerable progress has been made in identifying and dissecting the LPA-induced signal transduction events. To avoid undue overlap with recent reviews from our laboratory (Moolenaar, 1994, 1995; Jalink *et al.*, 1994a), only a brief summary is given here.

In mammalian fibroblasts, the LPA receptor couples to at least four G protein-mediated signaling pathways. These are (1) stimulation of phospholipase C (van Corven *et al.*, 1989; Jalink *et al.*, 1990) and phospholipase D (van der Bend *et al.*, 1992c; Ha *et al.*, 1994), (2) inhibition of adenylyl cyclase (van Corven *et al.*, 1989), (3) activation of the small GTP-binding protein Ras (van Corven *et al.*, 1993) and the downstream Raf/MAP kinase pathway, and (4) tyrosine phosphorylation of focal adhesion proteins in concert with remodeling of the actin cytoskeleton, which depends on the activity of the Ras-related Rho protein (Ridley and Hall, 1992; Jalink *et al.*, 1994b; Hordijk *et al.*, 1994; Seufferlein and Rozengurt, 1994). Of these pathways, inhibition of adenyly cyclase and stimulation of Ras signaling are sensitive to pertussis toxin. The Ras and Rho pathways are of particular importance because they account for long-term mitogenesis and rapid cytoskeletal events, respectively. But precisely how the LPA receptor couples to the Ras–MAP kinase pathway is still obscure. An as-yet-unidentified protein kinase has been implicated in the G_i-Ras activation route (van Corven *et al.*, 1993). Similarly, many details of the LPA receptor–Rho activation route are not understood, largely because the biochemical functions of Rho are unknown. For details of the LPA receptor signaling pathways, the interested reader is referred to the above-mentioned reviews.

8.6. Structure–Activity Relationship

From a structural point of view, the biological effects of LPA are highly specific. Maximal LPA activity is observed at long acyl chain lengths (C_{16} to C_{20})

and activity decreases with decreasing chain length. In fact, the short-chain species lauroyl- and decanoyl-LPA are almost devoid of activity (van Corven *et al.*, 1992; Jalink *et al.*, 1995). Human A431 carcinoma cells have recently been used for structure–activity analysis because of their high sensitivity to LPA in terms of Ca^{2+} mobilization (EC_{50} for oleoyl-LPA: 0.2 nM; Jalink *et al.*, 1995). The free phosphate group appears to be very critical: replacement of this moiety by either of three different phosphonates results in dramatic loss of activity. The methyl- and ethyl-esters of LPA are also inactive (Jalink *et al.*, 1995).

Ether-linked LPA can activate the LPA receptor, albeit with less potency than the corresponding ester-linked LPA, at least in A431 cells (Jalink *et al.*, 1995). In platelet aggregation assays, on the other hand, ether-linked LPA is the more potent species (Simon *et al.*, 1982). It might well be that platelets and A431 cells express pharmacologically distinct receptors for LPA. It appears that the *sn*-1 as well as the *sn*-3 enantiomers of ether-linked LPA have about equal activity (Jalink *et al.*, 1995; Simon *et al.*, 1982), with both isomers showing cross-desensitization. This apparent lack of stereospecificity is somewhat unexpected, since stereo-specificity is often a critical feature of physiological receptor ligands. It will be important to explore whether ether-linked LPA can occur extracellularly, either in physiological events or under pathological conditions.

Thus far, structure–activity studies have identified two promising LPA receptor antagonists (see Sugiura *et al.*, 1994). The best studied LPA antagonist is suramin; this polyanionic compound abolishes both early and late responses to LPA in a reversible manner (van Corven *et al.*, 1992; Jalink *et al.*, 1993a), apparently by inhibiting LPA–receptor binding (van der Bend *et al.*, 1992a). Suramin blocks LPA action at lower doses than those required for blocking the action of peptide growth factors (van Corven *et al.*, 1992).

In summary, many features of the LPA structure appear to be important for biological activity, but it is the phosphate group that matters most. Based on the available structure–activity data, we have proposed that the acyl chain may serve to anchor the LPA molecule in a hydrophobic cavity of the receptor in such a way that the glycerol-linked phosphate group is optimally oriented for direct interaction with positively charged residues in an extracellular region of the receptor (Jalink *et al.*, 1995). Such specific binding of the phosphate moiety would then trigger a conformational change that activates the receptor. Obviously, a direct test of this tentative model awaits cloning of the LPA receptor cDNA. It will then be possible to examine by site-directed mutagenesis how the LPA receptor interacts with its ligand.

8.7. Concluding Remarks

In conclusion, LPA has recently emerged as a unique intercellular phospho-lipid messenger that is secreted by certain activated cells, notably platelets, to influence target cells via activation of one or more specific G protein-coupled receptors. While much has already been learned about LPA receptor signaling, much work remains to be done. At the molecular level, cloning and characterization of the LPA receptor cDNA is one major challenge for future experiments. Other challenges relate to elucidating the *in vivo* role(s) of LPA and to further unraveling the receptor-mediated signaling events, particularly the biochemical cascades that underlie LPA-induced activation of the small GTP-binding proteins Ras and Rho. Furthermore, there is the question of whether LPA represents a family of related lipid mediators with different potencies and distinct biological activities. Hopefully, answers to these and other questions will come in the not too distant future.

References

Billah, M. M., Lapetina, E. G., and Cuatrecasas, P., 1981, Phospholipase A_2 activity specific for phosphatidic acid, *J. Biol. Chem.* **256:**5399–5403.

Bjerve, K. S., Daae, L. N., and Bremer, J., 1974, The selective loss of lysophospholipids in some commonly used lipid-extraction procedures, *Anal. Biochem.* **58:**238–245.

Eichholtz, T., Jalink, K., Fahrenfort, I., and Moolenaar, W. H., 1993, The bioactive phospholipid LPA is released from activated platelets, *Biochem. J.* **291:**677–680.

Fukami, K., and Takenawa, T., 1992, PA that accumulates in PDGF-stimulated 3T3 cells is a potential mitogenic signal, *J. Biol. Chem.* **267:**10988–10993.

Gerrard, J. M., and Robinson, P., 1989, Species of LPA produced in thrombin-activated platelets, *Biochim. Biophys. Acta* **1001:**282–285.

Ha, K.-S., Yeo, E.-J., and Exton, J. H., 1994, LPA activation of phosphatidylcholine-hydrolysing phospholipase D and actin polymerization by a pertussis toxin-sensitive mechanism, *Biochem. J.* **302:** 55–59.

Hii, C. S. T., Oh, S. Y., Schmidt, S. A., Clark, K. J., and Murray, A. W., 1994, LPA inhibits gap-junctional communication and stimulates phosphorylation of connexin-43 in WB cells, *Biochem. J.* **303:** 475–479.

Hordijk, P. L., Verlaan, I., van Corven, E. J., and Moolenaar, W. H., 1994, Protein tyrosine phosphorylation induced by lysophosphatidic acid in Rat-1 fibroblasts. Evidence that phosphorylation of MAP kinase is mediated by the G_i- p21^{ras} pathway, *J. Biol. Chem.* **269:**645–651.

Imamura, F., Horai, T., Mukai, M., Shinkai, K., Sawada, M., and Akedo, H., 1993, Induction of in vitro tumor cell invasion of cellular monolayers by LPA or phospholipase D, *Biochem. Biophys. Res. Commun.* **193:**497–503.

Jalink, K., van Corven, E. J., and Moolenaar, W. H., 1990, LPA, but not PA, is a potent calcium-mobilizing stimulus for fibroblasts, *J. Biol. Chem.* **265:**12232–12239.

Jalink, K., Eichholtz, T., Postma, F. R., van Corven, E. J., and Moolenaar, W. H., 1993a, LPA induces neuronal shape changes via a novel, receptor-mediated signaling pathway, *Cell Growth Differ.* **4:**247–255.

Jalink, K., Moolenaar, W. H., and van Duijn, B., 1993b, LPA is a chemoattractant for Dictyostelium amoebae, *Proc. Natl. Acad. Sci. USA* **90:**1857–1861.

Jalink, K., Hordijk, P. L., and Moolenaar, W. H., 1994a, Growth factor-like effects of lysophosphatidic acid, a novel lipid mediator, *Biochim. Biophys. Acta* **1198:**185–196.

Jalink, K., van Corven, E. J., Hengeveld, T., Morii, N., Narumiya, S., and Moolenaar, W. H., 1994b, Inhibition of LPA- and thrombin-induced neurite retraction and neural cell rounding by ADP ribosylation of the small GTP-binding protein Rho, *J. Cell Biol.* **126:**801–810.

Jalink, K., Hengeveld, T., Mulder, S., Postma, F. R., Simon, M.-F., Chap, H., van der Marel, G. A., van Boom, J. H., van Blitterswijk, W. J., and Moolenaar, W. H., 1995, LPA-induced calcium mobilization in human A431 cells: Structure–activity relationship, *Biochem. J.* **307:**609–616.

Kobayashi, T., Yamano, S., Murayama, S., Ishikawa, H., and Tokumura, A., 1994, Effect of LPA on the preimplantation development of mouse embryos, *FEBS Lett.* **351**:38–40.

Kolodney, M. S., and Elson, E. L., 1993, Correlation of myosin light chain phosphorylation with isometric contraction of fibroblasts, *J. Biol. Chem.* **268**:23850–23855.

Moolenaar, W. H., 1994, LPA: A novel lipid mediator with diverse biological actions, *Trends Cell Biol.* **4**:213–219.

Moolenaar, W. H., 1995, Lysophosphatidic acid signalling, *Curr. Opin. Cell Biol.* **7**:203–210.

Piazza, G. A., Ritter, J. L., and Baracka, C. A., 1995, Effects of LPA on keratinocyte proliferation, *Exp. Cell Res.* **216**:51–64.

Ridley, A. J., and Hall, A., 1992, The small GTP-binding protein Rho regulates the assembly of focal adhesions and actin stress fibers in response to growth factors, *Cell* **70**:389–399.

Seufferlein, T., and Rozengurt, E., 1994, Lysophosphatidic acid stimulates tyrosine phosphorylation of focal adhesion kinase, paxillin, and p130, *J. Biol. Chem.* **269**:9345–9351.

Simon, M. F., Chap, H., and Douste-Blazy, L., 1982, Ether-linked LPAs as activators of platelet aggregation, *Biochem. Biophys. Res. Commun.* **108**:1743–1750.

Sugiura, T., Tokumura, A., Gregory, L., Nouchi, T., Weintraub, S. T., and Hanahan, D. J., 1994, Action of phosphoric acid derivatives on platelet aggregation; structure activity relationship, *Arch. Biochem. Biophys.* **311**:358–368.

Thompson, F. J., and Clark, M. A., 1994, Purification of an LPA-hydrolysing lysophospholipase from rat brain, *Biochem. J.* **300**:457–461.

Thompson, F. J., and Clark, M. A., 1995, Characterization of PA-specific phospholipase A_2 activity in brain, *Biochem. J.* **306**:305–309.

Thompson, F. J., Perkins, L., Ahern, D., and Clark, M., 1994, Identification and characterization of a lysophosphatidic acid receptor, *Mol. Pharmacol.* **45**:718–728.

Thumser, A. E. A., Voysey, J. E., and Wilton, D. C., 1994, The binding of lysophospholipids to rat liver fatty acid-binding protein and albumin, *Biochem. J.* **301**:801–806.

Tigyi, G., and Miledi, R., 1992, Lysophosphatidates bound to serum albumin activate membrane currents in Xenopus oocytes and neurite retraction in PC12 cells, *J. Biol. Chem.* **267**:21360–21367.

Tigyi, G., Dyer, D. L., and Miledi, R., 1994, Lysophosphatidic acid possesses dual action in cell proliferation, *Proc. Natl. Acad. Sci. USA* **91**:1908–1912.

Tokumura, A., Iimori, M., Nishioka, Y., Kitahara, M., Sakashita, M., and Tanaka, S., 1994, LPAs induce proliferation of cultured vascular smooth muscle cells from rat aorta, *Am. J. Physiol.* **267**:C204–C210.

van Corven, E. J., Groenink, A., Jalink, K., Eichholtz, T., and Moolenaar, W. H., 1989, Lyso-phosphatidate-induced cell proliferation: Identification and dissection of signaling pathways mediated by G proteins, *Cell* **59**:45–54.

van Corven, E. J., van Rijswijk, A., Jalink, K., van der Bend, R. L., van Blitterswijk, W. J., and Moolenaar, W. H., 1992, Mitogenic action of LPA and PA on fibroblasts. Dependence on acyl-chain length and inhibition by suramin, *Biochem. J.* **281**:163–169.

van Corven, E. J., Hordijk, P. L., Medema, R. H., Bos, J. L., and Moolenaar, W. H., 1993, Pertussis toxin sensitive activation of Ras by G protein-coupled receptor agonists, *Proc. Natl. Acad. Sci. USA* **90**:1257–1261.

van der Bend, R. L., Brunner, J., Jalink, K., van Corven, E. J., Moolenaar, W. H., and van Blitterswijk, W. J., 1992a, Identification of a putative membrane receptor for the bioactive phospholipid lysophosphatidic acid, *EMBO J.* **11**:2495–2501.

van der Bend, R. L., de Widt, J., van Corven, E. J., Moolenaar, W. H., and van Blitterswijk, W. J., 1992b, Metabolic conversion of the biologically active phospholipid, lysophosphatidic acid, in fibroblasts, *Biochim. Biophys. Acta* **1125**:110–112.

van der Bend, R. L., de Widt, J., van Corven, E. J., Moolenaar, W. H., and van Blitterswijk, W. J., 1992c, The biologically active phospholipid, lysophosphatidic acid, induces phosphatidylcholine break-down in fibroblasts via activation of phospholipase D, *Biochem. J.* **285**:235–240.

Xie, M., and Low, M. G., 1994, An ecto-phosphohydrolase activity towards (L)PA in keratinocytes, *Arch. Biochem. Biophys.* **312**:254–259.

Zhang, Q., Checovich, W. J., Peters, D. M., Albrecht, R. M., and Mosher, D. F., 1994, Modulation of cell surface fibronectin assembly sites by lysophosphatidic acid, *J. Cell Biol.* **127**:1447–1459.

Chapter 9

Prostaglandins and Related Compounds
Lipid Messengers with Many Actions

Elizabeth A. Meade, David A. Jones, Guy A. Zimmerman, Thomas M. McIntyre, and Stephen M. Prescott

9.1. Introduction

Prostaglandins are oxidized derivatives of arachidonic acid that are produced by many tissues in response to specific stimuli. These biologically active lipids are not stored within cells but rather synthesized and released on stimulation to function in an autocrine or paracrine fashion. Prostaglandins serve diverse physiological and pathophysiological functions, affecting all major systems of the body, and mediating such physiological processes as vascular homeostasis, renal water absorption, ovulation, and parturition. In addition, prostaglandins have been implicated as pathological mediators in thrombosis, asthma, and inflammatory disorders such as arthritis.

Cells synthesize prostaglandins in response to a variety of hormones, autacoids, and growth factors (Smith and Borgeat, 1985; Smith and Marnett, 1991). These agonists bind to cell surface receptors to activate phospholipases, which results in the release of free arachidonic acid from phospholipids to initiate the process. The free arachidonic acid can be utilized as a substrate by cyclooxygenase, epoxygenase (a cytochrome P450 enzyme), or one of three lipoxygenase

Elizabeth A. Meade • The Eccles Program in Human Molecular Biology and Genetics and the Nora Eccles Harrison Cardiovascular Research and Training Institute, University of Utah, Salt Lake City, Utah 84112. *David A. Jones* • The Eccles Program in Human Molecular Biology and Genetics and the Nora Eccles Harrison Cardiovascular Research and Training Institute, University of Utah, Salt Lake City, Utah 84112. *Guy A. Zimmerman* • Department of Medicine, University of Utah School of Medicine, Salt Lake City, Utah 84112. *Thomas M. McIntyre* • Departments of Medicine and Biochemistry, University of Utah School of Medicine, Salt Lake City, Utah 84112. *Stephen M. Prescott* • Departments of Medicine and Biochemistry, the Nora Eccles Harrison Cardiovascular Research and Training Institute, and the Eccles Institute of Human Genetics, University of Utah School of Medicine, Salt Lake City, Utah 84112.

Handbook of Lipid Research, Volume 8: Lipid Second Messengers, edited by Robert M. Bell *et al.* Plenum Press, New York, 1996.

pathways. The products of these pathways are collectively known as eicosanoids, reflecting their origin from eicosatetraenoic acid—the proper name for arachidonic acid. Those compounds derived from the cyclooxygenase pathway are known as prostaglandins or, more broadly, prostanoids. The fate of free arachidonic acid is determined by the relative activity level of each of these pathways within the cell. This review focuses on the cyclooxygenase pathway.

The key enzyme committing arachidonic acid to prostaglandin biosynthesis is cyclooxygenase, which is more correctly called prostaglandin endoperoxide synthase or prostaglandin H synthase (PHS). PHS is a bifunctional enzyme that catalyzes the conversion of arachidonic acid to prostaglandin H_2. The first reaction is a bisoxygenase reaction whereby two molecules of oxygen are incorporated into arachidonic acid, resulting in formation of the unstable intermediate, prostaglandin G_2 (PGG_2). The second reaction catalyzed by PHS is to reduce PGG_2 to form PGH_2. This product also is an unstable intermediate and serves as the precursor for the terminal, biologically active prostaglandins. Specific synthases catalyze the conversion of PGH_2 to prostaglandins D, E, F, I and thromboxane A_2. Tissues often express only one of these enzymes and, therefore, synthesize a specific terminal prostaglandin. Vascular endothelial cells, for example, contain prostacyclin synthase and synthesize mainly prostacyclin (PGI_2). In contrast, platelets contain thromboxane synthase and convert PGH_2 to thromboxane A_2. Vascular tone and thrombogenesis are dependent on the proper balance of these two prostanoids. Prostacyclin possesses anti-inflammatory and antithrombogenic properties, while thromboxane A_2 is proinflammatory and prothrombogenic. Imbalance of these two eicosanoids can contribute to the risk of cardiovascular events (Belanger et al., 1988; Oates et al., 1988). The interplay of opposing prostanoids, in this case prostacyclin and thromboxane, to provide fine-tuning of an organ system is a theme found frequently in the actions of prostaglandins.

9.2. Regulation of Prostaglandin Formation

Prostaglandin biosynthesis often occurs in two phases: a rapid or acute phase followed by a sustained or chronic phase. The first phase involves rapid release of arachidonate resulting in a burst of prostaglandin synthesis. The sustained phase differs in that it can persist for hours and prostaglandin production coincides with increases in PHS enzyme levels. Initiation of the two phases depends on the agonist chosen for stimulation. Histamine, bradykinin, and calcium ionophore are examples of agents that elicit acute biosynthesis. In contrast, cytokines, growth factors, and phorbol esters activate both phases.

9.2.1. Availability of Substrate: Release of Arachidonic Acid

Arachidonic acid is not found as a free fatty acid within cells but is esterified in complex lipids such as phospholipids and triglycerides. However, PHS cannot act on the esterified form and an essential step is the release of arachidonic acid.

Eicosanoids are not produced constitutively, but only in response to an activation signal—when an appropriate agonist binds to its receptor, the target cell initiates prostanoid synthesis. The first step is the release of arachidonic acid and this requires a rise in free intracellular calcium. Thus, the agonists that induce eicosanoid production also stimulate phosphatidylinositol turnover, which results in a rise in calcium—mediated by inositol trisphosphate—and activation of protein kinase C by diacylglycerol. Both of these signals are needed for optimal release of arachidonic acid. Several pathways for release have been proposed, but the evidence supports a predominant role for phospholipase A_2 in most cells and tissues. These enzymes catalyze the hydrolysis of fatty acids at the sn-2 position of phospholipids, which is where arachidonic acid is found. Metabolic labeling studies have shown that the liberated arachidonic acid is derived from phospholipids, predominantly phosphatidylcholine or phosphatidylethanolamine—depending on the cell—and that the release is quite selective for this fatty acid even though others are in greater abundance. Several intracellular PLA_2's now have been characterized in detail and two, in particular, have been proposed to be responsible for arachidonic acid release. The enzyme that best fits the cellular studies is a PLA_2 of about 85 kDa found in many cells (Clark *et al.*, 1991; Leslie *et al.*, 1988; Sharp *et al.*, 1991). It requires phosphorylation to be active and the pathway involves PKC, which activates MAP kinase, which in turn phosphorylates the PLA_2 (Lin *et al.*, 1993). Moreover, in response to a rise in calcium it translocates from its resting location in the cytoplasm to a membrane location, which places it in an appropriate position to act on its substrates (Channon and Leslie, 1990). Finally, this enzyme is selective for arachidonic acid (Clark *et al.*, 1991; Leslie *et al.*, 1988). The alternate possible PLA_2 is a much smaller form that has been shown to be secreted from a variety of cells as they are stimulated (Crowl *et al.*, 1991; Hara *et al.*, 1989). The enzyme then binds to specific sites on the surface of the cell. This process corresponds well with the production of eicosanoids and the secreted enzyme is commonly found in exudates at sites of inflammation (Bomalaski *et al.*, 1991). Thus, a role for it has been sought. Experiments with antisense constructs and, conversely, overexpression of cDNAs have yielded conflicting results (Barbour and Dennis, 1993; Lin *et al.*, 1992), but it seems likely that the 85-kDa cytoplasmic form is the main catalyst for arachidonic acid release. The role of the smaller enzyme will undoubtedly be clarified in the intense studies under way in many laboratories.

9.2.2. Structure and Function of Prostaglandin H Synthase

The essential biochemical role of PHS and its prominence as a therapeutic target have made it the most well-characterized enzyme in the prostaglandin synthetic pathway. Prostaglandin synthase was originally purified from sheep vesicular gland (van der Ouderaa *et al.*, 1977). This organ was a rich source of the enzyme and proved useful in the initial characterization (Hemler *et al.*, 1978; Roth *et al.*, 1977; Egan *et al.*, 1976; Smith and Lands, 1972). Subsequent cDNA cloning of the ovine, murine, and human forms of PHS and expression of these proteins in

COS cells has yielded further information about the structure of PHS. The PHS protein is roughly 72 kDa in size (dependent on species) and exists functionally as a homodimer (DeWitt *et al.*, 1990; Picot *et al.*, 1994) associated with the endoplasmic reticulum and possibly with the nuclear envelope. Glycosylation at three asparagine residues is necessary for expression of synthase activity (Otto *et al.*, 1993). PHS has been crystallized and its structure determined (Picot *et al.*, 1994). The polypeptide chains of PHS have a highly helical content and very little organized beta sheet. PHS contains three independent folding units: an EGF domain, a membrane binding motif, and an enzymatic domain. As suggested by previous studies (Marshall and Kulmacz, 1988; Shimokawa and Smith, 1992), the active sites for cyclooxygenase and peroxidase activity are adjacent to one another, yet distinct. The cyclooxygenase active site is created by a long, hydrophobic channel that is located near the site where the PHS protein interacts with the membrane. The active site exists at the apex of the hydrophobic channel and the serine that is acetylated by aspirin lies just below the active site in this channel in a position where the acetate group would block access of arachidonic acid to the active site. Flurbiprofen, another nonsteroidal anti-inflammatory drug (NSAID), can also bind in this hydrophobic pocket to sterically hinder the interaction of arachidonic acid with the cyclooxygenase active site. The crystal structure of PHS suggests that it interacts with only one layer of the lipid bilayer, rather than passing entirely through the membrane as had been suggested by other studies (DeWitt and Smith, 1988; Merlie *et al.*, 1988). Based on this type of membrane association of PHS, arachidonic acid that had been liberated from membrane phospholipids would probably remain associated with the lipid bilayer as it interacted with PHS. The EGF domain of PHS is similar to that found in many extracellular proteins, blood clotting factors, and cell-surface proteins, and is located spatially adjacent to the membrane binding motif. This is one of the few regions of PHS to contain small regions of beta sheet structure. The function of this region is unknown, but it may serve as a structural building block to initiate or maintain protein–protein interactions.

9.2.3. Identification of a Second Isoform of Prostaglandin H Synthase

Over the past few years, investigators have advanced the understanding of the transcriptional and translational regulation of PHS expression. This work employed reagents specific for the original form of PHS. In some studies the observations were easily interpreted in the context of a single PHS. The results of other studies, however, were confusing and raised skepticism that a single PHS could account for all prostaglandin biosynthesis. Wong and Richards examined the distribution of PHS in rat ovarian tissue using two polyclonal antibodies prepared against ovine seminal vesicle PHS. Although both antibodies recognized the ovine PHS, they appeared to differentially detect PHS in the rat ovarian tissue. One of the antibodies recognized a protein 69 kDa in size that ran as a singlet on SDS-PAGE and had a wide tissue distribution. This protein was produced constitutively at low levels and was not induced by luteinizing hormone. The presence of this protein was associated with a low level of indomethacin-sensitive prostaglandin production. The second antibody recognized a protein of around 72 kDa that ran as a doublet. This protein was undetectable in unstimulated cells, but high-level expression was induced following treatment with luteinizing hormone and human chorionic gonadotropin (hCG). Stimulation of prostaglandin formation by hCG correlated with increases in expression of this protein (Wong and Richards, 1991). These experiments provided evidence that the ovary might contain two immunologically distinct forms of PHS which were present in a cell-specific manner and selectively regulated by hormones.

In experiments using cultured tracheal epithelial cells, Rosen *et al.* (1989) detected increases in prostaglandin formation that could not be accounted for by the changes in PHS message or protein levels that they measured. By reprobing their Northern blots at low stringency, they detected a novel mRNA that hybridized with their PHS probe at low, but not high, stringency. They hypothesized that they were detecting a novel form of PHS, and that the alterations in the expression of this PHS could account for the changes in prostanoid formation that they found in cultured tracheal cells.

The existence of a second PHS was confirmed by the discovery of cDNA clones encoding proteins similar to prostaglandin endoperoxide synthase. Two groups reported the isolation of cDNAs with predicted proteins homologous to PHS. Interestingly, both groups identified this second PHS as a growth-related transcript. Xie *et al.* (1991) reported a PHS homologue from chicken fibroblasts transformed with the Rous sarcoma virus. Nonproliferating cells contained this homologue at low levels as a transcript containing an unspliced intron. src-Transformed cells expressed a fully processed transcript at high abundance. At the same time, Kujubu *et al.* (1991) isolated a murine cDNA that was rapidly induced in quiescent 3T3 cells stimulated with TPA. The cDNA was designated TIS10 (TPA-inducible sequence 10) and the protein it encoded displayed cyclooxygenase activity following expression in COS cells (Fletcher *et al.*, 1992).

The protein sequences predicted by these cDNAs share 65% identity with the previously described PHS. Analysis of the protein sequence for the novel PHS revealed that important functional residues are conserved between PHS-1 and PHS-2 including the region with homology to EGF, histidines thought to be involved in the binding of heme, the aspirin binding site, the putative active site, and glycosylation sites (Kujubu *et al.*, 1991). The major differences between the two cDNAs are near the amino and carboxyl termini. The amino terminus of PHS-1 contains a hydrophobic sequence 17 amino acids in length which is absent in PHS-2. PHS-2, on the other hand, contains an 18-amino-acid insert near the carboxyl terminus that is not found in PHS-1. The role of the 18-amino-acid insert is unclear, but it has allowed the generation of antibodies specific for PHS-2 (Kujubu *et al.*, 1993; DeWitt and Meade, 1993; Ristimaki *et al.*, 1994). Human cDNAs for PHS-2 have also been isolated and contain features similar to the murine and chicken forms (Hla and Neilson, 1992; Jones *et al.*, 1993). At the level of protein, Sirois and Richards (1992) partially purified the novel form of PHS from hCG-stimulated rat granulosa cells. The partially purified protein exhibited peroxidase activity and sequencing of the bands immunoreactive with an antibody to the ovine PHS revealed N-terminal sequence with high homology of TIS10.

Like PHS-1, PHS-2 is a membrane-associated protein. Studies to date suggest that PHS-2 is associated with the endoplasmic reticulum and perinuclear membrane (Regier *et al.*, 1993; Kujubu *et al.*, 1993; Otto *et al.*, 1993; Otto and Smith, 1994). There is some evidence that PHS-2 may initially be localized in the nucleus and only later translocate to the perinuclear membrane and endoplasmic reticulum (Rimarachin *et al.*, 1994). PHS-2 is N-glycosylated at three sites analogous to those in PHS-1 and can be glycosylated at a fourth site about 50% of the time, making the native protein 72 and 74 kDa (Otto *et al.*, 1993) in size. N-glycosylation of the fourth site in PHS-2 is not necessary for maintenance of cyclooxygenase or peroxidase activity and the functional significance of the two differently glycosylated forms of PHS-2 is not known (Otto *et al.*, 1993).

9.2.4. *Enzymatic Properties of Prostaglandin H Synthases*

Prostaglandin endoperoxide synthases are bifunctional enzymes that display bisoxygenase and hydroperoxidase activities. The first activity of PHS, the bisoxygenase activity incorporates two molecules of oxygen into free arachidonic acid at carbon positions 9 and 11, and 15 to yield the unstable endoperoxide PGG_2. The 15-hydroperoxyl group of PGG_2 then undergoes a two-electron reduction, resulting in the formation of PGH_2, the precursor for terminal prostaglandins.

Traditionally, arachidonic acid has been considered to be the only substrate of importance in prostaglandin formation. Other substrates, such as the fish oils, can theoretically be metabolized by PHS-1 to form prostaglandins of the trienoic series, but actual formation of these compounds by PHS-1 is very limited. PHS-2, however, utilizes fatty acids other than arachidonic acid more readily than does PHS-1. It is able to oxidize eicosapentaenoic acid (EPA, 20:5) at about 50% of the efficiency of arachidonic acid, and actually appears to prefer eicosatrienoic acid (ETA, 20:3) to either EPA or AA (Meade, 1992). Although no studies have been reported to date, it is possible that PHS-2 could also utilize linoleic acid as a substrate. Since linoleic acid is the major fatty acid in a normal Western diet, the possibility that it could be utilized directly to form biologically active products, rather than forming products only following its conversion to arachidonic acid, is intriguing.

PHS undergoes an unusual process termed "suicide" inactivation during catalysis. The enzyme catalyzes its own destruction during the conversion of arachidonate to PGH_2. Approximately one in every 4000–5000 catalytic events of PHS results in the inactivation of the enzyme (Egan *et al.*, 1976; Ple and Marnett, 1989). Suicide inactivation is characteristic of several eicosanoid biosynthetic enzymes including thromboxane synthase (Jones and Fitzpatrick, 1990) and prostacyclin synthase (DeWitt and Smith, 1983). Incubation of PHS with arachidonic acid results in loss of cyclooxygenase but not peroxidase activity, concurrent with

PGH_2 production. The chemical basis for inactivation is unknown, but may involve covalent modification of the enzyme by a reaction intermediate. Suicide inactivation of PHS provides an additional regulatory mechanism for limiting prostaglandin formation. The susceptibility of PHS-2 to autoinactivation has not been determined, although based on the sequence similarity between the two isoforms, it is expected to autoinactivate. The balance between the formation of two synthases and subsequent inactivation adds to the complexity of control of prostaglandin formation.

9.2.5. Regulation of Expression of Prostaglandin H Synthases

The regulation of PHS-1 has been extensively studied, but the discovery of PHS-2 raised questions about the interpretation of previous results. Most of the experimental data collected have suggested a simple model for the regulation of expression of the two synthases. PHS-1 is regarded as a constitutive synthase, maintaining prostaglandin levels for routine housekeeping functions, while PHS-2 is proposed to be the inducible synthase, under strict transcriptional and translational regulation in response to mitogens and inflammatory mediators. This model accounts for most observations to date, but is probably oversimplified. Regulation of PHS-2 has been studied in fibroblasts, monocytes/macrophages, endothelial cells, vascular smooth muscle, ovarian follicles, and neurons. Diverse stimuli ranging from serum and other mitogens to inflammatory stimuli to synaptic activity induce expression of PHS-2. In fibroblast cell lines, PHS-2 mRNA levels are induced by mitogenic factors such as serum, phorbol esters, TGF-β, oncogenes and PDGF (Kujubu et al., 1991; Xie et al., 1991; Pilbeam et al., 1993). In monocytic cell lines, endotoxin (LPS) causes an upregulation of PHS-2 mRNA and protein with little effect on PHS-1 levels (Riese et al., 1994). Similarly, endothelial cells express PHS-2 in response to inflammatory mediators including LPS, TNF, and IL-1 (Hla and Neilson, 1992; Jones et al., 1993). PHS-2 is also induced in vascular smooth muscle cells by PDGF, thrombin, EGF, and PMA (Rimarachin et al., 1994). Ovarian follicles express PHS-2 following stimulation with hCG, LH, FSH, or gonadotropin-releasing hormone (Sirois et al., 1992). Expression of PHS-2 in neurons can be stimulated by seizures or by NMDA-induced synaptic activity (Yamagata et al., 1993). PHS-2 is an immediate early gene and the appearance of the transcript is quite rapid following exposure of the cells to an appropriate stimulus. For example, PHS-2 mRNA appears within 30 min in monocytes stimulated with LPS, and in various cell types, the mRNA level peaks at 2–6 hr and returns to basal levels by 8–12 hr. The relatively long duration of PHS-2 mRNA following stimulation suggests that posttranscriptional modification of the message to increase its half-life may also occur. Several groups have presented compelling evidence that the PHS-1 mRNA also accumulates in response to cytokines, although the level of induction is much lower than seen with PHS-2 and generally occurs at later times than PHS-2 (Maier et al., 1990; Diaz et al., 1992; DeWitt and Meade, 1993). Thus, both synthases may be inducible, but with the responses separated temporally—an acute rise in PHS-2 and a chronic elevation of PHS-1.

Interestingly, the protein synthesis inhibitor cycloheximide has been reported to induce accumulation of mRNA for both PHS-1 and PHS-2 (Maier *et al.*, 1990; O'Banion *et al.*, 1991; Rimarachin *et al.*, 1994). This response is a characteristic shared with growth-related, immediate-early genes (Herschman, 1991), and would be expected for PHS-2 but not PHS-1. Additionally, the PHS-2 transcript contains the AUUUA motif (Kujubu *et al.*, 1991) found in many immediate-early genes and which has been shown to confer instability to the transcript (Shaw and Kamen, 1986). Prolonged expression of the mRNA of immediate-early genes may be dependent on posttranscriptional processing to attenuate the effects of the reiterated AUUUA instability sequence (Piechaczyk *et al.*, 1985), and PHS-2 may similarly be dependent on posttranslational processing (Ristimaki *et al.*, 1994). The PHS-1 transcript does not share this degradation motif and appears much less labile than the PHS-2 transcript.

Glucocorticoids suppress prostaglandin formation in response to inflammatory stimuli. Two steps in the prostaglandin biosynthetic pathway appear sensitive to glucocorticoids. First, glucocorticoids inhibit the release of arachidonic acid by phospholipase A_2. Second, glucocorticoids suppress the expression of PHS. Dexamethasone, a clinically useful anti-inflammatory, attenuates the accumulation of the PHS-2 transcript in cultured fibroblasts, monocytes, and neurons (Kujubu and Herschman, 1992; DeWitt and Meade, 1993; O'Banion *et al.*, 1991, 1992; Yamagata *et al.*, 1993). O'Banion *et al.* first observed a glucocorticoid-regulated PHS homologue in murine fibroblasts. The partial mRNA sequence they obtained is identical to the murine PHS-2 obtained by Kujubu *et al.* (1991). PHS-1 levels are only slightly decreased or remain unchanged on the addition of dexamethasone (Kujubu and Herschman, 1992; O'Banion *et al.*, 1992; DeWitt and Meade, 1993). Convincing data showing dexamethasone suppression of PHS-2 in endothelial cells have not been presented. This suppressive effect may be limited to certain cell types, particularly cells with important inflammatory roles.

Masferrer *et al.* (1992) demonstrated the importance of endogenous glucocorticoids in regulating prostaglandin biosynthesis *in vivo*. They demonstrated that peritoneal macrophages from adrenalectomized mice, which lack endogenous glucocorticoids produced two to three times as much prostaglandin as control mice or adrenalectomized mice supplemented with dexamethasone. Injection of LPS further enhanced prostanoid biosynthesis and initiated a lethal inflammatory reaction in glucocorticoid-deficient mice. Coincident with this enhanced sensitivity to LPS was the upregulation of PHS protein. Dexamethasone suppressed LPS induction of PHS levels and eliminated the LPS toxicity. Although the reagents employed were not specific for PHS-2, the data suggest that PHS-2 is under strict glucocorticoid suppression in normal physiology.

The isolation of genomic clones for PHS-1 and-2 has made further study of gene regulation feasible. The genes for the two synthases are organized in a surprisingly similar fashion: PHS-1 contains 11 exons, while PHS-2 contains 10, and many intron–exon boundaries are conserved (Kosaka *et al.*, 1994). The PHS-1 gene is 22 kb in length (Yokoyama and Tanabe, 1989), while the PHS-2 gene is only 8 kb (Fletcher *et al.*, 1992). The smaller size of PHS-2 is consistent with its role as an immediate-early gene. Immediate-early genes are generally smaller than 10 kb, supposedly to allow rapid transcription and processing following stimulation (Herschman, 1991). The two synthases are located on separate chromosomes. In humans, PHS-1 maps to chromosome 9 (Funk *et al.*, 1991) whereas PHS-2 maps to chromosome 1 (Kosaka *et al.*, 1994; Jones *et al.*, 1993). The physical separation of these two genes within the genome suggests that different regulatory mechanisms control expression of each gene. The promoter regions of PHS-1 and -2 are providing additional insight into the regulation of prostaglandin production. The 5' regulatory sequence of PHS-1 does not contain the traditional TATA or CAAT boxes, but contains two SP-1 sites, an AP-2 site, three AP-1 sites, a sequence with similarity to a negative glucocorticoid regulatory element, and a dioxin responsive element (Kraemer *et al.*, 1992). The role of these regions in driving transcription of PHS-1 has not been analyzed. The 5' regulatory region of PHS-2 contains similar elements, but additionally contains TATA and CAAT boxes (Inoue *et al.*, 1994). Fletcher *et al.* reported that the first 1000 bp of the 5' region is sufficient to initiate serum and TPA induction of a luciferase reporter gene (Fletcher *et al.*, 1992). Subsequent experiments have shown that a cyclic AMP response element (CRE) within the first 125 bp of the 5' UTR that is conserved in the human, mouse, and rat genes is essential for expression of this transcript in response to TPA (Inoue *et al.*, 1994). Induction of expression by pp60[v-src] is dependent on the ATF/CRE element located in the first 80 nucleotides of the murine 5' UTR (Xie *et al.*, 1994). Somewhat surprisingly, Xie *et al.* (1993) demonstrated that dexamethasone had no effect on a reporter gene fused to the PHS-2 regulatory region and we have the same observation using a reporter construct with promoter sequence from the human gene. Several scenarios might explain this observation. It is possible that the promoter region was incomplete in both cases—this will be resolved in nuclear run-on experiments. Another explanation is that glucocorticoids exert their effect on PHS-2 expression posttranscriptionally. More thorough promoter analysis should help to clarify the transcriptional and translational controls for PHS expression.

9.3. Actions of Prostaglandins

Prostanoids are essential in the function of many organ systems. An example of the complexity of prostaglandin function is seen by examining the roles that these compounds play in normal kidney function. In the kidney, prostaglandins assist in the regulation of renal blood flow, renin release, and water reabsorption. Both prostacyclin and PGE_2, produced within the kidney, cause vasodilation and thereby increase renal blood flow. These two prostaglandins counteract the vasoconstrictive effects of the circulating hormones norepinephrine and angiotensin II (Codde and Beilin, 1986). $PGF_{2\alpha}$, which is formed by the kidney in response to bradykinin, and the platelet product thromboxane A_2 are vasoconstrictors that oppose the actions of prostacyclin and PGE_2 to reduce renal blood flow. PGE_2 stimulates renin formation and, in part, mediates the AVP-induced water reabsorption in the collecting tubule; assisting in regulation of whole body salt and water balance. It has been suggested that prostaglandins may play a direct role in the development of hypertension, which was a reasonable hypothesis given that several prostanoids can cause vasoconstriction and alter salt and water retention. However, current evidence suggests that it is more likely that the elevated levels of prostaglandins seen in hypertension are secondary to the development of this condition rather than a primary cause (Codde and Beilin, 1986).

In addition to their many important roles in maintaining homeostasis in the body, prostaglandins are essential components of a number of pathophysiological conditions including fever and inflammation. Fever-inducing stimuli, such as endotoxin and interleukin-1, cause an elevation of PGE_2 in the brain. Inhibition of PGE_2 formation by NSAIDs attenuates the fever (Bishai *et al.*, 1987; Ferrari *et al.*, 1990; Sirko *et al.*, 1989). Prostaglandins, particularly PGE_2 which is produced by macrophages at the site of injury or infection, stimulate extravasation of fluid from blood vessels and enhance the pain response (Moncada *et al.*, 1973; Crunkhorn and Willis, 1969). Moncada *et al.* (1973) proposed that the role of prostaglandins in inflammation is to potentiate the effects of primary mediators of inflammation, such as bradykinin or histamine. Inhibition of prostaglandin formation by steroidal or nonsteroidal anti-inflammatory drugs can limit the extent of the inflammatory process. High levels of PGE_2 are detected in the synovial fluid of patients suffering from rheumatoid arthritis where this prostanoid is thought to be involved in the erosion of cartilage and bone that occurs in this form of arthritis (Sano *et al.*, 1992). Inhibitors of prostaglandin synthase are mainstays of therapy for inflammatory arthritides and other disorders involving inflammation.

9.3.1. *Prostanoids Act by Binding to Specific Receptors*

Prostaglandins and thromboxane A_2 interact with their target tissues by binding to G protein-linked cell surface receptors. In the past few years, receptors for most of the biologically active prostanoids have been cloned. The receptors share a high degree of sequence homology and all of the cDNAs code for polypeptides with seven putative transmembrane domains, a pattern typical of G protein-linked receptors. Each receptor contains two potential glycosylation sites which may be important in proper targeting to the cell surface for expression (Abramovitz *et al.*, 1994). The receptors also contain numerous serine and threonine residues in their carboxy termini and third cytoplasmic loops; targets for phosphorylation and potential downregulation of the receptors. Northern blot analysis of the expression patterns of the different prostanoid receptors is shedding new light on the potential roles for these eicosanoids. For example, the thromboxane receptor is expressed at high level in thymus, suggesting that thromboxane A_2 may be important in the immune system. Ushikubi *et al.* (1993) have recently shown that the thromboxane receptor is expressed at high levels in immature thymocytes and that thromboxane induces apoptotic death of immature thymocytes. Thymocytes themselves are capable of producing only limited amounts of cyclooxygenase products, suggesting that the thromboxane to which the thymocytes respond is produced by the surrounding stroma. Thromboxane appears to act as a paracrine signal in the thymus involved in the regulation of thymocyte selection and maturation.

Localization of the message for the $PGF_{2\alpha}$ receptor in the gastrointestinal tract and lung corresponds with PGF_α's well-known role as a mediator of smooth muscle contraction (Sakamoto *et al.*, 1994; Abramovitz *et al.*, 1994; Sugimoti *et al.*, 1994; Nakano and Cole, 1969). Similarly, the prostacyclin receptor is expressed at high level in the kidney, consistent with the important role prostacyclin plays in the regulation of renal blood flow, renin release, and glomerular filtration rate (Boie *et al.*, 1994; Codde and Beilin, 1986).

9.3.2. *Prostaglandin H Synthases as Targets in Cardiovascular Disease and Inflammation*

The discovery of a form of PHS that is highly induced by inflammatory stimuli has ignited a search for new, specific inhibitors of prostaglandin production and the pharmacological differences between the two synthases are being investigated. Conventional inhibitors of PHS include aspirin, indomethacin, and ibuprofen. The majority of these PHS inhibitors compete with arachidonic acid at the cyclooxygenase active site (DeWitt *et al.*, 1990; Rome and Lands, 1975), but do not inhibit the peroxidase activity of the enzyme (Mizuno *et al.*, 1982; van der Ouderaa *et al.*, 1980). Several NSAIDs result in irreversible inhibition of the synthases *in vitro*. Aspirin is of particular importance clinically because it irreversibly inactivates the synthase *in vivo*. Aspirin inhibits PHS through a two-part mechanism; initially it acts as a competitive inhibitor. Following this process, aspirin transfers an acetyl group to a serine residue in the cyclooxygenase active site (Roth *et al.*, 1983; DeWitt and Smith, 1988). No endogenous process appears to hydrolyze this acetyl group. Thus, regeneration of activity requires the synthesis of new enzyme. PHS-2 contains a serine analogous to the one acetylated by aspirin in PHS-1 and LeComte *et al.* (1994) demonstrated that PHS-2 is acetylated at this analogous residue, resulting in attenuation of prostaglandin formation by this isozyme as well. Clinical studies have identified low-dose aspirin as effective in preventing the recurrence of myocardial infarction. The necessity of regeneration of PHS is the basis for aspirin's antithrombotic properties. Since platelets do not synthesize proteins, they are incapable of replenishing active PHS, and thromboxane formation is limited. Vascular endothelial cell PHS is spared inactivation by two processes: (1) low doses of aspirin are largely metabolized during a single pass through the portal circulation, limiting vascular endothelial cell exposure, and (2) the endothelium retains the capacity to synthesize active PHS and therefore continues prostacyclin production.

In initial studies, PHS-2 has proven susceptible to inhibition by conventional NSAIDs. The synthases do, however, display different pharmacological profiles. Meade *et al.* (1993) compared the potency of several common PHS inhibitors on the two isoforms expressed in COS cells. They observed that indomethacin, sulindac sulfide, and piroxicam preferentially inhibited PHS-1. Other anti-inflammatory drugs, including ibuprofen, flurbiprofen, and meclofenamate, were equipotent toward both enzymes. One compound, 6-methoxy-2-naphthyl-acetic acid (6-MNA), selectively inhibited PHS-2. Interestingly, aspirin inhibited the formation of prostaglandins by PHS-2, but did not alter the rate of oxygen consumption by the enzyme. Aspirin shifted the product profile of PHS-2 catalysis so that 15-HETE was generated by the synthase. Previous studies on PHS activity in cultured tracheal epithelial cells support this observation. Holtzman *et al.* (1992) observed that aspirin inhibited prostaglandin production coincident with in-creased production of 15-HETE. They demonstrated that a PHS catalyzed the formation of 15-HETE by two lines of evidence. First, indomethacin abolished the production of prostaglandins and 15-HETE. Second, they immunoprecipitated the 15-HETE and prostaglandin catalyzing activities with a specific anti-PHS antiserum. The data indicate a pharmacologically distinct isoform of PHS with an activity profile parallel to the recombinant PHS-2 in the experiments of Meade *et al.* The shift of product formation by aspirin raises interesting questions about the pharmacology of this drug. The interaction of aspirin with PHS-2 implicates 15-HETE in the anti-inflammatory response. Feedback regulation by 15-HETE has been proposed to limit the formation of proinflammatory substances including LTC_4 and LTD_4 formed by the 5-lipoxygenase pathway, as well as inhibiting the formation of the chemotactic LTB_4 (Vanderhoek *et al.*, 1980; Goetzl, 1981).

Recent work has identified new inhibitors of prostaglandin synthase that are more specific for PHS-2 than those previously available. As mentioned above, 6-MNA displayed specificity for the inhibition of PHS-2 *in vitro* (Meade *et al.*, 1993). No studies have yet examined the ability of 6-MNA to inhibit prostaglandin formation *in vivo*. NS-398 is a second new NSAID that specifically blocks the formation of prostaglandins in response to inflammatory stimuli. When tested in two different rat models of inflammation, NS-398 attenuated the production of inflammatory prostaglandins while sparing constitutive prostaglandin formation in the stomach of these animals (Futaki *et al.*, 1993; Masferrer *et al.*, 1994). When these drugs become available for clinical use, they should provide a better-tolerated approach to treatment of the many ailments for which NSAIDs are currently prescribed.

9.3.3. Role of Prostaglandins in Cancer

Several clinical studies have suggested that NSAIDs can decrease the inci-dence of colorectal cancer, particularly in patients with colorectal polyps (Thun *et al.*, 1991; Giardiello *et al.*, 1993; Logan *et al.*, 1993). The epidemiological evidence was reviewed by Marnett (1992), who concluded that the prostaglandins likely played a role in colon carcinogenesis. Several reports have strengthened the argument that inhibition of the prostaglandin synthesis lowers the risk of colon cancer by about 50%. Also, inhibitors of prostaglandin synthesis, in particular sulindac, have been tested in patients with adenomatous polyps of the colon and shown to cause regression of the polyps, indicating that the effect of prostaglandins occurs early in the carcinogenesis pathway (Giardiello *et al.*, 1993). Since the most likely target for the NSAIDs is PHS, Eberhart *et al.* (1994) determined whether colorectal polyps and adenocarcinomas contained higher levels of PHS than normal colon. They found no alteration in the expression of PHS-1 in colon tumor samples as compared to normal colon. However, Northern blot analysis demonstrated that PHS-2 was increased in the tumors compared to normal colon. Increased expression of PHS-2 was also found in about 50% of the colon polyps examined. We have carried out similar studies that confirmed and extended their findings. We found increased expression of PHS-2, but not PHS-1, in colon cancer cell lines. Subsequently, we demonstrated, by *in situ* hybridization of mRNA, that colon tumors contained high levels of PHS-2, while none was apparent in the normal colon. These findings suggest that dysregulation of PHS-2 expression may be crucial in the development of colorectal cancer. Additionally, studies in animal models suggest that increased prostanoid formation (and potentially dysregulation of PHS) may also be involved in breast cancer (Bandyopadhyay *et al.*, 1987; Carter *et al.*, 1983). Eicosanoids have been implicated in other tumors as well, and given their ability to stimulate growth under some conditions, it is likely that future studies will define a role for eicosanoids in some cancers.

9.3.4. *Actions of Eicosanoids Derived from Alternative Substrates*

Although arachidonic acid is the normal substrate for the prostaglandin synthases, there is increasing evidence that other fatty acids will also be utilized by PHS. Early studies recognized the ability of the "fish oils" (eicosapentaenoic acid and docosahexaenoic acid) to interact with PHS. More recently, it has become apparent that linoleic acid could also serve as a substrate for PHS. Linoleic acid is the precursor of the hydroxy-octadecadienoic acids, very potent compounds involved in a variety of physiological processes.

Interest in the fish oils dates back to the studies of Bang *et al.* (1976) and Dyerberg *et al.* (1975) examining the dietary fat content of the Greenland Eskimos, a population known to have low levels of atherogenesis. These authors expected to find a lower fat intake in the Eskimo population compared to their control population, but instead they found that the total dietary fat of the two groups was the same and only the distribution of dietary fat intake varied between the populations. The high-fish diet of the Eskimo population resulted in a much higher proportion of eicosapentaenoic acid (EPA) and docosahexaenoic acid (DHA) and relatively less linoleic acid (LA) in their plasma than seen in control subjects. These studies suggested that the fish oils had a protective, antiatheromatous effect. Further clinical studies demonstrated that a diet high in fish oil could protect against coronary artery disease (Kromhout *et al.*, 1985) and was antihypertensive as well (Knapp and FitzGerald, 1989). During the search for a mechanism to explain the protective effects, fish oils were found to alter the production of eicosanoids. Generally speaking, consumption of high levels of fish oil led to a decreased formation of prostanoids. It was assumed that EPA and DHA would competitively inhibit PHS, binding to the arachidonic acid binding site of the enzyme but not being metabolized. Although competitive inhibition of PHS is the main mechanism by which DHA attenuates prostaglandin formation, the mechanism by which EPA acts is more complex. Some cell types, such as endothelial cells, use EPA to form the trienoic series of prostaglandins (i.e., PGI_3) (Bordet *et al.*, 1986). Other cell types, such as platelets, are unable to efficiently utilize PAE as a substrate (Needleman *et al.*, 1979). Why some tissues were capable of utilizing EPA while others were not has, until recently, remained a mystery. Characterization of the inducible PHS, PHS-2, has suggested a potential mechanism for the cell-specific utilization of EPA. While PHS-1 utilizes EPA very poorly as a substrate, PHS-2 can utilize it relatively effectively. Maximal oxygen consumption of PHS-2 using EPA as a substrate is one-third that seen when arachidonic acid is used as a substrate, a level high enough to expect formation of trienoic prostaglandins by PHS-2 (Meade, 1992). Endothelial cells, which make PHS-2 protein on stimulation, are capable of producing relatively high levels of PGI_3. By shifting the EPA to PHS-2, endothelial cells may be able to preserve the formation of PGI_2 by PHS-1, since less EPA will be available to inhibit PHS-1 actions. Platelets, which are unable to synthesize new proteins and lack PHS-2, are unable to utilize EPA as a substrate. EPA which is liberated in platelets on stimulation will merely serve to competitively inhibit TxA_2 formation. While fish oils are no longer considered the miracle preventive for the cardiovascular system, it is possible that the ability of PHS-2 to utilize the fish oils effectively, while PHS-1 cannot, may explain the improved cardiovascular health of the Eskimos that was observed in the early epidemiological studies.

Linoleic acid is an essential fatty acid that is only now beginning to be recognized for its own importance rather than just as a precursor to arachidonic acid. At least three distinct enzymes—PH_s-2, 15-lipoxygenase, and 12-lipoxygenase—are able to catalyze the formation of hydroxyoctadecadienoic acids (HODEs) from linoleic acid. Two HODEs, 9- and 13-HODE, are formed in a variety of cell types and have potent biological activities. These products are rarely formed independently of each other, although the ratio of 9- to 13-HODE formed varies from tissue to tissue, perhaps dependent on the enzyme responsible for HODE synthesis. In hair follicles, for example, 15-lipoxygenase is the primary enzyme responsible for HODE formation, and 13-HODE is the only detectable product (Baer and Green, 1993). Vascular smooth muscle produces more 13-HODE than 9-HODE, presumably via PHS-2 (Daret *et al.*, 1993). Endothelial cells, which form HODEs through the action of both prostaglandin synthase and a lipoxygenase, synthesize more 9-HODE than 13-HODE (Kaduce *et al.*, 1989).

Like the eicosanoids, HODEs are potent compounds that are known to be involved in an increasing number of biological responses. For example, 13-HODE may be important in the maintenance of vascular tone. As demonstrated by Stoll *et al.*, 13-HODE increases cGMP and intracellular calcium levels in vascular smooth muscle cells (Stoll *et al.*, 1994). The authors suggest that the 13-HODE to which vascular smooth muscle responds is released by the endothelial cells and acts as a vasodilator because it increases vascular smooth muscle cGMP levels. In support of this theory, DeMeyer *et al.* (1992) demonstrated that 13-HODE will relax preconstricted arterial rings.

13-HODE formed by endothelial cells may also limit the adhesion of platelets to endothelial cells. Setty *et al.* (1987) demonstrated that 13-HODE decreases thromboxane formed by thrombin-stimulated platelets, in part by attenuating the release of arachidonic acid. The decreased formation of thromboxane limits the adhesion of platelets to the endothelial cell surface. The ability of 13-HODE to regulate adhesion of cells to the endothelial surface may also be important in determining the metastatic potential of carcinomas. Buchanan and Bastida (1988) demonstrated that carcinomas with a high metastatic potential produced less 13-HODE relative to HETEs than carcinomas with a lower metastatic potential. They hypothesized that carcinomas producing relatively low levels of 13-HODE would easily adhere to endothelial cell surfaces and therefore metastasize throughout the body, while those producing relatively high levels of 13-HODE would be unable to adhere to the endothelium and would, therefore, be less likely to metastasize.

13-HODE may also be able to block epidermal hyperproliferation (Cho and Ziboh, 1994a,b). In porcine epidermal cells, 13-HODE is incorporated into phosphatidyl 4,5-bisphosphate to form 13-HODE diacylglycerol (13-HODE-DAG). This compound competes with di-olein for binding to PKCβ, causing inhibition of the enzyme. Inhibition of PKCβ results in attenuation of the hyperproliferation seen in response to docosahexaenoic acid. Although 13-HODE-DAG does not associate with PKCα, its ability to associate with the other forms of PKC has not been examined. Through incorporation into membrane phospholipids, 13- and 9-HODE may have many important functions which have not yet been elaborated.

While 9-HODE function generally coincides with 13-HODE function, recent studies have demonstrated a potential pathophysiological role of 9-HODE which does not appear to involve 13-HODE. In experiments examining atherogenesis, Ku *et al.* identified both 9- and 13-HODE as products formed in response to oxidation of LDL (Ku *et al.*, 1992). 9-HODE caused the release of IL-1β from monocyte-derived macrophages, but release of IL-1β in response to 13-HODE was minimal. (IL-1β increases vascular smooth muscle proliferation which is a trademark of atherosclerosis.) These investigators demonstrated that 9-HODE not only caused the release of IL-1β from macrophages but actually increased its expression, suggesting that 9-HODE may play an active role in atherogenesis.

References

Abramovitz, M., Boie, Y., Nguyen, T., Rushmore, T. H., Bayne, M. A., Metters, K. M., Slipetz, D. M., and Grygorczyk, R., 1994, Cloning and expression of a cDNA for the human prostanoid FP receptor, *J. Biol. Chem.* **269**:2632–2636.

Baer, A. N., and Green, F. A., 1993, Fatty acid oxygenase activity of human hair roots, *J. Lipid Res.* **34**:1505–1514.

Bandyopadhyay, G., Imagawa, W., Wallace, D., and Nandi, S., 1987, Linoleate metabolites enhance the in vitro proliferative response of mouse mammary epithelial cells to epidermal growth factor, *J. Biol. Chem.* **262**:2750–2756.

Bang, H. O., Dyerberg, J., and Hjorne, N., 1976, The composition of food consumed by Greenland Eskimos, *Acta Med. Scand.* **200**:69–73.

Barbour, S. E., and Dennis, E. A., 1993, Antisense inhibition of group II phospholipase A_2 expression blocks the production of prostaglandin E_2 by P388D$_1$ cells, *J. Biol. Chem.* **268:**21875–21882.

Belanger, C., Buring, J. E., Cook, N., Eberlein, K., Goldhaber, S. Z., Gordon, D., and Hennekens, C. H., 1988, Preliminary Report: Findings from the aspirin component of the ongoing Physicians' Health Study, *N. Engl. J. Med.* **318:**262–264.

Bishai, I., Dinarello, C. A., and Coceani, F., 1987, Prostaglandin formation in feline cerebral microvessels: Effect of endotoxin and interleukin-1, *Can. J. Physiol. Pharmacol.* **65:**2225–2230.

Boie, Y., Rushmore, T. H., Darmon-Goodwin, A., Grygorczyk, R., Slipetz, D. M., Metters, K. M., and Abramovitz, M., 1994, Cloning and expression of a cDNA for the human prostanoid IP receptor, *J. Biol. Chem.* **269:**12173–12178.

Bomalaski, J. S., Lawton, P., and Browning, J. L., 1991, Human extracellular recombinant phospholipase A_2, induces an inflammatory response in rabbit joints, *J. Immunol.* **146:**3904–3910.

Bordet, J.-C., Guichardant, M., and Lagarde, M., 1986, Arachidonic acid strongly stimulates prostaglandin I$_3$ (PGI$_3$) production from eicosapentaenoic acid in human endothelial cells, *Biochem. Biophys. Res. Commun.* **135:**403–410.

Buchanan, M. R., and Bastida, E., 1988, Endothelium and underlying membrane reactivity with platelets, leukocytes and tumor cells: Regulation by the lipoxygenase-derived fatty acid metabolites, 13-HODE and HETES, *Med. Hypotheses* **27:**317–325.

Carter, C. A., Milholland, R. J., Shea, W., and Ip, M. M., 1983, Effect of the prostaglandin synthetase inhibitor indomethacin on 7,12-dimethylbenz(a)anthracene-induced mammary tumorigenesis in rats fed different levels of fat, *Cancer Res.* **43:**3559–3562.

Channon, J. Y., and Leslie, C. C., 1990, A calcium-dependent mechanism for associating a soluble arachidonoyl-hydrolyzing phospholipase A_2 with membrane in the macrophage cell line RAW 264.7, *J. Biol. Chem.* **265:**5409–5413.

Cho, Y., and Ziboh, V. A., 1994a, 13-Hydroxyoctadecadienoic acid reverses epidermal hyperproliferation via selective inhibition of protein kinase C-β activity, *Biochem. Biophys. Res. Commun.* **201:** 257–265.

Cho, Y., and Ziboh, V. A., 1994b, Expression of protein kinase C isozymes in guinea pig epidermis: Selective inhibition of PKC-β activity by 13-hydroxyoctadecadienoic acid-containing diacylglycerol, *J. Lipid Res.* **35:**913–921.

Clark, J. D., Lin, L. L., Kriz, R. W., Ramesha, C. S., Sultzman, L. A., Lin, A. Y., Milona, N., and Knopf, J. L., 1991, A novel arachidonic acid-selective cytosolic PLA$_2$ contains a Ca^{2+}-dependent translocation domain with homology to PKC and GAP, *Cell* **65:**1043–1051.

Codde, J. P., and Beilin, L. J., 1986, Prostaglandins and experimental hypertension: A review with special emphasis on the effect of dietary lipids, *J. Hypertens.* **4:**675–666.

Crowl, R. M., Stoller, T. J., Conroy, R. R., and Stoner, C. R., 1991, Induction of phospholipase A_2 gene expression in human hepatoma cells by mediators of the acute phase response, *J. Biol. Chem.* **266:**2647–2651.

Crunkhorn, P., and Willis, A. L., 1969, Actions and interactions of prostaglandins administered intradermally in rat and in man, *Br. J. Pharmacol.* **36:**216P–217P.

Daret, D., Blin, P., Dorian, B., Rigaud, M., and Larrue, J., 1993, Synthesis of monohydroxylated fatty acids from linoleic acid by rat aortic smooth muscle cells and tissues: Influence on prostacyclin production, *J. Lipid Res.* **34:**1473–1482.

De Meyer, G. R. Y., Bult, H., Verbeuren, T. J., and Herman, A. G., 1992, The role of endothelial cells in the relaxations induced by 13-hydroxy- and 13-hydroperoxylinoleic acid in canine arteries, *Br. J. Pharmacol.* **107:**597–603.

DeWitt, D. L., and Meade, E. A., 1993, Serum and glucocorticoid regulation of gene transcription and expression of the prostaglandin H synthase-1 and prostaglandin H synthase-2 isozymes, *Arch. Biochem. Biophys.* **306:**94–102.

DeWitt, D. L., and Smith, W. L., 1983, Purification of prostacyclin synthase from bovine aorta by immunoaffinity chromatography. Evidence that the enzyme is a hemoprotein, *J. Biol. Chem.* **258:**3285–3293.

DeWitt, D. L., and Smith, W. L., 1988, Primary structure of prostaglandin G/H synthase from sheep vesicular gland determined from the complementary DNA sequence, *Proc. Natl. Acad. Sci. USA* **85**:1412–1416.

DeWitt, D. L., el-Harith, E. A., Kraemer, S. A., Andrews, M. J., Yao, E. F., Armstrong, R. L., and Smith, W. L., 1990, The aspirin and heme-binding sites of ovine and murine prostaglandin endoperoxide synthases, *J. Biol. Chem.* **265**:5192–5198.

Diaz, A., Reginato, A. M., and Jimenez, S. A., 1992, Alternative splicing of human prostaglandin G/H synthase mRNA and evidence of differential regulation of the resulting transcripts by transforming growth factor β1, interleukin 1β, and tumor necrosis factor α, *J. Biol. Chem.* **267**:10816–10822.

Dyerberg, J., Bang, H. O., and Hjorne, N., 1975, Fatty acid composition of the plasma lipids in Greenland Eskimos, *Am. J. Clin. Nutr.* **28**:958–966.

Eberhart, C. E., Coffey, R. J., Radhika, A., Giardiello, F. M., Ferrenbach, S., and Dubois, R. N., 1994, Upregulation of cyclooxygenase 2 gene expression in human colorectal adenomas and adenocarcinomas, *Gastroenterology* **107**:1183–1188.

Egan, R. W., Paxton, J., and Kuehl, F. A., Jr., 1976, Mechanism for irreversible self-deactivation of prostaglandin synthetase, *J. Biol. Chem.* **251**:7329–7335.

Ferrari, R. A., Ward, S. J., Zobre, C. M., van Liew, D. K., Perrone, M. H., Connell, M. J., and Haubrich, D. R., 1990, Estimation of the *in vivo* effect of cyclooxygenase inhibitors on prostaglandin E_2 levels in mouse brain, *Eur. J. Pharmacol.* **179**:25–34.

Fletcher, B. S., Kujubu, D. A., Perrin, D. M., and Herschman, H. R., 1992, Structure of the mitogen-inducible TIS10 gene and demonstration that the TIS10-encoded protein is a functional prostaglandin G/H synthase, *J. Biol. Chem.* **267**:4338–4344.

Funk, C. D., Funk, L. B., Kennedy, M. E., Pong, A. S., and FitzGerald, G. A., 1991, Human platelet/erythroleukemia cell prostaglandin G/H synthase: cDNA cloning, expression, and gene chromosomal assignment, *FASEB J.* **5**:2304–2312.

Futaki, N., Yoshikawa, K., Hamasaka, Y., Arai, I., Higuchi, S., Itzuka, H., and Otomo, S., 1993, NS-398, a novel non-steroidal anti-inflammatory drug with potent analgesic and antipyretic effects, which causes minimal stomach lesions, *Gen. Pharmacol.* **24**:105–110.

Giardiello, F. M., Hamilton, S. R., Krush, A. J., Piantadosi, S., Hylind, L. M., Celano, P., Booker, S. V., Robinson, C. R., and Offerhaus, G. J. A., 1993, Treatment of colonic and rectal adenomas with sulindac in familial adenomatous polyposis, *N. Engl. J. Med.* **328**:1313–1316.

Goetzl, E. J., 1981, Selective feed-back inhibition of the 5-lipoxygenation of arachidonic acid in human T-lymphocytes, *Biochem. Biophys. Res. Commun.* **101**:344–350.

Hara, S., Kudo, I., Chang, H. W., Matsuta, K., Miyamoto, T., and Inoue, K., 1989, Purification and characterization of extracellular phospholipase A_2 from human synovial fluid in rheumatoid arthritis, *J. Biochem.* **105**:395–399.

Hemler, M. E., Crawford, C. G., and Lands, W. E. M., 1978, Lipoxygenation activity of purified prostaglandin-forming cyclooxygenase, *Biochem. J.* **17**:1772–1779.

Herschman, H. R., 1991, Primary response genes induced by growth factors and tumor promoters, *Annu. Rev. Biochem.* **60**:281–319.

Hla, T., and Neilson, K., 1992, Human cyclooxygenase-2 cDNA, *Proc. Natl. Acad. Sci. USA* **89**:7384–7388.

Holtzman, M. J., Turk, J., and Shornick, L. P., 1992, Identification of a pharmacologically distinct prostaglandin H synthase in cultured epithelial cells, *J. Biol. Chem.* **267**:21438–21445.

Inoue, H., Nanayama, T., Hara, S., Yokoyama, C., and Tanabe, T., 1994, The cyclic AMP response element plays an essential role in the expression of the human prostaglandin-endoperoxide synthase 2 gene in differentiated U937 monocytic cells, *FEBS Lett.* **350**:51–54.

Jones, D. A., and Fitzpatrick, R. A., 1990, "Suicide" inactivation of thromboxane A_2 synthase. Characteristics of mechanism-based inactivation with isolated enzyme and intact platelets, *J. Biol. Chem.* **265**:20166–20171.

Jones, D. A., Carlton, D. P., McIntyre, T. M., Zimmerman, G. A., and Prescott, S. M., 1993, Molecular cloning of human prostaglandin endoperoxide synthase type II and demonstration of expression in response to cytokines, *J. Biol. Chem.* **268**:9049–9054.

Kaduce, T. L., Figard, P. H., Leifur, R., and Spector, A. A., 1989, Formation of 9-hydroxyoctadecadienoic acid from linoleic acid in endothelial cells, *J. Biol. Chem.* **264**:6823–6830.

Knapp, H. R., and FitzGerald, G. A., 1989, The antihypertensive effects of fish oil. A controlled study of polyunsaturated fatty acid supplements in essential hypertension, *N. Engl. J. Med.* **320:**1037–1043.

Kosaka, T., Miyata, A., Ihara, H., Hara, S., Sugimoto, T., Takeda, O., Takahashi, E., and Tanabe, T., 1994, Characterization of the human gene (PTGS2) encoding prostaglandin-endoperoxide synthase 2, *Eur. J. Biochem.* **221:**889–897.

Kraemer, S. A., Meade, E. A., and DeWitt, D. L., 1992, Prostaglandin endoperoxide synthase gene structure: Identification of the transcriptional start site and 5'-flanking regulatory sequences, *Arch. Biochem. Biophys.* **293:**391–400.

Kromhout, I., Bosschieter, E. B., and Coulander, C. L., 1985, The inverse relation between fish consumption and 20-year mortality from coronary heart disease, *N. Engl. J. Med.* **312:**1205–1209.

Ku, G., Thomas, C. E., Akeson, A. L., and Jackson, R. L., 1992, Induction of interleukin 1β expression from human peripheral blood monocyte-derived macrophages by 9-hydroxyoctadecadienoic acid, *J. Biol. Chem.* **267:**14183–14199.

Kujubu, D. A., and Herschman, H. R., 1992, Dexamethasone inhibits mitogen induction of the TIS10 prostaglandin synthase cyclooxygenase gene, *J. Biol. Chem.* **267:**1–4.

Kujubu, D. A., Fletcher, B. S., Varnum, B. C., Lim, R. W., and Herschman, H. R., 1991, TIS10, a phorbol ester tumor promoter-inducible mRNA from Swiss 3T3 cells, encodes a novel prostaglandin synthase/cyclooxygenase homologue, *J. Biol. Chem.* **266:**12866–12872.

Kujubu, D. A., Reddy, S. T., Fletcher, B. S., and Herschman, H. R., 1993, Expression of the protein product of the prostaglandin synthase-2/TIS10 gene in mitogen-stimulated Swiss 3T3 cells, *J. Biol. Chem.* **268:**5425–5430.

Lecomte, M., Laneuville, O., Ji, C., DeWitt, D. L., and Smith, W. L., 1994, Acetylation of human prostaglandin endoperoxide synthase-2 (cyclooxygenase-2) by aspirin, *J. Biol. Chem.* **269:**13207–13215.

Leslie, C. C., Voelker, D. R., Channon, J. Y., Wall, M. M., and Zelarney, P. T., 1988, Properties and purification of an arachidonoyl-hydrolyzing phospholipase A_2 from a macrophage cell line, RAW 264.7, *Biochim. Biophys. Acta* **963:**476–492.

Lin, L.-L., Lin, A. Y., and Knopf, J. L., 1992, Cytosolic phospholipase A_2 is coupled to hormonally regulated release of arachidonic acid, *Proc. Natl. Acad. Sci. USA* **89:**6147–6151.

Lin, L.-L., Wartmann, M., Lin, A. Y., Knopf, J. L., Seth, A., and Davis, R. J., 1993, cPLA$_2$ is phosphorylated and activated by MAP kinase, *Cell* **72:**269–278.

Logan, R. F. A., Little, J., Hawtin, P. G., and Hardcastle, J. D., 1993, Effect of aspirin and non-steroidal anti-inflammatory drugs on colorectal adenomas: Case–control study of subjects participating in the Nottingham faecal occult blood screening programme, *Br. Med. J.* **307:**285–289.

Maier, J. A. M., Hla, T., and Maciag, T., 1990, Cyclooxygenase is an immediate-early gene induced by interleukin-1 in human endothelial cells, *J. Biol. Chem.* **265:**10805–10808.

Marnett, L. J., 1992, Aspirin and the potential role of prostaglandins in colon cancer, *Cancer Res.* **52:**5575–5589.

Marshall, P. J., and Kulmacz, R. J., 1988, Prostaglandin H synthase: Distinct binding sites for cyclooxygenase and peroxidase substrates, *Arch. Biochem. Biophys.* **266:**162–170.

Masferrer, J. L., Seibert, K., Zweifel, B., and Needleman, P., 1992, Endogenous glucocorticoids regulate an inducible cyclooxygenase enzyme, *Proc. Natl. Acad. Sci. USA* **89:**3917–3921.

Masferrer, J. L., Zweifel, B. S., Manning, P. T., Hauser, S. D., Leahy, K. M., Smith, W. G., Isakson, P. C., and Seibert, K., 1994, Selective inhibition of inducible cyclooxygenase 2 *in vivo* is antiinflammatory and nonnuclerogenic, *Proc. Natl. Acad. Sci. USA* **91:**3228–3232.

Meade, E. A., 1992, Kinetic and Pharmacological Characterization of Murine PGH Synthase-1 and PGH Synthase-2, Ph.D. thesis, pp. 21–66, Michigan State University.

Meade, E. A., Smith, W. L., and DeWitt, D. L., 1993, Differential inhibition of prostaglandin endoperoxide synthase (cyclooxygenase) isozymes by aspirin and other non-steroidal anti-inflammatory drugs, *J. Biol. Chem.* **268:**6610–6614.

Merlie, J. P., Fagan, D., Mudd, J., and Needleman, P., 1988, Isolation and characterization of the complementary DNA for sheep seminal vesicle prostaglandin endoperoxide synthetase (cyclooxygenase), *J. Biol. Chem.* **263:**3550–3553.

Mizuno, K., Yamamoto, S., and Lands, W. E. M., 1982, Effects of non-steroidal anti-inflammatory drugs on fatty acid cyclooxygenase and prostaglandin hydroperoxidase activities, *Prostaglandins* **23:** 743–757.

Moncada, S., Ferriera, S. H., and Vane, J. R., 1973, Prostaglandins, aspirin-like drugs and the oedema of inflammation, *Nature* **246:**217–219.

Nakano, J., and Cole, B., 1969, Effects of prostaglandins E_1 and $F_{2\alpha}$ on systemic, pulmonary, and splanchnic circulations in dogs, *Am. J. Physiol.* **217:**222–227.

Needleman, P., Raz, A., Minkes, M. S., Ferrendelli, J. A., and Sprecher, H., 1979, Triene prostaglandins: Prostacyclin and thromboxane biosynthesis and unique biological properties, *Proc. Natl. Acad. Sci. USA* **76:**944–948.

Oates, J. A., FitzGerald, G. A., Branch, R. A., Jackson, E. K., Knapp, H. R., and Roberts, L. J., II, 1988, Clinical implications of prostaglandin and thromboxane A_2 formation, *N. Engl. J. Med.* **319:** 689–698.

O'Banion, M. K., Sadowski, H. B., Winn, V., and Young, D. A., 1991, A serum- and glucocorticoid-regulated 4-kilobase mRNA encodes a cyclooxygenase-related protein, *J. Biol. Chem.* **266:**23261–23267.

O'Banion, M. K., Winn, V. D., and Young, D. A., 1992, cDNA cloning and functional activity of a glucocorticoid-regulated inflammatory cyclooxygenase, *Proc. Natl. Acad. Sci. USA* **89:**4888–4892.

Otto, J. C., and Smith, W. L., 1994, The orientation of prostaglandin endoperoxide synthases-1 and -2 in the endoplasmic reticulum, *J. Biol. Chem.* **269:**19868–19875.

Otto, J. C., DeWitt, D. L., and Smith, W. L., 1993, N-glycosylation of prostaglandin endoperoxide synthases-1 and -2 and their orientations in the endoplasmic reticulum, *J. Biol. Chem.* **268:**18234–18242.

Picot, D., Loll, P. J., and Garavito, R. M., 1994, The x-ray crystal structure of the membrane protein prostaglandin H_2 synthase-1, *Nature* **367:**243–249.

Piechaczyk, M., Yang, J.-Q., Blanchard, J.-M., Jeanteur, P., and Marcu, K. B., 1985, Posttranscriptional mechanisms are responsible for accumulation of truncated *c-myc* RNAs in murine plasma cell tumors, *Cell* **42:**589–597.

Pilbeam, C. C., Kawaguchi, H., Hakeda, Y., Voznesensky, O., Alander, C. B., and Raisz, L. G., 1993, Differential regulation of inducible and constitutive prostaglandin endoperoxide synthase in osteoblastic MC3T3-E1 cells, *J. Biol. Chem.* **268:**25643–25649.

Ple, P., and Marnett, L. J., 1989, Alkylaryl sulfides as peroxidase reducing substrates for prostaglandin H synthase, *J. Biol. Chem.* **264:**13983–13993.

Regier, M. K., DeWitt, D. L., Schindler, M. S., and Smith, W. L., 1993, Subcellular localization of prostaglandin endoperoxide synthase-2 in murine 3T3 cells, *Arch. Biochem. Biophys.* **301:**439–444.

Riese, J., Hoff, T., Nordhoff, A., DeWitt, D. L., and Resch, K., 1994, Transient expression of prostaglandin endoperoxide synthase-2 during mouse macrophage differentiation, *J. Leukocyte Biol.* **55:** 476–482.

Rimarachin, J. A., Jacobson, J. A., Szabo, P., Maclouf, J., Creminon, C., and Weksler, B. B., 1994, Regulation of cyclooxygenase-2 expression in aortic smooth muscle cells, *Arterioscler. Thromb.* **14:**1021–1031.

Ristimaki, A., Garfinkel, S., Wessendorf, J., Macing, T., and Hla, T., 1994, Induction of cyclooxygenase-2 by interleukin-1α. Evidence for post-transcriptional regulation, *J. Biol. Chem.* **269:**11769–11775.

Rome, L. H., and Lands, W. E. M., 1975, Structural requirements for time-dependent inhibition of prostaglandin biosynthesis by anti-inflammatory drugs, *Proc. Natl. Acad. Sci. USA* **72:**4863–4865.

Rosen, G. D., Birkenmeier, T. M., Raz, A., and Holtzman, M. J., 1989, Identification of a cyclooxygenase-related gene and its potential role in prostaglandin formation, *Biochem. Biophys. Res. Commun.* **164:**1358–1365.

Roth, G. J., Stanford, N., Jacobs, J. W., and Majerus, P. W., 1977, Acetylation of prostaglandin synthetase by aspirin. Purification and properties of the acetylated protein from sheep vesicular gland, *Biochem. J.* **16:**4244–4248.

Roth, G. J., Machuga, E. T., and Ozols, J., 1983, Isolation and covalent structure of the aspirin-modified, active-site region of prostaglandin synthetase, *Biochem. J.* **22:**4672–4675.

Sakamoto, K., Ezashi, T., Miwa, K., Okuda-Ashitaka, E., Houtani, T., Sugimoto, T., Ito, S., and Hayashi, O., 1994, Molecular cloning and expression of a cDNA of the bovine prostaglandin $F_{2\alpha}$ receptor, *J. Biol. Chem.* **269:**3881–3886.

Sano, H., Hla, T., Maler, J. A. M., Crofford, L. J., Case, J. P., Maciag, T., and Wilder, R. L., 1992, In vivo cyclooxygenase expression in synovial tissues of patients with rheumatoid arthritis and osteoarthritis and rats with adjuvant and streptococcal cell wall arthritis, *J. Clin. Invest.* **89:**97–108.

Setty, B. N., Berger, M., and Stuart, M. J., 1987, 13-Hydroxyoctadeca-9,11-dienoic acid (13-HODE) inhibits thromboxane A_2 synthesis, and stimulates 12-HETE production in human platelets, *Biochem. Biophys. Res. Commun.* **148:**528–533.

Sharp, J. D., White, D. L., Chiou, X. G., Goodson, T., Gamboa, G. C., McClure, D., Burgett, S., Hoskins, J., Skatrud, P. L., Sportsman, J. R., Becker, G. W., Kang, L. H., Roberts, E. F., and Kramer, R. M., 1991, Molecular cloning and expression of human Ca^{++}-sensitive cytosolic phospholipase A_2, *J. Biol. Chem.* **266:**14850–14853.

Shaw, G., and Kamen, R., 1986, A conserved AU sequence from the 3′ untranslated region of CM-CSF mRNA mediates selective mRNA degradation, *Cell* **46:**659–667.

Shimokawa, T., and Smith, W. L., 1992, Prostaglandin endoperoxide synthase. The aspirin acetylation region, *J. Biol. Chem.* **267:**12387–12392.

Sirko, S., Bishai, I., and Coceani, F., 1989, Prostaglandin formation in the hypothalamus in vivo: Effect of pyrogens, *Am. J. Physiol.* **256:**R616–R624.

Sirois, J., and Richards, J. S., 1992, Purification and characterization of a novel, distinct isoform of prostaglandin endoperoxide synthase induced by human chorionic gonadotropin in granulosa cells of rat preovulatory follicles, *J. Biol. Chem.* **267:**6382–6388.

Sirois, J., Simmons, D. L., and Richards, J. S., 1992, Hormonal regulation of messenger ribonucleic acid encoding a novel isoform of prostaglandin endoperoxide H synthase in rat preovulatory follicles. Induction in vivo and in vitro, *J. Biol. Chem.* **267:**11586–11592.

Smith, W. L., and Borgeat, P., 1985, The eicosanoids: Prostaglandins, thromboxanes, leukotrienes, and hydroxy-eicosaenoic acids, in: *Biochemistry of Lipids and Membranes* (D. E. Vance and J. E. Vance, eds.), pp. 325–360, Benjamin/Cummings, Menlo Park, CA.

Smith, W. L., and Lands, E. M., 1972, Oxygenation of polyunsaturated fatty acids during prostaglandin biosynthesis by sheep vesicular gland, *Biochem. J.* **11:**3276–3285.

Smith, W. L., and Marnett, L. J., 1991, Prostaglandin endoperoxide synthase: Structure and catalysis, *Biochim. Biophys. Acta* **1083:**1–17.

Stoll, L. L., Morland, M. R., and Spector, A. A., 1994, 13-HODE increases intracellular calcium in vascular smooth muscle cells, *Am. J. Physiol.* **266:**C990–C996.

Sugimoti, Y., Hasumoto, K.-y., Namba, T., Irie, A., Katsuyama, M., Negishi, M., Kakiszuka, A., Narumiya, S., and Ichikawa, A., 1994, Cloning and expression of a cDNA for mouse prostaglandin F receptor, *J. Biol. Chem.* **269:**1356–1360.

Thun, M. J., Namboodiri, M. M., and Heath, C. W., Jr., 1991, Aspirin use and reduced risk of fatal colon cancer, *N. Engl. J. Med.* **325:**1593–1596.

Ushikubi, F., Aiba, Y.-I., Nakamura, K.-I., Namba, T., Hirata, M., Mazda, O., Katsura, Y., and Narumiya, S., 1993, Thromboxane A_2 receptor is highly expressed in mouse immature thymocytes and mediates DNA fragmentation and apoptosis, *J. Exp. Med.* **178:**1825–1830.

Vanderhoek, J. Y., Bryant, R. W., and Bailey, J. M., 1980, Inhibition of leukotriene biosynthesis by the leukocyte product 15-hydroxy-5,8,11,13-eicosatetraenoic acid, *J. Biol. Chem.* **255:**10064–10065.

van der Ouderaa, F. J., Buytenhek, M., Nugteren, D. H., and van Dorp, D. A., 1977, Purification and characterisation of prostaglandin endoperoxide synthetase from sheep vesicular glands, *Biochim. Biophys. Acta* **487:**315–331.

van der Ouderaa, J., Buytenhek, M., Nugteren, D. H., and van Dorp, D. A., 1980, Acetylation of prostaglandin endoperoxide synthetase with acetylsalicyclic acid, *Eur. J. Biochem.* **109:**383–389.

Wong, W. Y. L., and Richards, J. S., 1991, Evidence for two antigenically distinct molecular weight variants of prostaglandin H synthase in the rat ovary, *Mol. Endocrinol.* **5:**1269–1279.

Xie, W., Chipman, J. G., Robertson, D. L., Erikson, R. L., and Simmons, D. L., 1991, Expression of a mitogen-responsive gene encoding prostaglandin synthase is regulated by mRNA splicing, *Proc. Natl. Acad. Sci. USA* **88:**2692–2696.

Xie, W., Merrill, J. R., Bradshaw, W. S., and Simmons, D. L., 1993, Structural determination and promoter analysis of the chicken mitogen-inducible prostaglandin G/H synthase gene and genetic mapping of the murine homolog, *Arch. Biochem. Biophys.* **300**:247–252.

Xie, W., Fletcher, B. S., Anderson, R. D., and Herschman, H. R., 1994, *v-src* induction of the TIS10/PGS2 prostaglandin synthase gene is mediated by an ATF/CRE transcription response element, *Mol. Cell. Biol.* **14**:6531–6539.

Yamagata, K., Andreasson, K. I., Kaufmann, W. E., Barnes, C. A., and Worley, P. F., 1993, Expression of a mitogen-inducible cyclooxygenase in brain neurons: Regulated by synaptic activity and glucocorticoids, *Neuron* **11**:371–386.

Yokoyama, C., and Tanabe, T., 1989, Cloning of human gene encoding prostaglandin endoperoxide synthase and primary structure of the enzyme, *Biochem. Biophys. Res. Commun.* **165**:888–894.

Index